Between Quantum and Cosmos

# Between Quantum and Cosmos

## Studies and Essays in Honor of John Archibald Wheeler

Edited by
Wojciech Hubert Zurek
Alwyn van der Merwe
Warner Allen Miller

Princeton University Press
Princeton, New Jersey

Published by Princeton University Press, 41 William Street,
Princeton, New Jersey 08540
In the United Kingdom: Princeton University Press, Guildford, Surrey

The contents of this book were first published in six special issues of *Foundations of Physics*, Volume 16, Numbers 2 through 7 (February 1986 through July 1986), comprising invited papers in honor of John Wheeler. 

Library of Congress Cataloging in Publication Data

Between quantum and cosmos : studies and essays in honor of John Archibald Wheeler / edited by Wojciech Hubert Zurek, Alwyn van der Merwe, and Warner Allen Miller.
p. cm.
"The contents . . . were first published in six special issues of Foundations of physics"—T.p. verso.
ISBN 0-691-08490-4 (alk. paper)
1. Physics. 2. Quantum theory. 3. Wheeler, John Archibald, 1911– .
I. Wheeler, John Archibald, 1911– . II. Zurek, Wojciech Hubert. III. Van der Merwe, Alwyn. IV. Miller, Warner Allen.
QC71.B45 1988
539—dc19 88-1497

Clothbound editions of Princeton University Press books are printed on acid-free paper, and binding materials are chosen for strength and durability

Printed in the United States of America
by Princeton University Press,
Princeton, New Jersey

# Contents

# Preface

The forty papers collected here honor one of the great twentieth-century scientists—John Archibald Wheeler. Preceding each paper is an original drawing by Wheeler (Wheeler's descriptions are provided at the end of the book). The papers and interwoven drawings capture and illuminate Wheeler's many contributions to physics, from his work with Niels Bohr in atomic and nuclear physics (the scattering matrix and liquid drop model) through his influential contributions to Albert Einstein's theory of gravity (black holes), his deep insights into quantum theory and measurement (the elementary quantum phenomenon), and his efforts to explain the origins of the quantum postulate and quantum gravity (the meaning circuit and the Wheeler-DeWitt Equation).

The majority of the papers contained here are a reflection and sharpening of Wheeler's original and potentially earthshaking ideas. Many scientists are convinced that Wheeler's insights into the foundation of modern-day physics will spur a revolution in our perception of the universe. This book attempts to capture one man's rendering of the "big picture" by providing a glimpse of it through the eyes of his many colleagues. Rather than talk about John in this preface we have decided to let him speak for himself: quotations and, especially, the drawings illustrate the essence of the physics that forms the framework of his deep insight into the inner workings of nature. The contributed papers reveal this framework in a somewhat different, but no less dramatic, manner. The measure of the impact a scientist has on his field is best provided by the influence his ideas have on his colleagues. The contents of this volume—viewed in this light—speak for themselves.

We wish to extend our gratitude to John Archibald Wheeler, not only for inspiring the papers in this book, but also for providing the drawings: they are intended by Professor Wheeler as an expression of his gratitude to his colleagues who have helped him celebrate his seventy-fifth birthday with this volume. We would also like to thank Zelda Davis for her indispensable help, the editors of Princeton University Press for the care with which this book was prepared, and the publishers of *Foundations of Physics*, in which the papers appeared in 1986, for their cooperation. We are particularly grateful to all the contributors.

Los Alamos
November 5, 1987

WOJCIECH HUBERT ZUREK
ALWYN VAN DER MERWE
WARNER ALLEN MILLER

# Between Quantum and Cosmos

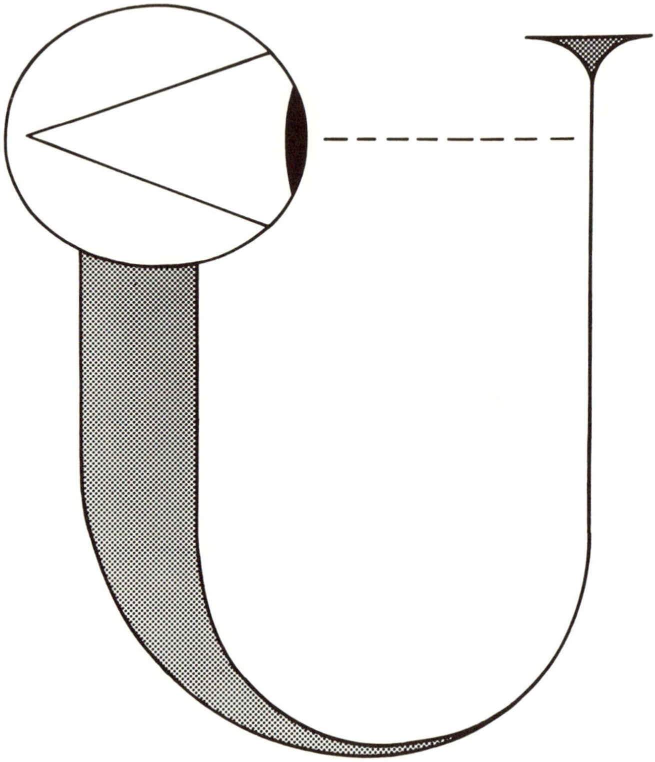

*Plate 1*

# John Archibald Wheeler: A Few Highlights of His Contributions to Physics

**Compiled and edited by Kip S. Thorne[1] and Wojciech H. Zurek[2]**

*Received October 28, 1985*

*The following quotations describe in "nutshells" a few highlights of John Archibald Wheeler's contributions to physics. The contributions are arranged in roughly the following order: (i) concrete research results, (ii) innovative ideas that have become foundations for the research of others, (iii) insights that give guidance for the development of physics over the coming decades. Since most of Wheeler's work contains strong elements of two or even all three of these characteristics, the editors have not attempted to delineate the dividing lines between the three categories.*

*A description of the nucleus which regards the neutrons and protons as spending part of their time in configurations corresponding, for example, to interacting alpha-particles, part of their time in other groupings, already takes into account to a large extent that intimate interaction between nuclear particles which is so entirely different from the situation in atomic structure ... The method of "resonating group structure" builds up a wave function for the whole nucleus out of partial wave functions which describe the close interactions within the individual groups.*—Wheeler (1937).

[Wheeler's method of "resonating group structure," introduced in this paper, was called "clustering theory" by some later researchers, and was the direct ancestor of the "method of generator coordinates" developed later by Wheeler and D. L. Hill.]

[1] Division of Physics and Astronomy, California Institute of Technology, Pasadena, California 91125.

[2] Theoretical Astrophysics, Los Alamos National Laboratory, Los Alamos, New Mexico 87545.

*The c's form a unitary [scattering] matrix ... The equality which must exist between the numbers of incoming and outgoing groups in the [given] state ... for arbitrary choices of the a's ... is just the necessary and sufficient condition for unitary character. ... [Time-reversal invariance] shows that* $\|c_{mn}\|$ *is a symmetrical matrix, and ... each element of the matrix is determined up to a factor* $\pm 1$.—Wheeler (1937).

[The matrix $\|c_{mn}\|$ introduced for the first time in this paper was later given the name "scattering matrix" or "S-matrix" and has come to play a major role in nuclear physics and elementary particle physics.]

*On the basis of the liquid drop model of atomic nuclei, an account is given of the mechanism of nuclear fission. In particular, conclusions are drawn regarding the variation from nucleus to nucleus of the critical energy required for fission, and regarding the dependence of the fission cross section for a given nucleus on the energy of the exciting agency."*—Bohr and Wheeler (1939).

[This paper, written only a few months after the discovery of nuclear fission, gave the definitive theory of fission. It even enabled Bohr and Wheeler to predict that the yet-to-be-discovered nucleus plutonium-239 would have a very large cross section for fission when bombarded by slow neutrons—a cross section that would make it the chosen fuel for the first American atomic bomb.]

*[Describing the first, abortive attempt on September 27, 1944 to operate at full strength the Hanford Washington reactor that was to produce Plutonium-239, Wheeler says]: It had been one of my jobs to consider every possible way that things might go wrong. I was, therefore, very aware that a fission product, when it decayed, could give rise to another one which could absorb the neutrons. When [the reactivity of the reactor mysteriously began to fall] and then, a few hours later, ... began rising again, I was sure that this was what had happened. Now the second nucleus had decayed into a third one which did not absorb the neutrons; ... the culprit had to be Xenon-135.*—Wheeler (1985).

[Thanks to Wheeler, the Hanford reactor had already been designed to deal with the possibility of such "poisoning" by fission product nuclei with high neutron capture cross sections. Some months earlier he had started to warn of the possibility of cross sections far higher than the typical 10 barns that had been measured up to that time; perhaps even as high as 1 to 25 million barns. The reactor had been redesigned at the last minute to allow for this possibility; and this permitted it, with the read-justement of the safety rods, to handle Xenon-135 whose cross section turned out to be 3 million barns.]

*A quantitative formulation of [the absorber as the mechanism of radiation] is given here on the basis of the following postulates: (1) An accelerated charge in otherwise charge-free space does not radiate energy. (2) The fields which act on a given particle arise only from other particles. (3) These fields are represented by one-half the retarded plus one-half the advanced Lienard–Wiechert solutions of Maxwell's equations. In a system containing particles sufficient in number ultimately to absorb all radiation [these postulates produce the standard radiation-reaction phenomena in the emitter] ... Radiation is [thereby] concluded to be a phenomenon as much of statistical mechanics as of pure electrodynamics.*—Wheeler and Feynman (1945).

[Feynman (1966) in his Nobel Prize lecture describes the genesis of this work, his interactions with Wheeler during its development, and its considerable influence on Feynman's subsequent Nobel-prize-winning contributions to quantum electrodynamics.]

*... by far the dominating type of annihilation is that in which the positron combines with an electron whose spin forms a singlet state with respect to the positron. Associated with this selection of pairs which have zero relative angular momentum, before the annihilation process, is an analogous polarization phenomenon in the two quanta which are left at the end of the process. According to the pair theory, if one of these photons is linearly polarized in one plane, then the photon which goes off in the opposite direction with equal momentum is linearly polarized in the perpendicular plane.*—Wheeler (1946).

[This paper, in addition to developing the theory of electron–positron atoms, also laid the foundations (in the above quote) for the now-most-familiar variant of the Einstein–Podolsky–Rosen paradox.]

*Ignorance of the deeper relation between nucleons and mesons makes it especially appropriate at this time to investigate those features of the behavior of mesons which are largely independent of uncertainties about the nature of elementary particles. Fortunately, a number of conclusions may be drawn about the interaction between the meson and the nucleus when we assume little more than the laws of electrodynamics, elementary notions of nuclear structure, the principle of microscopic reversibility, and the simplest ideas of quantum theory. Thus, it has become clear not only that the meson possesses characteristic Bohr orbits of its own around the nucleus, but also that trapping into these orbits via ordinary atomic interactions is the precursor of any specific reaction with the nucleus. ... Also interesting are the energy levels and interlevel transitions of the meson, the first experimental evidence for which is given by W. Y. Chang in a following paper. Among the possible transitions are not only those in which a photon is emitted or an atomic electron is ejected, but also processes of pair creation and meson-induced fission. ... An attempt is made to analyze these questions in the present paper.*—Wheeler (1949).

[This paper developed the definitive theory of mu-mesic atoms.]

*Analyzing different formulations of conditions for validity of the dispersion formula relating absorption and refraction of light, especially the condition that no wave be scattered before arrival of the primary wave (Kronig) we conclude: it is reasonable to apply the dispersion formula for arbitrarily high frequencies.*—Toll and Wheeler (1951).

[This work showed for the first time that dispersion theory could be extended into the relativistic domain by a proper application of the principle of causality.]

*Associated with an electromagnetic disturbance is a mass, the gravitational attraction of which under appropriate circumstances is capable of holding the disturbance together for a time long in comparison with the characteristic periods of the system. Such gravitational-electromagnetic entities, or "geons," are analyzed via classical relativity theory. They furnish for the first time a completely classical, divergence-free self-consistent picture of the Newtonian concept of body...*—Wheeler (1955).

[This paper introduced the concept of "geon" into physics, and in analyzing geons it introduced the concept of two-lengthscale perturbative expansions into general relativity. Such expansions, in the hands of Wheeler's subsequent students and their students, have become the key foundation for the theory of the interaction of propagating gravitational waves with the "background" curvature of space-time.]

*We consider a number of mass concentrations so distributed in space and of such relative magnitudes, that the zone of influence of each can be reasonably approximated by a sphere. Inside each cell [sphere] we replace the actual gravitational potentials by the expressions of Schwarzschild. ... The mass concentrations on either side of a cell boundary accelerate toward that boundary at such a rate as to nullify the discontinuity in matching of the normal derivative of the gravitational potentials that would otherwise occur. ... [This] expresses the equation of motion of the mass at the center of a cell as a dynamic condition on the boundary of the cell. ... The whole of the expansion and subsequent contraction [of the universe] is derived [in this way] from the elementary static Schwarzschild solution.*—Lindquist and Wheeler (1957).

[This paper introduced into general relativity a technique for matching together two different space-time geometries. This matching technique, in the hands of Wheeler's students and colleagues, has become a foundation for modern analyses of the laws of motion and precession of black holes and other relativistic bodies.]

*Space tells matter how to move, ... and matter tells space how to curve.*—Misner, Thorne, and Wheeler (1973).

*... we examined the consequences of an equation of state that attempts to span the whole region from normal densities to supranuclear densities. We assume: (1) Cold matter ideally catalyzed to the end point of thermonuclear evolution. ... (2) General relativity equations of hydrostatic equilibrium. ... The numerical integrations show for the first time both crushing points [white dwarf and neutron star] on a single curve for mass as a function of central density. [The form of this curve emphasizes that] assembly of an amount of mass that exceeds in order of magnitude the mass of the sun and catalysis of this matter to the endpoint of energy evolution results in a condition which lies at the untamed frontier between elementary particle physics and general relativity. Of all the implications of general relativity for the structure and evolution of the universe, this question of the fate of great masses is one of the most challenging.*—Harrison, Wakano, and Wheeler (1958).

[With these words Wheeler began the quest to understand the fates of great masses, a quest that led him a decade later to formulate the concept of a black hole.]

*Energy of rotation [of a central neutron star] appears not yet to have been investigated as a source of power [for the Crab nebula]. Presumably this mechanism can only be effective—if then—when the magnetic field of the residual neutron star is well coupled to the surrounding ion clouds.*"—Wheeler (1966).

[This statement, published one year before the discovery of pulsars, was the closest anybody ever came before pulsars to the correct explanation of what powers the Crab nebula.]

*We write down ... the linear differential equations for small first-order departures from the Schwarzschild metric. ... It is shown that a Schwarzschild singularity ["black hole" in the terminology of the 1980s] ... will undergo small vibrations about the spherical form and will therefore remain stable if subjected to a small nonspherical perturbation.*—Regge and Wheeler (1957).

[This was the pioneering analysis of the pulsations and stability of black holes, performed a decade before the concept of a black hole was fully understood.]

*... by reason of its faster and faster infall [the surface of a collapsing star] moves away from the observer more and more rapidly. The light is shifted to the red. It becomes dimmer millisecond by millisecond, and in less than a second too dark to see... [The star] like the Cheshire cat fades from view. One leaves behind only its grin, the other, only its gravitational attraction. Gravitational attraction, yes; light, no. No more than light do any particles emerge. Moreover, light and particles incident from outside ... go down the black hole only to add to its mass and increase its gravitational attraction.*—Wheeler (1968).

[With these sentences, Wheeler coined the name "black hole" thereby catalyzing a major change in viewpoint on something that previously had gone under such names as "Schwarzschild singularity," "collapsed star," and "frozen star." Note that, in typical Wheelerian style, he does not tell us that he is coining a new name for and viewpoint on a fundamental concept in physics; rather, he writes as though this name and viewpoint had always been used—and almost overnight all other researchers in "black-hole physics" embrace them.]

*A black hole has no hair.*—Ruffini and Wheeler (1971).

[This phrase epitomized a result of which hints existed at the time Wheeler coined the phrase; and like so many of Wheeler's pithy phrases, it acted as a stimulus for further research by Wheeler's colleagues, former students and students of students—research that ultimately gave strong validity to the phrase.]

*On an atomic scale the metric appears flat, as does the ocean to an aviator far above. The closer the approach, the greater the degree of irregularity. Finally, at distances of the order* $L^*[=(\hbar G/c^2)=1.6\times10^{-33}$ *cm*], *the fluctuations in the typical metric component,* $g_{\mu\nu}$, *become of the same order as* $g_{\mu\mu}$ *themselves. Then the character of the space undergoes an essential change ... Multiple connectedness develops, as it does on the surface of an ocean where waves are breaking.*—Wheeler (1957).

[Wheeler's intuitive insight that space-time may be multiply connected (foam-like) on the scale $L^*$ has received considerable substantiation from recent research on the quantum theory of gravity. The lengthscale $L^*$ is called by Wheeler the "Planck length" and by others the "Planck–Wheeler length," in recognition that Wheeler was the first to understand clearly its significance.]

*Along with the fluctuations of the metric there occur fluctuations in the electromagnetic field. In consequence the typical multiply connected space ... has a net flux of electric lines of force passing through the "wormhole." These lines are trapped by the topology of space. These lines give the appearance of a positive charge at one end of a wormhole and a negative charge at the other.*"—Wheeler (1957).

*The quantum mechanical state of the metric field is a functional of the three-dimensional geometry $\mathcal{G}$ intrinsic to a spacelike surface $\sigma$*

$$\Psi = \Psi(\mathcal{G})$$

*The probability amplitude $\Psi$ typically falls off rapidly for* large scale *departures from a smoothly behaved metric, but falls off very little for departures from smoothness which are of the scale $L^* = (\hbar G/c^2) = 1.6 \times 10^{-33}$ cm or less.*—Wheeler (1960).

[This line of thought led, within the next few years, to the DeWitt–Wheeler formulation of quantum gravity, which in the hands of Hartle, Hawking, and others has recently achieved considerable success in understanding the initial conditions of the universe.]

*... one feels that one has, at last in gravitational collapse, a phenomenon where general relativity dramaticaly comes into its own, and where its fiery marriage with quantum physics will be consummated.*—Wheeler (1964a).

*The law of conservation of baryons must be abandoned for a star that exceeds the critical baryon number.*—Wheeler (1964b).

[This conclusion, forced on Wheeler by a long struggle to understand the fate of great masses, became a central issue in the 1970s with the discovery by Hawking that black holes can evaporate. This conclusion also led Wheeler to speculate, as an inspiration for future research, about the mutability of all conservation laws—and indeed of all the laws of physics—in the big bang and big crunch of the universe.]

*... These advances in cosmology and black hole physics lend new emphasis to some of the most remarkable consequences of Einstein's standard theory: ... nature conserves nothing; there is no constant of physics that is not transcended; or, in one word, mutability is a law of nature.*"—Wheeler (1979a).

Law without Law. "*Physical space-time is not mathematical space-time" is the one lesson of mutability; the other, "Physical law is not ideal mathematical law." Law that comes into being at the beginning of time and fades away at the end of time cannot be forever 100% accurate. Moreover, it must have come into being without anything to guide it into being.*—Wheeler (1980).

*Not only particles and the fields of force have to come into being at the big bang, but the laws of physics themselves, and this by a process of higgledy-piggledy as genetic mutation or the second law of thermodynamics.*—Wheeler (1982).

*Time is defined so that motion looks simple.*—Misner, Thorne, and Wheeler (1973).

*Time ends. That is the lesson of the "big bang." It is also the lesson of the black hole.*—Wheeler (1981a).

*Time is not a primary category in the description of nature. It is secondary, approximate, and derived.*—Wheeler (1982).

[Wheeler is also fond of the phrase "Time is God's way of keeping things from happening all at once" (anonymous graffiti, discovered by him on the wall of Pecan Street Cafe in Austin, Texas).]

*No account of existence that presupposes the concept of time can ever account for either time or existence.*—Wheeler (1982).

*No elementary phenomenon is a phenomenon until it is an observed phenomenon.*—Wheeler (1979b).

[This phrase, and variations on its theme, paraphrase the central point of Niels Bohr. They serve Wheeler and his colleagues both as a point of departure and as a motto in the struggle to understand the relation between the quantum and measurements.]

*... if one really understood the central point and its necessity in the construction of the world, one ought to be able to state it in one clear, simple sentence. Until we see the quantum principle with this simplicity we can well believe that we do not know the first thing about the universe, about ourselves, and about our place in the universe.*—Wheeler (1979a).

*The final story of the relation between the quantum and the universe is unfinished business. We can well believe that we will first understand how simple the universe is when we will recognise how strange it is.*—Wheeler (1979a).

[Wheeler often quotes Gertrude Stein on modern art: "It looks strange and it looks strange and it looks very strange, and then it suddenly does not look strange at all and you can't understand what made it look strange in the first place."]

*The elementary quantum phenomenon is the strangest thing in this strange world. It is strange because it has no localisation in space and time. It is strange because it has a pure yes–no character—one bit of meaning. It is strange because it is more deeply dyed with an information-theoretic flavor than anything in all physics.*—Wheeler (1984).

*Quantum mechanics promotes the mere "observer of reality" to the "participator in the defining of reality." It demolishes the view that the universe exists out there.*—Wheeler (1979a).

*The universe is a self-excited circuit. As it expands, cools and develops it gives rise to observer participancy. Observer participancy in turn gives what we call tangible reality to the universe.*—Wheeler (1980).

[A graphic summary of this point can be found, for instance, in Fig. 22.13 of Wheeler (1980).]

*... The wave function is within the present incomplete marriage of quantum theory and relativity not a very happy device for calculating the probability of real coincidences. Only when the counters have gone off has the reality of the situation declared itself.*—Wheeler (1979a).

*... It is wrong to think of that past [ascribed to a quantum phenomenon] as "already existing" in all detail. The "past" is theory. The past has no existence except as it is recorded in the present. By deciding what questions our quantum registering equipment shall put in the present we have an undeniable choice in what we have the right to say about the past.*—Wheeler (1980).

[The delayed-choice experiment described by Wheeler in this paper forces one to recognize this unexpected and yet inescapable conclusion about the quantum Universe.]

*What we call reality consists ... of a few iron posts of observation between which we fill an elaborate papier-mache of imagination and theory.*—Wheeler (1980).

*Spacetime in the prequantum dispensation was a great record parchment. This sheet, this continuum, this carrier of all that is, was and shall be, had its definite structure with its curves, waves and ripples; and on this great page every event, like a glued down grain of sand, had its determinate place. In this frozen picture a far-reaching modification is forced by the quantum. What we have the right to say of past spacetime, and past events, is decided by choices—of what measurements to carry out—made in the near past and now.*—Wheeler (1981b).

*Planck's discovery of the quantum in 1900 drove a crack in the armor that still covers the deep and secret principle of existence. In the exploitation of that opening we are at the beginning, not the end.*—Wheeler (1982).

*Beyond the probability interpretation of quantum mechanics, beyond all the standard analysis of idealised experiments, beyond the principle of indeterminacy and the limits it imposes, lie deep issues on which full agreement has not yet been reached in the physics community. They include questions like these: Does observation demand an irreversible act of amplification such as takes place in a grain of photographic emulsion or in the electron avalanche of a Geiger counter? And if so, what does one mean by "amplification"? And by "irreversible"? Does the quantum theory of observation apply in any meaningful way to the "whole universe"? Or is it restricted, even in principle, to the light cone? And if so, whose light cone? How are the observations made by different observers to be fited into a single consistent picture in space-time?*—Wheeler and Zurek (1983).

*... the most revolutionary discovery in science is yet to come! And come, not by questioning the quantum, but by uncovering that utterly simple idea that demands the quantum.*—Wheeler (1984).

*WHEELER'S FIRST MORAL PRINCIPLE. Never make a calculation until you know the answer. Make an estimate before every calculation, try a simple physical argument (symmetry! invariance! conservation!) before every derivation, guess the answer to every puzzle.*—Taylor and Wheeler (1966).

## REFERENCES

Bohr, Niels, and Wheeler, John A. (1939), "The Mechanism of Nuclear Fission," *Phys. Rev.* **56**, 426.

Feynman, Richard P. (1966), "The Development of the Space-Time View of Quantum Electrodynamics," *Phys. Today*, August 1966, p. 31.

Harrison, Kent, Wakano, Masami, and Wheeler, John A. (1958), "Matter-Energy at High Density; End Point of Thermonuclear Evolution," in *La Structure et L'Evolution de l'Univers*, Proceedings of Eleventh Solvay Congress, Brussels, 9–13 June, 1958 (Stoops, Brussels), p. 124.

Lindquist, Richard W., and Wheeler, John A. (1957), "Dynamics of a Lattice Model Universe by the Schwarzschild-Cell Method," *Rev. Mod. Phys.* **29**, 432.

Misner, Charles W., Thorne, Kip S., and Wheeler, John A. (1973), *Gravitation* (Freeman, San Francisco), p. 23.

Regge, Tullio, and Wheeler, John A. (1957), "Stability of a Schwarzschild Singularity," *Phys. Rev.* **108**, 1063.

Ruffini, Remo, and Wheeler, John A. (1971), "Relativistic Cosmology and Space Platforms," in *Proceedings of Conference on Space Physics* (ESRO, Paris).

Taylor, Edwin F., and Wheeler, John A. (1966), *Spacetime Physics* (Freeman, San Francisco).

Toll, John, and Wheeler, John A. (1951), "Some Pair-Theoretic Applications of the Dispersion Relation," *Phys. Rev.* **81**, 654.

Wheeler, John A. (1937), "On the Mathematical Description of Light Nuclei by the Method of Resonating Group Structure," *Phys. Rev.* **52**, 1107.

Wheeler, John A. (1946), "Polyelectrons," *Ann. N.Y. Acad. Sci.* **48**, 219–238.

Wheeler, John A. (1949), "Some Consequences of the Electromagnetic Interaction between $\mu^-$-Mesons and Nuclei," *Rev. Mod. Phys.* **21**, 133.

Wheeler, John A. (1955), "Geons," *Phys. Rev.* **97**, 511.

Wheeler, John A. (1957), "On the Nature of Quantum Geometrodynamics," *Ann. Phys.* **2**, 604–614.

Wheeler, John A. (1960), "Neutrinos, Gravitation, and Geometry," in *Rendiconti della Scuola Internazionale di Fisica "Enrico Fermi," XI Corso* (Zanichelli, Bologna, 1960); reprinted in J. A. Wheeler, *Geometrodynamics* (Academic Press, New York, 1962).

Wheeler, John A. (1964a), "Geometrodynamics and the Issue of the Final State," in *Relativity, Groups, and Topology*, C. deWitt and B. DeWitt, eds. (Gordon and Breach, New York), pp. 317–522.

Wheeler, John A. (1964b), "The Superdense Star and the Critical Nucleon Number," in *Gravitation and Relativity*, H.-Y. Chiu and W. F. Hoffmann, eds. (Benjamin, New York).

Wheeler, John A. (1966), "Superdense Stars," *Ann. Rev. Astron. Astrophys.* **4**, 393.

Wheeler, John A. (1968), "Our Universe: The Known and the Unknown," *Am. Sci.* **56**, 1.

Wheeler, John A. (1979a), "The Quantum and the Universe," in *Relativity, Quanta and Cosmology*, Vol. II, M. Pantaleo and F. de Finis, eds. (Johnson Reprint Corp., New York).

Wheeler, John A. (1979b), *Frontiers of Time* (North-Holland Publishing Co., Amsterdam); also in *Rediconti della Scuola Internazionale di Fisica "Enrico Fermi,"* N. Toraldo di Francia and Bas van Fraassen, eds. (North-Holland Publishing Co., Amsterdam), pp. 395–497.

Wheeler, John A. (1980), "Beyond the Black Hole," Chapter 22 in *Some Strangeness in Proportion*, Harry Wolf, ed. (Addison-Wesley, Reading, Massachusetts).

Wheeler, John A. (1981a), "The Lesson of the Black Hole," *Proc. Am. Philos. Soc.* **125**, 25.

Wheeler, John A. (1981b), "Delayed-Choice Experiments and the Bohr–Einstein Dialogue," The American Philosophical Society and the Royal Society, pp. 25–37, papers read at the Meeting, June 5, 1980 (American Philosophical Society, Philadelphia).

Wheeler, John A. (1982), "The Computer and the Universe," *Int. J. Theor. Phys.* **21**, Nos. 6/7.

Wheeler, John A. (1984), "Bits, Quanta and Meaning," in *Theoretical Physics Meeting: Commemorative Volume on the Occasion of Eduardo Caianiello's Sixtieth Birthday*, A. Giovanni, M. Marinaro, F. Mancini, and A. Rimini, eds. (Edizioni Scientifici Italiani, Naples), pp. 121–134.

Wheeler, John A. (1985), quoted in Jeremy Bernstein, "Retarded Learner," *Princeton Alumni Weekly*, October 9, 1985, p. 28.

Wheeler, John A., and Feynman, Richard P. (1945), "Interaction with the Absorver as the Mechanism of Radiation," *Rev. Mod. Phys.* **17**, 157.

Wheeler, John A., and Zurek, Wojciech H., eds. (1983), *Quantum Theory and the Measurement* (Princeton University Press, Princeton), Preface, p. i.

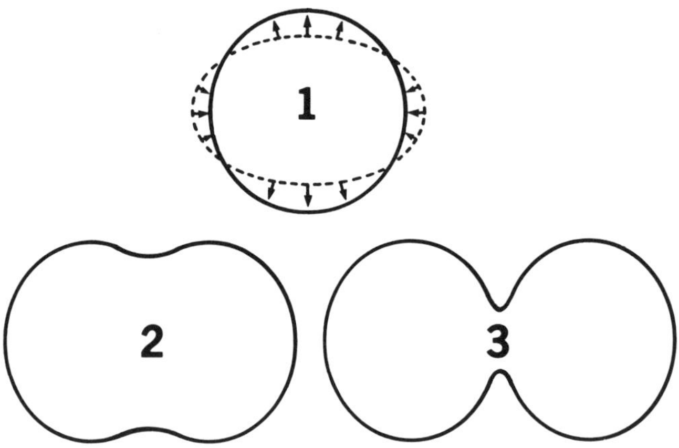


*Plate 2*

# Early Estimates of the Strength of the Nuclear Spin–Orbit Force

**H. H. Barschall**[1] **and Louis Brown**[2]

*Received September 4, 1985*

*Before the development of the nuclear shell model estimates of the strength of the nuclear spin–orbit interaction varied widely. Wheeler was the first to conclude that the nuclear spin–orbit interaction produces splittings of several MeV. This conclusion appeared, however, to be inconsistent with some experimental results that later turned out to be faulty.*

In 1949 Mayer[(1)] and Haxel, Jensen, and Suess[(2)] obtained the "Magic Numbers" in nuclei by assuming strong spin–orbit coupling. Maria Mayer in her first paper on the subject thanks Enrico Fermi for having asked her whether there is any indication of spin–orbit coupling. None of these physicists appear to have known that John Wheeler[(3)] had concluded in 1940 that the nuclear spin–orbit interaction was very strong and might produce splittings of several MeV.

In atomic physics the spin–orbit interaction had been understood since the 1920's, i.e., that it is due to the interaction of the magnetic moment of the electron and the magnetic field generated by the orbital motion of the electron. The resulting fine structure of atomic states is very small.

The earliest mention of nuclear spin–orbit coupling was in a 1936 paper by Inglis.[(4)] On the basis of the Thomas relativistic effect he showed that the nuclear spin–orbit force would produce an inverted doublet, i.e., that the $l+1/2$ state would have lower energy than the $l-1/2$ state, the opposite of the order in atomic physics. Inglis proposed that the excited state of $^7$Li that had just been found by Rumbaugh and Hafstad[(5)] was the

[1] University of Wisconsin, Madison, Wisconsin 53706.
[2] Carnegie Institution, Washington, D. C. 20015.

$^2P_{1/2}$ component of a doublet, the ground state being the $^2P_{3/2}$ component. His calculation yielded a splitting of 0.2 MeV, which, considering the rather general assumptions, appeared consistent with the observed value of 0.4 MeV. In the same issue of the *Physical Review*, Furry[(6)] derived the same result from the Dirac equation.

Inglis' explanation of the splitting of the ground state of $^7$Li turned out to be correct, but the 1936 theory was cast in general terms and did not consider the exceptional structure of $^7$Li. The theory greatly underestimated the splitting for the general case, as was recognized by Inglis[(7)] in 1951. The apparent experimental confirmation of the early theory of the nuclear spin–orbit coupling strongly affected the development of nuclear theory for the following thirteen years during which time faulty experiments appeared to support the theory, while experiments and their analysis that showed a much larger spin–orbit coupling had little impact.

The measurements that provided the first evidence that spin–orbit splitting in nuclei was much larger than previously assumed were not originally intended to address this question. Instead the intent was to develop a method for measuring spectra of fast neutrons. In 1938 Baldinger, Huber, and Staub[(8)] in Zürich had attempted to obtain such spectra by observing the distribution in energy of recoiling alpha particles in an ionization chamber. They deduced the neutron spectrum from the derivative of the recoil energy distribution. Their neutron source was the $^2$H$(d, n)$ $^3$He reaction. From the observed recoil energy distribution they concluded that, in addition to the main group of 2.6-MeV neutrons, many neutrons of lower energy were present, presumably degraded by inelastic scattering.

At Princeton where a similar neutron source had been constructed under the direction of Rudolf Ladenburg the Zürich experiments were repeated soon afterwards.[(9)] Even though an effort was made to minimize scattering material that might degrade the neutrons, the pulse height distribution was very similar to that observed in Zürich. When the experiment was repeated with hydrogen instead of helium, the low-energy pulses that the Zürich group had interpreted as caused by degraded neutrons were not observed. The relationship which the Zürich group had used to deduce the neutron spectrum was correct only if the scattering was isotropic in the center-of-mass system. Since the recoil proton spectrum observed at Princeton was consistent with the expected isotropic scattering of monoenergetic neutrons, the observed alpha-particle recoil spectrum could serve to determine the angular distribution of neutron-alpha scattering. Wheeler[(3)] pointed out that there was a surprisingly simple relationship between these distributions, i.e., the energy distribution of the recoils in the laboratory system is the same as the angular distribution of the scattering in the center-of-mass system, if this distribution is plotted against the cosine of the center-of-mass scattering angle. This simple relation holds also at relativistic energies and for inelastic processes.[10]

The Princeton measurements[9] were performed at two neutron energies, 2.5 and 3.1 MeV, and are shown in Fig. 1. The results were similar at the two energies and consistent with the Zürich measurements. The anisotropy in the neutron-alpha scattering at neutron energies around 3 MeV was not unexpected because in 1939 Staub and Stephens,[11] work-

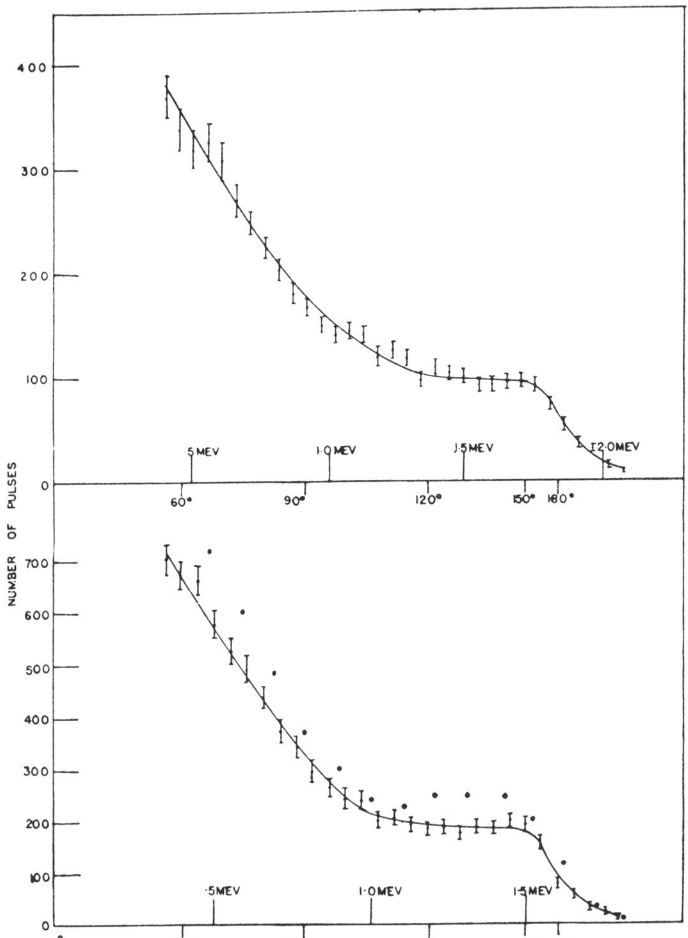


Fig. 1. Energy distribution of recoiling alpha particles. Upper curve for incident 3.1-MeV neutrons, lower curve for 2.5-MeV neutrons (from Ref. 9). Small circles are from Ref. 8. The energy distribution is also the CM angular distribution plotted against the cosine of the CM scattering angle.

ing at the California Institute of Technology, had measured the backscattering of neutrons as a function of energy by observing alpha recoils in a cloud chamber. They found an anomalously high scattering cross section near 1 MeV neutron energy which they attributed to the effect of a level in $^5$He. From the width and magnitude of the peak they concluded that it must be a *P*-wave resonance. The neutron source used in these measurements did not really allow a measurement as a function of energy. In fact, no source available at the time produced monoenergetic neutrons of variable energy around 1 MeV. In order to study the energy dependence of the neutron-alpha scattering cross section in more detail Staub and Tatel,[(12)] working at Stanford, used *d–d* neutrons that were slowed down with paraffin. They tested this method with *n–p* scattering, which was expected to show only *s*-wave scattering, essentially independent of neutron energy. The result was a uniform distribution of neutron energies. The comparison of *n–p* and $n$–$\alpha$ scattering showed the resonance in $n$–$\alpha$ scattering clearly and is reproduced in Fig. 2. Staub and Tatel saw in these data two peaks, separated by 0.3 MeV. Whether they would have seen two peaks, if they had not expected to find a $P_{3/2}$–$P_{1/2}$ splitting between 0.2 and 0.4 MeV, is an open question. They interpreted their data in terms of a resonance theory developed by Bloch,[(13)] but were not able to decide whether the $P_{1/2}$ level was above or below the $P_{3/2}$ level.

Staub and Tatel[(12)] suggested that the low-energy alpha recoils in the Zürich experiments were due to relatively few neutrons of energies near 1 MeV, which were detected with high probability because of the peak in the scattering cross section near that energy.

That there was a *P*-wave resonance in neutron-alpha scattering at 1 MeV neutron energy was known when the Princeton measurements were performed, hence an anisotropy in $n$–$\alpha$ scattering at 2.5 MeV was not entirely unexpected. Although the resonance at 1 MeV did not appear to extend to 2.5 MeV, there might be unknown *P*-wave resonances at higher energies. On the other hand, the observed angular distribution could not be fitted with *S*-waves and only one *P*-wave. Although at this energy *D*-waves were not expected to be important, introducing a *D*-wave phase shift still did not give a fit to the measurements either. If the resonance at 1 MeV is a closely spaced doublet, the $P_{1/2}$ and $P_{3/2}$ phase shifts at 2.5 MeV would be expected to be almost the same. Wheeler[(3)] showed, however, that the observations required a difference in the $P_{1/2}$ and $P_{3/2}$ phase shifts of the order of 60° although there was a substantial uncertainty in the phase shifts because of the uncertainty in the possible effect of *D*-waves and because there was no information about the magnitude of the *S*-wave phase shift. The analysis was inconsistent with the small magnitude of spin–orbit splitting found by Staub and Tatel.

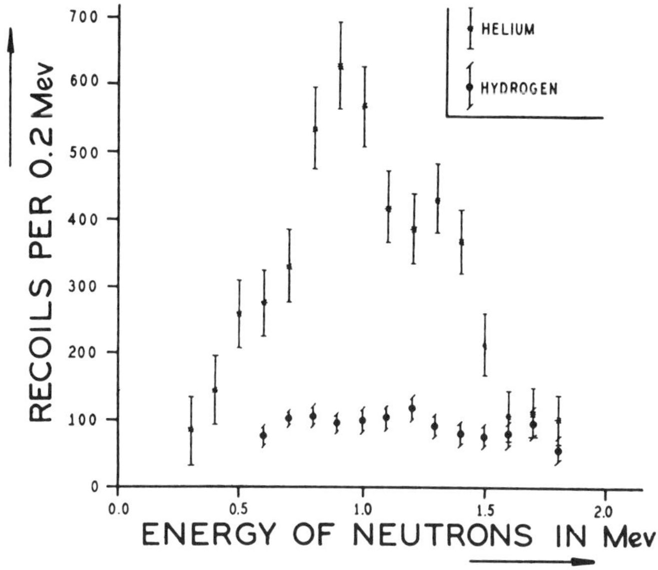


Fig. 2. Backward scattering of neutrons by hydrogen and helium (from Ref. 12). The helium data were interpreted as showing two peaks separated by 0.3 MeV.

In addition to the experiments performed in California and at Princeton on neutron scattering, Heydenburg and Ramsey,[14] at the Department of Terrestrial Magnetism of the Carnegie Institution in Washington, measured the differential scattering cross section for proton-alpha scattering from 1.2 to 3.0 MeV in steps of 0.2 MeV and observed a strong resonance at 2 MeV. Measurements of charged-particle scattering were much easier than neutron measurements and yielded more reliable results, because of much higher count rates and because of fewer problems with degraded projectiles. The DTM group had developed this technique to high precision for studying the proton–proton interaction. The analysis of the $p$–$\alpha$ data was, however, more difficult than that of $n$–$\alpha$ scattering because of the interference between nuclear and Coulomb scattering. The analysis would ordinarily have been performed by Breit, who had been a long-time collaborator of the DTM group, but in 1941 Breit and most other physicists had turned to other tasks. The $p$–$\alpha$ data were analyzed by Landau and Smorodinsky[15] in 1944, and they write: "Wheeler and Barschall have shown that for neutrons with an energy of 2.5 MeV the phases corresponding to the waves $P_{1/2}$ and $P_{3/2}$ differ approximately by 60°. Similar results have been obtained by us for the scattering of protons by $\alpha$-particles, according to the data of Heydenburg and Ramsey. It turned out that in this case for the description of the scattering it is necessary to take into account three states: $S$, $P_{1/2}$, and $P_{3/2}$, the phases of the latter differing strongly. These data show that the conclusion of Staub and Tatel that a narrow doublet exists is wrong.... The existence of such a narrow doublet is even quite unexpected if one takes into account the fact that the spin–orbit interaction in the case of nuclear particles is not small. This follows from the presence of a relatively large quadrupole moment of the deuteron.... Bloch's formula, which has been used by Staub and Tatel,... was based on the incorrect assumption that the neutron width does not depend on energy."

In 1945 the experimenters who had been involved in the contradictory $n$–$\alpha$ experiments that had been performed at Princeton and Stanford were working at Los Alamos. It was now possible to perform measurements as a function of neutron energy around 1 MeV by using the reaction $^7$Li$(p, n)$ $^7$Be. Therefore it appeared easily possible to resolve the discrepancy by observing the energy distribution of alpha particle recoils in an ionization chamber in the region around 1 MeV. Such distributions were measured at five neutron energies from 0.8 to 1.6 MeV by Hall and Koontz.[(16)] The distributions looked rather strange, but the total cross section as a function of energy looked very similar to that observed by Staub and Tatel. Hall and Koontz tried to analyze their data in terms of Bloch's theory, taking into account, however, the dependence of level width on energy, but they did not obtain a fit to the data. The problem with these experiments was that the neutrons were not really monoenergetic, as the experimenters had assumed. The first information about spin–orbit coupling had come from the first excited state of $^7$Li. It should not have come as a surprise that the mirror nucleus $^7$Be had a similar state, which was responsible for the production of a group of lower-energy neutrons in the source reaction $^7$Li$(p, n)$ $^7$Be. Although the presence of this second group of neutrons was actually clearly visible in the $\alpha$ particle spectra, it was not recognized until 1950 when Hall[(17)] reexamined the data. The experiment that was intended to settle the argument had given what later was shown to be a wrong answer.

Neither Wheeler's work nor that of Landau and Smorodinsky appears to have been convincing. In 1948 the well-known book on nuclear theory entitled *Nuclear Forces* was published by Rosenfeld.[(18)] The entire Chapter 17 is devoted to the effect of spin–orbit coupling. One section of the chapter is entitled "Fine Structure of $^5$He and $^5$Li." In this section Rosenfeld discusses the California, Princeton, and Los Alamos experiments, as well as the analyses performed by Wheeler, and by Landau and Smorodinsky. He concludes: "It would seem that Staub and Tatel's interpretation of the $^5$He resonance level, while in serious need of further check, may provisionally be retained." In the following section Rosenfeld discusses another derivation of the spin–orbit coupling in $^5$He by Dancoff[19] who found it to be a consequence of the nucleon–nucleon tensor force. Dancoff found that this theory gave a narrow doublet, but with the $P_{1/2}$ level at lower energy than the $P_{3/2}$ level. This may have influenced Rosenfeld's conclusion.

Schwinger[20] also believed that there was a closely spaced $P_{1/2}$–$P_{3/2}$ doublet in $^5$He and suggested that neutron polarization could be demonstrated by double scattering in He because the energy loss in elastic collisions would be comparable with the doublet separation.

Even though Schwinger's suggestion to use the $^5$He doublet to observe neutron polarization was based on an erroneous measurement, it was in fact equally applicable to what turned out to be a single $^5$He level. Polarization effects were soon observed and provided the most important tool for studying not only the spin–orbit interaction, but nuclear structure in general. The large polarization produced in $n$–$\alpha$ and $p$–$\alpha$ scattering has remained the most useful polarization analyzer, since the analyzing power can be calculated reliably from the known scattering phase shifts.

A direct measurement of the $n$–$\alpha$ total cross section in a transmission experiment[21] was performed at Wisconsin in 1951. It showed a very broad peak at 1.15 MeV with no evidence for any splitting. Also at Wisconsin and at about the same time Adair[22] repeated the measurement of the angular distribution of $n$–$\alpha$ scattering at energies from 0.4 to 2.7 MeV using monoenergetic neutrons from the $^3$H$(p, n)$ $^3$He reaction. The results were consistent with the old Princeton measurements. Adair's analysis led to the conclusion that the peak in the cross section near 1 MeV is due to a $P_{3/2}$ level with the $P_{1/2}$ member of the doublet located 5 MeV above the $P_{3/2}$ level. Figure 3 shows the measured $n$–$\alpha$ total cross section as a function of neutron energy compared with Adair's calculation of the contributions of the three partial waves that contribute to the cross section. Adair showed that the ground state of $^5$He, which produces the large peak shown in Fig. 3, has a reduced width about equal to the single-particle width. This means that the interaction can be represented as the motion of a neutron in the potential of the ground state of the alpha particles. This confirmed the relevance of the large spin–orbit coupling observed in nucleon-alpha scattering to that postulated in the shell model.

In the meantime the measurement of the elastic scattering of protons by alpha particles had been extended to higher energies by Freier *et al.*[23] at Minnesota, and a phase shift analysis by Critchfield and Dodder[24] had confirmed the conclusion by Landau and Smorodinsky that the phase shift difference between the $P_{1/2}$ and $P_{3/2}$ waves was large, although the order of the $P_{1/2}$–$P_{3/2}$ doublet could not be deduced from this analysis. In 1952

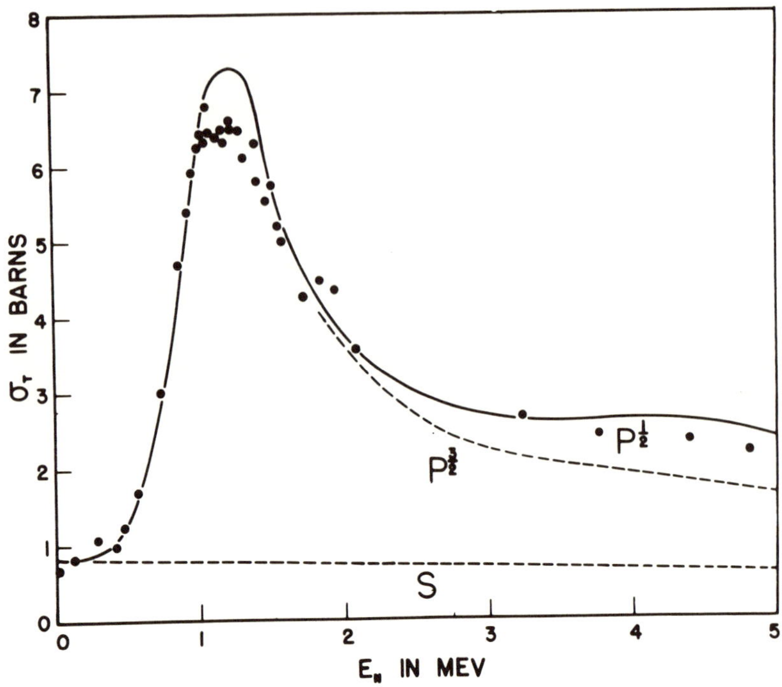


Fig. 3. The total cross section of helium as a function of neutron energy. The data points are from Ref. 21; the curves (from Ref. 22) show the results of the analysis of the contributions of various partial waves.

polarization measurements,[25] also carried out at Minnesota, showed that the doublet was inverted, as was assumed in the shell model and had been deduced by Adair from the analysis of neutron-alpha particle scattering.

In spite of the evidence from nucleon-alpha scattering and from the shell model the importance of the large spin–orbit interaction in the nucleon–nucleus interaction was not yet generally accepted. In 1954 Feshbach, Porter, and Weisskopf[26] proposed an optical model to explain the observed energy dependence of the total interaction cross section of neutrons with nuclei.[27] The potential describing the interaction did not contain a spin–orbit coupling term. Figure 4 shows the comparison of the results of the calculation with the measurements. The surprising feature are the letters *S*, *P*, *D*, *F*, *G* on the calculated broad maxima. If these maxima represented single-particle states, spin–orbit coupling should split, for example, a *P*-state into widely separated $P_{3/2}$ and $P_{1/2}$ states. That the spin–orbit interaction is important in neutron–nucleus scattering was shown by polarization measurements performed at Wisconsin by Adair,

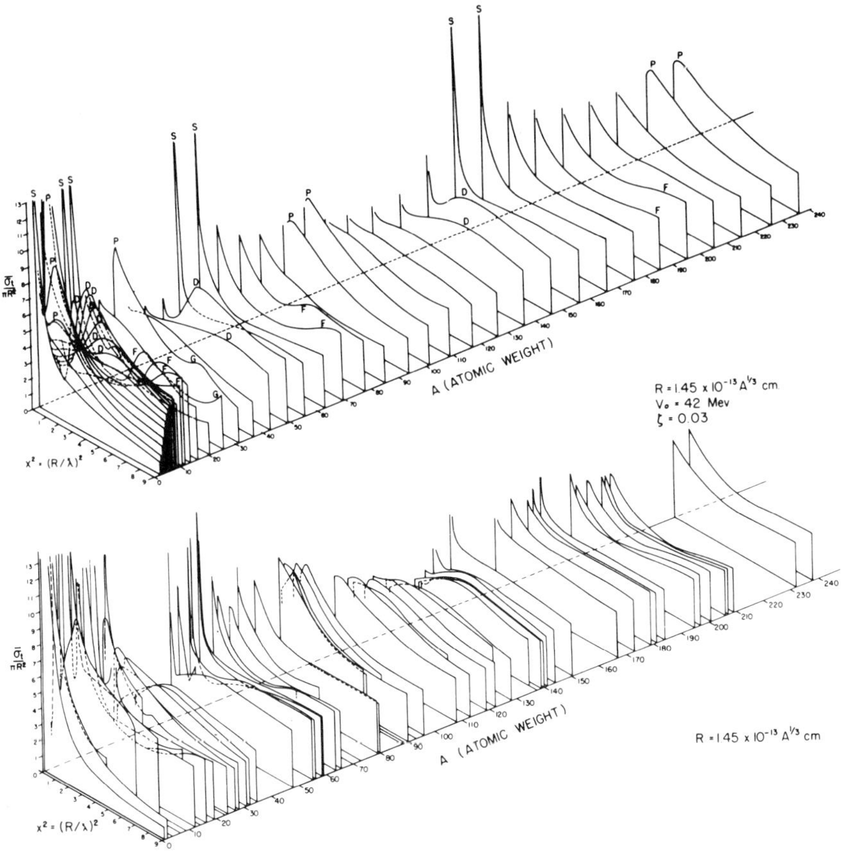


Fig. 4. Neutron total cross sections as functions of neutron energy and as functions of atomic weight. The upper portion is calculated from an optical model (Ref. 26). The lower portion is a collection of measurements (from Ref. 27). The capital letters on the broad maxima in the upper part appear to imply the absence of spin–orbit splitting.

Darden, and Fields[28] and published in the same issue of the *Physical Review* as the theoretical results shown in Fig. 4. The reason why the broad peaks shown in Fig. 4 were not split turned out to be that the peaks are not resonances caused by neutrons of definite angular momentum, but nonresonant interference maxima.[29]

Contemporary textbooks on nuclear physics usually give the impression that the importance of the spin–orbit interaction in nuclei was first recognized in 1949 in the discovery of the shell model. We have tried to review some of the earlier investigations of the nuclear spin–orbit interaction, which, however, did not have any direct impact on the development of the shell model.

## REFERENCES


1. M. G. Mayer, *Phys. Rev.* **75**, 1969 (1949).
2. O. Haxel, J. H. D. Jensen, and H. E. Suess, *Phys. Rev.* **75**, 1766 (1949).
3. J. A. Wheeler and H. H. Barschall, *Phys. Rev.* **58**, 682 (1940).
4. D. R. Inglis, *Phys. Rev.* **50**, 783 (1936).
5. L. Rumbaugh and L. Hafstad, *Phys. Rev.* **50**, 681 (1936).
6. W. Furry, *Phys. Rev.* **50**, 784 (1936).
7. H. H. Hummel and D. R. Inglis, *Phys. Rev.* **81**, 910 (1951); D. R. Inglis, *Phys. Rev.* **85**, 492 (1952).
8. E. Baldinger, P. Huber, and H. Staub, *Helv. Phys. Acta* **11**, 245 (1938).
9. H. H. Barschall and M. H. Kanner, *Phys. Rev.* **58**, 590 (1940).
10. H. H. Barschall and J. L. Powell, *Phys. Rev.* **96**, 713 (1954).
11. H. Staub and W. E. Stephens, *Phys. Rev.* **55**, 131 (1939).
12. H. Staub and H. Tatel, *Phys. Rev.* **58**, 820 (1940).
13. F. Bloch, *Phys. Rev.* **58**, 829 (1940).
14. N. P. Heydenburg and N. F. Ramsey, *Phys. Rev.* **60**, 42 (1941).
15. L. Landau and J. Smorodinsky, *J. Phys. USSR* **8**, 154 (1944).
16. T. A. Hall and P. G. Koontz, *Phys. Rev.* **72**, 196 (1947).
17. T. A. Hall, *Phys. Rev.* **77**, 411 (1950).
18. L. Rosenfeld *Nuclear Forces* (North-Holland, Amsterdam, 1948).
19. S. M. Dancoff, *Phys. Rev.* **58**, 326 (1940).
20. J. Schwinger, *Phys. Rev.* **69**, 681 (1946).
21. S. Bashkin, F. P. Mooring, and B. Petree, *Phys. Rev.* **82**, 378 (1951).
22. R. K. Adair, Phys, Rev. **82**, 750 (1951); **86**, 155 (1952).
23. G. Freier, E. Lampi, W. Sleator, and J. H. Williams, *Phys. Rev.* **75**, 1345 (1949).
24. C. L. Critchfield and D. C. Dodder, *Phys. Rev.* **76**, 602 (1949).
25. M. Heusinkveld and G. Freier, *Phys. Rev.* **85**, 80 (1952).
26. H. Feshbach, C. E. Porter, and V. F. Weisskopf, *Phys. Rev.* **96**, 448 (1954).
27. H. H. Barschall, *Phys. Rev.* **86**, 431 (1952).
28. R. K. Adair, S. E. Darden, and R. E. Fields, *Phys. Rev.* **96**, 503 (1954).
29. K. W. McVoy, *Ann. Phys. (N.Y.)* **43**, 91 (1967).

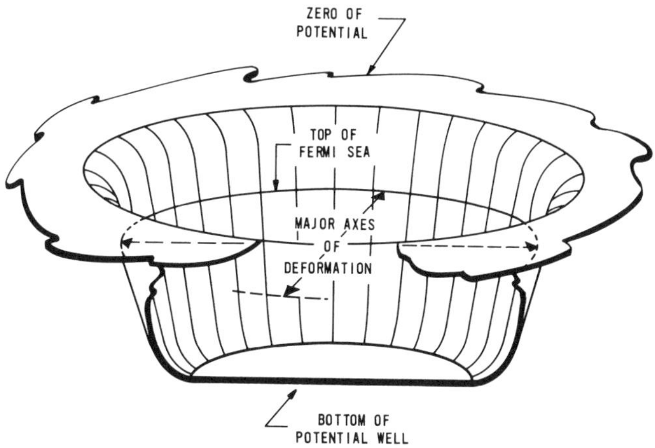
ZERO OF
POTENTIAL
TOP OF
FERMI SEA
MAJOR AXES
OF
DEFORMATION
BOTTOM OF
POTENTIAL WELL

*Plate 3*

# Momentum Projection of Solitons Including Quantum Corrections

**Lawrence Wilets**[1]

*Received August 8, 1985*

*The method of projection is applied to a relativistic field theory of fermions interacting with a nonlinear scalar field, specifically the Friedberg–Lee soliton model. Projection is effected by operating on a localized "bag" state with the translation operator* $\exp(i\mathbf{P}\cdot\mathbf{Z})$*, and integrating over* $\mathbf{Z}$*. The resulting state is an eigenstate of zero momentum. The energy and the expectation value of other physical operators can be expressed as Gaussian moments of the Hamiltonian or the physical operator times powers of the momentum operator taken with respect to the bag state. Renormalization in the one-loop approximation is discussed in detail for the boson sector, and briefly for the fermion sector. The method can be tested for convergence against nonexpansion techniques. The latter, however, cannot so easily handle distortion of the Bose modes or the distortion of the Dirac sea.*

## 1. INTRODUCTION

Hill and Wheeler[(1)] invented the method of generator coordinates in order to handle collective phenomena in many-body systems without the introduction of redundant variables. The method has proved to be very useful for nonrelativistic systems of fermions interacting through two-body potentials. It is especially appropriate for relativistic systems of interacting Fermi and Bose fields which cluster into localized structures such as quark bags, since then there is no tractable center-of-mass coordinate.[(2)]

In the present work, we address one facet of the generator coordinate method, namely that of projection of a localized state onto a state of zero momentum. This removes spurious center-of-mass motion from the wave

[1] Institute for Nuclear Theory, Department of Physics, FM-15, University of Washington, Seattle, Washington 98195.

function and permits the calculation of energies, moments of the electromagnetic form factors, and other system properties without further center-of-mass corrections.

A method of handling projection energy variation after projection, and subsequent projection to a state of finite momentum, has been discussed by Lübeck, Birse, Henley, and Wilets[3] (LBHW) in the context of the (nonlinear) soliton bag model. They expanded the Bose (sigma) field in plane waves in order to preserve translational invariance of the field operators. The Fermi (quark) fields were expanded in distorted bag states. The quark vacuum is thus not relativistically invariant. The translation of only the valence quarks was considered. The neglect of the (distorted) quark vacuum caused no numerical difficulties, but left open the possibility of further corrections.

In the present work, the sigma field is written as the sum of a $c$-number mean field function plus distorted quantum modes.[4] Quark states are calculated in the localized sigma field. Projection is effected by a moment expansion of the translation operator $P$ about the localized state. This involves evaluation of the $n$-even moments, $\langle P^n \rangle$ and $\langle HP^n \rangle$, where $P$ is the sum of quark and sigma field momenta.

Quantum corrections to classical solutions of field theories have been discussed by a number of authors.[5] Most make use of collective coordinate approaches. An exception is the work of Callan and Gross[6] which uses generator coordinates in a path-integral formalism.

In order to evaluate quantum corrections consistently, it is necessary to remove divergences by renormalization. We exhibit explicitly the renormalization procedure for the Bose fields through the one-loop approximation for the lowest moment ($n = 0$) of the expansion. The first quantum corrections to the mean field approximation are displayed. There still remain finite terms after renormalization, analogous to those which appear in vacuum polarization in QED. The procedure for renormalization of the fermion sector is not given explicitly, but the techniques are similar.

The matrix elements of the next moment ($n = 2$) are given. This is the leading order for the projection (recoil) correction, and is probably as far as one would wish to carry the moment expansion. Renormalization in this order has not been explored thoroughly yet, but any further divergences should be identifiable with renormalization of the coupling constants as in the lowest order; indeed, it does not appear to us that there are further divergences.

Since the LBHW calculations are equivalent to all orders in the moment expansion but plane waves in the sigma field expansion, the present program presents an opportunity to test the two methods: (a) How good is the moment expansion (using plane waves)? (b) How important is the distortion of the sigma modes (for, say, the lowest moment)? (c) How important is the distortion of the Dirac sea? The numerical implementation of this program is a subject of future research.

## 2. THE GAUSSIAN-MOMENTS EXPANSION

We begin with a localized "bag" state $|B\rangle$ centered at the origin and normalized to $\langle B|B\rangle = 1$. A state translated by $\mathbf{Z}$ can be constructed by

$$|\mathbf{Z}\rangle = e^{i\mathbf{P}\cdot\mathbf{Z}}\,|B\rangle \tag{1}$$

where $\mathbf{P}$ is the total momentum operator and hence the generator of infinitesimal translations. The un-normalized state obtained by projecting $|B\rangle$ onto a state of zero momentum is

$$|\Psi\rangle = V^{-1/2}\int |\mathbf{Z}\rangle\, d^3Z \tag{2}$$

The normalization is given by the integral

$$\begin{aligned}\langle\Psi|\Psi\rangle &= V^{-1}\int d^3Z'\int d^3Z\,\langle\mathbf{Z}'|Z\rangle = \int d^3Z\,\langle -\tfrac{1}{2}\mathbf{Z}\,|\tfrac{1}{2}\mathbf{Z}\rangle\\ &= \int d^3Z\,\langle B|\,e^{i\mathbf{P}\cdot\mathbf{Z}}\,|B\rangle = \int d^3Z\,\langle B|\,j_0(PZ)\,|B\rangle\\ &= \int d^3Z[1-\tfrac{1}{6}\langle P^2\rangle Z^2 + \tfrac{1}{120}\langle P^4\rangle Z^4 - \cdots]\end{aligned} \tag{3}$$

where we have used the observation that $|B\rangle$ is a spherically symmetric state and so the angle average can be done immediately. We use the notation $\langle P^2\rangle = \langle B|\,P^2\,|B\rangle$, etc. Note that $j_0(PZ)$ is a function of $P^2$ and $Z^2$.

Following the experience of nuclear physics, we re-expand the final square bracket in (3) in terms of a Gaussian times a power series:

$$\begin{aligned}&[1-\tfrac{1}{6}\langle P^2\rangle Z^2 + \tfrac{1}{120}\langle P^4\rangle Z^4 - \cdots]\\ &\quad = e^{-1/6\langle P^2\rangle Z^2}[1+(\tfrac{1}{120}\langle P^4\rangle - \tfrac{1}{72}\langle P^2\rangle^2)\,Z^4 + \cdots]\end{aligned} \tag{4}$$

The Gaussian factor was chosen to eliminate the quadratic term in the power series. The $\mathbf{Z}$-integration can now be performed to yield

$$\langle\Psi|\Psi\rangle = 2\left(\frac{6\pi}{\langle P^2\rangle}\right)^{3/2}\left[1+\frac{9\langle P^4\rangle - 15\langle P^2\rangle^2}{8\langle P^2\rangle^2}+\cdots\right] \tag{5}$$

A similar procedure can be used to calculate $\langle \Psi | H | \Psi \rangle$. We note that **P** commutes with the Hamiltonian, so we can write immediately

$$\langle \Psi | H | \Psi \rangle = \int d^3 Z [\langle H \rangle - \tfrac{1}{6} \langle HP^2 \rangle Z^2 + \tfrac{1}{120} \langle HP^4 \rangle Z^4 + \cdots] \quad (6)$$

It is convenient and appropriate to extract the same Gaussian from the integrand of (6) as was done for the normalization integral. We find

$$\langle \Psi | H | \Psi \rangle = \int d^3 Z \, e^{-1/6 \langle P^2 \rangle Z^2} [\langle H \rangle - \tfrac{1}{6} (\langle HP^2 \rangle - \langle H \rangle \langle P^2 \rangle) Z^2 + (\tfrac{1}{120} \langle HP^4 \rangle - \tfrac{1}{36} \langle HP^2 \rangle \langle P^2 \rangle + \tfrac{1}{72} \langle H \rangle \langle P^2 \rangle^2) Z^4 + \cdots] \quad (7)$$

Performing the **Z**-integration and taking the ratio to $\langle \Psi | \Psi \rangle$, we find

$$\frac{\langle \Psi | H | \Psi \rangle}{\langle \Psi | \Psi \rangle} = \left[ \langle H \rangle - \frac{3}{2} \frac{\langle HP^2 \rangle - \langle H \rangle \langle P^2 \rangle}{\langle P^2 \rangle} + \frac{9 \langle HP^4 \rangle - 30 \langle HP^2 \rangle \langle P^2 \rangle + 15 \langle H \rangle \langle P^2 \rangle^2}{8 \langle P^2 \rangle^2} + \cdots \right] \div \left[ 1 + \frac{9 \langle P^4 \rangle - 15 \langle P^2 \rangle^2}{8 \langle P^2 \rangle^2} + \cdots \right] \quad (8)$$

## 3. REPRESENTATION OF THE FIELD OPERATORS; VARIATIONAL EQUATIONS

In the present work, we obtain the one-loop approximation using a variational approach based on the Hamiltonian formulation of a field theory. This will provide wave functions which can be projected onto momentum eigenstates. We will see that for the vacuum and for localized states the results are identical to those obtained with the more conventional effective Lagrangian methods.

To be specific, we consider the soliton bag model.[4,7,8] This is defined by the Hamiltonian

$$H = \int d^3 r \left[ \psi^\dagger(\mathbf{r}) (-i \boldsymbol{\alpha} \cdot \nabla + g \beta \sigma(\mathbf{r})) \psi(\mathbf{r}) + \tfrac{1}{2} (Z_\sigma^{-1} \pi(\mathbf{r})^2 + Z_\sigma |\nabla \sigma(\mathbf{r})|^2) + d_0 \sigma(\mathbf{r}) + \frac{a_0}{2} \sigma(\mathbf{r})^2 + \frac{b_0}{3!} \sigma(\mathbf{r})^3 + \frac{c_0}{4!} \sigma(\mathbf{r})^4 \right] \quad (9)$$

where the coefficients $d_0$, $a_0$, $b_0$, and $c_0$ are bare values and $Z_\sigma$ is a wave function renormalization factor. We work in the Schrödinger picture.

To describe a stationary state, we expand the $\sigma$ field about a time-independent, $c$-number function of $\mathbf{r}$,

$$\sigma(\mathbf{r})=\sigma_0(\mathbf{r})+\sum_n \sqrt{\frac{1}{2\Omega_n}}\,[a_n+a_n^\dagger]\,u_n(\mathbf{r})\equiv\sigma_0+\sigma_1 \tag{10}$$

and the conjugate momentum as

$$\pi(\mathbf{r})=-i\sum_n \sqrt{\frac{\Omega_n}{2}}\,[a_n-a_n^\dagger]\,u_n(\mathbf{r}) \tag{11}$$

where the $u_n(\mathbf{r})$ are a complete, orthonormal set and have been chosen to be real.

Similarly, the quark field may be expanded as

$$\psi(\mathbf{r})=\sum_m c_m q_m(\mathbf{r}) \tag{12}$$

where the $q_m(\mathbf{r})$ are a complete set of Dirac spinors. A trial "bag"state $|B\rangle$ is constructed by defining

$$\begin{aligned} a_n\,|B\rangle&=0, && \text{for all } n\\ c_m\,|B\rangle&=0, && \text{for } m \text{ unoccupied}\\ c_m^\dagger\,|B\rangle&=0, && \text{for } m \text{ occupied} \end{aligned} \tag{13}$$

The energy of this state can be evaluated by taking the expectation value of $H$, and using (10)–(13) along with the standard commutation relations for creation and annihilation operators. This gives

$$\begin{aligned}\langle B|\,H\,|B\rangle=\int d^3r\Bigg[&\sum_{m(occ)} q_m^\dagger(\mathbf{r})(-i\boldsymbol{\alpha}\cdot\boldsymbol{\nabla}+g\beta\sigma_0)\,q_m(\mathbf{r})\\ &+\tfrac{1}{4}\sum_n u_n(\mathbf{r})\Bigg(Z_\sigma^{-1}\Omega_n+\Omega_n^{-1}\\ &\times\Bigg(-Z_\sigma\nabla^2+a_0+b_0\sigma_0+\tfrac{1}{2}c_0\sigma_0^2+\frac{c_0}{8}\sum_{n'}\frac{u_{n'}(\mathbf{r})^2}{\Omega_{n'}}\Bigg)\Bigg)\,u_n(\mathbf{r})\\ &+Z_\sigma\tfrac{1}{2}\,|\boldsymbol{\nabla}\sigma_0|^2+d_0\sigma_0+\frac{a_0}{2}\sigma_0^2+\frac{b_0}{3!}\sigma_0^3+\frac{c_0}{4!}\sigma_0^4\end{aligned} \tag{14}$$

In obtaining (14), we have used the following relations:

$$\langle B|\, \sigma_1^n \,|B\rangle = 0 \qquad \text{for } n \text{ odd} \tag{15a}$$

$$\langle B|\, \sigma_1^4 \,|B\rangle = 3\langle B|\, \sigma_1^2 \,|B\rangle^2 \tag{15b}$$

$$\langle B|\, \sigma_1^2 \,|B\rangle = \tfrac{1}{2}\sum_n \frac{u_n(\mathbf{r})^2}{\Omega_n} \tag{15c}$$

$$\langle B|\, \pi^2 \,|B\rangle = \tfrac{1}{2}\sum_n \Omega_n u_n(\mathbf{r})^2 \tag{15d}$$

We now vary $\langle B|\, H \,|B\rangle$ with respect to the various functions (parameters) of the state vector. Variation with respect to $\bar{q}_m$ yields

$$(\boldsymbol{\alpha}\cdot\mathbf{p} + g\beta\sigma_0 - \varepsilon_m)\, q_m = 0 \tag{16a}$$

with respect to $\sigma_0$ yields

$$-Z_\sigma \nabla^2 \sigma_0 + +\tfrac{1}{4} b_0 S(\mathbf{r}) + d_0 + [a_0 + \tfrac{1}{4} c_0 S(\mathbf{r})]\, \sigma_0 + \tfrac{1}{2} b_0 \sigma_0^2 + \tfrac{1}{6} c_0 \sigma_0^3$$
$$+ g \sum_{m(occ)} \bar{q}_m q_m = 0 \tag{16b}$$

with respect to $u_n$ yields

$$[-Z_\sigma \nabla^2 + a_0 + b_0\sigma_0 + \tfrac{1}{2} c_0 \sigma_0^2 + \tfrac{1}{4} c_0 S(\mathbf{r}) - \lambda_n]\, u_n = 0 \tag{16c}$$

where $\lambda_n$ is a Lagrangian multiplier constraining the normalization of the $\mu_n$; and with respect to $\Omega_n$ yields

$$Z_\sigma^{-1}\Omega_n^2 = \int d^3r\, u_n[-Z_\sigma \nabla^2 + a_0 + b_0\sigma_0 + \tfrac{1}{2} c_0 \sigma_0^2 + \tfrac{1}{4} c_0 S(\mathbf{r}) - \lambda_n]\, u_n \tag{16d}$$

where

$$S(\mathbf{r}) = \sum_n \frac{|u_n|^2}{\Omega_n} \tag{17}$$

Equation (16d) shows that the Lagrange multiplier $\lambda_n$ in (16c) is $Z_\sigma^{-1}\Omega_n^2$.

## 4. THE ONE-LOOP APPROXIMATION AND RENORMALIZATION

We will presently show that $S(\mathbf{r})$ contains divergences which multiply certain powers of $\sigma_0$. These divergent terms can be identified with a renormalization of the coupling constants $a_0$, $b_0$, $c_0$, and $d_0$ to the "physical" values $a$, $b$, $c$, and $d = 0$.

Let us define

$$W(\mathbf{r}) = a_0 + b_0\sigma_0 + \tfrac{1}{2}c_0\sigma_0^2 + \tfrac{1}{4}c_0 S(\mathbf{r}) - m_\sigma^2 \tag{18}$$

with $m_\sigma$ defined so that $W(\mathbf{r}) \to 0$ as $r \to \infty$; it is the renormalized sigma-field mass in vacuum. The $u_n(\mathbf{r})$ are then determined by the equation

$$[-Z_\sigma \nabla^2 + m_\sigma^2 + W(\mathbf{r}) - Z_\sigma^{-1}\Omega_n^2]\, u_n(\mathbf{r}) = 0 \tag{19}$$

We can now write the energy as

$$\begin{aligned}\langle B|\, H\, |B\rangle = \sum_{m(occ)} \varepsilon_m + \int d^3r \\ \times \left[ \tfrac{1}{2}Z_\sigma\, |\nabla\sigma_0(\mathbf{r})|^2 + d_0\sigma_0 \frac{a_0}{2}\,\sigma_0(\mathbf{r})^2 + \frac{b_0}{3!}\,\sigma_0(\mathbf{r})^3 + \frac{c_0}{4!}\,\sigma_0(\mathbf{r})^4 \right] \\ + \tfrac{1}{2}Z_\sigma^{-1} \sum_n \Omega_n - \tfrac{1}{32}c_0 \int S^2(\mathbf{r})\, d^3r \end{aligned} \tag{20}$$

But we can also set

$$\begin{aligned} \tfrac{1}{2}Z_\sigma^{-1} \sum_n \Omega_n &\equiv E_\Omega \\ &= \tfrac{1}{4} \sum_n \int u_n [Z_\sigma^{-1}\Omega_n + Z_\sigma \Omega_n^{-1}(-\nabla^2 + a_0 + b_0\sigma_0 + \tfrac{1}{2}c_0\sigma_0^2 + \tfrac{1}{4}c_0 S(\mathbf{r}))]\, u_n\, d^3r \\ &= \tfrac{1}{2} \int \left[ (-Z_\sigma \nabla^2 + m_\sigma^2 + W(\mathbf{r})) \sum_n \frac{u_n(\mathbf{r})\, u_n(\mathbf{r}')}{\Omega_n} \right]_{\mathbf{r}'=\mathbf{r}} d^3r \end{aligned} \tag{21}$$

We evaluate the sum by a contour integral:

$$\begin{aligned} \sum_n \frac{u_n(\mathbf{r})\, u_n(\mathbf{r}')}{\Omega_n} &= \frac{1}{\pi i} \oint_C dz \sum_n \frac{u_n(\mathbf{r})\, u_n(\mathbf{r}')}{z^2 - \Omega_n^2} \\ &= \frac{i}{\pi} \int_{-\infty}^{\infty} d\omega \sum_n \frac{u_n(\mathbf{r})\, u_n(\mathbf{r}')}{\omega^2 - \Omega_n^2 + i\varepsilon} \end{aligned} \tag{22}$$

The contour $C$ encloses the poles along the positive real axis. The second form is obtained by deforming the contour so that it runs below the negative real axis and above the positive real axis; note the minus sign from the direction reversal. Other contours will be considered later.

Let

$$G(\mathbf{r}, \mathbf{r}'; z) \equiv \sum_n \frac{u_n(\mathbf{r})\, u_n(\mathbf{r}')}{z^2 - \Omega_n^2} \tag{23}$$

Then

$$[z^2 + Z_\sigma \nabla^2 - m_\sigma^2 - W(\mathbf{r})]\, G(\mathbf{r}, \mathbf{r}'; z) = \delta^3(\mathbf{r} - \mathbf{r}') \tag{24}$$

Similarly we define $G^0$ by

$$[z^2 + Z_\sigma \nabla^2 - m_\sigma^2]\, G^0(\mathbf{r}, \mathbf{r}'; z) = \delta^3(\mathbf{r} - \mathbf{r}') \tag{25}$$

As a matrix (integral) equation in $(\mathbf{r}, \mathbf{r}')$, $G$ satisfies

$$\begin{aligned} G &= G^0 + GWG^0 = G^0 + G^0WG^0 + GWG^0WG^0 = \cdots \\ &\equiv G^0 + G^1 + G^2 + G^3 + \cdots \end{aligned} \tag{26}$$

where the superscript on $G^n$ denotes the order in $W$.

Using the above equations, we can write

$$\begin{aligned} E_\Omega &= \frac{1}{2\pi i} \int d^3r \oint_C dz [-W(\mathbf{r})\, G^0(\mathbf{r}, \mathbf{r}; z) + z^2 G(\mathbf{r}, \mathbf{r}; z)] \\ &\equiv E_\Omega^0 + E_\Omega^1 + E_\Omega^2 + E_\Omega^3 + \cdots \end{aligned} \tag{27}$$

where again the superscripts denote the order in $W$.

The terms in $E_\Omega$ through order $W^2$ are infinite; the higher-order terms are finite. The infinite terms can be treated by the introduction of a cut-off $\Lambda$ in $k$-space. Using

$$G^0(\mathbf{k}, \mathbf{k}'; z) = \frac{\delta_{\mathbf{k}\mathbf{k}'}}{z^2 - k^2 - m_\sigma^2} \tag{28}$$

the leading-order terms can be evaluated to yield

$$\begin{aligned} &E_\Omega^0 + E_\Omega^1 + E_\Omega^2 \\ &\quad = \frac{1}{64\pi^2} \int d^3r \left[ 4\Lambda^4 + 4\Lambda^2(m_\sigma^2 + W(\mathbf{r})) - (m_\sigma^2 + W(\mathbf{r}))^2 \ln \frac{4\Lambda^2}{m_\sigma^2} \right] + \tilde{E}_\Omega^2 \end{aligned} \tag{29}$$

where $\tilde{E}_\Omega^2$ is the finite part of $E_\Omega$ through second order and can be evaluated by quadrature using the explicit form of $G^0$, Eq. (28).

We now consider an expansion of the energy in powers of $\hbar$. To first order in $\hbar$ (the one-loop approximation) we may drop $S(\mathbf{r})$ wherever it is written explicitly in the equations for $\sigma_0(\mathbf{r})$, $u_n(\mathbf{r})$ and the energy. The infinite terms which arise can be regrouped, where they occur, to renormalize the coefficients of $\sigma_0$ through fourth order. The resultant effective Hamiltonian, however, is not quartic in $\sigma_0$ and is nonlocal.

Renormalization in the fermi-sector also requires consideration. The quarks are imbedded in a sigma field, and this distorts the negative energy Dirac sea, similar to the problem in QED.[(10)] For a uniform sigma field, this leads to divergent terms which involve powers of $\sigma_0$ to fourth order plus a finite term proportional to $\sigma_0^4 \ln \sigma_0$. The divergent terms can be regrouped to further renormalize the coefficients of powers of $\sigma_0$, now denoted by $a$, $b$, $c$ with $d$ chosen to be zero. For the nonuniform case, nonlocality introduces a divergent term in the gradient of sigma. Regrouping of this term can, in effect, remove $Z_\sigma$ from the equations. Other nonlocal terms are finite.

We now summarize the equations through the one-loop approximation:

$$(\boldsymbol{\alpha}\cdot\mathbf{p}+g\beta\sigma_0-\varepsilon_m)\,q_m=0 \tag{30a}$$

$$-\nabla^2\sigma_0+U'(\sigma_0(\mathbf{r}))+g\sum_{m(valence)}\bar{q}_m q_m=0 \tag{30b}$$

$$[-\nabla^2+U''(\sigma_0(\mathbf{r}))-\Omega_n^2]\,u_n=0 \tag{30c}$$

with

$$U(\sigma)=\tfrac{1}{2}a\sigma^2+\tfrac{1}{3!}b\sigma^3+\tfrac{1}{4!}c\sigma^4$$

and the primes denote differentiation with respect to $\sigma$.

There are several methods which can be used to evaluate the terms higher order in $W$ in the Bose sector, within the one-loop approximation. One method which has been effective in other contexts, for spherically symmetric $W$, goes as follows:

(1) Deform the $z$-contour to the imaginary axis.

(2) Do an angular momentum decomposition of $G$:

$$G(\mathbf{r},\mathbf{r}';z)=(rr')^{-1}\sum_{l,m}\mathscr{G}_l(r,r';z)\,Y_{lm}(\Omega)\,Y^*_{lm}(\Omega') \tag{30d}$$

where

$$\left[z^2+\frac{d^2}{dr^2}-\frac{l(l+1)}{r^2}-W(r)\right]\mathscr{G}_l(r,r';z)=\delta(r-r') \tag{31}$$

which can be solved by standard methods. Since $z^2<0$, $\mathscr{G}(r,r';z)$ is not oscillatory, at least as $|z|\to\infty$. The numerical methods are thus stable.

(3) For each $l$ and discrete $z$, evaluate and subtract the terms through second order in $W$:

$$E_\Omega - E_\Omega^0 - E_\Omega^1 - E_\Omega^2 = \frac{1}{\pi i}\int_0^\infty dr \int_0^{i\infty} z^2 dz \sum_l (2l+1) \times [\mathcal{G}_l(r, r'; z) - \mathcal{G}_l^0(r, r'; z) - \mathcal{G}_l^1(r, r'; z) - \mathcal{G}_l^2(r, r'; z)] \tag{32}$$

The $\mathcal{G}_l^n(r, r'; z)$ are finite for all $n$ and may be obtained either analytically or numerically. For example, one could replace $W \to \lambda W$ and solve for $\mathcal{G}_l(\lambda)$ for several $\lambda$ and extract the various orders in $\lambda$; this approach was used by Rinker and Wilets[11] in a QED fermion problem.

(4) The integrations in Eq. (32) can be performed over a discrete grid in $r$ and $z$; the sum over $l$ can be truncated at a reasonable value.

Similar techniques are available for the fermion one-loop corrections.

## 5. THE MOMENTUM OPERATOR AND MATRIX ELEMENTS

The momentum operator contains two parts, the quark and the soliton contributions:

$$\mathbf{P} = \mathbf{P}_q + \mathbf{P}_\sigma \tag{33}$$

with

$$\mathbf{P}_q = -i\int d^3r\, \psi^\dagger \nabla \psi = \sum_{k,l} c_k^\dagger c_l \int d^3r\, q_k^\dagger \mathbf{p} q_l \tag{34}$$

and

$$\mathbf{P}_\sigma = -\tfrac{1}{2}\int d^3r[\pi \nabla \sigma + (\nabla \sigma)\, \pi] \tag{35}$$

The vacuum is a zero-eigenstate of $\mathbf{P}$; hence all moments of $\mathbf{P}$ vanish in vacuum, In the bag state only even moments of $\mathbf{P}$ occur, and in our approximation of a factored $q$-$\sigma$ state only even moments of $\mathbf{P}_q$ and $\mathbf{P}_\sigma$ occur.

### 5.1. Calculation of $\langle B|P_q^2|B\rangle$

In what follows, it is convenient to use first-quantized notation for the quarks, and take into account the Dirac sea explicitly. We use $i$ and $j$ for quark coordinates and $k$ and $l$ for quantum numbers. Then

$$\mathbf{P}_q = \sum_i \mathbf{p}_i \tag{36}$$

$$P_q^2 = \sum_i p_i^2 + \sum_{i \neq j} \mathbf{p}_i \cdot \mathbf{p}_j \tag{37}$$

$$\langle B|P_q^2|B\rangle \equiv \langle P_q^2\rangle = \sum_{k(occ)} \langle k|p^2|k\rangle - \sum_{\substack{k(occ)\\ l(occ)}} |\langle k|\mathbf{p}|l\rangle|^2 \tag{38}$$

There is no direct term for the two-body operator $\mathbf{p}_i \cdot \mathbf{p}_j$ because $\langle k|\mathbf{p}|k\rangle = 0$. In the second (exchange) term in (38), there is no contribution from terms where $k$ and $l$ are both valence states (in the case of the baryon) since all the valence quarks are in the same spatial state. Thus $k$ can be restricted to negative energy states, This gives

$$\langle P_q^2\rangle = \sum_{k(occ)} \langle k|p^2|k\rangle + \Bigg[ - \sum_{\substack{k(\varepsilon_k<0)\\ l(all)}} + \sum_{\substack{k(\varepsilon_k<0)\\ l(unocc)}} - \sum_{\substack{k(valence)\\ l(occ)}} \Bigg] |\langle k|\mathbf{p}|l\rangle|^2$$

$$= \sum_{k(valence)} \langle k|p^2|k\rangle + \Bigg[ \sum_{\substack{k(\varepsilon_k<0)\\ l(unocc)}} - \sum_{\substack{k(valence)\\ l(occ)}} \Bigg] |\langle k|\mathbf{p}|l\rangle|^2 \tag{39}$$

The second term can be identified as arising from polarization of the vacuum.

### 5.2. Evaluation of $\langle B|P_\sigma^2|B\rangle$

Since the $u_n$ are taken to be real, the matrix elements

$$-i\int u_m \nabla \sigma_0 \, d^3r \equiv \mathbf{p}_{m\sigma_0} = -\mathbf{p}_{\sigma_0 m} \tag{40a}$$

and

$$-i\int u_m \nabla u_n \, d^3r \equiv \mathbf{p}_{mn} = -\mathbf{p}_{nm} \tag{40b}$$

are pure imaginary and $\mathbf{p}_{mm}=0$. Then (35) can be written

$$\begin{aligned} P_\sigma = & -\sum_m \sqrt{\frac{\Omega_m}{2}}\, \mathbf{p}_{m\sigma_0}(a_m - a_m^\dagger) \\ & -\tfrac{1}{4}\sum_{mn}\left[\left(\frac{\Omega_m}{\Omega_n}\right)^{1/2} - \left(\frac{\Omega_n}{\Omega_m}\right)^{1/2}\right] \mathbf{p}_{mn}(a_m a_n - a_m^\dagger a_n^\dagger) \\ & +\tfrac{1}{2}\sum_{mn}\left[\left(\frac{\Omega_m}{\Omega_n}\right)^{1/2} + \left(\frac{\Omega_n}{\Omega_m}\right)^{1/2}\right] \mathbf{p}_{mn} a_m^\dagger a_n \end{aligned} \tag{41}$$

We can readily find

$$\langle B\,|P_\sigma^2|\,B\rangle = \langle P_\sigma^2\rangle = \tfrac{1}{2}\sum_m \Omega_m\,|\mathbf{p}_{m\sigma_0}|^2 + \tfrac{1}{8}\sum_{mn}\frac{(\Omega_m-\Omega_n)^2}{\Omega_m\Omega_n}\,|\mathbf{p}_{mn}|^2 \tag{42}$$

Note that in vacuum the first term on the right-hand side vanishes because $\mathbf{p}_{m\sigma_0}=0$; the second term vanishes because $\mathbf{p}_{mn}=0$ unless $\Omega_m=\Omega_n$.

## 6. EVALUATION OF $\langle B\,|HP^2|\,B\rangle$

Because $\mathbf{P}$ commutes with $H$, we can write $HP^2$ in any of the alternate forms

$$HP^2 = P^2H = \mathbf{P}\cdot H\cdot \mathbf{P} \tag{43}$$

The cross terms between $P_q$ and $P_\sigma$ in the expectation value vanish. Thus we can write

$$\langle B\,|HP^2|\,B\rangle = \langle B\,|HP_q^2|\,B\rangle + \langle B\,|HP_\sigma^2|\,B\rangle \tag{44}$$

We note that $H_q+H_{q\sigma}$ is a one-body operator with respect to the quarks, and we work in a representation where the quark state is diagonal in $\langle\sigma_0\,|H_q+H_{q\sigma}|\,\sigma_0\rangle$, where $|\sigma_0\rangle$ is the sigma part of the state vector. Then we can write

$$\langle B|\,(H_q+H_{q\sigma})\,P_q^2\,|B\rangle \equiv \langle (H_q+H_{q\sigma})\,P_q^2\rangle = \langle H_q+H_{q\sigma}\rangle\langle P_q^2\rangle \tag{45}$$

where expectation values are with respect to $|B\rangle$; $\langle P_q^2\rangle$ is given by (39). Thus the first term on the right-hand side of (44) reduces to

$$\langle HP_q^2\rangle = \langle H\rangle\langle P_q^2\rangle \tag{46}$$

In evaluating the last two terms in (9), we note that $H$ contains a $c$-number part and a "quantum" part with respect to $|\sigma_0\rangle$. Let us define

$$H \equiv E_c + H' \tag{47}$$

where $H'$ contains only finite powers of $\sigma_1$ (through order 4) and $\pi$ (through order 2). Then

$$\langle HP_\sigma^2 \rangle = \langle H \rangle \langle P_\sigma^2 \rangle + \langle H' P_\sigma^2 \rangle \tag{48}$$

and for the first correction (8) we are left with

$$\langle HP^2 \rangle - \langle H \rangle \langle P^2 \rangle = \langle \mathbf{P}_\sigma \cdot H' \cdot \mathbf{P}_\sigma \rangle \tag{49}$$

## 7. CALCULATION OF OTHER PROPERTIES

The projected state vector (2) can be used to calculate the expectation value (or transition matrix elements) of other physical operators. Let us denote by $\mathcal{O}$ an operator of interest. Then the expectation value of $\mathcal{O}$ is given by (8) with $\mathcal{O}$ replacing $H$ and symmetrized with respect to $P^2$:

$$\frac{\langle \Psi | \mathcal{O} | \Psi \rangle}{\langle \Psi | \Psi \rangle} = \langle \mathcal{O} \rangle - \frac{3}{2} \frac{\langle \frac{1}{2}\{\mathcal{O}, P^2\} \rangle - \langle \mathcal{O} \rangle \langle P^2 \rangle}{\langle P^2 \rangle} \cdots \tag{50}$$

We illustrate this by the magnetic moment operator, which depends only on the quark fields:

$$\boldsymbol{\mu} = \tfrac{1}{2} \int d^3r \, \bar{\psi} (\mathbf{r} \times \boldsymbol{\gamma}) \, \tfrac{1}{2}(\tau_3 + \tfrac{1}{3}) \, \psi \tag{51}$$

and

$$\frac{\langle \Psi | \boldsymbol{\mu} | \Psi \rangle}{\langle \Psi | \Psi \rangle} = \langle \boldsymbol{\mu} \rangle - \frac{3}{2} \frac{\langle \frac{1}{2}\{\mathbf{u}, P_q^2\} \rangle - \langle \boldsymbol{\mu} \rangle \langle P_q^2 \rangle}{\langle P_q^2 + P_\sigma^2 \rangle} \cdots \tag{52}$$

The magnetic moment is the expectation value of $\mu_z$ in the state $M = J$. The indicated expectation values can be carried out in a straightforward manner.

## 8. SUMMARY AND OUTLOOK

The method of projection is used to construct an eigenstate of zero linear momentum in the case of a nonlinear relativistic field theory. The expectation value of the energy and physical operators can be evaluated in terms of the various moments of the physical operators times $P^n$. Renormalization was addressed in the one-loop approximation, and was discussed in detail in the Bose sector, including a program to evaluate corrections to all orders within the framework of the one-loop approximation. The finite one-loop corrections in the Fermion sector are left for a future paper.

The program as outlined here can be used to test at least three important aspects of the projection method.

(1) By using plane waves for $u_n$ in the moment expansion to a given order, say $n=2$, and comparing with the calculations of LBHW,[3] which is equivalent to all orders in $n$, one can test the convergence of the moment expansion—which is easier to use than the complete projection.

(2) The plane wave expansion of LBHW can already be checked in order $n=0$ by calculating $E_\Omega$, Eq. (21), as indicated in Sec. 4 with distorted $u_n$, and comparing the results with an evaluation using plane waves.

(3) I have not discussed in detail here the distortion of the Dirac sea. An important part of it is included in the $n=2$ terms, but it would be useful to calculate the finite (vacuum polarization) correction terms in the Fermi sector. The LBHW paper neglects the Dirac sea.

The LBHW calculations exhibited the importance of variation of the state vector functions $(\psi_m, u_n, \sigma_0)$ *after* projection. Those authors further presented virial theorem tests for the "goodness" of the state vector. Such variations and tests should be included in the implementation of the present program. However, the renormalization problem has not yet been studied for the case of variation after projection.

The utility of the proposed program is, ultimately, the calculation of physical quantities in a relativistic field theory with proper accounting for recoil effects.

## ACKNOWLEDGMENTS

I am deeply grateful to Drs. M. C. Birse and R. Perry for in-depth discussions on all aspects of this problem, for what they have taught me about renormalization, and for a full reading of the manuscript. This work was supported in part by the U.S. Department of Energy.

## REFERENCES

1. D. L. Hill and J. A. Wheeler, *Phys. Rev.* **89**, 1106 (1953); J. J. Griffin and J. A. Wheeler, *Phys. Rev.* **108**, 311 (1957); R. E. Peierls and J. Yoccoz, *Proc. Phys. Soc. London A* **70**, 381 (1957).

2. J-L. Dethier, R. Goldflam, E. M. Henley, and L. Wilets, *Phys. Rev. D***27**, 2193 (1983).
3. G. Lübeck, M. C. Birse, E. M. Henley, and L. Wilets, *Phys. Rev. D***33**, 234 (1986).
4. R. Goldflam and L. Wilets, *Phys. Rev. D***25**, 1951 (1982).
5. R. Jackiw, *Rev. Mod. Phys.* **49**, 681 (1977); R. Rajaraman, *Solitons and Instantons* (North-Holland, Amsterdam, 1982).
6. C. G. Callan and D. J. Gross, *Nucl. Phys. B***93**, 29 (1975).
7. R. Friedberg and T. D. Lee, *Phys. Rev. D***15**, 1694 (1977); **16**, 1096 (1977); **18**, 2623 (1978); T. D. Lee, *Particle Physics and Introduction to Field Theory* (Harwood Academic, New York, 1981).
8. L. Wilets, "Quark Models of Hadronic Interactions" in *Hadrons and Heavy Ions* (Springer-Verlag, New York, 1985).
9. J. M. Cornwall, R. Jackiw, and E. Tomboulis, *Phys. Rev. D***10**, 2428 (1974); S. Coleman, R. Jackiw, and H. Politzer, *Phys. Rev. D***10**, 2491 (1974); S.-J. Chang, *Phys. Rev. D***12**, 1071 (1975); T. Barnes and G. I. Ghandour, *Phys. Rev. D* **22**, 924 (1980).
10. E. H. Wichmann and N. M. Kroll, *Phys. Rev.* **101**, 843 (1956); M. Gyulassy, *Nucl. Phys. A* **244**, 497 (1975).
11. G. A. Rinker and L. Wilets, *Phys. Rev. A* **12**, 748 (1975).

Therefore, given the Fredholm determinant, we can determine the scattering matrix $\|c_{mn}\|$ uniquely up to a transformation by an arbitrary phase matrix and a possible interchange of rows and columns.

*Plate 4*

# An $S$ Matrix Theory for Classical Nonlinear Physics[1]

**J. Glimm**[2–5] **and D. H. Sharp**[6,7]

*Received June 28, 1985*

*The basic concepts appropriate for an* S *matrix theory for classical nonlinear physics are formulated here. These concepts are illustrated by a discussion of shock wave diffraction patterns. Other information concerning solutions of nonlinear conservation laws is surveyed, so that a coherent picture of this theory can be seen. Within this* S *matrix framework, a number of open problems as well as a few solved ones will be discussed.*

## 1. INTRODUCTION

In 1937, John Wheeler[(1)] introduced the technique of the $S$ matrix into physics. This idea owes its success to its ability to describe the outcome of a collision process in a fashion which suppresses the transient details of this process. This methodology can be usefully carried over to classical physics when studying problems characterized by localized disturbances or coherent nonlinear waves. In fact localized disturbances and coherent waves are generic in problems governed by nonlinear conservation laws. In this paper we formulate the basic concepts appropriate for an $S$ matrix theory in this context.

---

[1] Dedicated to John A. Wheeler on the occasion of his 75th birthday.

[2] Courant Institute of Mathematical Sciences, New York University, New York, New York 10012.

[3] Supported in part by the National Science Foundation, grant DMS-831229.

[4] Supported in part by the Aplied Mathematical Sciences subprogram of the Office of Energy Research, U.S. Department of Energy, under contract DE-AC02-76ER03077.

[5] Supported in part by the Army Research Office, grant DAAG29-83-K-0007.

[6] Los Alamos National Laboratory, Los Alamos, New Mexico, 87545.

[7] Work supported by the U.S. Department of Energy.

The equations of physics, in their simplest form, often express conservation of certain quantities. Assuming that there are $s$ such independent conservations laws, the solution $u=u(x. t)$ take its values in a state space which is some subset of $R^s$. The lowest-order dynamical equations for $u$ then have the form

$$u_t+\nabla\cdot\mathbf{F}(u)=0 \tag{1}$$

and state that $\mathbf{F}(u)$ is the flux associated with the conserved quantity $u$. Further terms, not discussed in this article, express nonconservation (external sources) and higher-order effects such as diffusion. In an $S$ matrix theory, one is concerned with the asymptotic behavior at large times to the solutions of these full equations. To leading order, the large time behavior is governed by the infinite scaling limit of the original equations. Under scaling, each term in the equation has a dimension, and in the infinite scaling limit, only terms of the maximal dimension survive. This process generally eliminates diffusion terms, yielding the above conservation law. We assume that the source terms are of bounded extent in space in the original problem. Under scaling, they do survive, and go into a multiple of a delta function, i.e., a point source at the origin.

Let $\mathbf{A}=\partial\mathbf{F}/\partial u$ be the Jacobian matrix of $\mathbf{F}$. Then (1) is hyperbolic if the $s\times s$ matrix $\mathbf{A}\cdot\xi$ has real eigenvalues $\lambda_1,\dots,\lambda_s$ for each $\xi$, and it is strictly hyperbolic if the $\lambda_j$ are distinct. We assume real eigenvalues (hyperbolicity) throughout this article; the critical values of $u$ where some of the eigenvalues coincide are of considerable importance in cases where they occur.

The eigenvalues $\lambda_j(u,\xi)$ are characteristic speeds, or wave speeds, for propagation of disturbances of infinitesimal size in the direction $\xi$: that is, they are the characteristic speeds of the system

$$\mathsf{w}_t+(\mathbf{A}\cdot\xi)(\xi\cdot\nabla)\mathsf{w}=0 \tag{2}$$

which is the linearization of (1) about the constant solution $u(x,t)=u\in R^n$. Similarly, the right eigenvectors $e_j=e_j(u,\xi)$ of $\mathbf{A}(u)\cdot\xi$ are the normal modes for the linearized system (2).

The $S$ matrix is the product of two wave operators, each of which expresses the large time asymptotics for large positive and negative times respectively. We focus on the outgoing wave operator $\mathsf{W}^+$. Its data are in the range of the incoming wave operator $\mathsf{W}^-$. Because of the occurrence of shocks in the solutions of (1), this equation (when supplemented by the necessary entropy condition to select physical shocks and reject nonphysical ones) is not reversible in time and $\mathsf{W}^-$ is not well defined. As a substitute for the range of $\mathsf{W}^-$, which is still needed for an $S$ matrix theory, we restrict to scale-invariant data. More precisely, the range of $W^-$ is defined to be the infinite scaling limit about the origin (in space alone) of an arbitrary solution of (1). For some purposes, it is convenient to restrict further to generic data, which means that one starts with a generic solution of (1), and it may also be useful to retain certain nonleading terms in the infinite scaling limit. $W^+$ thus solves a Riemann problem, which is simply the initial-value problem for (1) with scale-invariant initial data.

In one space dimension, the Riemann problem has data which are piecewise constant, with two constant states, one defined for $x$ positive and the other defined for $x$ negative, and a simple jump discontinuity at the origin. In higher dimensions, the data are constant on rays through the origin, and is otherwise unrestricted, except perhaps by some regularity assumption. There is in general also a source term which is a multiple of the delta function concentrated at the origin. The Riemann problem is of fundamental importance in a study of the phenomena associated with (1); it describes both the large time asymptotic behavior of a general solution and the instantaneous interaction of pairs of nonlinear waves, when other disturbances are missing or can be neglected. In one spatial dimension there have been mathematical proofs[(2,3)] of the convergence of a general solution to the solution of a Riemann problem as $t \to \infty$. The solution of Riemann problems with delta function sources has also been analyzed mathematically.[(4)]

The Riemann problem is not only fundamental; it is more elementary than the full equation (1). In fact the imposition of scale invariance leads to a reduction by one in the number of independent variables. Thus in one space dimension, the Riemann problem is roughly equivalent in scientific difficulty to the theory of ordinary differential equations. It is important to study data in the large, so that the jump at the origin is not restricted to be small. In this case the comparison theory is ordinary differential equations in the large. As with this comparison subject, global geometry, topology, and computer experimentation all play a role in the study of Riemann problems in one space dimension. We will discuss this below, as an introduction to the study of Riemann problems with higher co-dimension.

A generic phenomenon in scattering theory is the decomposition of a scattering center ($W^+$ data as defined above) into localized coherent disturbances, which then separate and propagate independently of one another. We call these localized disturbances elementary waves or, to be more precise, $d$-dimensional elementary waves, if the equation (1) is in $d$ space dimensions. For one-dimensional gas dynamics, the elementary waves are shock waves, infinitesimal portions of rarefaction waves, and contact discontinuities. In two dimensions, the elementary waves are located in physical space at the points where one-dimensional elementary waves meet or cross, such as a Mach triple point. In three dimensions, an elementary wave is defined by the meeting or crossing of lower-dimensional elementary waves. Thus we see that the dimensional label is actually a co-dimension, and an elementary wave in $d$ dimensions has a discontinuity of co-dimension $d$, and is defined by the intersection of $d$ discontinuities of co-dimension 1. Each elementary wave moves with a definite velocity, and assuming (1) to be invariant under Galilean boosts, we define an elementary wave to be a scale-invariant solution to (1) which is Galilean equivalent to a stationary solution.

Elementary waves are both more fundamental and more elementary than solutions to Riemann problems. They are the building blocks out of which Riemann problem solutions are built. Because they are both stationary and scale invariant, they have two fewer independent variables in their solutions. Thus the elementary waves in three dimensions have (or should have) a scientific theory comparable to the conservation law with general data in one space dimension, or to the solution of Riemann problems in two space dimensions, while the theory of elementary waves in two dimensions is comparable in scientific difficulty to the solution of a Riemann problem in one space dimension. In fact the basic method for investigation of elementary waves in two dimensions, as with Riemann problems in one dimension, is the interaction of the shock and rarefaction wave curves defined by the various modes of equation (1).

The solution of a Riemann problem is completely determined by its values on the hyperplane $t=1$ in space-time. In this hyperplane, time can be eliminated from the Eq. (1), which then becomes

$$-\frac{\mathbf{x}}{t}\nabla u+\nabla\mathbf{F}(u)=0 \tag{3}$$

This reduced equation will always be hyperbolic at large radii (large $\mathbf{x}/t$), with the positive time-like direction the inward-directed radius and a space-like surface given by a sphere of large radius. Assuming a knowledge of solutions of lower-dimension Riemann problems, the data are known on this sphere of large radius, because the finite propagation speed of solutions to the hyperbolic equation (1) allows only lower-dimensional wave interactions to influence the solution at large radii and $t=1$. Depending on the system (1), there may be regions inside this sphere where some of the characteristic speeds of the reduced equation (3) become complex. Thus the Riemann problem in its formulation (3) is in general transonic. For a scalar hyperbolic equation, the situation is somewhat different. Because a real $1\times 1$ matrix has only real eigenvalues, the scalar verion of (1) is hyperbolic in some direction globally except at singular points of the characteristic vector field, which occur at shocks (co-dimension 1 elementary waves) and 2- and 3-co-dimensional elementary waves. See also Refs. 5 and 6.

## 2. RIEMANN PROBLEMS WITH CO-DIMENSION 1

There is a complete theory for Riemann problems of co-dimension 1 in three situations only. If the jump discontinuity is small enough, the wave speeds distinct, and the system convex in a suitable sense, then a theory of Lax[(7)] says that the nonlinear waves correspond exactly to the linear modes found by eigenfunction analysis of the corresponding linearized system. In this case the waves in each mode are fully nonlinear and exhibit features characteristic of nonlinear waves with variable sound speed, such as breaking of nonlinear waves to form shocks and spreading of nonlinear waves. If the equation is scalar, then a general theory of Oleinik[(8)] says that the nonlinear wave structure can be understood in terms of the convex envelope of the associated flux function. In this case a typical wave is composite and in general contains several regions in which the characteristics are spreading. These regions touch, and at their point of meeting, there is in general a shock. However, all such nonlinear pieces define a single wave, and represent the response of the system to a single mode. The number of composite pieces is exactly the number of pieces (line segments and curve segments) in the convex envelope of the flux function, taken between the left and right states of the Riemann problem. If the equation belongs to a very small number of specific examples (including a gamma law gas,[(9)] adsorption with a Langmuir isotherm,[(10)] and water and polymer displacement of oil without adsorbtion[(11,12)]), then again the general theory of the Riemann problem (and thus of the nonlinear waves and their interactions) is known. Outside of these cases, there is no theory.

However, there are tips to the iceberg. Emerging results from the example of three-phase flow show that critical points (where the eigenvalues and, what is the same, the nonlinear wave speeds, coincide) introduce nontrivial topology into the picture. Specifically, the Riemann problem in this example has a critical point in the interior of its state space which is described by the butterfly singularity of bifurcation theory. The local theory has been developed, and is different from the Lax local theory cited above. The difference reflects the fact that the linearized problem is different, due to the occurrence of coinciding wave speeds. Related examples have critical points with a nontrivial Jordan form. The global theory gives rise to surprising and new mathematical phenomena, such as stable shocks whose admissibility condition is distinct from that introduced by

Lax and Oleinik. Each mode has composite waves with more structure (more pieces) than would be inferred from the above rule applied to each mode separately; i.e., than occur in the associated scalar Buckley–Leverett equation with a corresponding but scalar flux function. In other words, each of the modes in the system has less convexity or behaves as if it had more inflection points than occur in each of the two equations defining the system. Also the shock curves, which are by definition the locus of solutions of the shock relations (Rankine–Hugoniot conditions), have global branches not predicted by the local theory, i.e., which do not connect continuously to shocks of zero strength through a fixed (left or right) end point.[(13)] Also a critical point and the associated eigenvalue play an important role in detonation modeling.[(14)] In contrast, we observe that gas dynamics has so much convexity that no critical points occur except at the vacuum, on the border of state space, even in the large. Critical points on the border of the state space introduce no topology into the problem, as is also reflected in the fact that the qualitative nature of the nonlinear waves and their interactions is the same in the large as in the small for gas dynamics.

**Problem 1. Global Analysis for Co-dimension 1 Riemann Problems**

Here we pose a general problem: to obtain an understanding of the global theory of co-dimension 1 Riemann problems, starting from specific applications of intrinsic interest, and going from there toward general concepts and results. This problem should be approached from two avenues: experimental numerical computations and applied analysis, including ordinary differential equations, geometry, and topology.

## 3. ELEMENTARY WAVES WITH CO-DIMENSION 2

The co-dimension 2 and 3 Riemann problems are of even more interest and fundamental importance. We begin with some definitions. A Riemann problem of co-dimension $j$ is the Cauchy problem for a nonlinear conservation law in $d$ space dimensions which is scale invariant in $j$ coordinates and independent of the remaining $d$–$j$ coordinates. We also speak of a Riemann solution as the solution of a Riemann problem; it has the same invariance properties. Now let us suppose that the conservation law is Galilean invariant. An elementary wave of co-dimension $j$ is a Riemann solution of co-dimension $j$ which is also Galilean equivalent to a stationary solution. The elementary waves of co-dimension 2 are formed by the intersection of jump discontinuities (elementary waves of co-dimension 1).

Again a Riemann problem and its solution define an outgoing wave operator, and the elementary waves are the various localized disturbances (of co-dimension $j$) which can arise in its solution. Examples include the familiar Mach triple point, the point of regular reflection, the point of crossing or overtake of two shocks, and the point of diffraction of an incident shock meeting a contact wave. On the basis of a slightly more restrictive definition, the elementary waves of co-dimension 2 for a gamma law gas have been classified and yield the above list of examples.[(15)] For a scalar conservation law in two dimensions, the elementary waves and the corresponding Riemann problem have been studied by several authors.[(5,6,16,17)] This seems to be the extent of the exact mathematical knowledge of this problem.

There is an extensive lore, backed up by laboratory experiments, concerning the solution of co-dimension 2 Riemann problems.[(18–22)] However, there are also open questions and controversies here. Shock polar analysis leads to nonunique solutions. In general the choice among these solutions is given by boundary conditions. In the case of a Riemann problem, there are no freely prescribable boundary conditions. For example, one principle states that in case of apparently nonunique solutions, the Riemann problem selects the one with the lowest pressure behind the incident shock. However, there are other proposed principles, distinct from and inconsistent with the foregoing, to resolve the apparent nonuniqueness of solutions.

### Problem 2. Elementary Waves and Riemann Problems with Two Independent Variables

From a mathematical point of view, a Riemann problem is equivalent to a differential equation in one fewer independent variable. Similarly, elementary waves are equivalent to problems in two fewer independent variables. There is a general theory for conservation laws in one space and one time dimension for general data, valid at least if the data are sufficiently close to constant. On this basis we propose our second problem: to study Riemann problems of co-dimension 2 and elementary waves of co-dimension 3, at least in favorable cases. Here we propose nearly constant data as an example of a restriction which would make the problem favorable. Restriction to a regular reflection configuration is another example of such a restriction. The above-cited controversies still occur within this regime. The appropriate methods are again a mixture of computer experiments and applied analysis, but because of the occurrence of two independent variables, the appropriate mathematical disciplines are now partial differential equations and functional analysis.

## 4. LOSS OF PIECEWISE SMOOTHNESS

Examination of laboratory photographs suggests that many fluid flows are piecewise smooth. However, mathematical analysis shows that this class of flows is not universal or even generic. We discuss three mechanisms for the breakdown of piecewise smoothness and argue that two of these three mechanisms can arise in the solution of co-dimension 2 Riemann problems. In one-dimensional gas dynamics, one can consider initial data with compression regions. These compression regions will steepen due to the nonlinearities in the equations, and eventually form shocks. By adjusting the slope and location of the initial compression waves, it can be arranged to initiate a shock at an arbitrary location in space-time. This mechanism is not very convincing from a physical point of view, as the carefully adjusted initial data would be difficult to construct in a laboratory and perhaps not likely to arise spontaneously in a flow. A second mechanism for loss of piecewise smoothness, proposed by Rauch and Reed,[(23)] is the multiple reflection of wave signals. In Ref. 23, a semilinear problem with four modes and four wave speeds was considered, and it was shown that the points where the problem failed to be $C^\infty$ were dense in an open set in space-time, on the basis of multiple wave reflections. Specifically there were an infinite number of reflections of fast signals between singularities carried on slow wave mode characteristics. In Ref. 6 a third mechanism for loss of piecewise smoothness of solutions was proposed. For a nonconvex equation of state, it is possible for shocks of a single mode to split into pieces and also to recombine. This mechanism requires only one mode, and so arises even for a scalar hyperbolic equation, but requires a sufficient number of inflection points in the flux function.

We consider specifically Riemann problems with co-dimension 2 (or more). In this case the Cauchy data are very restricted and the first mechanism of complex initial data seems unlikely to cause a breakdown of piecewise smoothness. However, both of the other two mechanism discussed above can lead to a breakdown of piecewise smoothness, provided either that there are enough modes or that there is enough nonconvexity in the problem. In Ref. 6, a construction was indicated to show that four inflection points in a scalar equation would allow loss of piecewise smoothness for a co-dimension 2 Riemann problem. The same phenomena should occur for systems with nonconvex flux functions.

Here we argue that the reverberating shock (multiple reflection) mechanism should cause an unbounded or perhaps an infinite number of discontinuities in the solution of a co-dimension 2 Riemann problem for compressible gas dynamics. The phenomenon is marginal in compressible gas dynamics in the sense that there seem to be just enough modes present to generate the reverberating shocks. For compressible isentropic potential flow in a region exterior to a convex obstacle, it seems likely that the mechanism fails and that solutions can be found within the class of piecewise smooth functions. Compressible isentropic potential flow has two fewer modes than gas dynamics.

Consider a weak shock wave moving into a narrow wedge-shaped opening between two reflecting walls. The shock will generate regular reflections off the wall boundaries. These reflected shocks will cross, and since they are in general not of equal strength, will generate a contact discontinuity at their point of crossing. After crossing, the reflected shocks continue to the opposite boundary, where they will undergo another reflection. This process continues until a subsonic region is created, and each further transit of a reflected shock across the opening requires crossing all previously generated contacts and shocks from the opposite side. In this way a large number of shocks and contacts will be created. As the strength of the initial weak shock is decreased, there will be more reflections before the subsonic region is created. In this way an unbounded number of discontinuities will be created. The resulting configuration is data for a Riemann problem defined by a weak initial shock reaching the vertex of the narrow opening. We conclude that the range of $W^-$ can be arbitrarily complex. Similarly the solutions of the Riemann problem, that is, the range of $W^+$, can be arbitrarily complex. Here we imagine a contact inclined obliquely to a wall, and a shock normal to the wall and approaching the contact. Suppose that the gas below the contact is colder, so that its sound speed is slower. After the shock meets the contact, a transmitted wave is formed in the region below the contact. As it is slow, it is inclined backward, toward the point where the contact meets the wall. There will be multiple reflections again between the contact and the wall. See Refs. 18 and 19.

In spite of the negative conclusions of the present section, we do not believe the reverberating shocks are of great significance in most problems. In fact the waves reflected off of the contacts are likely to be weak, and most of the shocks will have little importance. In order to have a set of equations in which this phenomenon does not occur at all, and so to obtain a qualitative picture of what seem to be the important waves and wave intersections for compressible gas dynamics, we consider isentropic potential flow.

For isentropic potential flow there are no contacts or slip lines, and the only allowed discontinuities are shocks. Making the plausible assumption that the classification of elementary waves given in Ref. 15 applies in this case also with the simple change of deleting all contacts, we see that the only allowed elementary waves are the cross, overtake, and

Mach triple-point configurations. There are at most two outgoing waves. The importance of this simplification for the study of Riemann problems is that in this case we believe they can be analyzed without the complicated phenomenon of reverberating shocks, and hence their solutions can be sought in the space of piecewise smooth functions.

## 5. ELEMENTARY WAVES WITH CO-DIMENSION 3

An elementary wave is by definition a solution of (1) which is stationary and scale invariant. As such, it is completely determined by its values on the surface of the unit sphere $S^2$ in $R^3$. As with the Riemann problem of co-dimension 2, we expect the reverberating shock phenomena to produce elementary waves with an unbounded or infinite number of pieces, leading to a loss of piecewise smoothness. To avoid this possibility, we restrict our attention to compressible isentropic potential flow (or we do not consider the reflection of shocks off of contacts in the case of compressible gas dynamics.) Then we introduce a distinction between the incoming and outgoing portions of the sphere $S^2$. In the incoming portion of the sphere we would expect that more or less arbitrary wave configurations are possible, restricted principally by the condition that they be generic, as was the case in Ref. 15. These incoming waves then determine data for the solution of the problem in the outgoing portion of $S^2$. The reduced equations considered on $S^2$ should be transonic, i.e., containing both hyperbolic and elliptic regions, and data provided by the incoming waves will be Cauchy data in the hyperbolic regions and elliptic boundary data for the elliptic regions. In a different terminology, there are regions where the flow fields restricted to $S^2$ are respectively supersonic or subsonic.

In order to clarify these ideas, we refer the reader to Figs. 1 and 2 which depict an example of a co-dimension 3 elementary wave formed by the intersection of three incident shocks.

On the basis of the above discussion, we see that the classification of three-dimensional elementary waves will consist mainly or entirely in listing the incoming wave configurations which are generically allowed, as the remaining outgoing waves are then determined as a consequence. Because of the assumed scale invariance, all shocks will be planar except insofar as curvature results from interaction with other waves in the problem. Specifically, a shock bordering a subsonic region, or running either through a simple wave (i.e., where the solution gradient has rank 1) or a region with general solution variation (rank 2), will have a curvature which is determined dynamically. The generic configuration will then be a co-

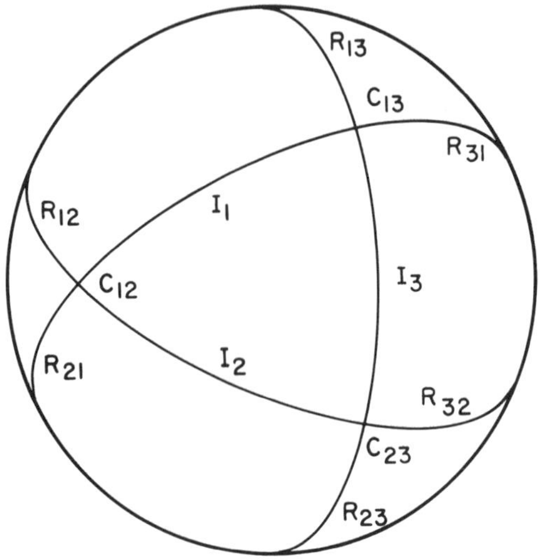


Fig. 1. Three planar incident shocks $I_1$, $I_2$, and $I_3$ meet at the origin. The locus of pairwise crossings $C_{12}$, $C_{13}$, and $C_{23}$ meets the sphere $S^2$ in three distinct points, as illustrated. In a neighborhood of the pairwise crossings $C_{ij}$, the configuration is locally a co-dimension 2 elementary wave, namely a cross, with incident shocks $I_i$ and $I_j$ and reflected shocks $R_{ij}$ and $R_{ji}$. Not shown (or considered) here is the contact between $R_{ij}$ and $R_{ji}$, which occurs for the full Euler equations.

dimension 1 wave crossing a co-dimension 2 wave or, in other words, a shock crossing an overtake, cross, or Mach triple-point configuration.

We now turn to the definition of incoming and outgoing regions and the boundary conditions at the transition between them. Following Ref. 15, we define a shock wave in $R^3$ to be incoming if it turns the flow toward the origin; otherwise it is outgoing. This definition can be applied also to infinitesimal waves, which in effect are points in the cotangent bundle of $S^2$. For an infinitesimal shock wave the definition of incoming is as above; for an infinitesimal rarefaction wave the definition is reversed, i.e., it is incoming when it turns the flow away from the origin. However, the definition is inapplicable if the flow is subsonic, and subsonic regions are defined to be outgoing. For the supersonic case, this notion can be projected onto $S^2$; in effect the flow is incoming where the flow velocity is supersonic and has a radial component which is directed inwards, and otherwise it is outgoing.

The dynamical equations projected onto $S^2$ will be hyperbolic if two conditions are met: (a) the flow in $R^3$ must be hyperbolic (supersonic), and

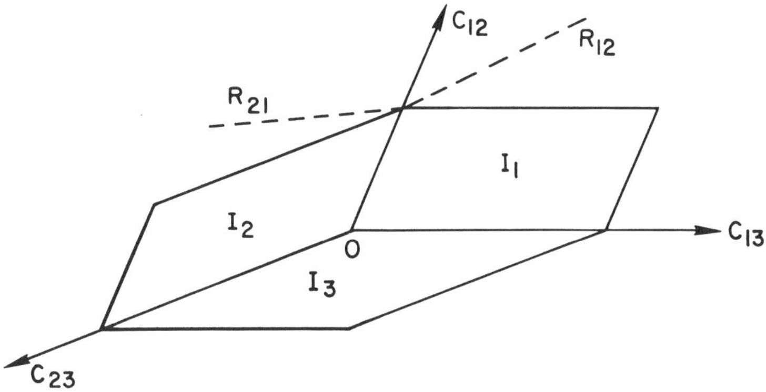


Fig. 2. The intersection of three planar shocks $I_1$, $I_2$, and $I_3$ at the origin 0, shown in a perspective drawing, with pairwise crossing lines $C_{12}$, $C_{23}$, and $C_{13}$. These lines define an oblique set of coordinate axes, and in these coordinates, the incident shocks $I_i$ are the coordinate planes. Notice that the reflected shocks are bent. For clarity we have drawn only $R_{12}$ and $R_{21}$. These shocks are inclined relative to $I_1$ and $I_2$ respectively. They are drawn with dashed lines, to suggest their location behind $I_1$ and $I_2$. To complete this picture in the outgoing region, we note that the reflcted shocks $R_{ij}$ also have pairwise crossings. These crossings define lines which could be regarded as extensions of the axes $C_{ij}$ in the negative directions. In this picture, the extended axis bends on passage through the origin.

(b) the Mach cone defined for the $R^3$-flow when projected to the tangent space of $S^2$ must still define a cone. This cone is the Mach cone for the reduced flow on $S^2$. A transition can occur between ingoing and outgoing portions of $S^2$, either at a shock or in a smooth region. We exclude the possibility of a transition at a shock. If the flow is incoming ahead of the shock, it is turned toward the origin by the shock, and hence the flow remains incoming behind the shock. The transition in a smooth region has two subcases. If the transition line is a sonic line, then the outgoing side is elliptic and data on the transition line are elliptic boundary data. Otherwise the flow is tangential to $S^2$ at the transition point, and supersonic. Because the flow is tangential, it is hyperbolic for the reduced equations on $S^2$. Furthermore the transition line is normal to the particle streamlines, and thus is space-like with respect to the hyperbolic equations on $S^2$.

Having defined boundary data for the outgoing region, we observe that the reduced equations on $S^2$ are transonic, having in general both hyperbolic and elliptic regions. The hyperbolic boundary data along the incoming/outgoing transition curves determine a solution throughout the entire hyperbolic region, and provide the data necessary to determine a solution in the elliptic regions.

The preceding framework is too narrow in that it excludes some interesting phenomena depending on next to leading order scaling behavior, such as the initial point of contact between curved shocks and the point of bifurcation from regular reflection to Mach stem for a shock hitting a curved ramp. With next to leading order scaling data included, the incoming shocks are no longer restricted to be planar.

## 6. FRONT TRACKING

Singularities in solutions are often the points of greatest interest and the points of greatest numerical difficulty. As we have seen in this paper, information concerning solution singularities is sometimes available, even when a general solution is not. The ability to incorporate knowledge of solution singularities into practical numerical computations holds considerable promise for increasing the overall quality of the computation. Specifically, for nonlinear conservation laws and shock wave problems, the ability of Riemann problems to resolve nonlinear wave interactions has been exploited by several authors.[24–27] This principle has received its furthest development in the method of front tracking.

Front tracking is a computational method distinguished by the use of a lower-dimensional grid (interface) to represent the location of solution discontinuity surfaces. Because the conservation laws (1) are correct only in their weak formulation, special methods must be used to formulate the equations on the level of finite differences. In the case of front tracking, these methods consist of using the exact answer. Specifically, for jump discontinuities the exact solution is given by the solution of jump relations known as Rankine–Hugoniot conditions. Because the problem will in general contain modes other than the discontinuity in question, it is important to select the desired discontinuity from more general data, i.e., data not constrained by the jump condition identity. This objective is accomplished by the solution of Riemann problems.

The first major development of this method was due to Richtmyer.[28] The work was carried forward by Moretti and co-workers[29] and by Roger Lazarus.[30] This work has characteristically produced solutions of high resolution and high quality for specific problems.

The current effort[15,31–35] has as its scientific goal the production of comparably high-quality solutions within the context of a general purpose two-dimensional fluid dynamics code. This code currently supports four application areas: gas dynamics, petroleum reservoirs, incompressible fluid flow, and detonations. In each of these applications the code has been tested on certain benchmark problems where the solutions are available, either from closed-form mathematical expressions, independently validated spcialized numerical codes, or laboratory experiments. A sample calculation performed with this code is shown in Fig. 3, which shows the instability and associated shock wave structures of an accelerated interface between a dense and a light fluid.[36]

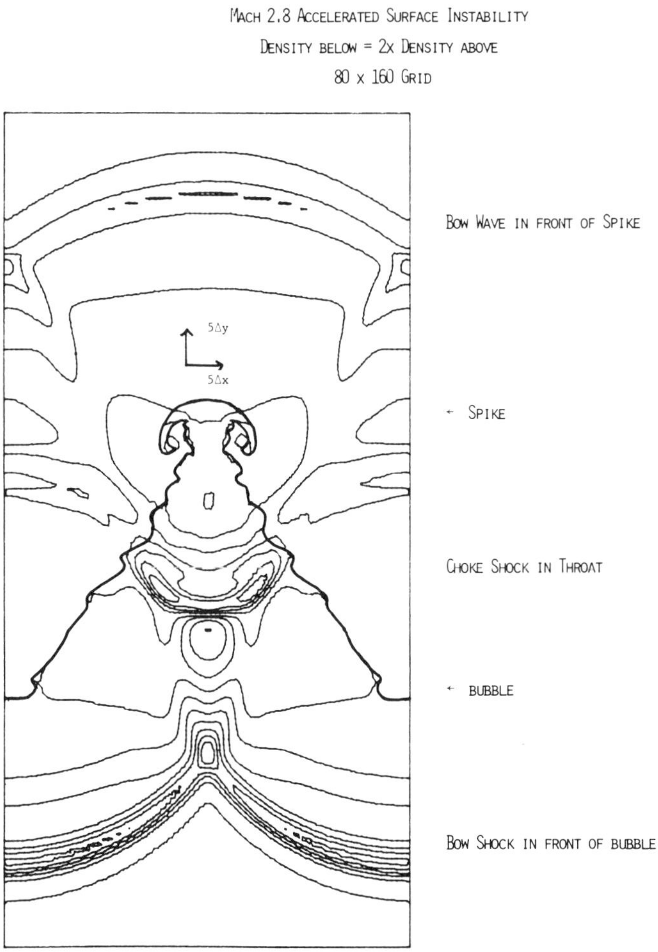


Fig. 3. Instability of an accelerated interface. The heavy fluid is below, and has a density twice that of the upper fluid. The velocities are Mach 2.8 in the dense fluid. The figure shows the density contours and the fluid interface as computed on an 80 × 160 grid. The left/right symmetry observed was not imposed, and therefore provides a check on the accuracy of the computation. The result shown here is due to Dr. Carl Gardner (Ref. 36). The authors express their appreciation to Dr. Gardner for permission to display this result.

The current focus of effort is on analysis of bifurcations in the interface topology and on the numerical implementation of the solutions of the Riemann problems to which these bifurcations give rise. This is one of the key capabilities needed to enable the front tracking method to be applied to problems with a level of complexity typical of engineering applications. Clearly, the solutions of problems (1) and (2) stated above have a direct bearing on this program, over and above their intrinsic interest.

It is evidently important to consider extension of these methods to three dimensions. Such an extension has many aspects, and will face substantial scientific as well as software development problems. The major scientific problems concern the analysis of Riemann problems and elementary waves in higher dimension, as explained above. To give a small indication of some of the software development issues involved, we discuss in the next section data structures for interfaces imbedded in $R^n$ from the point of view of the topological information contained in the boundary operator. Such data structures might be of interest to mathematicians concerned with computational solutions to problems in geometry and topology.

## 7. DATA STRUCTURES FOR MANIFOLDS WITH BOUNDARIES IMBEDDED IN $R^n$

We define an interface to be manifold with boundary imbedded in $R^n$. In this section we describe the type of data structures which would be useful for a computational representation of interfaces. There are two separate issues which must be addressed. One is the description of a smooth surface and the other is the manner in which smooth surfaces intersect along curves and at points to define a singular surface. The specification of a smooth surface can be accomplished by use of parallelepipeds and simplices. The fact that we allow an interface to contain the intersections of or joining of smooth surfaces means that the surfaces will in general have boundaries, which in turn will have boundaries, etc. Thus the data structure for a numerical description of an interface will consist of a list of smooth surfaces of various dimensions together with the boundary relations among these surfaces. Along with each surface of dimension $j$ is stored a knowledge of its boundary surfaces of dimensions $j-1$ and those surfaces of dimension $j+1$ which it bounds. Hypersurfaces of dimension $n-1$ are exceptional, and bound regions or domains in $R^n$. In fact an interface of dimension $n-1$ divides $R^n$ into connected components, and each hypersurface in the interface bounds at most two of these components. The zero-dimensional surface (a node) is also exceptional in that it has no boundary. We consider oriented surfaces, so that there is a sign defined at each boundary relation, to be recorded in the data structure. Thus a surface $\Sigma$ of dimension $j$ will contain pointers to lists of those $j+1$ dimensional oriented surfaces which it bounds and whose orientation relative to $\Sigma$ is positive; it will also contain pointers to lists of those $j+1$ dimensional oriented surfaces which it bounds and whose orientation relative to $\Sigma$ is negative. In two dimensions, this means that nodes have incoming and outgoing curve lists. The boundary information must also go the other way. A surface of dimension $j$ must know its boundary of dimension $j-1$. This is an interface of highest dimension $j-1$, also imbedded in $R^n$. Not every $j-1$-dimensional interface is the boundary of a $j$-dimensional interface or, even more restrictively, of a smooth, connected $j$-dimensional surface imbedded in $R^n$. In two dimensions the boundaries of a curve are just two points, namely its two endpoints, while in three dimensions the boundary of a (smooth, connected) surface is an arbitrary number of unknotted simple closed curves. In general the constraints are summarized by the identity $\partial^2=0$ satisfied by the topological boundary operator $\partial$, together with further constraints depending on $n$ which restrict the possibility of imbedding an abstract manifold with boundary in $R^n$. These ideas in their two-dimensional formulation were originally developed and implemented in terms of a computer library of data structures and support routines as described in Ref. 37.

## ACKNOWLEDGMENTS

It is a pleasure for the authors to thank Dr. John Grove for helpful discussions.

## REFERENCES

1. J. A. Wheeler, *Phys. Rev.* **52**, 1107 (1937).
2. T.-P. Liu, *Commun. Math. Phys.* **55**, 163 (1977).
3. T.-P. Liu, *Commun. Pure Appl. Math.* **30**, 585 (1977).
4. D. Marchesin and P. Paes-Leme, to appear, *Adv. Hyperbolic Partial Differential Equations, Computers and Math. with Applications.*
5. B. Lindquist, to appear, *Adv. Hyperbolic Partial Differential Equations, Computers and Math. with Applications.*
6. B. Lindquist, To appear, *SIAM J. Anal.*
7. P. Lax, *Commun. Pure Appl. Math.* **7**, 159 (1954).
8. O. Oleinik, *Usp. Mat. Nauk (N.S.)* **12**, 3 (1957); English Translation: *Am. Math. Soc. Transl. Ser. 2* **27**, 95 (1963).

9. J. Smoller, *Shock Waves and Reaction-Diffusion Equations* (Springer-Verlag, New York, 1983).
10. F. Helfferidge and G. Klein, *Multicomponent Chromatography: Theory of Interference* (Marcel Dekker, New York, 1970).
11. B. Keyfitz and H. Kranser, *J. Differ. Equ.* **27**, 444 (1978).
12. Eli Isaacson, To appear, *J. Comput. Phys.*
13. D. Marchesin, P. Paes-Leme, D. Schaeffer, and M. Shearer, work in progress.
14. W. Fickett and W. C. Davis, *Detonation* (University of California Press, Berkeley, 1979).
15. J. Glimm, C. Klingenberg, O. McBryan, B. Plohr, D. Sharp, and S. Yaniv, *Adv. Appl. Math.* **6**, 259 (1985).
16. J. Guckenheimer, *Arch. Ration. Mech. Anal.* **59**, 281 (1975).
17. D. Wagner, *J. Math. Anal.* **14**, 534 (1983).
18. A. M. Abd-el-Fattah and L. F. Henderson, *J. Fluid Mech.* **86**, 15 (1978).
19. A. M. Abd-el-Fattah and L. F. Henderson, *J. Fluid Mech.* **89**, 79 (1978).
20. G. Ben-Dor and I. I. Glass, *AIAA J.* **16**, 1146 (1978).
21. R. Deschambault and I. I. Glass, *J. Fluid Mech.* **131**, 22 (1983).
22. R. G. Jahn, *J. Fluid Mech.* **1**, 457 (1956).
23. J. Rauch and M. Reed, *Commun. Math. Phys.* **81**, 203 (1981).
24. J. Glimm, *Commun. Pure Appl. Math.* **18**, 697 (1965).
25. B. Engquist and S. Osher, *Math. Comput.* **36**, 321 (1981).
26. P. Woodward and P. Colella, *J. Comput. Phys.* **54**, 115 (1984).
27. P. Collela and P. Woodward, *J. Comput. Phys.* **54**, 174 (1984).
28. R. D. Richtmyer and K. W. Morton, *Difference Methods for Initial-Value Problems*, 2nd edn. (Interscience Publishers, New York, 1967).
29. G. Moretti, Report No. PIBAL-72-37, Polytechnic Institute of Brooklyn, 1972.
30. Roger Lazarus, private communication.
31. J. Glimm, E. Isaacson, D. Marchesin, and O. McBryan, *Adv. Appl. Math.* **2**, 91 (1981).
32. I. L. Chern, J. Glimm, O. McBryan, B. Plohr, and S. Yaniv, to appear, *J. Comput. Phys.*
33. J. Glimm, B. Lindquist, O. McBryan, and L. Padmanabhan, "A Front Tracking Reservoir Simulator, Five-Spot Validation Studies and the Water Coning Problem," in *The Mathematics of Reservoir Simulation*, R. E. Ewing, ed. (Society for Industrial and Applied Mathematics, Philadelphia, 1983), pp. 107–135.
34. J. Glimm, E. Isaacson, B. Lindquist, O. McBryan, and S. Yaniv, "Statistical Fluid Dynamics: the Influence of Geometry on Surface Instabilities," in *The Mathematics of Reservoir Simulation*, R. E. Ewing, ed. (Society for Industrial and Applied Mathematics, Philadelphia, 1983), pp. 137–160.
35. J. Glimm, O. McBryan, R. Menikoff, and D. H. Sharp, to appear, *SIAM J. Sci. Stat. Comput.*
36. C. Gardner, work in progress.
37. J. Glimm and O. McBryan, in preparation.

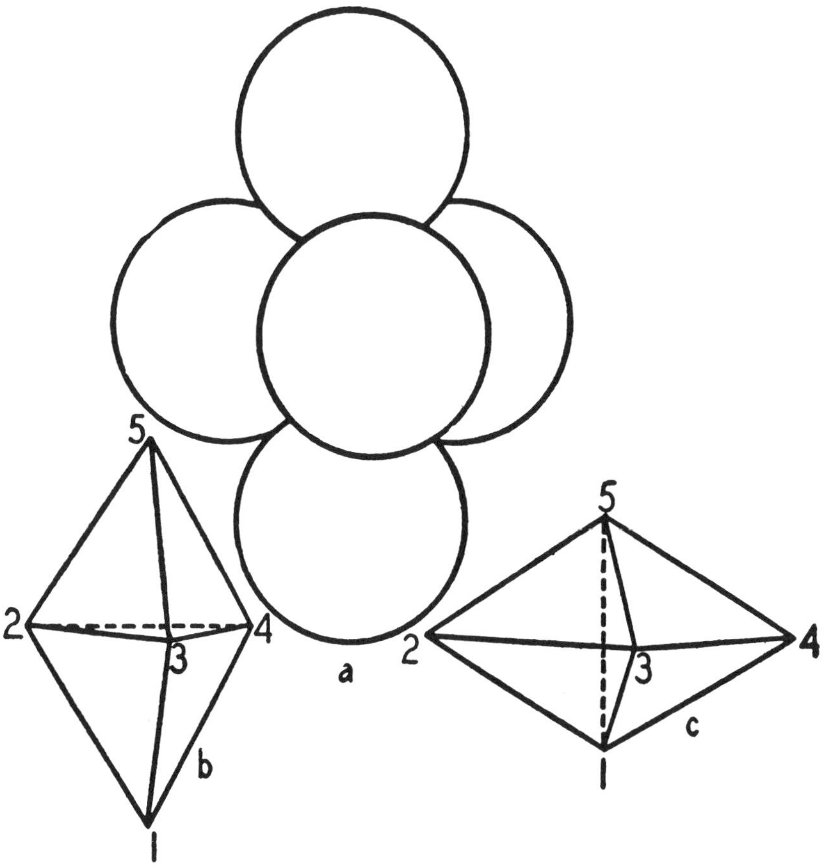


*Plate 5*

# Some Generalizations of the Virial Theorem

**Edward Teller**[2]

*Received August 20, 1985*

*Generalizations of the virial theorem are derived: In atomic physics, in systems including electromagnetic radiation, in Newtonian gravitation, and in general relativity and also some types of nuclear forces. The cases discussed are limited to potentials which can be produced by the exchange of one particle, which include potentials of the form* $1/r$. *The method used is to set equal a change in energy produced by an infinitesimal similarity transformation to a change of energy obtained by a first-order perturbation.*

## 1. THE SIMPLE VIRIAL THEOREM IN ATOMIC STRUCTURES

A primitive derivation of the virial theorem in atomic physics can be obtained by considering an infinitesimal increase of all masses by a factor $1+\mu$ (with $\mu \ll 1$):

$$m \rightarrow m(1+\mu)$$

while leaving all other relevant constants $e$, $\hbar$, $c$ unchanged. A similarity transformation, in which distances $r$, momentum values $p$, velocities $v$, energies $E$, and times $t$ are replaced by

$$r \rightarrow r(1-\mu)$$

$$p \rightarrow p(1+\mu)$$

$$v \rightarrow v$$

$$E \rightarrow E(1+\mu)$$

$$t \rightarrow t(1-\mu)$$

[1] Work performed under the auspices of the U.S. Department of Energy by Lawrence Livermore National Laboratory under contract #W-7405-Eng-48.

[2] University of California, Lawrence Livermore National Laboratory, Livermore, California 94550.

will leave the Schrödinger equation unchanged. The change in energy $\mu E$ can be written in terms of potential energy $E_{pot}$ and kinetic energy $E_{kin}$

$$\mu E = \mu E_{pot} + \mu E_{kin}$$

The same change in energy may be derived by first-order perturbation which leaves the potential energy which is propertional to $e^2/r$ unchanged, while the kinetic energy $p^2/m$ is perturbed:

$$\frac{p^2}{2m} \to \frac{p^2}{2m}(1-\mu)$$

It is to be noted that in first-order perturbation $p$ remains unchanged, while $m$ is changed into $m(1+\mu)$. Setting equal the change in energy $\Delta E$ obtained from the similarity transformation and from first-order perturbation, we obtain

$$\Delta E = \mu E_{pot} + \mu E_{kin} = -\mu E_{kin}$$

which leads to the most simple form of the virial theorem[3]

$$E_{pot} + 2E_{kin} = 0$$

If the system, which may be macroscopic, is under pressure $P$ and is confined to a volume $V \sim r^3$, then in the similarity transformation the volume will change as

$$V \to V(1-3\mu)$$

In this similarity transformation the energy will still change in the same way as above, though the volume is changed. To the previous result of first-order perturbation we should now add the work by the pressure $P$ due to the volume contraction $3\mu V$, which gives

$$\mu E_{pot} + \mu E_{kin} = -\mu E_{kin} + 3\mu PV$$

or

$$E_{pot} + 2E_{kin} = 3PV \tag{1}$$

which is the usual form of the virial theorem. In this equation, as well as in the following analogous ones, all terms on the left-hand side must be summed or integrated over the components of the system.

[3] A similar derivation is based on the variational theorem. If all distances $r$ are replaced by $r(1-\mu)$, where $\mu$ is infinitesimal, then the change in the potential and in the kinetic energy will be respectively $\mu E_{pot}$ and $2\mu E_{kin}$. According to the variational principle, the sum of these two terms must be zero.[(1)]

This derivation is valid not only for the actual system but also for the results of approximate calculations such as obtained from the Hartree method, the Thomas–Fermi equation, or the refined form of that equation in which exchange forces are included.[2] It is equally valid in statistical mechanics, and can be applied to quasi-ergodic systems which return in a finite time within arbitrarily close limit to any earlier location. In this case the time average of each quantity $E_{pot}$, $E_{kin}$ is considered.

## 2. IS THE RESULT OF THE SIMILARITY TRANSFORMATION UNIQUE?

Instead of changing $m$, we may have changed $e$, $\hbar$, or $c$. If the latter quantities are changed, however, one should conserve the fine-structure constant $e^2/\hbar c$, otherwise the nature of our problem will no longer be the same. We shall now show that any of the permissible changes (leaving $e^2/\hbar c$ unaltered) will lead to the same virial theorem.

It is clear that if several of the constants $m$, $e$, $\hbar$, $c$ are changed by infinitesimal amounts, the results are additive, both in the energy change due to the similarity transformation and in the energy change due to first-order perturbation. Instead of these changes, one may construct another three in which of the three dimensions—length, time, and mass—only one changes. We shall now show that of these three only one leads to a non-trivial result: if we change the lengths $r$.

It is to be noted that in the discussion given in the first section all three dimensions changed. It is also obvious that if the length $r$ is unchanged, the volume $V$ will not change and the term $3PV$ cannot appear. Furthermore, if the distances and, therefore, the gradients of the wave function remain unchanged in the wave function of the Schrödinger (or Dirac) equation during the similarity transformation, only the constants $m$, $e$, $\hbar$, $c$ can be changed. But in the perturbation procedure of the first order the change of the same constants is taken into account exclusively; thus, the two approaches give identical results. Nothing new can be derived.

By contrast, if the distances are changed, this has a direct effect on the similarity transformation but not on the first-order perturbation. The perturbation theory is affected only by the change of the constants, while in the similarity transformation both the change of the constants and the change of the variables play a role. In this way only one relation can be derived: the virial theorem.

## 3. MAGNETIC ENERGY

The virial theorem is applicable in its usual form only to central forces having a $1/r^m$ dependence. We consider in this paper $1/r$ which is derivable from a $1/r$ potential. Magnetic forces between mowing charged particles are not central and spin–orbit or spin–spin forces have a different dependence on $r$. The method used here is, however, readily applicable.

In the similarity transformation, currents remain unchanged and the magnetic moments of spins change as $1/m$ and are, therefore, multiplied by $(1-\mu)$. Since $r$ is also multiplied by $(1-\mu)$ and since orbit–orbit, spin–orbit, and spin–spin interactions vary as $1/r$, $1/r^2$, and $1/r^3$, respectively, all these interactions change as $(1+\mu)$, that is, like all other energies. The infinitesimal change in energy in the similarity transformation will therefore be

$$\mu E_{\text{pot}} + \mu E_{\text{kin}} + \mu E_{\text{m}}$$

where $E_{\text{m}}$ is the magnetic energy and $E_{\text{pot}}$ is limited to the Coulomb interaction and no externally imposed fields are assumed.

In first-order perturbation theory we must leave $r$ values, gradients, and momentum values unchanged, but velocities $(\sim p/m)$ and currents change as $(1-\mu)$ and magnetic moments of spins change in the same way. Therefore, the change in $E_{\text{m}}$ is proportional to $1/m^2$ and can be written $-2\mu E_{\text{m}}$.

The generalization of Eq. (1) is

$$E_{\text{pot}} + 2E_{\text{kin}} + 3E_{\text{m}} = 3PV \tag{2}$$

This equation may be applied to ferromagnets or superconductors.

If external electromagnetic field and potentials are imposed, the usual term $3PV$ should be replaced by a more involved expression.

One may then apply Eq. (3) to more complicated situations including a plasma held together by external electromagnetic forces (in most of the research on controlled fusion only external magnetic fields are applied). In that case the right-hand side of Eq. (3) is to be replaced by

$$\int P\mathbf{r}\cdot d\boldsymbol{\sigma} + \int \rho_{\text{el}}\mathscr{E}_{\text{ext}}\cdot\mathbf{r}\,d\tau + \int (\mathbf{i}\times\mathscr{H})\cdot\mathbf{r}\,d\tau$$
$$+\int [\nabla(\mathscr{H}_{\text{ext}}\cdot\boldsymbol{\mu}_{\text{mag}})]\cdot\mathbf{r}\,d\tau \tag{3a}$$

Here the first term is an integral to be taken over the surface of the finite volume to which the virial theorem is applied. The vector $\mathbf{r}$ is drawn from the surface element over which we integrate to an arbitrary fixed point and $d\boldsymbol{\sigma}$ is a vector perpendicular to the surface element and pointing inward; it has the absolute value $d\sigma$. It is easy to see that this integral is equal to $3PV$. In the second term, $\rho_{el}$ is the electric charge density and $\mathscr{E}_{ext}$ is the electric field imposed by external sources. The vector $\mathbf{r}$ points from the volume element to a fixed position. In the third term the current density is $\mathbf{i}$ and $\boldsymbol{\mathscr{H}}_{ext}$ is the independently given magnetic field, usually the confining agent acting on the plasma. Finally, the last term is due to the force of the inhomogeneous magnetic field and $\mu_{mag}$ is the magnetic moment of the spin. Here the gradient operator $\boldsymbol{\nabla}$ is to be applied to the magnetic interaction $(\boldsymbol{\mathscr{H}}_{ext} \cdot \mu_{mag})$. All integrations have to be carried out over the volume of the plasma. In experiments involving a high-temperature plasma, the pressure $P$ is usually zero since the plasma is not supposed to be in contact with the container. The whole right-hand side is the sum of all the forces imposed on the plasma volume under consideration.

The modified form of Eq. (3) can be applied to a part of a system where the left-hand side is summed or integrated over the part of the system under consideration while the right-hand side describes the forces due to all other portions of the whole system. Our approach to the virial theorem would then involve an application of the similarity transformation to the total system. In the perturbation theory the similarity changes still apply to the "external" parts, while in the "internal" part only $m$ is replaced by $m(1+\mu)$ and the work accomplished by the motion of the "internal" part relative to the "external" parts must be added. This procedure could be useful in numerical calculations in that more crude zoning may become permissible.

If one wants to verify that application of the virial theorem to parts add up to the virial theorem for the whole system, orbit–orbit, spin–orbit, and spin–spin interactions must be separately considered. For instance, the factor 3 in the term $3E_m$ for the spin–spin case is due to the $1/r^3$ interaction and its influence on the term due to the work considered in our procedure. The other terms in the $3E_m$ expression have a more complex origin.

## 4. ELECTROMAGNETIC RADIATION

So far we have included only stationary, or slowly varying, electric and magnetic fields. It seems reasonable that the energy of electromagnetic radiation behaves like the potential energy. This is actually the case.

Indeed, the electromagnetic radiative energy density

$$\frac{1}{8\pi}(\mathscr{E}^2 + \mathscr{H}^2)$$

changes in the similarity transformation like $(1+4\mu)$ since $\mathscr{E}$ and $\mathscr{H}$ change due to

$$r \to r(1-\mu)$$

as

$$\mathscr{E} \to \mathscr{E}(1+2\mu)$$

$$\mathscr{H} \to \mathscr{H}(1+2\mu)$$

If we multiply the energy density by the volume element $d\tau$, which changes as $(1-3\mu)$, and integrate, we obtain for the electromagnetic radiation energy

$$E_{\text{em}} = \int \frac{1}{8\pi} (\mathscr{E}^2 + \mathscr{H}^2)\, d\tau$$

and the similarity change

$$E_{\text{em}} \to E_{\text{em}}(1+\mu)$$

That is, the radiation energy changes like all other energies.

In the perturbation calculation, $E_{\text{em}}$ remains unchanged. Therefore, Eq. (2) may be replaced by

$$E_{\text{pot}} + E_{\text{em}} + 2E_{\text{kin}} + 3E_{\text{m}} = 3PV \tag{3}$$

This equation may be applied to stars in which radiation may account for a considerable fraction of the energy in a steady state (or quasi-ergodic state, i.e., a variable star), though gravitational energy has to be included. This will be done further below.

## 5. RELATIVISTIC EQUATIONS

If we want to apply our equation to cases where special relativity must be taken into account, the similarity transformation remains unchanged. But, in the first-order perturbation calculation $E_{\text{kin}}$ and $E_{\text{m}}$ must be reexamined.

For the first of these we may write

$$E_{\text{kin}} = [(mc^2)^2 + c^2p^2]^{1/2}$$

where $m$ is the rest mass and $mc^2$ has been included in $E_{\text{kin}}$. The change in kinetic energy $\Delta E_{\text{kin}}$ due to an infinitesimal change in $m$ will become

$$\Delta E_{\text{kin}} = [(\{1+\mu\}\, mc^2)^2 + c^2p^2]^{1/2} - [(mc^2)^2 + c^2p^2]^{1/2} = \mu E_{\text{kin}} \left(\frac{mc^2}{E_{\text{kin}}}\right)^2$$

In considering the portion of $E_{\mathrm{m}}$ which is the current–current interaction, this quantity is not simply proportional to $1/m^2$ but to $(c^2/E_{\mathrm{kin}})^2$. For the change of the relevant part of $\Delta E_{\mathrm{m}}$ due to first-order perturbation (where the momentum $p$ is kept constant) we may, therefore, write

$$\Delta E_{\mathrm{m}} = E_{\mathrm{m}}\left[\left(\frac{E_{\mathrm{kin}}}{E_{\mathrm{kin}} + \Delta E_{\mathrm{kin}}}\right)^2 - 1\right] = -2\mu E_m \left(\frac{mc^2}{E_{\mathrm{kin}}}\right)^2$$

Using the calculated values of $\Delta E_{\mathrm{kin}}$ and $\Delta E_{\mathrm{m}}$, one obtains, for the relativistic form of the virial theorem,

$$E_{\mathrm{pot}} + E_{\mathrm{em}} + E_{\mathrm{kin}}\left[1 - \left(\frac{mc^2}{E_{\mathrm{kin}}}\right)^2\right] + E_{\mathrm{m}}\left[1 + 2\left(\frac{mc^2}{E_{\mathrm{kin}}}\right)^2\right] = 3PV \quad (4)$$

What has been said so far about the relativistic case relates to macroscopic situations. The most discussed case in quantum mechanics is the Dirac equation for a single particle. Straightforward application of our procedure gives

$$E = mc^2\bar{\beta} = [(\psi_1|\psi_1) + (\psi_2|\psi_2) - (\psi_3|\psi_3) - (\psi_4|\psi_4)]\, mc^2$$

where $\beta$ is the Dirac operator which is positive for the first pair of components and negative for the second pair. The left-hand side is obtained from the similarity transformation while the right-hand side is derived by the perturbation theory where only the term $\beta mc^2$ is changed.

If we subtract this relation from the Dirac equation averaged over a stationary state, we obtain the more usual form of the virial theorem[(1)]

$$E_{\mathrm{pot}} + c\boldsymbol{\alpha}\cdot\mathbf{p} = 0$$

Here $c\boldsymbol{\alpha}$ stands for the velocity and $\mathbf{p}$ for the momentum.

One may get a different generalization of the virial theorem by deriving the Klein–Gordon equation from the Dirac equation. For the sake of simplicity, this will be done in the absence of magnetic fields. The result is

$$(E - E_{\mathrm{pot}})^2\,\psi = (p^2c^2 + m^2c^4)\psi$$

This holds, of course, for any single-particle problem. If the average is taken over a stationary state and if, as is permissible for $E_{\mathrm{pot}} \approx 1/r$, the change $m \to m(1+\mu)$ leads to $E \to E(1+\mu)$, then using the perturbation theory we get

$$E(E - \overline{E_{\mathrm{pot}}}) = m^2c^4$$

where $\overline{E_{\text{pot}}} = (\psi\,|E_{\text{pot}}|\,\psi)$. The solution of this equation yields, for $\overline{E_{\text{pot}}} < 0$,

$$E = \tfrac{1}{2}[\overline{E_{\text{pot}}} + (\overline{E_{\text{pot}}}^2 + 4m^2c^4)^{1/2}]$$

where the positive square root should be taken. The reason for this relatively simple result is that the term $\overline{E_{\text{pot}}^2}$ drops out when taking the difference of energies after and before the similarity transformation. This would not have been possible if higher power of $E - \overline{E_{\text{pot}}}$ had been taken.

An interesting though bizarre consequence of our potential-theory result is the situation arising if $\overline{E_{\text{pot}}}^2 \gg m^2c^4$. In this case the approximate result is

$$E \approx \frac{m^2c^4}{|\overline{E_{\text{pot}}}|}$$

It is obvious that any stable solutions must have a positive energy. For $\overline{E_{\text{pot}}}^2 \gg m^2c^4$ it is by no means obvious that stable solutions exist. It may be possible to assume a continuum of such solutions with infinitely many modes near $r = 0$. We see, in addition, that the values of $E$ will be inversely proportional to $|\overline{E_{\text{pot}}}|$.

The quantum-mechanical, many-body problem is essentially unsolved. To approximate solutions, like the Hartree method, generalizations of the virial theorem can be applied.[(3)]

## 6. GRAVITATION AND GENERAL RELATIVITY

In our derivation of the virial theorem for atomic structures we assumed that the charge of the electron, $e$, remained unchanged, but we replaced all masses $m$ by $m(1+\mu)$. To retain the argument in the same form, we must assume that the gravitational interaction of two masses $Gm^2$ must not change. Hence, we have

$$G \to G(1 - 2\mu)$$

Designating the potential energy of gravitation by $E_{\text{grav}}$, one obtains for isolated systems in the nonrelativistic approximation

$$E_{\text{pot}} + E_{\text{em}} + E_{\text{grav}} + 2E_{\text{kin}} = 3PV \tag{5}$$

where $E_{\text{kin}}$ includes macroscopic as well as atomic and subatomic motions. Nuclear energies are not taken into account and for the sake of simplicity energy $E_{\text{m}}$ is neglected[4] in Eq. (5). In the obvious applications to stars, globular clusters, and galaxies, $P = 0$, and the right-hand side of Eq. (5)

[4] Its inclusion would be routine.

vanishes. The same conclusions also apply to periodic and quasi-ergodic motion such as cepheid variables and rotating stars or galaxies. In this case all terms must be replaced by their time averages.

One can also apply the virial theorem to a part of an astronomical object, for instance, to an arbitrarily defined volume within a star. Then the right-hand side of Eq. (5) will not be zero but will have the form

$$\int P\mathbf{r}\cdot\mathbf{d}\sigma+\int(\mathbf{g}_{\mathrm{ext}}\cdot\mathbf{r})\rho\,d\tau \tag{5a}$$

where the first term becomes $3PV$ provided the pressure is a constant over the surface which in the present case is not always true. In the second terms, $g_{\mathrm{ext}}$ is the gravitational acceleration due to external masses and $\rho$ the density. The integration is to be carried out over the volume.

It is interesting to note that if a star is subdivided into two regions, an internal sphere and an external shell, then the interaction potentials occur twice. This wrong factor 2 is corrected when the second term in the expression (5a) is taken into account. Similar considerations apply to Eq. (3a).

The application to general relativity is remarkably simple. Actually the similarity transformation may be used without change.

To see this one has to consider the definition of space curvature. The relevant point is the angular change of a vector carried around an infinitesimal surface in four-dimensional space, divided by the area of that surface. The curvature is proportional to the gravitational constant $G$ and the energy-momentum-tension tensor which is an energy per unit volume. Thus, we get

$$G\frac{E}{r^3}\sim\frac{(1-2\mu)(1+\mu)}{(1-3\mu)}=1+2\mu$$

On the other hand, the angular change of the vector is the curvature times the area. Since $r$ and $t$ change as $1-\mu$ the area is proportional $1-2\mu$. Thus, the change in the direction remains unaltered by the similarity transformation if the vector is carried around a scaled-down area.

The first-order perturbation theory may be applied in an unchanged fashion. Thus, one is led to the equation for an isolated system

$$E_{\mathrm{pot}}+E_{\mathrm{em}}+E_{\mathrm{kin}}\left[1-\left(\frac{mc^2}{E_{\mathrm{kin}}}\right)^2\right]=E_{\mathrm{pot}}+E_{\mathrm{em}}+\frac{p^2c^2}{E_{\mathrm{kin}}}=0 \tag{6}$$

In Eq. (6), $E_{\mathrm{pot}}$ includes the Coulomb energy as well as all forms of the gravitational energy. We again use $E_{\mathrm{kin}}^2=(mc^2)^2+c^2p^2$ which leads to the rewriting of the third term in Eq. (6). This term is to be summed over all particles in the isolated systems (star, cluster, or galaxy) under consideration.

If the system is not isolated but is under the influence of external pressure and externally imposed gravitational forces, the right-hand side of Eq. (6) will not be zero but will be of the same form as stated in the case of Eq. (5). It should be recognized, however, that in general relativity there is no obvious way in which to separate out the gravitation forces $g_{\text{ext}}$ due to the external masses. Indeed, the solutions in general relativity are supposed to be self-consistent.

All energies should be given the value as seen by one selected observer who is at rest relative to the observed system or, in case of periodic motion, relative to the average position of the system. For different observers terms in Eq. (6) will change due to a change of the gravitational red-shift which will amount to a common factor in all terms.

Equation (6) does not apply to black holes, not even to very massive black holes (with a correspondingly big radius) in which high densities such as those occurring in atomic nuclei need not be considered. On the one hand, the interior of a black hole is unobservable. On the other, the formation of a black hole needs infinite time; though configurations closely approaching a black hole can be obtained in a short time, systematic, non-ergodic velocities approaching light velocity are necessarily involved. All our considerations apply only to systems which remain in or return to the same state, at least in good approximation.

## 7. APPLICATION TO NUCLEAR FORCES

There is no indication that forces which exclude the separation of the interacting particles such as occur in the description of quarks in quantum chromodynamics can be treated by any method resembling the virial theorem. On the other hand, older theories describing nuclear forces based on the exchange of mesons might be treated in analogy with the procedures used in this paper.

We again perform a similarity transformation in which all masses are changed by a factor $1+\mu$. The masses include those of the Yukawa particles which are exchanged. The potentials due to these particles are of the Coulomb type with an added exponential factor $\exp(-r/r_Y)$, where $r_Y$ is the Yukawa distance and is inversely proportional to the mass of the Yukawa particle. The similarity transformation holds with all masses changed by a factor $1+\mu$ and all distances changed by a factor $1-\mu$. The masses include $m_Y$, the Yukawa mass, and the distances the Yukawa radius

$r_Y$, which changes by $1-\mu$. The result is an equation similar to Eq. (1) except that on the left-hand side a term $-(r/r_Y)\,E_Y$ will appear, where $E_Y$ is the Yukawa potential and $r_Y$ is again the Yukawa radius. In this expression, $r$ is the distance between a pair of nucleons, and a sum over all the pairs has to be taken.

Actually, several terms of this kind will have to be introduced, depending on various Yukawa masses that can be exchanged between nucleons. In this way Yukawa-type repulsive potentials may be introduced which, of course, have in general a shorter range than the attractive interactions.

The virial theorem can be further generalized to the older approaches of nuclear forces when one introduces orbit–orbit interactions, spin–orbit interactions, and spin–spin interactions or tensor forces. The appearance of these various forces will be similar to what we discussed in connection with Eq. (3) and in relation to the magnetic term $E_m$. The discussion here is a simple translation of what has been said in connection with Eq. (3). In every case, the similarity transformation will occur as usual with the only exception being that $m_Y$ and $r_Y$ have to be changed as described. In every case, therefore, additional terms of the form $(r/r_Y)\,E_Y$ will make their appearance, where $E_Y$ stands for the various kinds of Yukawa interactions describing plain interactions, orbit–orbit interactions, spin–orbit interactions, and spin–spin interactions. The latter three will furthermore carry a factor 3 as did $E_m$ in Eq. (3).

For clarity we repeat the reasons for their occurrence. Every one of these terms will occur with a factor 1 due to the similarity transformation. In addition, in the case of the orbit–orbit interactions, perturbation theory will add the term multiplied by a factor 2 due to the perturbation originating from the changed mass of the nucleons which in perturbation theory corresponds at a constant momentum to a decrease of the velocity of the interacting particles. Since, however, it is this velocity which gives rise to the orbit–orbit interaction, the perturbation theory will give a factor $-\mu$ on the right-hand side of the equation which can be transferred to the left-hand side with a positive sign and will furnish two more terms of the orbit–orbit type due to the two interacting particles.

In the case of the spin–orbit and spin–spin type of interactions, the exponential factors are preceded by a $1/r^2$ or $1/r^3$ dependence, respectively. In order to maintain the similarity transformation, the forces due to the spin as well as the forces due to the velocity of the motion have to be reduced by a factor $1-\mu$. This will result in all cases in a further factor $2\mu$ due to the perturbation theory which will in the end give rise to a factor 3 in analogy with the one obtained for $E_m$ in Eq. (3).

It might be mentioned that the spin–orbit forces have played a particularly strong role in the shell theory of nuclear structure. This is probably due to the circumstance that the spin–orbit forces are strong when the gradients of the potentials become pronounced. The Yukawa-type behavior increases these gradients.

The end result is an expression similar to Eq. (3) in which, however, on the left side Yukawa-type additions of the appearance $(r/r_Y)\,E_Y$ have to be added[5] for every Yukawa-type force whether it be a simple type interaction or one analogous to the terms designated as $E_m$ in Eq. (3).

It would be tempting to apply these considerations to the interior of neutron stars. This would appear to be possible by using virial-type expressions on volume elements within the neutron star and deriving sound velocities by deriving expressions for $dP/d\rho = (1/c^2)(dP/dE)$ where $\rho$ is the mass density and $E$ the energy density. In some approximations, one can see how a sound velocity $c/\sqrt{3}$ is approached; this sound velocity is, indeed, characteristic for ideal gases at very high temperatures.

The appearance of strong repulsive forces and the corresponding Yukawa potentials, which can become predominant at high pressures, can give rise in a formal sense to velocities higher than the velocity of light. One of my old memories is that Fermi asked me in 1951 or 1952 how probable it was that within ten years a velocity would be discovered exceeding light velocity. My answer was one in a million. Fermi then told me he thought it was 10%, the well-known value for a Fermi miracle. Unfortunately, I remain convinced that the probability for the next ten years is even now $10^{-6}$.

The derivation of sound velocity discussed in the previous paragraphs seems actually to be inapplicable for velocities approaching the velocity of light. The expression $dP/d\rho$ is defined in a static way, but is based on the exchange of mesons. The circumstance that these mesons themselves cannot move faster than light should make the derivation based on static compressibilities inapplicable. We seem to have reached in this regard the limits of the usefulness of the methods that are applied in this paper.

## 8. CONCLUSION

Many of the results obtained in this paper are repetitions of the obvious. Others approach the absurd. I am deeply indebted to my friend, Mort Weiss, who prevented me from going over the edge and also

[5] It is noted that these extra terms should in no case be multiplied by a factor 3 since they are obtained from the similarity transformation and no contribution is made by the perturbation theorem.

encouraged me by his interest in what appeared almost obvious. I hope, particularly with his help, to have touched upon some subjects which are neither obvious nor absurd.

## REFERENCES

1. M. Brack, *Phys. Rev. D* **27**, 1950 (1983).
2. R. P. Feynman, N. Metropolis, and E. Teller, *Phys. Rev.* **75**, 1561 (1949).
3. Y. K. Kim, *Phys. Rev. D* **154**, 17 (1967).

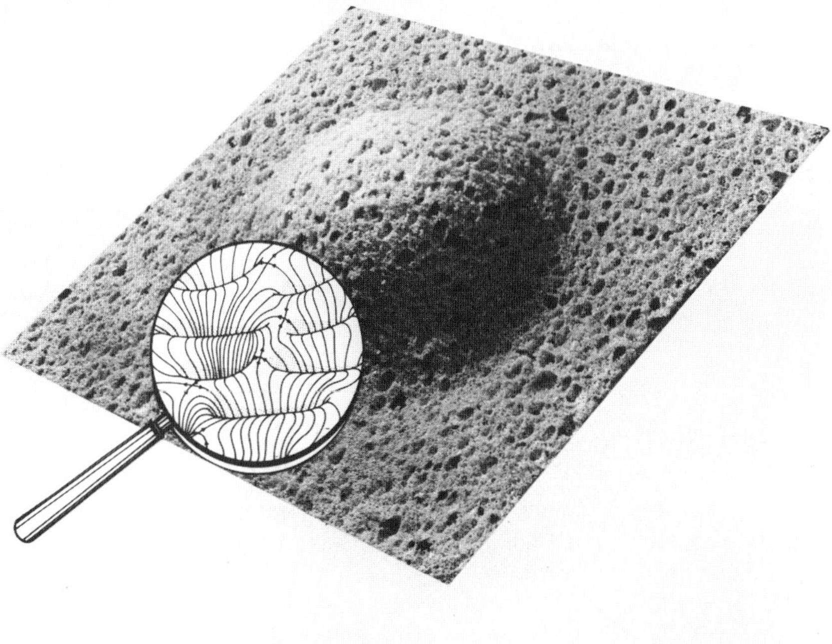

Plate 6

# Does the Topology of Space Fluctuate?[1]

**Arlen Anderson**[2] **and Bryce DeWitt**[2]

*Received February 15, 1985*

*Evidence is presented that the singularities induced in causal Lorentzian spacetimes by changes in 3-space topology give rise to infinite particle and energy production under reasonable laws of quantum field propagation. In the case of the gravitational field, if 3-space is compact the total energy must vanish. A topological transition therefore induces a violent collapse that effectively aborts the transition, since the collapse mode is the only mode carrying the negative energy needed to compensate the associated infinite energy production. The existence of the Hamiltonian constraint of general relativity suggests that topological stability is a local property of the quantum theory that is maintained even when 3-space is noncompact.*

## 1. INTRODUCTION

No one has had a greater influence on the development of the quantum theory of gravity than John Wheeler. He has taught many of us not only how to ask the right questions of the theory but also to understand some of its fantastic implications. In a memorable conference held in Chapel Hill in January, 1957,[3] John told the following fairy story: A great scientific competition is underway between two nations. The leaders of one of these nations confront their scientists with a vital piece of intelligence: The scientists of the other nation have recently discovered that everything can be explained in terms of geometry. These leaders demand of their own scien-

[1] Supported in part by National Science Foundation Grant No. PHY-8318635.

[2] The University of Texas at Austin, Austin, Texas 78712.

[3] *Conference on the Role of Gravitation in Physics*, Chapel Hill, North Carolina, January 18–23, 1957 (WADC technical report 57-216, ASTIA document No. AD 118180). Nowadays this conference is referred to as GRG 1. The origin of coordinates was the 1955 Jubilee of Relativity conference held in Bern, now known as GRG 0.

tists, under threat of torture, that they formulate a similar theory of their own. How do they go about it?

One of the suggestions that John made in answer to this question was to take seriously the idea that the topology of 3-space might undergo quantum fluctuations. He reasoned as follows: If one attempts to measure the average electric field in a space-time volume $L^4$ in a vacuum at 0°K (e.g., by observing during a time $L$ the motion of a charged test body of volume $L^3$), one will not generally obtain a vanishing result. In a sequence of such measurements one will obtain a statistical distribution of values with magnitudes peaked around $1/L^2$ ($\hbar = c = 1$).

Similarly, quantization of the linearized theory of gravity suggests that if one attempts to measure the Riemann tensor averaged over a space-time volume $L^4$ in a vacuum at 0°K (e.g., by observing during a time $L$ the change of strain of a gravitational antenna[4] of volume $L^3$), one will probably not obtain zero but rather a value of magnitude in the neighborhood of $\lambda_P/L^3$ ($\lambda_P$ = Planck length) corresponding to a space-time radius of curvature of order $(L^3/\lambda_P)^{1/2}$. John noted that this radius of curvature becomes comparable to $L$ itself when $L \approx \lambda_P$. Although this is precisely the realm in which the linearized theory fails, the result suggests that the curvature at the Planck scale fluctuates so violently as practically to cause portions of space to "pinch off" or to become multiply connected, developing structures such as "wormholes." John proposed that this is indeed what happens.

## 2. DIFFICULTIES

John's proposal faces two immediate difficulties. First of all, if wormholes can develop, then wormholes themselves can acquire wormholes *ad infinitum*, and space can take on the character of a "foam." But, on a coarse scale ($\gg \lambda_P$), a foam can appear to have a dimensionality arbitrarily greater than that of the underlying manifold out of which it is built. The observed stability of the dimensionality of space is thus called into question, and any theory that predicts quantum fluctuations in the topology of space must also explain why nature seems so rigidly to prefer three dimensions.

Secondly, if a wormhole forms, the two entrances to it can move arbitrarily far apart before the wormhole dissolves. Indeed, classically, if time is reckoned by means of a foliation of space-time into maximal

[4] For the design of such a measurement, see B. S. DeWitt in *Gravitation: An Introduction to Current Research*, L. Witten, ed. (Wiley, 1962).

spacelike hypersurfaces, then, once the two entrances get far enough apart the tunnel that connects them will "never" dissolve, and the distance between the entrances can become vastly greater on the "outside" than through the tunnel itself. Of course, one must expect quantum fluctuations to violate classical laws. But if a wormhole can fluctuate out of existence when its entrances are far apart—a transition that seemingly *ought* to take place readily enough—then, by the principle of microscopic reversibility, the fluctuation *into* existence of a wormhole having widely separated entrances ought to occur equally readily. This means that every region of space must, through the quantum principle, be potentially "close" to every other region, something that is certainly not obvious from the operator field equations which, like their classical counterparts, are strictly local.

It is difficult to imagine any way in which widely separated regions of space can be "potentially close" to each other unless space-time itself is embedded in a convoluted way in a higher dimensional manifold. Additionally, a dynamical agency in that higher-dimensional manifold must exist which can transmit a sense of that closeness. A model of this kind could well be called a *string theory* of space-time. A major defect of a string theory of space-time is that it is anti-Riemann, anti-Clifford, and anti-Einstein. It was Riemann who first insisted that a curved manifold needs no higher-dimensional embedding space for its definition. Clifford and Einstein went even further, insisting that space-time itself is *all.* In a string theory of space-time the Universe is not everything; there is something physical *outside.*

Even if one embraces the string idea, one is by no means free of difficulties. The idea by itself gives no clue to the nature of the "closeness transmitting" agency in the embedding space. The trouble with changing the rules of a game (in this case the standard rules of quantum gravity) is that there are no rules about changing the rules. The same difficulty arises with other suggestions that have been put forward from time to time, such as the proposal that one should cease to regard space-time as a differentiable manifold at scales below the Planck length. If not a differentiable manifold, then *what*? Since speculation is generally profitless we believe that the only useful procedure is to challenge the theory on its own ground, to see where its real weaknesses and strengths lie. One might then learn something.

## 3. CONTINUITY AND THE EUCLIDEAN OPTION

It is a general rule in the quantum theory that the classically forbidden regions into which a system can fluctuate can always be connected by continuous paths (in configuration space) to classically allowed regions. There is no reason to suppose that quantum gravity is an exception to this rule. But this means that if one wishes to calculate, by means of a Feynman functional integral, the amplitude for a transition between states having different topologies, then one must have available a set of continuous field histories that connect those topologies. Now it is well known that the metric of any Lorentzian manifold that has no closed timelike curves, or other causal anomalies, necessarily becomes singular at points where its spatial sections undergo topological change. How does one handle these singularities?

One answer is to regard Lorentzian space-time as an analytic continuation of a Euclidean manifold. In a Euclidean manifold the metric need not become singular at points where cross sections undergo topological change. The search for interpolating histories becomes a standard *cobordism* problem. This "Euclidean option" thus converts the problem of topological change into what looks like a standard quantum tunneling problem. However, it has difficulties of its own. The first is the well-known failure of the Euclidean action to be positive definite: Conformal changes in the metric contribute negatively to it. The second is the seldom emphasized and much less well-known fact that in applications of the Euclidean option one generally avoids asking for the amplitude for a topological change to be *observed.* It seems to be very difficult to ask for this, and hence, because physicists like to solve easy problems, a real test of John's idea has never been achieved.

In our view it is scandalous that, after more than a quarter of a century, John's idea remains untested. The aim of the present work is to take a tentative step towards improving the situation. Euclidean techniques will be avoided. Historically, most problems in quantum mechanics that can be illumined by analytic continuation were first solved by other more straightforward methods. The advantage of proceeding in a direct manner is that the physics remains in the foreground and our understanding is enhanced. The results described in the following paragraphs turn out to suggest that John's topological transitions may be both energetically and dynamically forbidden.

## 4. A (1+1)-DIMENSIONAL MODEL. THE TROUSERS TOPOLOGY

The full range of topological varieties that can be connected to a given 3-manifold by interpolating 4-manifolds is daunting. Where shall we begin? It seems obvious that we should begin by making drastic simplifications, for example, by reducing the dimensionality of space-time from 4 to 2. The trouble with this is that the gravitational field in a 2-dimensional space-time has $-1$ degrees of freedom! A direct analog of Einstein's theory simply does not exist in two dimensions.

We shall avoid this difficulty for the moment by replacing the gravitational field by a linear massless scalar field. This field will not be regarded as describing a dynamics of the underlying space-time; the metric of space-time will be taken as given *a priori*. The topology we shall choose is shown in Fig. 1. This may be called the *trousers topology*: Space, which is initially diffeomorphic to the circle, splits into two circles: $S^1 \to S^1 + S^1$. Our initial task will not be to ask how this topological transition can come about but rather to determine reasonable laws of propagation for the scalar field on such a background.

The only serious problem confronting us is how to deal with the metric singularity that inevitably occurs in the crotch region. A possible solution is to allow the metric to become complex (in a suitable patch) in this region. The metric then need not be singular and one may study the propagation problem as an ordinary problem in the theory of partial differential equations. A difficulty with this, however, is that the scalar field, if initially real in the trunk region, will not generally remain real in the legs. Moreover, if the field is quantized, the coefficients connecting "in" mode

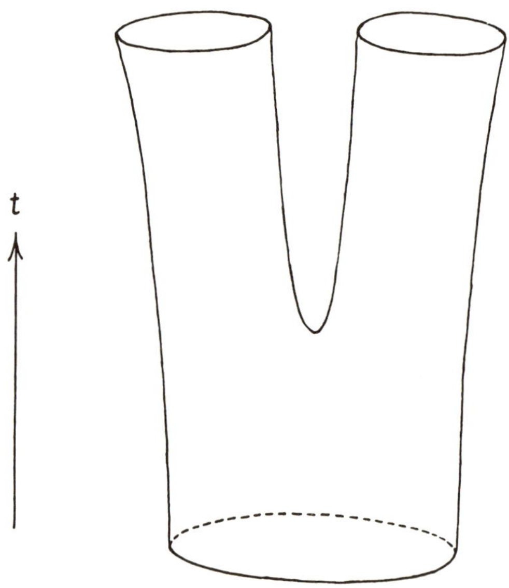


Fig. 1. The trousers topology, corresponding to the topological transition $S^2 \to S^1 + S^1$.

functions (in the trunk) to "out" mode functions (in the legs) will not satisfy the standard Bogoliubov relations, and unitarity will be violated.

Unitarity could, of course, be regained by allowing the metric in the end to become real and the singularity to reappear. But this would be equivalent to simply excising the singular point from space-time and specifying how waves are to propagate around it (different laws corresponding to different ways in which the imaginary part of the metric is allowed to tend to zero). This is in fact the procedure that will be adopted. In two dimensions there is also no need to introduce nonzero curvature. The purely topological aspect of the effects to be studied can therefore be cleanly isolated.

Figure 2 shows a convenient way to display the flat trousers manifold. The metric is everywhere Minkowskian and static except at the singular point, which is indicated by crosses. Below the crosses in the trunk region the two sides of the figure are to be identified. Similar identifications occur

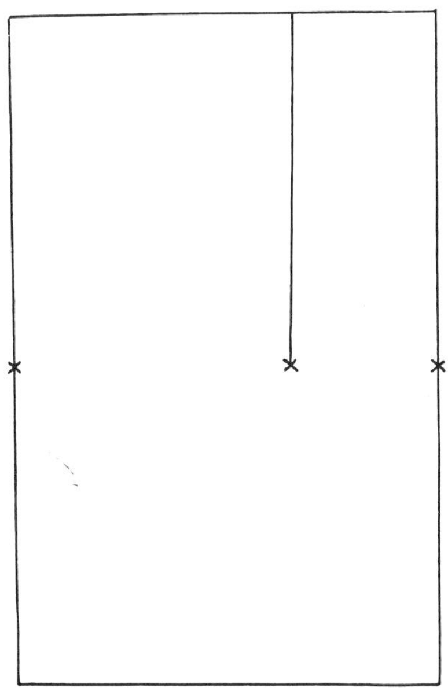

Fig. 2. Representation of the flat trousers topology. The crosses indicate the singular point. Below the crosses, opposite sides of the figure are to be identified. Above the crosses, identifications occur across each "leg."

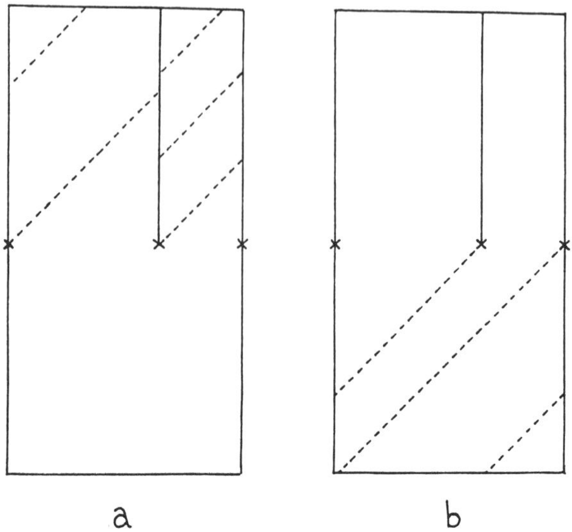


Fig. 3. Right-propagating mode functions: (a) "in" mode functions; (b) "out" mode functions. The dotted lines show the location of discontinuities.

across each leg above the crosses. The circumferences of the legs need not be equal, but their sum must equal the circumference of the trunk.

Figures 3a and 3b show one possible law of propagation past the singular point, which may be called the *shadow law*. In Fig. 3a an "in" mode function propagating to the right splits into components propagating to the right in each leg. Although continuous in the trunk region, such mode functions generally have discontinuities (located along the dotted lines) in the legs. Figure 3b shows the corresponding discontinuities possessed by right-propagating "out" mode functions associated with either leg. Every "out" mode function is continuous in each leg (vanishing in one of them) but has discontinuities in the trunk region. Mirror-image figures can be drawn for left-propagating mode functions.

## 5. THE "IN" AND "OUT" STATE VECTOR SPACES

The scalar field can be expressed in terms of either "in" or "out" mode functions as follows:

$$\phi(x) = \sum_{\mathbf{p}} [u_{\text{in}}(x, \mathbf{p})\, \mathbf{a}_{\text{in}}(\mathbf{p}) + u^*_{\text{in}}(x, \mathbf{p})\, \mathbf{a}^*_{\text{in}}(\mathbf{p})] + u_{0\text{in}}(x)\, \mathbf{Q}_{\text{in}} + v_{0\text{in}}(x)\, \mathbf{P}_{\text{in}} \tag{1a}$$

$$
\begin{aligned}
&= \sum_{\mathbf{p}'} [u_{\text{out}}(x, \mathbf{p}', R)\, \mathbf{a}_{\text{out}}(\mathbf{p}', R) + u^*_{\text{out}}(x, \mathbf{p}', R)\, \mathbf{a}^*_{\text{out}}(\mathbf{p}', R)] \\
&\quad + u_{0\text{out}}(x, R)\, \mathbf{Q}_{\text{out}}(R) + v_{0\text{out}}(x, R)\, \mathbf{P}_{\text{out}}(R) \\
&\quad + \sum_{\mathbf{p}''} [u_{\text{out}}(x, \mathbf{p}'', L)\, \mathbf{a}_{\text{out}}(\mathbf{p}'', L) + u^*_{\text{out}}(x, \mathbf{p}'', L)\, \mathbf{a}^*_{\text{out}}(\mathbf{p}'', L)] \\
&\quad + u_{0\text{out}}(x, L)\, \mathbf{Q}_{\text{out}}(L) + v_{0\text{out}}(x, L)\, \mathbf{P}_{\text{out}}(L)
\end{aligned} \tag{1b}
$$

In addition to standard Fock-space modes, massless fields in compact universes always possess zero-frequency modes. Here the $u_{\text{in}}(x, \mathbf{p})$ are the Fock-space mode functions associated with the "in" region (the trunk) and $u_{0\text{in}}(x)$ and $v_{0\text{in}}(x)$ are zero-frequency mode functions. ($u_{0\text{in}}$ is actually constant in space and time while $v_{0\text{in}}$ is constant in space and varies linearly with time.) The $\mathbf{a}_{\text{in}}(\mathbf{p})$ and $\mathbf{a}^*_{\text{in}}(\mathbf{p})$ satisfy the standard commutation relations for annihilation and creation operators for particles of momentum $\mathbf{p}$. They commute with the operators $\mathbf{Q}_{\text{in}}$ and $\mathbf{P}_{\text{in}}$, which are Hermitian and canonically conjugate. The momentum $\mathbf{p}$ is quantized by the periodic boundary conditions in the trunk region, except that the value $\mathbf{p} = 0$ is excluded.

The ground-state vector in the trunk region is defined by

$$
\mathbf{a}_{\text{in}}(\mathbf{p})\, |\text{in, vac}\rangle = 0, \qquad \mathbf{P}_{\text{in}}\, |\text{in, vac}\rangle = 0 \tag{2}
$$

Although the symbol "vac" is used here, $|\text{in, vac}\rangle$ is not a true Fock-space vacuum state vector, for the second of equations (2) prevents it from being normalizable. In fact, the "in" state vector space is not a true Fock space but rather the direct product of a Fock space and an ordinary Hilbert space of complex functions of a single real variable $Q_{\text{in}}$.

The operators $\mathbf{a}_{\text{out}}(\mathbf{p}', R)$, $\mathbf{a}^*_{\text{out}}(\mathbf{p}', R)$, $\mathbf{Q}_{\text{out}}(R)$, $\mathbf{P}_{\text{out}}(R)$, $\mathbf{a}_{\text{out}}(\mathbf{p}'', L)$, $\mathbf{a}^*_{\text{out}}(\mathbf{p}'', L)$, $\mathbf{Q}_{\text{out}}(L)$, $\mathbf{P}_{\text{out}}(L)$ have analogous properties and interpretations in the leg regions. The momenta $\mathbf{p}'$ and $\mathbf{p}''$ are quantized by the periodic boundary conditions in the right and left legs respectively. Again the values $\mathbf{p}' = 0$ and $\mathbf{p}'' = 0$ are excluded. The ground-state vector in the leg regions is defined by

$$
\begin{aligned}
\mathbf{a}_{\text{out}}(\mathbf{p}', R)\, |\text{out, vac}\rangle &= 0, \qquad \mathbf{P}_{\text{out}}(R)\, |\text{out, vac}\rangle = 0 \\
\mathbf{a}_{\text{out}}(\mathbf{p}'', L)\, |\text{out, vac}\rangle &= 0, \qquad \mathbf{P}_{\text{out}}(L)\, |\text{out, vac}\rangle = 0
\end{aligned} \tag{3}
$$

Alternatively, one may introduce ground-state vectors associated with each leg and regard $|\text{out, vac}\rangle$ as their Cartesian product. The "out" state vector space is the direct product of the Fock spaces and Hilbert spaces for the two legs.

## 6. GROUND-STATE ENERGY AND UNITARY INEQUIVALENCE

The renormalized energy operator (i.e., with the vacuum energy of infinite Minkowski space subtracted off) for the scalar field in the trunk region is

$$\mathbf{E}_{\text{in}} = \sum_{\mathbf{p}} |\mathbf{p}| \, \mathbf{a}^*_{\text{in}}(\mathbf{p}) \, \mathbf{a}_{\text{in}}(\mathbf{p}) + \tfrac{1}{2}\mathbf{P}^2_{\text{in}} + C \tag{4}$$

where $C$ is the Casimir energy in the trunk. The corresponding operator for the legs is

$$\begin{aligned}\mathbf{E}_{\text{out}} = &\sum_{\mathbf{p}'} |\mathbf{p}'| \, \mathbf{a}^*_{\text{out}}(\mathbf{p}', R) \, \mathbf{a}_{\text{out}}(\mathbf{p}', R) + \tfrac{1}{2}\mathbf{P}_{\text{out}}(R)^2 + C_R \\ &+ \sum_{\mathbf{p}''} |\mathbf{p}''| \, \mathbf{a}^*_{\text{out}}(\mathbf{p}'', L) \, \mathbf{a}_{\text{out}}(\mathbf{p}'', L) + \tfrac{1}{2}\mathbf{P}_{\text{out}}(L)^2 + C_L\end{aligned} \tag{5}$$

where $C_R$ and $C_L$ are the Casimir energies in the right and left legs respectively. Evidently

$$\begin{aligned}\mathbf{E}_{\text{in}} \, |\text{in, vac}\rangle &= C \, |\text{in, vac}\rangle \\ \mathbf{E}_{\text{out}} \, |\text{out, vac}\rangle &= (C_R + C_L) \, |\text{out, vac}\rangle\end{aligned} \tag{6}$$

$C$, $C_R$, and $C_L$ are generally all negative (if the scalar field is "untwisted") and both $C_R$ and $C_L$ are more negative than $C$. Therefore the Casimir energy is not conserved.

The situation is actually more dramatic than this. Because of the topological change the trousers manifold is not stationary, and there is no reason for $\mathbf{E}_{\text{in}}$ and $\mathbf{E}_{\text{out}}$ to be equal. Their expectation values, in fact, turn out to differ from one another, in all states, by infinite amounts. One way (the hard way) to show this is to compute the Bogoliubov coefficients that connect the "in" and "out" mode functions (or alternatively the operators $\mathbf{a}_{\text{in}}$, $\mathbf{a}^*_{\text{in}}$, $\mathbf{Q}_{\text{in}}$, $\mathbf{P}_{\text{in}}$ with the operators $\mathbf{a}_{\text{out}}$, $\mathbf{a}^*_{\text{out}}$, $\mathbf{Q}_{\text{out}}$, $\mathbf{P}_{\text{out}}$) and to reexpress $\mathbf{E}_{\text{out}}$ in terms of the "in" operators, or vice versa. This has been done, although the details, which are not particularly illuminating, will not be presented here.

Another result that can be obtained directly from the Bogoliubov coefficients is

$$\langle \text{out, vac} | \text{ in, vac} \rangle = 0 \tag{7}$$

It is a corollary of this result that the "in" and "out" state-vector spaces are unitarily inequivalent: A state having a finite number of "in" particles has an infinite number of "out" particles with an infinite amount of "out" energy, and vice versa.

The easy way to demonstrate the above results is to look at ground-state expectation values of the renormalized stress-energy tensor. These may be obtained from the Wightman functions of the theory by suitable limiting procedures. Actually, Wightman functions do not exist for massless fields in $1 + 1$ dimensions. However, expectation values of the stress-energy tensor involve space-time derivatives of Wightman functions, and these do exist.

Figure 4 shows the distribution of energy in the "in" ground state, i.e., the structure of $\langle \text{in, vac} | \mathbf{T}^{00}_{\text{ren}} | \text{in, vac} \rangle$. In the trunk region the energy density is simply the Casimir energy density $C/L$, where $L$ is the circumference of the trunk. In the legs the energy density is also equal to $C/L$ everywhere, except along the 45° lines emanating from the crotch singularity. Along these lines the energy density behaves like the square of the delta function

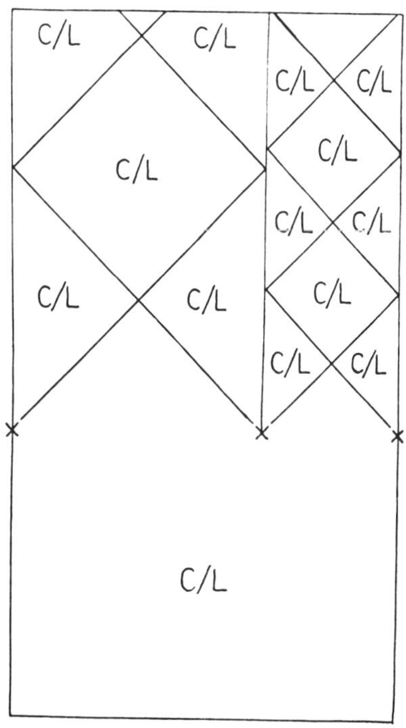


Fig. 4. Distribution of energy in the "in" ground state. The lines at 45° represent the infinitely bright flash emitted by the crotch singularity.

and hence is not only infinite there but also corresponds to infinite *total* energy. That is to say, the crotch singularity produces an infinitely bright flash that propagates forward in time into the legs.

The reason for the infinitely bright flash is easily seen if one expresses $\langle \text{in, vac} | \mathbf{T}^{00}_{\text{ren}} | \text{in, vac} \rangle$ as a sum over "in" mode functions. In the legs the "in" mode functions have discontinuities. When these functions are differentiated the discontinuities give rise to delta functions. Since the terms of the mode sum for $\langle \text{in, vac} | \mathbf{T}^{00}_{\text{ren}} | \text{in, vac} \rangle$ are bilinear in differentiated mode functions, the square of the delta function automatically appears.

## 7. LINEARIZED GRAVITY IN A 3-TORUS

One can make the above results bear directly on the real problem of dynamically induced topological change by increasing the dimensionality of space-time from two to four and replacing the scalar field by the gravitational field. We shall work in a 3-torus universe and write the metric operator in the form

$$\mathbf{g}_{\mu\nu} = \eta_{\mu\nu} + \boldsymbol{\phi}_{\mu\nu} \tag{8}$$

where $\eta_{\mu\nu}$ is the Minkowski metric. In linear approximation one expands $\boldsymbol{\phi}_{\mu\nu}$ in the form

$$\begin{aligned} \boldsymbol{\phi}_{\mu\nu} = & \sum_{\mathbf{p},s} [u_{\mu\nu}(x, \mathbf{p}, s)\, \mathbf{a}(\mathbf{p}, s) + u^*_{\mu\nu}(x, \mathbf{p}, s)\, \mathbf{a}^*(\mathbf{p}, s)] \\ & + \sum_A [u_{0\mu\nu}(x, A)\, \mathbf{Q}_A + v_{0\mu\nu}(x, A)\, \mathbf{P}_A] \\ & + \text{gauge terms} \end{aligned} \tag{9}$$

Here $s$ is a helicity label, which can assume the values $+2$ and $-2$, and $\mathbf{p}$ is a 3-momentum quantized by the boundary conditions of the 3-torus, the value $\mathbf{p} = 0$ being excluded. The $u_{\mu\nu}(x, \mathbf{p}, s)$ are Fock-space mode functions having the form

$$u_{\mu\nu}(x, \mathbf{p}, \pm 2) = \mu_{\text{P}}^{-1} V^{-1/2} (2\omega)^{-1/2} e_\mu(\mathbf{p}, \pm 1)\, e_\nu(\mathbf{p}, \pm 1)\, e^{ip \cdot x} \tag{10}$$

where $\mu_{\text{P}}$ is the Planck mass, $V$ is the volume of the 3-torus, $\omega = \text{p}^0 = |\mathbf{p}|$, and the $e_\mu$ are complex vectors satisfying

$$\begin{gathered} e_0(\mathbf{p}, \pm 1) = 0, \qquad \mathbf{e}(\mathbf{p}, \pm 1) = \mathbf{e}(\mathbf{p}, \mp 1)^*, \qquad \mathbf{p} \cdot \mathbf{e}(\mathbf{p}, \pm 1) = 0 \\ \mathbf{e}(\mathbf{p}, \pm 1) \cdot \mathbf{e}(\mathbf{p}, \pm 1)^* = 1, \qquad \mathbf{e}(\mathbf{p}, \pm 1) \cdot \mathbf{e}(\mathbf{p}, \pm 1) = 0 \end{gathered} \tag{11}$$

In addition to the Fock-space mode functions there are six pairs of independent zero-frequency mode functions given by

$$\begin{aligned} u_{0\mu\nu}(x, A) &= \mu_{\mathrm{P}}{}^{-1} V^{-1/2} \chi_{\mu\nu}(A) \\ v_{0\mu\nu}(x, A) &= \mu_{\mathrm{P}}^{-1} V^{-1/2} \chi_{\mu\nu}(A)\, x^0 \end{aligned} \qquad A = 0, 1, 2, 3, 4, 5 \tag{12}$$

where

$$\chi_{\mu\nu}(0) = \frac{1}{\sqrt{6}} (\eta_{\mu\nu} + \eta_{\mu 0} \eta_{\nu 0}) \tag{13}$$

and the $\chi_{\mu\nu}(A)$ with $A = 1, \ldots, 5$ are any five other constant real symmetric tensors satisfying

$$\begin{aligned} \chi_{\mu 0}(A) &= 0, & \chi^{\mu}_{\mu}(A) &= 0 \\ \chi_{\mu\nu}(0)\, \chi^{\mu\nu}(A) &= 0, & \chi_{\mu\nu}(A)\, \chi^{\mu\nu}(A') &= \delta_{AA'} \end{aligned} \qquad A, A'' = 1, \ldots, 5 \tag{14}$$

$\chi_{\mu\nu}(0)$ describes pure dilation of the 3-metric, whereas the $\chi_{\mu\nu}(A)$ with $A = 1, \ldots, 5$ describe pure shear.

The **a**'s and **a***'s are standard annihilation and creation operators and the **Q**'s and **P**'s form canonically conjugate Hermitian pairs. Since the static 3-torus possesses a timelike Killing vector field, there is a gauge-invariant conserved total energy operator. Its expression in terms of the above operators may be shown to be[5]

$$\begin{aligned} \mathbf{E} = &\sum_{\mathbf{p}, s} \omega \mathbf{a}^*(\mathbf{p}, s)\, \mathbf{a}(\mathbf{p}, s) \\ &+ \tfrac{1}{2}(-\mathbf{P}_0^2 + \mathbf{P}_1^2 + \mathbf{P}_2^2 + \mathbf{P}_3^2 + \mathbf{P}_4^2 + \mathbf{P}_5^2) \\ &+ \text{Casimir energy} \end{aligned} \tag{15}$$

The noteworthy feature of this expression is that the dilation mode contributes negatively to the total energy whereas all other modes contribute positively.

## 8. LINEARIZATION CONSTRAINT

If one goes beyond the linearized theory one finds that the total energy, through the operator field equations, can be expressed as an integral over a 2-dimensional surface at spatial infinity. Since in a compact

[5] B. S. DeWitt in *Relativity, Groups and Topology II*, DeWitt and Stora, eds. (North-Holland, 1984). See especially pp. 454–457 and problem 95 on pp. 672–674.

universe such as the 3-torus there is no spatial infinity, it follows that the total energy must vanish. In the classical theory this fact may be restated as follows: Let $\Phi$ be the space of solutions of the classical field equations and let $\eta$ be the point in $\Phi$ corresponding to the flat 3-torus. Let $\eta+\phi$ be a point in $\Phi$ close to $\eta$. Then $\phi$ is a solution of the linearized field equations and may be regarded as the tangent vector at $\eta$ to a curve in $\Phi$, i.e., to a one-parameter family of solutions of the nonlinear field equations. Not *all* solutions of the linearized equations are tangent to curves in $\Phi$. The only solutions of the linearized equations that are tangent to curves in $\Phi$ are those for which the total energy vanishes. Because of the spatial homogeneity of the present model, one may show that the total 3-momentum too must vanish. This means that the coefficients in the decomposition (9) cannot all be chosen independently but must be subjected to quadratic constraints, one for each independent Killing vector field that the 3-torus possesses.

The existence of the energy constraint implies that if any Fock modes or shearing modes are excited, then, to the extent that their energy content together with the Casimir energy is positive, the dilation mode must be excited too. This phenomenon is well known in relativistic cosmology. The universe as a whole reacts to the energy that is in it by expanding or contracting. It cannot stand still.

Quantum mechanically the only way to impose the energy constraint is as a condition on the state vector $\Psi$:

$$\mathbf{E}\Psi=0 \tag{16}$$

There are also three other similar constraints corresponding to the condition of vanishing total 3-momentum.

## 9. THE TRANSITION $T^3 \rightarrow T^3 + T^3$

Suppose now that the 3-torus undergoes a topological transition in which it splits into two 3-tori. One may imagine this split occurring along one of the principal axes of the 3-torus so that the transition may be expressed more precisely as $T^2 \times S^1 \rightarrow T^2 \times (S^1 + S^1)$. In this case there will be a submanifold of singular points, having the topology $T^2$. Again one must propose laws by which the (linearized) gravitational field propagates around these points.

Here a naive shadow law is no longer adequate since not all plane waves propagate parallel to the splitting axis. However, there is a simple alternative that is applicable also to the $(1+1)$-dimensional trousers model previously considered. Consider an "in" mode function along a line $t = \text{constant}$ just above the singularity point in Fig. 3a. The Cauchy data are $\phi$ and $\partial\phi/\partial t$ along this line. The derivative $\partial\phi/\partial t$ contains a delta function in each leg, introduced by the discontinuities along the dotted lines. The alternative proposal is simply to eliminate these delta functions at the entrances to each leg. Initially $\phi$ and $\partial\phi/\partial t$ then possess only discontinuities but no delta functions, and this rule may be applied to the transition $T^3 \to T^3 + T^3$ also.

In the case of the trousers topology although $\partial\phi/\partial t$ possesses no delta function initially (according to the new rule) it immediately thereafter acquires one—four, in fact; two in each leg. Even though the "in" mode function was propagating to the right in the trunk region, it now acquires, in the legs, discontinuities that propagate to *both* the left and the right from the singular point.

## 10. DYNAMICAL SUPPRESSION OF TOPOLOGICAL CHANGE

Examination of yet more complicated topological transitions suggests that discontinuities in the Cauchy data will always emanate from the singular points under almost any conceivable law of propagation. These discontinuities will immediately give rise to delta functions in the derivatives of the fields and to squares of delta functions in the stress-energy tensor (or pseudo-tensor in the case of gravity). From this it follows that the singular points will always be the source of infinitely bright flashes and that a change of topology will always be accompanied by infinite particle and energy production.

In the case of the gravitational field the energy constraint must hold both before and after the topological change. The production of an infinite amount of energy will cause the dilation mode to come instantly into play. Indeed the entire burden of compensating for the infinite energy falls on the dilation mode alone. This means that the universe must either expand or contract infinitely rapidly immediately after the change. It cannot expand infinitely rapidly because of causality requirements. It therefore collapses, aborting the topological change.

The collapse, in fact, will occur locally. In the full nonlinear theory the condition of vanishing total energy and momentum is replaced by an infinity of local constraints: one Hamiltonian constraint and three momentum constraints at each point of space. These constraints control not merely the total energy and momentum but the distribution of energy and momentum. The infinite energy density associated with field discontinuities is thus locally suppressed, which is to say that the Hamiltonian and momentum constraints are not merely manifestations of the diffeomorphism invariance of the theory but also serve locally to maintain everywhere the integrity of the topology of 3-space. Because these constraints are local it is obvious that the topology of 3-space is dynamically preserved even if 3-space is noncompact and its volume is infinite.

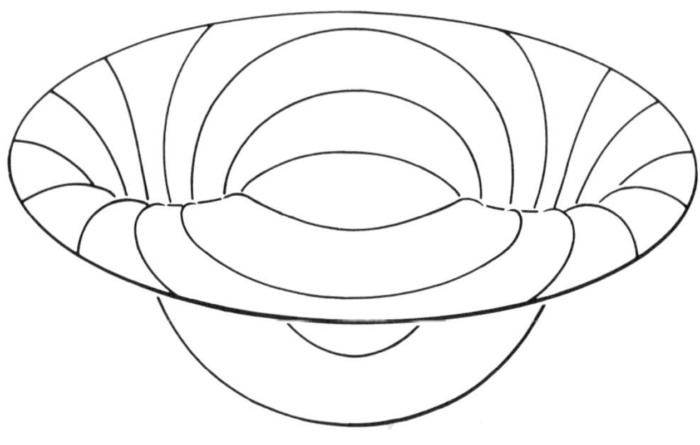

*Plate 7*

# Barrier Penetration and Initial Values in Kaluza–Klein Theories[1]

**Dieter R. Brill**[2,3]

*Received September 30, 1985*

Topology change is discussed as a barrier penetration process and illustrated by explicit calculations in Witten's Kaluza–Klein bubble geometry. Initial values are given for other bubble geometries, including ones with negative total mass.

## 1. THE ROLE OF TOPOLOGY IN SPACE-TIME PHYSICS

Our view of four-dimensional space-time today was shaped in large measure by two men, Einstein and Wheeler. Einstein introduced into physics the richness of curvature; we owe to Wheeler the recognition that a similar richness is provided by topology.[(1)] He first emphasized that localized topological peculiarities of empty space-time can represent features usually attributed to matter—a well-known example is the description of "charge without charge" as field lines trapped in a "wormhole." He also emphasized that such descriptions, while possible in the classical theory, can be appropriate to our world only in the quantum version of general relativity.

Einstein's classical theory of general relativity not only predicts and requires that space-time be cuved, but it also gives curvature a dynamical role. By contrast it provides no similar dynamical role for topology: a space-time with a given topology on an initial three-dimensional spacelike surface cannot change that topology in the course of time within the

[1] Work supported in part by the National Science Foundation, under grant No. PHY82-05659, and by the Humboldt Foundation.

[2] Max-Planck Institut für Astrophysik, 8046 Garching, Federal Republic of Germany.

[3] Permanent address: Department of Physics, University of Maryland, College Park, Maryland 20742.

domain of dependence in which Einstein's classical equations can predict that development.[2] The dynamics of one topology changing into another therefore appeared to be a typical quantum process of the barrier penetration type; but aside from the problems of formulating a consistent quantum gravity, even the kinematics of such a process as the pinch-off or formation of a wormhole offered formidable difficulties, contributing to motivate Wheeler's search for a foundation of space-time physics more fundamental than geometry.[3]

Considerable progress[4] in calculating processes of quantum gravity has been made by an alternative approach, which circumvents both the kinematical and the quantum mechanical difficulties: both processes can be more simply, albeit approximately, described in terms of analytic continuation to imaginary or "Euclidean" time. Barrier penetration appears much like classical propagation in Euclidean time; and four-dimensional nonsingular Euclidean spaces can admit a sequence of three-dimensional nonintersecting slices whose topology changes. A familiar two-dimensional example is a pair of trousers, which have some slices (near the belt) that are single circles, and others (near the cuffs) that are a pair of circles. (Of course at least one of the sequence of one-dimensional slices in between these is singular, but the two-dimensional trousers as a whole—if made by a good tailor—are everywhere smooth.)

In quantum gravity then we might expect that we can start with an initial wave packet concentrated about a classical state of one topology, say the Minkowski vacuum, and find with some probability a packet concentrated about a different classical state, with a different topology in the subsequent developement. Fortunately such a "vacuum decay" by barrier penetration is not possible in four-dimensional quantum gravity. The reason is that the total energy is conserved in such a process, and that the Minkowski vacuum is the unique classical state of lowest energy[5]—vacuum decay is energetically impossible. It may come as a relief to know that at least the vacuum is stable, but in order really to enjoy this fortune one would like to know what fate one has escaped—what might have happened if the vacuum were unstable. For this reason, and also as one simple concrete example of a calculation of topology change, it is instructive to consider five-dimensional Kaluza–Klein theory where the vacuum does decay.[6]

In the following we shall compare the barrier penetration formalism for particles and geometries; point out the special role of time-symmetric geometries; consider Witten's process and some properties of the classical decay products; find other time-symmetric initial data in five-dimensional Kaluza–Klein theories; and discuss possible generalizations to higher dimensions.

## 2. BARRIER PENETRATION FOR PARTICLES AND FOR SPACE-TIMES

The semiclassical description of barrier penetration is familiar, e.g., from the elementary treatment of alpha decay. Although an essential quantum process, barrier penetration can be discussed in terms that are closely related to classical dynamics. Classical dynamics can be recovered from quantum dynamics whenever a wave packet can be formed that holds together for a relatively long time, so that it can be said to be concentrated about one trajectory—that of the classical motion. For example, Einstein's equations can be recovered as such a limit of quantum geometrodynamics.[7] By contrast, even if a packet is incident on a classically forbidden barrier region, the wave function that penetrates into that region is not a packet. However, typical wave functions in a (one-dimensional) barrier region can be analytically continued into packets, by rotating the time variable to purely imaginary values.[8] Thus barrier penetration can be said to be described approximately by classical propagation in imaginary "Euclidean" time.

A particle that propagates in Euclidean time has a negative kinetic energy. It can therefore conserve energy while moving in a classically forbidden region, where the potential $V$ exceeds the total energy $E$. Such a particle also has a purely imaginary action. Its contribution to the probability amplitude is therefore a decreasing exponential. In general, many Euclidean trajectories will contribute to the amplitude for crossing from one side of a barrier to another, because the Euclidean motion may bounce back and forth many times inside the barrier. These contributions are particularly important for energies near the top of the barrier, and lead to such effects as resonances. In the following we shall confine attention to barrier penetration processes that are dominated by just one Euclidean trajectory. This is particularly true for Euclidean trajectories that start from a classically locally stable point,[9] for such "bounces" take an infinite amount of Euclidean time, and hence cannot happen more than once.

Taking a giant leap from particles to space-times, one has reason to believe that the dominant contribution to barrier penetration of geometries is given by "Euclidean"—that is, Riemannian or positive-definite metric—solutions of Einstein's field equations. As for particles, the connection with the classical motion is made at the turning points where the kinetic energy vanishes: these configurations lie both on the classical and on the Euclidean trajectories; they are the transition points between the two kinds of motion. In geometrodynamics the kinetic energy is measured by the extrinsic curvature, or second fundamental form, of a three-dimensional spacelike surface. Hence the simplest kind of turning point, at which the motion reverses exactly, is a totally geodesic subspace, or surface of time symmetry, at which the extrinsic curvature vanishes.[(10)] Therefore, *time-symmetric solutions are candidates for products of barrier penetration processes.* Interpreted in this way, the classical solution only applies for times after the time-symmetric moment $t_0$—at that moment it "emerged" from the barrier.[4] (Of course the classical solution for times before $t_0$ is an equally possible past, which is in fact more likely, in the sense of retrodiction.)

What is the physical interpretation of these Euclidean solutions? We know[(13)] how wave packets describe classical (Lorentzian) solutions of Einstein's equations: in the representation where the intrinsic geometry $^3g$ is diagonal, the wave functional $\Psi$ depends only $^3g$; and in the classical limit, $\Psi(^3g)$ is large only for those 3-geometries that fit as subspaces into one and the same classical four-dimensional space-time. For the barrier penetration processes described by Euclidean four-dimensional spaces, the wave functional is a packet only when continued to imaginary times; however, the main effect of this continuation is to change exponential decay into harmonic dependence, and the actual wave functional is still relatively larger for those 3-geometries in the Euclidean packet than for other 3-geometries. Thus, *wave functionals* $\Psi(^3g)$ *that describe a simple barrier penetration* (with one dominant history) *are large for all those* 3-*geometries* $^3g$ *that fit into one and the same Euclidean space-time.*

A concept of time only makes (approximate) sense for (nearly) classical wave packets.[(14)] The wave functionals corresponding to barrier penetration are examples of states in which there is *no* notion of time. But in a field theory, unlike in the case of a one-dimensional particle, there are always other degrees of freedom that might be excited during barrier penetration—in fact, must be excited in a realistic description where there is room for observers with clocks in an asymptotic region at infinity. In these cases the total energy of the system need not be less than the potential energy: there may be enough energy available to allow a classical transition, but a part of the system is sufficiently isolated that it cannot share in this energy, and the most likely way of making the transition is by barrier penetration. Clearly the well-isolated observers at infinity can have enough energy so that a classical notion of time makes sense there. How are we to interpret the Euclidean solution in this context? As McLaughlin writes (for the particle case)[(8)] "This result may be interpreted as a particle traveling in

[4] Could our whole universe have arisen in this way? Not unless the cosmological constant $\Lambda$ was positive, for it is known[(11)] that cosmological solutions with $\Lambda \leqslant 0$ do not expand from a moment of time-symmetry. Barrier penetration for a universe with $\Lambda > 0$ has been suggested[(12)] as a creation "from nothing."

real time until it strikes the turning point where its real time "freezes." It then penetrates the barrier by taking an excursion into complex time. Upon striking the second turning point, its complex time freezes, and it leaves the barrier with increasing real time." Thus it takes "no" (classically observable) real time to penetrate the barrier: the uncertainty principle demands that this time be ill defined. In the case of space-time, however, this freezing of the real time of the observers at infinity means that it must be possible to describe the Euclidean 4-geometry by a sequence of three-dimensional slices (of constant Euclidean time) in which the asymptotic geometry is indeed frozen, so that it connects with consistent observers at infinity. Hence the demand that the Euclidean solution be asymptotically flat. (The next section will exhibit an example of such a slicing.)

A word of warning is appropriate concerning the turning points (moments of time symmetry): where as any classical time-symmetric solution of a certain energy may, with some probability, have been produced by barrier penetration or tunneling from any other time-symmetric solution of that energy, the most probable decay product corresponds to very special time-symmetric solutions. This becomes clear as soon as we consider tunneling in more than one dimension. For example, in a typical two-dimensional barrier situation the solution for turning points (points where $E = V$) consists of two disjoint curves. Typically there is just one path of absolute minimum Euclidean action between these curves. Correspondingly for space-time one would expect just one time-symmetric initial state (in each topological class) to be produced with maximum probability in the decay by barrier penetration of a given state. We note, however, that in order to compute the exponential factor in the production probability we do not need to know the exact Euclidean solution. We need to evaluate the classical action, which consists of a four-dimensional volume term and a three-dimensional surface term. The volume term, $\int {}^4R\, d^4V$, vanishes for solutions of the sourceless (Euclidean) Einstein equations since they imply ${}^4R = 0$. The surface term needs only to be computed in the asymptotically flat regions, where the weak-field equations are valid. Thus if we can find the weak field version of a Euclidean solution relevant to our problem, we can evaluate the action and hence the tunneling probability.[5]

[5] The tunneling probability must be multiplied by a frequency to obtain a transition rate (or, in a field theory, the corresponding quantities per volume). For vacuum decay this frequency can be computed as shown by Coleman[(9)] and is nonzero if the operator corresponding to the action's second variation has one negative mode.

## 3. AN EXAMPLE—WITTEN'S BUBBLE

One of the simplest examples of barrier penetration with topology change is given by the instability of vacuum in five-dimensional Kaluza–Klein theory.[(6)] The sourceless field equations are quite analogous to the four-dimensional case, namely $^5\text{Ricci} = 0$, and a Euclidean solution relevant to vacuum decay is the analog of the Schwarzschild solution

$$ds^2 = (1 - (R/r)^2)\, d\phi^2 + (1 - (R/r)^2)^{-1}\, dr^2 + r^2\, d\Theta^2 \tag{1}$$

Here, $r$, $\theta_1$, $\theta_2$, $\theta_3$ are four-dimensional polar coordinates; $d\Theta^2 = d\theta_1^2 + \sin^2\theta_1(d\theta_2^2 + \sin^2\theta_2\, d\theta_3^2)$ is the line element on a unit 3-sphere; and $\phi$ is the "internal" Kaluza–Klein dimension with period $2\pi R$. It is well known that, with this periodicity, there is no geometrical singularity at $r = R$; instead this locus (for fixed $\theta_i$) represents a point, the origin of two-dimensional polar coordinates $r^* = (r^2 - R^2)^{1/2}$ and $\phi$.

Because this solution approaches flat space to order $1/r^2$, it describes a transition from the vacuum state, and from its action we can compute the penetration factor. However, the solution also tells us *to* what state the transition is made: we only have to look for the "turning point," that is, a totally geodesic subspace with the same asymptotic dependence as the initial vacuum. (Physically the latter demand means that this new classical state can be observed from infinity by the original observers.) Such subspaccs arc all equivalent under rotation to the "equatorial plane" $\theta_1 = \pi/2$. The metric on this subspace

$$d\sigma^2 = (1 - (R/r)^2)\, d\phi^2 + (1 - (R/r)^2)^{-1}\, dr^2 + r^2\, d\Omega^2 \tag{2}$$

(with $d\Omega^2$ the line element on a unit 2-sphere) describes the time-symmetric initial data for the subsequent classical development; this development has also been given by Witten, by an analytic continuation of (1).

As remarked, we may consider (1) as an approximate description of the quantum wave function $\Psi(^4g)$ in the sense that $\Psi(^4g)$ is relatively large for those 4-geometries[6] $^4g$ that are subspaces of (1). To obtain a typical barrier penetrating history of 4-geometries corresponding to this semiclassical $\Psi(^4g)$, we only need to slice the 5-geometry (1) by 4-surfaces leading from flat space to the surface described by (2). For example, we may choose surfaces of constant imaginary time given by

$$\tau = r \cos\theta_1 = \text{const}$$

[6] In five-dimensional Kaluza–Klein theory, in the representation chosen, $\Psi$ depends on spacelike 4-geometries.

If we also choose a new radial coordinate[7]

$$\rho = r \sin \theta_1$$

the sequence of 4-metrics labelled by $\tau$ is

$$d\sigma_\tau^2 = \left(1 - \frac{R^2}{\rho^2 + \tau^2}\right) d\phi^2 + \left(1 - \frac{\rho^2 R^2}{(\rho^2 + \tau^2)^2 - R^2\tau^2}\right)^{-1} d\rho^2 + \rho^2\, d\Omega^2 \qquad (3)$$

For $\tau \to \pm\infty$ this becomes flat 4-space; for $\tau = 0$ it reduces to (2); it has a conical singularity for $\tau = \pm R$; and it is nonsingular for all other $\tau$, $\rho^2 + \tau^2 = R^2$ being the origin of polar coordinates in the $\rho$, $\tau$ subspace as can be seen by expanding that neighborhood in terms of the coordinate $\rho^* = (\rho^2 + \tau^2 - R^2)^{1/2}$. For $\tau^2 > R^2$ the topology is $R^3 \times S^1$, the same as that of a spacelike surface in five-dimensional Kaluza–Klein vacuum, but for $\tau^2 < R^2$ a bubble appears at the center, and the topology is $R^2 \times S^2$. Thus this sequence of geometries, changing from one topology to another via a singular configuration, is very similar to the pinch-off process envisaged several decades ago by Wheeler. Of course, the singularity, although bound to happen for some $\tau$, is due to the slicing—the five-dimensional geometry (1) is nowhere singular.

For $\tau^2 = 0$ we reach the turning point, since (3) reduces to (2). The subsequent classical behavior can be obtained by analytically continuing $\theta_1$ as Witten does, or by replacing $\tau$ by *it* in Eq. (3). The metric coefficients are then singular at $\rho^2 - t^2 = R^2$, but as before this is not a geometrical singularity but an origin of two-dimensional polar coordinates $\rho^* = (\rho^2 - t^2 - R^2)^{1/2}$ and $\phi$. The size of the minimum 2-sphere $(\theta_1, \theta_2)$ on a slice $t = \text{const}$ is $\rho_{\min} = (R^2 + t^2)^{1/2}$, and in this sense the bubble expands with increasing $t$.

The time-dependent 3-metric obtained from Eq. (3) by the usual dimensional reduction, where the small size in the $\phi$-direction is set to zero, is of course singular at $\rho_{\min}$. However, it is our view that the extra dimension must be taken seriously in this context; as $\rho_{\min}$ is approached, the curvature of the reduced metric becomes large and comparable to $1/R^2$, indicating that the dimensional reduction is no longer appropriate. Regarded as a dimensionally reduced object, the bubble cannot be analyzed in interaction with matter because of the singularity, but as an unreduced, nonsingular object it yields definite predictions for the behavior of test particles and fields incident on it: Because the geometry is invariant under

[7] This goes over into the usual three-dimensional radial distance in the asymptotic region. It is not the coordinate orthogonal to $\tau$, $(r^2 - R^2)^{1/2} \sin \theta_1$, in terms of which the intrinsic metric is too unwieldy.

rotations in the $\phi$ direction, the corresponding conjugate momentum is conserved. For example, uncharged massless matter, such as light, can be described by rays pointing at the $r^*$, $\phi$ origin while incident on the bubble. As they reach the bubble, they cross the origin and become rays pointing away from the origin. Thus the bubble acts as a mirror at least at the time-symmetric moment. Now we recall that the four-dimensional rotational symmetry of (1) becomes a Lorentz invariance after an analytic continuation to the classical bubble geometry.[8] Hence any point on the bubble surface can be Lorentz transformed to lie in a surface of time symmetry. Thus *the expanding bubble reflects light like a moving mirror*. For similar reasons other test particles or fields incident on the bubble will also be reflected. Particles with finite rest mass can collide several times with a bubble as the latter accelerates, eventually moving with the bubble surface. Of course the back reaction on the bubble cannot be calculated in this test particle approximation.

## 4. TIME-SYMMETRIC INITIAL DATA

In the last section we exhibited one classical solution with zero energy and momentum. Is this an isolated solution or are there others? Are there solutions of negative energy, or of other topology, say corresponding to several bubbles? To show the existence of such solutions it is often simplest to exhibit the initial data only. In particular, we saw that the most probable products of tunneling are described by time-symmetric initial data.

As in four-dimensional general relativity,[10,15] the time-symmetric constraints of five-dimensional Kaluza–Klein theory demand simply that the scalar curvature of the (four-dimensional) initial spacelike surface vanish,

$$^4R = 0 \tag{4}$$

(Time symmetry, of course, implies that the second fundamental form $K_{ij}$ vanishes, so that the spacelike constraints are identically satisfied.) Taking a hint from the four-dimensional theory, we try to choose a form of the metric in which one scalar function is determined by Eq. (4) in terms of the other arbitrarily metric coefficients. Since we want metrics that are

[8] A bubble has zero energy and momentum, as it must since it is the product of vacuum decay. It remains unchanged, and acquires no momentum, when observed from a moving frame.

independent of the "internal" coordinate $\phi$, we choose the initial metric (at $x^0 = t = 0$)

$$ds^2 = {}^3g_{ij} dx^i \, dx^j + V^2(x^i) \, d\phi^2, \qquad i = 1, 2, 3 \tag{5}$$

The constraint (4) written as an equation for $V$ becomes

$$\nabla^2 V + \tfrac{1}{2} \, {}^3RV = 0 \tag{6}$$

Here the Laplacian $\nabla^2$ and the scalar curvature ${}^3R$ are formed from the 3-metric $g_{ij}$. We note that, except for the factor multipling ${}^3R$, this equation is the same as the Lichnerowicz–York equation[15] in the time-symmetric four-dimensional theory, $\hat{\nabla}^2\psi - \frac{1}{8}\hat{R}\psi = 0$, which determines the conformal factor needed in that theory to produce a solution $g = \psi^4 \hat{g}$ from an arbitrary metric $\hat{g}$. The existence theory of solutions of the latter therefore also applies to (6); qualitatively solutions $V > 0$ exists provided the scalar curvature $R$ is not too positive over large regions; a conservative estimate analogous to that of Cantor and Brill[16] gives the condition

$$\int |{}^3R|^{3/2} \sqrt{{}^3g} \, d^3x \leqslant C \tag{7}$$

where $C$ is a positive constant that can be estimated, and the integral extends over the support of the positive part of ${}^3R$.

Unlike the factor $\psi$ in the three-dimensional case, a solution $V$ that vanishes on a 2-sphere (or a disjoint union of 2-spheres) in the coordinates $x^i$, and is not defined in the interior, can still describe a nonsingular 4-geometry by Eq. (5). Let $\partial_n$ denote the normal derivative evaluated at, and $r$ the geodesic distance from, the 2-sphere in a 3-space $\phi = \text{const}$. Then near the 2-sphere the 4-geometry has the form

$$ds^2 = d\sigma^2 + dr^2 + (r \, \partial_n V)^2 d\phi^2$$

where $d\sigma^2$ is the metric on the 2-sphere. Thus we can interpret $r$, $\phi$ as polar coordinates provided $\partial_n V$ is constant on the 2-sphere and equal to $1/R$, where $2\pi R$ is the periodicity of $\phi$. The result is an initial state of a bubble similar to that of Eq. (2) in the sense that the 2-spheres is minimal and has no interior. For a given base metric ${}^3g$ such spheres cannot be arbitrarily placed: if we think of $V$ as an electrostatic potential, the condition that $\partial_n V$ be constant implies that no net electrostatic force can act on the spheres.

Examples of solutions of Eq. (6) can be given analytically in several simple cases. Suppose first that ${}^3g$ is the metric of flat Euclidean 3-space. The only everywhere regular solution of (6) is then $V = \text{const}$.

corresponding to the initial state of flat Kaluza–Klein vacuum. An example of a bubble type initial value solution is

$$V = 1 - (R/r), \qquad ds^2 = dr^2 + r^2 d\Omega^2 + (1 - R/r)^2\, d\phi^2 \tag{8}$$

To generalize this example somewhat we may suppose that ${}^3g$ is a time-symmetric initial state of four-dimensional Einstein theory that is conformally flat,

$${}^3g_{ij} = \psi^4 f_{ij}, \qquad {}^3R = 0 \text{ (hence } \nabla_f^2 \psi = 0) \tag{9}$$

where $f_{ij}$ is the flat metric and $\nabla_f^2$ is the flat Laplacian. The constraint (6), $\nabla^2 V = 0$, can then be written as

$$\nabla_f^2 (\psi V) = 0 \tag{10}$$

The only everywhere regular solution of Eq. (10) is $V = \text{const}$. Bubble-type solutions can be generated by adding the solution $V = 1/\psi$. For example, choose

$$\psi = 1 + (m/2r)$$

the three-dimensional Schwarzschild initial state. The corresponding Kaluza–Klein solution (without excitation in the extra dimension) of course corresponds to $V = 1$,

$$ds^2 = \left(1 + \frac{m}{2r}\right)^4 (dr^2 + r^2 d\Omega^2) + d\phi^2$$

and bubble type solutions are obtained from

$$\begin{aligned} V &= 1 - \frac{a}{r + \frac{1}{2}m}, \\ ds^2 &= \left(1 + \frac{m}{2r}\right)^4 (dr^2 + r^2\, d\Omega^2) + \left(1 - \frac{a}{r + \frac{1}{2}m}\right)^2 d\phi^2 \end{aligned} \tag{11}$$

The bubble surface is located at the coordinate $r = a - \frac{1}{2}m$ and has a radius (as measured by its area)

$$r_B = a^2/(a - \tfrac{1}{2}m).$$

To avoid a conical singularity there, the constant $a$ must be related to $R$ and $m$ via

$$R = a^3 (a - \tfrac{1}{2}m)^{-2} \tag{12}$$

Consider the periodicity $2\pi R$ of $\phi$ fixed (say, by a "cosmological" boundary condition at $R \to \infty$). What time-symmetric bubbles are then possible about a Schwarzschild-type wormhole characterized by $m$? Equation (12) has solutions only if $m < 8R/27$. If $m > 0$ the bubble radius $r_B$ always exceeds the Schwarzschild radius $2m$, the smallest size that any sphere $r = \text{const}$ can have in the geometry (11). When $r_B = 2m$ (hence $a = m = R/4$, bubble at coordinate $r = m/2$), Eq. (11) is the Euclidean four-dimensional Schwarzschild solution, which has vanishing Ricci tensor. Its five dimensional time development is therefore static.[9] For other values of $r_B$ one can compute the initial acceleration of the bubble from the time development equations. Bubbles with $R > m/2$ expand, and those with $R < m/2$ contract (as would be expected from the gravitational collapse that occurs when the bubble fails to cover up the event horizon).

If $m > 0$, $r_B$ is less than $R$; if $m = 0$ (Eq. 8), $r_B$ equals $R$. If $m < 0$, the solution (11) (with positive $a$) still describes a regular geometry because the bubble always "covers up" the singularity of ${}^3g$ at[10] $R = -m/2$. Here Eq. (12) can be solved for $a$ no matter what the (negative) value of $m$. The bubble radius $r_B$ always exceeds $R$ and can be macroscopic if $-m$ is sufficiently large, even if $R$ is small as prescribed by dimensional reduction. These solutions are examples of negative-energy bubble initial states, whose existence is to be expected from the absence of a positive-energy theorem in the bubble topology.[(6)]

The *ansatz* (9) is of course much too specialized to describe the general time-symmetric initial state. We mention some further but also special classes of solutions. Any Euclidean solution of the four-dimensional sourceless Einstein equations with periodicity appropriate to the coordinate $\phi$ satisfies not only the constraint (4) but has a static time development. Any Euclidean solution of the four-dimensional Einstein–Maxwell equations (or any other source with $T^\alpha_\alpha = 0$) with appropriate periodicity satisfies (4) but will in general not be static. For example, the Euclidean Reissner–Nordstrom solution

$$\begin{aligned} ds^2 &= f(r)\, d\phi^2 + f(r)^{-1}\, dr^2 + r^2\, d\Omega^2 \\ f(r) &= (1 - A/r)(1 + B/r), \qquad 2A^2/(A+B) = R \end{aligned} \tag{13}$$

can be viewed as Kaluza–Klein initial data. For $A = B = R$ it is the Witten bubble initial state (2), and for $0 < A < B$ it represents a similar bubble but with nonvanishing total mass. As in the previous examples, the mass

[9] This solution is known, e.g., to Chodos and Detweiler.[(17)]

[10] The $r$ used in (11) is the isotropic coordinate for the Schwarzschild part of the metric, so $r = -m/2$ corresponds to the zero (origin) of the Schwarzschild radial coordinate.

parameter $2m = A - B$ cannot be too large (it is limited by $R/2$) and the bubble radius $r_B = A$ is comparable to $R$ unless $m$ is negative.

The methods mentioned in this section to construct explicit solutions of the initial value equations do not yield the initial state of two or several bubbles, because the unusual boundary conditions $V = 0$, $\partial_n V = \text{const}$ (by choice of the bubble's shape) cannot be satisfied by regular solutions of Eq. (10). In addition, in order to be created with finite probability by vacuum decay, $\psi$ and $V$ have to fall off as $1/r^2$. (Of course, a weak-field, linear superposition of solutions of type (13) with the proper fall-off is easily written down and could be used to evaluate the action and hence the tunneling probability, but since it is not reliable at $V = 0$ the relation to the internal size $R$ would be lacking.)

Can bubble-type solutions be generalized to higher-dimensional Kaluza–Klein theories? An important uncertainty here is the description of the vacuum. If Einstein's equations should hold, and the dimensionally reduced vacuum should be Minkowski space, then we either need additional, nongeometrical fields or the extra dimension should be Ricci flat. In some of the latter cases one can have a positive-energy theorem that would exclude zero-energy bubbles. A simple example of the former type has an internal space homogeneous under a group $G$ and an additional 4-form $F$ that satisfies equations of the Maxwell type[18], $dF = 0 = \delta F$. Suppose there were a bubble solution in a space-time of dimension $n$. Draw a curve from the bubble surface to infinity and let the group $G$ act on this curve. This generates an $n-3$-dimensional surface $S$ that bounds a fiber $H$ at infinity. In order to allow the proper dimensional reduction of the vacuum at infinity, there must be flux of the dual of $F$ through $H$,

$$\int_H {*F} = 0 \tag{14}$$

But since $H$ bounds, $H = \partial S$, we have

$$\int_H {*F} = \int_{\partial S} {*F} = \int_S {*\delta F} = 0 \tag{15}$$

by the field equations for $F$. The contradiction between (14) and (15) shows that there can be no bubble solutions in this case. (This argument generalizes a proof by Young[19] for the spherically symmetric case.)

## REFERENCES

1. J. A. Wheeler, *Geometrodynamics* (Academic Press, New York, 1962).
2. R. Geroch, *J. Math. Phys.* **8**, 782 (1967).

3. G. M. Patton and J. A. Wheeler, "Is Physics Legislated by Cosmogeny?" in *Quantum Gravity, An Oxford Symposium*, C. J. Isham, R. Penrose, and D. W. Sciama, eds. (Clarendon Press, Oxford, 1975).
4. S. Hawking, "The path-integral approach to quantum gravity," in *General Relativity: An Einstein Centenary Survey*, S. Hawking and W. Israel, eds. (Cambridge University Press, Cambridge, 1979).
5. D. Brill and S. Deser, *Ann. Phys.* **50**, 548 (1968); R. Schoen and S. Yau, *Comm. Math. Phys.* **65**, 47 (1979); **79**, 47 (1981); **79**, 231 (1981); E. Witten, *Comm. Math. Phys.* **80**, 381 (1981).
6. E. Witten, *Nucl. Phys. B* **195**, 481 (1982).
7. U. Gerlach, *Phys. Rev.* **177**, 1929 (1969).
8. D. McLaughlin, *J. Math. Phys.* **13**, 1099 (1972).
9. S. Coleman, *Phys. Rev. D* **15**, 2929 (1977); C. Callan and S. Coleman, *Phys. Rev. D* **16**, 1762 (1977).
10. C. W. Misner and J. A. Wheeler, *Ann. Phys.* **2**, 525 (1957); D. Brill, *Ann. Phys.* **7**, 466 (1959); C. W. Misner, K. S. Thorne, and J. A. Wheeler, *Gravitation* (Freeman, San Francisco, 1973), Chap. 21.
11. D. Brill and F. Flaherty, *Commun. Math. Phys.* **50**, 157 (1976), Theorem 4.2; D. Brill and F. Flaherty, *Ann. Inst. H. Poincaré* **28**, 335 (1978).
12. A. Vilenkin, *Phys. Rev. D* **27**, 2848 (1983).
13. J. A. Wheeler, in *Batelle Rencontres 1967 Lectures in Mathematics and Physics* (W. Benjamin, New York, 1968), pp. 242–307; G. M. Patton and J. A. Wheeler, in *Quantum Gravity*, C. W. Isham, R. Penrose, and D. Sciama, eds. (Oxford, University Press, Oxford, 1975).
14. C. W. Misner, K. S. Thorne, and J. A. Wheeler, *Gravitation* (Freeman, San Francisco, 1973), Chap. 43.
15. J. York, "The initial value problem and dynamics," in *Gravitation Rediation*, N. Deruelle and T. Piran, eds. (North-Holland, Amsterdam, 1983).
16. M. Cantor and D. Brill, *Compos. Math.* **43**, 317 (1981).
17. A. Chodos and S. Detweiler, *Gen. Relativ. Gravit.* **14**, 885 (1982).
18. P. Freund and M. Rubin, *Phys. Lett. B.* **79**, 233 (1980).
19. R. Young, *Phys. Lett. B* **142**, 149 (1984).

*Plate 8*

# Obvious, Incredible ...

**V. N. Lukash and I. D. Novikov**

*Received July 30, 1985*

*The role of neutrinos and other weakly interacting particles in cosmology is considered in view of the evolution of concepts on the initial stages of the universe's expansion.*

## 1. INTRODUCTION

When John A. Wheeler wrote his book *Neutrinos, gravitation and geometry*(1) a quarter century ago, he began and closed it with the sections of the same name, "Central role of neutrinos in the physics of elementary particles." It shows how essential, in his view, was the role of weakly interacting particles in nature already in those years. The same idea of the fundamental role of neutrino is repeated in numerous papers by Wheeler.

Now, 25 years later, we are as much impressed by Wheeler's vision and never cease to admire his insight into the essence of this and many other fields of science, this "magic without magic."

The neutrino was hypothesized by Wolfgang Pauli in 1931 to save the energy conservation law. Twenty-five years had to pass before the neutrino was detected in a direct experiment. In this period, however, none of the physicists ever doubted its existence.

The situation with the neutrino seems to repeat these days. We mean relic neutrinos, whose existence has been reliably predicted but will probably take much time to be confirmed in a direct experiment. But if a half century ago the idea was to save the energy conservation law at the scale of a decaying neutron, now the relic neutrinos are needed to "save" the law of mass (energy) conservation at the scale of the universe, that is,

[1] Institute for Space Research, Academy of Sciences of the USSR, Moscow 117810, USSR.

to explain the "dark mass," whose only manifestation is gravity, in the intergalactic space.

The extraordinary penetrability of neutrinos that Pauli predicted, a most essential factor in the physics of elementary particles, should be equally decisive in cosmology. Indeed, it is perhaps only the existence of relic neutrinos or other similar particles, with a nonzero rest mass and quite easily penetrating any thicknesses of matter, that may explain the absence of appreciable fluctuations in the relic radiation intensity in the contemporary universe. And this is against the background of gigantic fluctuations in the large-scale matter distribution. The fact is that nonrelativistic neutrinos do begin clustering gravitationally still in the epoch when the hot-matter pressure impedes the development of inhomogeneities therein. This hindered the development of perturbations in the microwave background when the matter was not yet transparent for the relic quanta.

It is now believed that, along with neutrinos, other weakly interacting relic particles may be candidates for the "dark cosmic mass," particles which have similar properties (e.g., photinos, gravitinos, etc.); sometimes they are called "... inos." For brevity we shall refer as neutrinos all weakly interacting particles which, when being still relativistic, decoupled from ordinary matter during the universe's expansion. In addition, there may also be cold relic particles. They either originated due to nonequilibrium processes (monopoles, axions, primordial black holes, etc.) or decoupled after they had already become nonrelativistic.

In this paper we would like to discuss some effects involving weakly interacting particles, which may be dominant in various cosmological models. Not all models considered here may be explicitly related with the real universe, part of them being probably of purely of academic interest now. However, a long way had to be covered before such a conclusion could be made. Moreover, if we are to follow Wheeler's method, not only is it necessary to study the universe as it is, but it is also necessary to try and understand what it could have been under differing circumstances.

Energy units used below are $c = \hbar = 8\pi G = 1$.

## 2. OPPOSING NEUTRINO FLUXES

We proceed to consider some unusual effects involving neutrinos in homogeneous anisotropic cosmological models.

Let us recall the properties of the simplest homogeneous anisotropic models with metric

$$ds^2 = dt^2 - a^2(t)\, dx^2 - b^2(t)\, dy^2 - c^2(t)\, dz^2 \qquad (1)$$

Models with matter at isotropic pressure and the equation of state $p=\varepsilon/3$ have a Kasner[(2)] solution near the singularity ($t \to 0$):

$$a=a_0 t^p, \qquad b=b_0 t^{p_2}, \qquad c=c_0 t^{p_3}$$
$$p+p_2+p_3=p^2+p_2^2+p_3^2=1, \qquad p \leqslant p_2 \leqslant p_3 \tag{2}$$
$$\varepsilon=t_{\rm I}^{-2/3} t^{-4/3}, \qquad t_{\rm I} \gg t_{\rm Pl}=1$$

The inequality $-1/3 \leqslant p \leqslant 0$ is valid for parameter $p$. Compression occurs along the x axis; expansion, along other axes. At this stage the presence of matter does not affect the expansion. At $\sim t_{\rm I}$ the asymptotic solution (2) transforms to the isotropic expansion:

$$a \sim b \sim c \sim t^{1/2}, \qquad \varepsilon=3(2t)^{-2} \tag{3}$$

Parameters that determine the model are, for example, $t_{\rm I}$ and $p$. If, however, neutrinos are present in the model, its dynamics strongly changes.[(3)] Near the singularity, all particles, including neutrinos, are in thermodynamic equilibrium. The neutrino relaxation time is denoted as $\tau$; it is energy dependent. The instant of neutrino decoupling is determined by the $t \simeq \tau$ condition and may be written using the moment of neutrino decoupling $\tau_0$ in the isotropic model ($\tau_0 \sim 0.1$ s for $\nu_e$):

$$\tau=\tau_0(\tau_0/t_{\rm I})^{5/4} < t_{\rm I} \tag{4}$$

After neutrino decoupling in the anisotropic model,[2] the neutrino momentum component along the x axis grows at stage (2) due to contraction along this axis. The interaction cross-section $\sigma$ increases correspondingly and the annihilation reaction of the $\nu+\tilde{\nu} \to e^+ + e^-$ type predominates. For energies lower than $E_0 \sim 100$ GeV, the reaction effective cross-section is $\sigma \sim E^2$. For $E_\nu \gg E_0$ the cross-section $\sigma$ decays.

At the anisotropic stage after $t=\tau$ and at $E_\nu < E_0$ the number of neutrinos becomes smaller owing to the $\nu$ and $\tilde{\nu}$ annihilation, and their energy irreversibly converts into $e^-, e^+$ pairs, which are then termolized. The entropy of the matter is growing. The energy densities of neutrinos, photons, and $(e^+, e^-)$ pairs always remain of the same order.

The energy of oppositely moving neutrinos grows as $E_\nu \sim t^{-p/3}$, and stage (2) is completed at $t^*=t_{\rm I}(\tau_0/t_{\rm I})^{9(1+p)/4(3+p)}$. The annihilation process continues either until the neutrino energy reaches $E_0$ or up to the end of the anisotropic stage. Stage (2) is followed by the isotropization of the expansion.

[2] Note that the assumption of an important role of neutrino viscosity in the anisotropic model[(4)] does not seem valid even for the least realistic parameters.[(3)]

As the papers of the late 60s suggested (see, e.g., Ref. 5), the result of the processes considered above should be a sharply anisotropic distribution of relic neutrino momenta in the present universe. Their energy should be much higher and the density lower by the same factor than that of relic photons. The presence of anisotropic fluxes of relativistic particles should affect the expansion dynamics at the almost-isotropic stage. All this resulted in a considerable predicted anisotropy of the microwave background $\Delta T/T \sim 10^{-3}$–$10^{-4}$.[3,6]

Later, however, it became clear that the model considered did not account for some important questions.

## 3. WHY WERE THERE NO KASNER STAGES?

Kasner solutions are unstable against the quantum process of particle production.

Physically, cosmological stages (2) are tidal gravitational compression and stretching (in the perpendicular directions) of space. Kasner stages are generalized by oscillatory solutions[7,8] which consist of chains of Kasner stages transfering one to another. Classical oscillatory solutions are stable in the case of cosmological collapse; they describe a general type of space-like singularity.[9] A physical source of oscillatory solutions is gravitational waves or vortical matter motions.[10]

Kasner and oscillatory solutions are the vacuum solutions, which means that the matter gravity does not affect their dynamics. At the vacuum stage, with the volume increasing, there is always a direction along which the universe is contracting. In the presence of such an intense tidal gravitational field, quantum creation of matter particles from vacuum takes place.[11] Gravitation of particles so produced rapidly halts the cosmological compression along the specified axis and sets up instead expansion over all directions. It is this effect that is responsible for the quantum instability of vacuum solutions.

In the simplest anisotropic model with a flat space (1) the back reaction of the produced particles on the metric causes rapid isotropization of cosmological expansion. Thus, if the particle production is "turned on" at $t_0 > t_{\mathrm{Pl}}$ (the only condition for which we can make calculations), then the moment of isotropization $t_{\mathrm{I}} = t_{\mathrm{I}}(t_0)$ meets the following limit relation[12]:

$$\lim_{t_0 \to t_{\mathrm{Pl}}} t_{\mathrm{I}}(t_0) = t_{\mathrm{Pl}} \tag{5}$$

In arbitrary inhomogeneous models, the gravity of the created matter tends to isotropize cosmological expansion as well. The same relation (5) holds for the general case if the moment $t_I$ is understood as the beginning of gravitational influence of the matter on the cosmological evolution. This tendency, however, is not accompanied by metric inhomogeneity smoothing over large scales, or by space curvature isotropization, so the model remains inhomogeneous globally.

Hence, the intense particle production in the gravitational field forbids all vacuum stages,[3] and the expansion goes in any direction since the very beginning (i.e., since $\sim t_{Pl}$).

Can we suppose that the universe starts expanding isotropically? Generally, the answer is no. While vorticity is heavily suppressed by the particle creation efect, it is not the case with shear deformation. Even if the scale of inhomogeneity is much larger than the horizon at $t = t_{Pl}$, the anisotropy of space curvature induced by gravitational waves may be sufficiently strong for the curvature (along with the matter gravity) to determine the dynamics of expansion. This is the case when the so-called "damping" stages are realized during which the curvature and shear anisotropy decrease by a power law, with expansion over all directions. Damping stages are stable not only against the quantum process of particle production. Generally they appear either from the very beginning or later during the expansion, to replace the quasi-isotropic solution[4] near the singularity. In classical cosmology, the damping stage is a link to the changeover from the chaotic oscillatory singularity to a quasi-Friedmann expansion.(13,14) It is only degenerate initial conditions that makes it possible to avoid damping stages (e.g., for small initial inhomogeneities the quasi-isotropic solution turns into a Friedmann solution with the "growing" mode of perturbations).

The next section deals with the most general damping stage.

## 4. NEUTRINOS AT THE DAMPING STAGE

For arbitrary initial conditions specified at the moment when matter gravity becomes important for dynamics ($\varepsilon \sim t^{-2}$), one of the eigenvalues of the space curvature tensor is always larger than the other two. We denote this maximum curvature as $Q$. If $Q \ll \varepsilon \sim t^{-2}$, where $\varepsilon$ is the total density of matter, the expansion starts with the quasi-isotropic stage during which, however, $Q$ decreases less rapidly than $\varepsilon$. Hence, at

$$Q \sim \varepsilon \sim t^{-2} \tag{6}$$

[3] Except the degenerate Taube solution $p = p_2 = 0,\ p_3 = 1$.

[4] Cosmological isotropic expansion with anisotropic space curvature tensor is called the quasi-isotropic solution.(15)

the quasi-isotropic stage is replaced by a damping stage. If the condition (6) is observed from the very beginning, the damping stage is a singularity asymptote.

Expansion at the damping stage is the slowest in the maximum curvature direction $\boldsymbol{l}=\{l_\alpha(\mathbf{x})\}$ which is a function of space coordinates $\mathbf{x}=\{x^\alpha\}$. Let the second-rank tensor $a_{\alpha\beta}(\mathbf{x})$ specify a space structure in the plane orthogonal to $\boldsymbol{l}$. The countervariant vector $l^\alpha$ is defined as the only solution of the equation[5]

$$a_{\alpha\beta}l^\beta=0, \qquad l_\alpha l^\alpha=1 \tag{7}$$

The metric tensor and the 3-tensor of the curvature (Ricci tensor) are as follows[(14)]:

$$\begin{aligned} ds^2 &= dt^2-\gamma_{\alpha\beta}\,dx^\alpha\,dx^\beta, \qquad \gamma_{\alpha\beta}=\nu\tilde{l}_\alpha\tilde{l}_\beta+\mu a_{\alpha\beta} \\ P^\beta_\alpha &= Q(l_\alpha(l^\beta-h^\beta)-\tfrac{1}{2}\delta^\beta_\alpha), \qquad Q=k^2\nu\mu^{-2} \end{aligned} \tag{8}$$

where

$$k=(\det\{l_\alpha l_\beta+a_{\alpha\beta}\})^{1/2}\cdot\boldsymbol{l}\cdot\operatorname{rot}\boldsymbol{l}=\text{const}$$

is the wave vector magnitude and $\lambda\!\!\!^{-}=\mu^{1/2}k^{-1}$ is the gravity wavelength. The solution (8) is written in the leading order of the small parameter $t/\lambda\!\!\!^{-}\lesssim(\mu/\nu)^{1/2}\ll 1$, which increases with expansion. We have, to accuracy $\sim(\mu/\nu)^3$,

$$\tilde{l}_\alpha=l_\alpha+\frac{\mu}{\nu}h_\alpha \tag{9}$$

where

$$h_\alpha=h_\alpha(\mathbf{x})=-(\mathbf{l}\cdot\operatorname{rot}\mathbf{l})^{-1}\cdot a_{\alpha\beta}\operatorname{rot}^\beta\mathbf{l}$$

$\nu$ and $\mu$ are time functions that obey the equations

$$\begin{aligned} V^{-1}\left(V\frac{\dot{\nu}}{\nu}\right)^{\cdot}+Q &= \varepsilon+p_\nu-2p_\mu \\ V^{-1}\left(V\frac{\dot{\mu}}{\mu}\right)^{\cdot}-Q &= \varepsilon-p_\nu \end{aligned} \tag{10}$$

where $(\dot{\ })=\partial/\partial t$ and $V=\nu^{1/2}\mu$; $\varepsilon$ and $p_{\nu,\mu}$ are the energy density and matter pressure in the $\pm\mathbf{l}$ direction and in the orthogonal plane. To the leading order, the reference system (8) is comoving with matter and the functions $\varepsilon$ and $p_{\nu,\mu}$ are homogeneous over space.

[5] All manipulations with vector $\mathbf{l}$ and other objects that depend on $\mathbf{x}$ are performed as if the $\mathbf{x}$ space were Euclidean.

The solution (8) incorporates four physically arbitrary functions of the spatial coordinates.[6] The quasi-isotropic stage is described by Eqs. (10) for $Q \ll \varepsilon$ and $p_\nu = p_\mu$. In this case, $\mu/\nu = \text{const}$, and the number of independent functions reduces to three.

For the damping stage, with a hydrodynamical equation of state $p_\nu = p_\mu = (\gamma - 1)\varepsilon$, we have

$$\begin{gathered} \varepsilon = (6-\gamma)(2\gamma t)^{-2}, \qquad Q = (3\gamma - 2)(2-\gamma)(2\gamma t)^{-2} \\ \nu \sim t^{-1+2/\gamma}, \qquad \mu \sim t^{1/2+1/\gamma} \end{gathered} \tag{11}$$

where $\frac{2}{3} < \gamma = \text{const} < 2$. A most remarkable test of damping stages is weakly interacting massless particles (say, gravitions and neutrinos). Due to the slow expansion along the $\nu$ axis, opposite fluxes of free particles in the $\pm$l directions start forming and dominating energy density. In this case, the expansion rate changes:

$$\begin{gathered} \varepsilon = p_\nu = \tfrac{2}{3} t^{-2} \gg p_\mu, \qquad Q = \tfrac{8}{9} t^{-2} \\ \nu \sim t^{2/3}, \qquad \mu \sim t^{4/3} \end{gathered} \tag{12}$$

As a result of such evoluation in the past, the density of massless free particles with a sharply anisotropic momentum distribution may be very high today. However, before astrophysical estimates are given as to whether such detection is possible, we consider one more isotropization mechanism so ardently discussed in the literature of the 80s. It is now referred to as *inflation*. If inflation occurred in the most distant past of our universe, this rules out all damping stages, the last stronghold of anisotropic cosmology.

## 5. CHAOTIC COSMOLOGY AND INFLATION

The development of any idea obeys the dialectic triad: thesis, antithesis, synthesis. The beginning of the 1970s saw isotropization as the central problem of the early universe's evolution: If there was chaos initially, then how did the ordered Friedmann expansion originate from the chaotic initial conditions? The early 80s were marked by the progress of the opposite tendency: If there was complete order initially, then how could "primordial" perturbations originate in a strictly isotropic and homogeneous universe, of which later galaxies have formed? The primordial cosmological perturbations, small as they are in amplitude, are enormously large compared to natural fluctuations in a homogeneous medium

[6] As to the general solution of the gravity equations at the stage (8), the missing arbitrary functions may be obtained from the subsequent expansion terms.

(quantum and thermal). And it is only in recent years that it has become clear that the problems of isotropization and generation of perturbations should be considered together. The inflation concept is an example—and the only one until now—of such an approach.

To realize inflation, the initial state of matter should correspond to a high-density metastable vacuum ($\varepsilon_{\mathrm{vac}} \sim 10^{15}$ GeV). The ideal properties of the vacuum state, homogeneity and isotropy, constitute a physical source of isotropization. During the expansion, the vacuum density remains constant due to the negative-pressure action ($p_{\mathrm{vac}} = -\varepsilon_{\mathrm{vac}}$); this leads to "false" vacuum domination over all other matter components and over the space curvature. At the stage of false vacuum domination and before its disintegration into particles (vacuum phase transition), the universe is expanding very rapidly, and it takes only several tens of cosmological times for a causally-connected (at the beginning of expansion) region to become equal in size to the visible universe. Because of this inflation the curvature is effectively ironed out so much that the total density of matter is now equal to the critical one. During inflation and phase transition into the "true" vacuum state quantum-statistical fluctuations, which could be the cause of primordial perturbations, are developed. The parameters of these inhomogeneities strongly depend on the kinetics of phase transitions. The spectra of the perturbations so produced are almost flat in the first generation of inflationary models. Here models in which the total density of all kinds of matter is critical ($\Omega_{\mathrm{tot}} = 1$), and primordial perturbations are adiabatic (metric perturbations) with flat spectrum, are referred to as *standard inflationary* theories.

The question about the total matter density $\Omega_{\mathrm{tot}}$ and the spectrum of primordial perturbations is of great significance to the processes of formation of the observed structure of the universe—clusters and supercusters. Furthermore, the growth of perturbations in the early universe depends upon the parameters of cosmological missing mass, whatever that is: massive neutrinos, primordial black holes, axions, gravitinos, or any other hypothetical weakly interacting particles.

What are the predictions of the modern theory of the very early universe, as far as primordial perturbations are concerned? First of all, the perturbations must have random normal distribution. The reason is that primordial perturbations are generated by quantum (or thermal) fluctuations, which obey, by definition, the Gaussian law of distribution. The question about the spectrum of primordial perturbations is still less clear.

We have already mentioned quantum-statistical perturbations, which are generated by vacuum phase transitions. These perturbations develop from the quantum fluctuations of the Higgs fields at scales less than the horizon size. At greater scales, an alternative mechanism dominates[(18)]: the parametric amplification of metric vacuum fluctuations due to the fast "nonadiabatic" expansion of the universe [more precisely, due to the breaking of the $a''=0$ condition; here $a$ is the scale factor factor and $(')=a(\partial/\partial t)$]. In the case where incomplete (shortened in duration) inflation occurs, one obtains a spectrum of primordial metric perturbations decreasing with growth of scale.[19,20] Moreover, the primordial perturbations may not have been metric perturbations but density perturbations in the component of missing mass which dominates at present time (isothermal or isocurvature perturbations). Isocurvature perturbation spectra predicted by inflationary models of the universe may be flat[21] or have a cutoff on the long wavelength side.[22] In the latter case, as well as in parametric amplification models mentioned above, the theory has a free parameter—the scale, from which the spectrum decays with growing wavelength.

Does theory inevitably predict inflation in the very early universe? Not yet. The possibility of inflation (e.g., in grand unified theories) is based on the extreme extrapolation of elementary particle physics to high energies. And, even if current theoretical ideas prove to be valid, it is necessary, in addition, that the initial state of quantum fields (before the phase transition) provide an inflationary stage with duration enough to account for homogeneity and isotropy of the universe.

Thus, one cannot yet exclude the possibility of anisotropic damping stages in the past. Besides, we do not know the nature of the weakly interacting particles that the cosmic dark mass consists of. In such a situation the observational astrophysical restrictions on possible parameters of anisotropy and missing mass are most significant. Before presenting the observational predictions of various cosmological models, we will breifly summarize the modern view of possible types of the missing mass in an isotropic universe.

## 6. WEAKLY INTERACTING PARTICLES IN THE FRIEDMANN MODELS

We have seen above what peculiar role may be played by neutrinos or some other "... inos" in the various cosmological models, including exotic ones.

It is worth emphasizing that these particles play a decisive role in most popular modern cosmological models.

Thus, if one admits an inflationary model of the universe, then the inevitable conclusion arises that the main part of matter in the universe is in the form of invisible mass consisting of "... inos." Indeed, theory predicts in this case that $\Omega_{\text{tot}}=1$. Together with data on light element abundances

(H, $^3$He, $^4$He, D, $^7$Li) and on possible anisotropy of the microwave background, this means that the main part of matter in the universe does not consist of baryons. There must be missing mass, perhaps in the form of weakly interacting particles.[23,24]

It is the dynamical mass of "... inos" that, probably, rules the rate of the universe's expansion and controls the processes of growth of primordial perturbations, which have led to formation of the structure of the universe.

The so-called standard neutrino model is the most attractive one from the aesthetic point of view. According to it, there is one or several types of nonrelativistic stable neutrinos in the present universe. This model has no free parameters and gives a natural cutoff in the perturbation spectrum at scales of superclusters. On the other hand, the standard $\nu$ model faces some difficulties. One of those, as we shall see in Sec. 6, is that theory predicts too high fluctuations of the microwave background. Among the other difficulties, let us enumerate the following[25–28]:

(1) The high rate of the nonlinear evolution, which leads to a contradiction with observations of distant quasars.

(2) The too low part of matter that gets into galaxies, which contradicts observations as well.

(3) The prediction of theory that the correlation scale of missing mass structure is $\sim 20$ Mpc, which probably exceeds the observational value $r_c \sim 10$ Mpc.

(4) The high peculiar velocities of galaxies.

Although these difficulties of the standard $\nu$ model are not crucial because of our poor knowledge of galaxy formation process, nevertheless they give an impetus to the development of alternative models.

We shall discuss here those alternative models which try to overcome the difficulties mentioned in the following manner. The mass carriers in these models are not usual neutrinos with nonzero rest mass, but some or other hypothetic "... inos" with *a priori* postulated properties. This introduction of some extra parameters into the theory certainly makes it less attractive. The most important theories of this kind have the missing mass in the form of decaying neutrinos or cold matter (axions or very heavy relic particles, $m_R \gtrsim 1$ KeV). In the theories of the first type, the missing mass is suggested to be formed of massive unstable neutrinos which decay just after the formation of the structure of the universe. The relativistic decay products escaping out of the structure clumps, the gravity of the latter weakens, and hence their nonlinear evolution slows down. But the inevitable fit of the neutrino lifetime to the required value $10^{16}$–$5 \times 10^{16}$ s seems to be artificial.

In this sense, theories of the second type seem to be more attractive. These theories deal with hypothetical superheavy particles, which become nonrelativistic at an early stage of evolution of the universe. Hence, the gravitational growth of clusters of these particles begins very early, far before the hydrogen recombination,(29) which enables one to avoid certain difficulties. Unfortunately this approach is not free of problems as well. One has to admit some additional hypotheses (e.g., biasing—galaxies form only in a very dense clusters of "... inos").

It must be emphasized that none of the modern theories is free from serious drawbacks.

What is the way to overcome these difficulties? Should one choose some more sophisticated combinations of the known theories or follow an essentially different approach? The answer is not yet clear.

## 7. WHAT WAS THE UNIVERSE IN REALITY?

In the previous sections we tried to show possible options for the early universe from the viewpoint of theorists. The models discussed are so numerous mainly because what we know about the microcosm (that is, the physics of elementary particles) is so uncertain. It is in the early universe that the microcosm and the macrocosm overlaps, that is, elementary processes predetermine future features of the observable Metagalaxy. Calculations based on theoretically predicted possibilities (in different models suggested) and their comparison with observational data are still the only reliable research tool in cosmology, and not only in cosmology. Cosmological predictions may considerably restrict the class of possible physical theories which are consistent with the present observations or sometimes help verify the theories of a very early universe.

What are then the observational aspects most efficiently reflecting the processes that occurred in the early universe? Above all it is the observable structure of the large-scale universe. However, to detect fundamental parameters of missing mass from the observed structure is extremely difficult, owing to complex, nonlinear gasdynamical processes accompanying the formation of galaxies.

An independent and more accurate test of an early universe is the observations of the primordial chemical composition of the matter and especially of the anisotropy of relic radiation. The $\Delta T/T$ fluctuations are reliably and readily calculated in the framework of the linear theory of perturbations. The anisotropy amplitude $\Delta T/T$ is unambiguously determined by primordial inhomogeneities, by the dynamics of their growth in the early universe (which is explicitly related with missing-mass parameters), and by the global spatial curvature of today's universe.

The abundance of light elements in the prestellar matter is consistent with the predictions of Friedmann's cosmology. What are now the options for anisotropic models?

Current studies show that far from all nonhomogeneous anisotropic models become isotropic in the expansion after the damping stage. Hence, here we restrict ourselves only to most general homogeneous anisotropic models whose evolution is well known. Let $t_D$ be the moment the damping stage starts (see Sec. 4). If $t_D > 1$ s, the chemical elements are forming at the stage of quasi-isotropic expansion and the outcome of nuclear reactions is the same as in the Friedmann model. About 25% $He^4$ by mass from. If $t_D < 1$ s, chemical elements are forming at the anisotropic stage (12). For $He^4$ to be less than 50% (the contrary would drastically contradict observations), $t_D > 10^{-3}$ s is required.[(30)] The result of such evolution would be a sharply anisotropic momentum distribution of free particles (e.g., neutrinos or gravitons). As to the energy density of free particles, today it may be of the order of the matter density in galaxies and still not contradict the observations of the chemical composition. (Restrictions imposed by $\Delta T/T$ are discussed below.) Note that in the isotropic model the free-particle density is not more than 3 or 4 times higher than the density of the background radiation.[(31)] In the anisotropic model the energy of free particles gets larger at the damping stage that follow the nucleosynthesis stage.[(30)]

Let us now turn to restrictions arising from the observations of microwave background. At present the experiment is close to $\Delta T/T \sim 10^{-5}$; however, no angular variations of temperature, except the dipole component[(32,33)] due to the Earth's motion relative to the background radiation, have yet been detected. This high degree of background isotropy imposes stringent restrictions on the anisotropy parameters.

For arbitrary initial conditions, the isotropization moment seems to be of the order of the Planckian, which implies that since the very beginning the Universe was in essence a Friedmann Universe. Still, not all models have been studied; neither can we draw the final conclusion that the damping stage was impossible in the past.[7]

We now consider possible isotropic models with different carriers of missing mass.

As theory puts it, the $\Delta T/T(\mathbf{e})$ amplitude is a function of the unit vector $\mathbf{e}$ along the line of sight. The hypothesis on randomly distributed

[7] The damping stage might be possible for a certain set of initial conditions that ensure rapid isotropization.[(10)] Here we do not deal with this question.

primordial perturbations (see Sec. 4) allows comparison with observations in terms of the correlation function

$$\xi(\theta) = \left\langle \frac{\Delta T}{T}(\mathbf{e}) \frac{\Delta T}{T}(\mathbf{e}') \right\rangle^{1/2} \tag{13}$$

or by employing harmonic analysis

$$\left(\frac{\Delta T}{T}\right)_l = \left( \sum_{m=-1}^{l} a_{lm}^2 \right)^{1/2} \tag{14}$$

where

$$\frac{\Delta T}{T}(\mathbf{e}) = \sum_{l=2}^{\infty} \sum_{m=-1}^{l} a_{lm} \Psi_{lm}(\mathbf{e})$$

$\Psi_{lm}(\mathbf{e})$ is a spherical function, $\langle \cdots \rangle$ means the average over the state of the primordial perturbation field, $\theta$ is the angle between the observation directions, and $\cos\theta = \mathbf{e}\cdot\mathbf{e}'$. For comparison of (13) with the observed distribution, where the average is taken over all directions on the celestial sphere with the fixed angle $\theta$, the antenna beam width $\theta_a$ should be taken into account. For an estimate we have (dipole is subtracted)

$$\xi(\theta_a) = \left( \sum_{l=2}^{\theta_a^{-1}} \left(\frac{\Delta T}{T}\right)_l^2 \right)^{1/2} \tag{15}$$

Table I presents the expected $\Delta T/T$ anisotropies ($H_0 = 50\ \mathrm{km\,s^{-1}\,Mpc^{-1}}$, $T_\gamma = 2.7$ K) for the flat primordial adiabatic[8] perturbation spectrum in different cosmological dark matter models ($\Omega_{\mathrm{tot}} = 1$). For the thermodynamic massive neutrino models, the density contrast is supposed to reach $\sim 1$ at the redshift $z_s = 3$. If one chooses $z_s \simeq 1$ (see Sec. 6; quasars at $z \sim 3$ are accounted for by random pick of density[(34)]) the given numbers must be reduced by twice as much. In the decaying neutrino models ($\tau_d \simeq 1$–$5\times10^{16}$ s, $z_\tau \simeq 4$–$10$, $m_\nu \simeq 40-100$ eV),[(35,36)] large-scale harmonics $(\Delta T/T)_l$ increase for $l \lesssim z_\tau$, which may provide a test for these models.[(37)] Normalization of the primordial spectrum in the axion models is taken by the coorelation scale $r_c = 10$ Mpc.[(38)] Increase of the correlation scale of galaxies[(39)] leads to growth in the predicted $\Delta T/T$ anisotropy. For $r_c \sim 30$ Mpc, large-scale $\Delta T/T$ fluctuations appear of the same order of

[8] Flat spectrum of the isocurvature perturbations seems to disagree with the large-scale $\Delta T/T$ limit, since the predicted $\Delta T/T$ amplitude turns out to be much larger than that for the adiabatic perturbations.

**Table I.** $\Delta T/T$ Anisotropy in Different Models of Dark Matter ($\Omega_{tot} = 1$, Flat Spectrum of Primordial Perturbations)[a]

| | $\Delta T/T$ | | | |
|---|---|---|---|---|
| | Stable neutrino | Decaying neutrino | Axions or heavy "... inos" | $\Lambda > 0$ |
| Quadrupole, $(\Delta T/T)_{l=2}$ $(2 \times 10^{-5})$ | $2 \cdot 10^{-5}$ | $\gtrsim 2 \cdot 10^{-5}$ | $4 \cdot 10^{-6}$ | $10^{-5}$ |
| Large scales $\xi(6^\circ)$ $(5 \times 10^{-5})$ | $5 \cdot 10^{-5}$ | $\gtrsim 5 \cdot 10^{-5}$ | $10^{-5}$ | $2 \cdot 10^{-5}$ |
| Smale scales $\xi(20')$ $(3 \times 10^{-5})$ | $10^{-4}$ | $2 \cdot 10^{-5}$ | $2 \cdot 10^{-5}$ | $8 \cdot 10^{-5}$ |

[a] Observational data are given in parentheses: large-scale restrictions for the quadrupole and dispersion ($\Theta_a \simeq 6^\circ$, $\Theta_a^{-1} \simeq 10^{(50)}$) and small-scale limit for the ideal antenna.[53]

magnitude in both neutrino and axion models. In models with the $\Lambda$ term the density of homogeneous missing mass component is taken as $\Omega_\Lambda \simeq 0.8$, with the density of dynamical missing mass $\Omega_m \simeq 0.2$.[40]

Small-scale $\Delta T/T$ fluctuations may well be essentially weaker due to a possible secondary ionization of cosmic plasma by the products of decay of unstable particle. According to Ref. 41, the secondary ionization reduces $\Delta T/T$ $(\theta < 1^\circ)$ to the level $\lesssim 8 \times 10^{-6}$. A test for secondary ionization is to measure polarization, which may reach here several ten percents with respect to $\Delta T/T$ anisotropy.[42]

Large-scale anisotropy $\Delta T/T$ $(\theta > 6^\circ)$ does not depend at all on the hydrogen recombination and secondary ionization dynamics; it is directly linked to the primordial cosmological perturbations. The flat spectrum is a crucial test in the $\Delta T/T$ problem. White noise spectrum (growing to large scales) practically contradicts present-day observations, while decaying spectra reduce the expected level of large-scale anisotropy. A very important effect in the large-scale $\Delta T/T$ structure—so-called "spottiness"—occurs if $\Omega_{tot} < 1$. In this case, the universe 3-space curvature results in the correlation angular scale $\theta_c \simeq \Omega_{tot}$,[43–46] which gives a unique opportunity to detect directly $\Omega_{tot}$ by the distribution $\xi(\theta)$.

The confrontation of predicted $\Delta T/T$ fluctuations with modern observational limits leads to the following conclusions:

1. The standard inflationary model of the universe with massive neutrinos as missing mass contradicts the large-scale $\Delta T/T$ limits.[47] The mismatching factor here is $\sim 2$, since we deal with the single universe (i.e., with one of all possible realizations of the primordial perturbation field). Thus, large-scale $\Delta T/T$ anisotropy provides a good test of the neutrino-dominated universe.

2. The minimum predicted anisotropy level $\Delta T/T$ $(\theta > 6°)$ for the standard inflationary model, which can be obtained if axions or very heavy particles are the missing mass,[38,47–49] may be detected if the RELIC experiment[50] sensitivity is improved to $\sim 6 \times 10^{-6}$.

3. Models with $\Lambda > 0$ and baryon or neutrino dynamical dark mass disagree with current observations. All other models do not contradict the Uson and Wilkinson[51] limit; however, stable neutrino models and $\Lambda \neq 0$ models contradict the data of Refs. 52 and 53.

4. The optimal range of search for $\Delta T/T$ anisotropy is $20' < \theta < 90°$. If $\Delta T/T$ is detected at the angular scales $\theta > 6°$, the spectrum of primordial perturbations and the total matter density in the universe can be determined.[54]

The world, so obvious and so incredible, is around us. The ways to learn its mysteries are as obvious and as incredible. We may hope that a theory most adequately describing the exploding universe will be gradually crystalized from the chaos of theoretical models after they are compared with more and more advanced observations. No easy approaches are possible here. John Wheeler[55] once wrote that if it is sometimes said that chaos is a fundamental feature of nature let it be so .... It is not new for physics to meet with chaos and make it obey the laws of reason. We would like to conclude our paper as Wheeler ended his. Maybe we shall only then understand for the first time how simple the universe is when we learn how strange it is.

## REFERENCES

1. J. A. Wheeler, "Neutrinos, Gravitation and Geometry," Rendiconti della Scuola Internazionale di Fisica "Enrico Fermi," Corso XI (Varenna sul Lago di Como, 1959), Bologna.
2. E. Kasner, *Am. J. Math.* **43**, 217 (1921).
3. A. G. Doroshkevich, Ya. B. Zeldovich, and I. D. Novikov, *Problems of Theoretical Physics* (dedicated to N. N. Bogolubov) (Nauka, Moscow, 1969).
4. C. W. Misner, *Phys. Rev. Lett.* **19**, 533 (1967).
5. A. G. Doroshkevich, Ya. B. Zeldovich, and I. D. Novikov, *Sov. Phys. JETP* **53**, 644 (1967).

6. A. G. Doroshkevich, V. N. Lukash, and I. D. Novikov, *Astron. Zh.* **51**, 940 (1974).
7. V. A. Belinskii, I. M. Khalatnikov, and E. M. Lifshitz, *Adv. Phys.* **19**, 525 (1970).
8. A. G. Doroshkevich and I. D. Novikov, *Astron. Zh.* **47**, 948 (1970).
9. V. A. Belinskii, I. M. Khalatnikov, and E. M. Lifshitz, *Adv. Phys.* **31**, 639 (1982).
10. V. N. Lukash, *Nuovo Cimento B* **35**, 268 (1976).
11. Ya. B. Zeldovich, *JETP Lett.* **12**, 443 (1970).
12. V. N. Lukash and A. A. Starobinskii, *Sov. Phys. JETP* **66**, 1515 (1974).
13. A. G. Doroshkevich, V. N. Lukash, and I. D. Novikov, *Sov. Phys. JETP* **64**, 1457 (1973).
14. V. N. Lukash, in *Contributed Papers, Eighth International Conference on GRG* (Waterloo University, 1977), p. 236.
15. E. M. Lifshitz and I. M. Khalatnikov, *Usp. Fiz. Nauk* **80**, 391 (1963).
16. A. H. Guth, *Phys. Rev. D* **32**, 347 (1981).
17. A. D. Linde, in *The Very Early Universe*, G. W. Gibbons *et al.*, eds. (Cambridge University Press, Cambridge, 1983), p. 251.
18. V. N. Lukash, *Sov. Phys. JETP* **79**, 1601 (1980).
19. D. A. Kompaneets, V. N. Lukash, and I. D. Novikov, *Astron. Zh.* **59**, 424 (1982).
20. V. N. Lukash and I. D. Novikov, in *Contributed Papers, Tenth International Conference on GRG*, B. Bertotti *et al.*, eds. (Consiglio Nazionale Delle Ricerche—Roma, 1983), p. 844.
21. A. D. Linde, *JETP Lett.* **40**, 496 (1984).
22. L. A. Kofman, *Phys. Lett. B* **165**, 100 (1985).
23. G. S. Bisnovatyi-Kogan and I. D. Novikov, *Astron. Zh.* **57**, 899 (1980).
24. A. G. Doroshkevich, M. Yu. Khlopov, Ya. B. Zeldovich, and R. A. Sunyaev, *Pis'ma Astron. Zh.* **6**, 457 (1980).
25. A. A. Klypin and S. F. Shandarin, *Mon. Not. R. Astron. Soc.* **204**, 891 (1983).
26. C. S. Frenk, S. D. M. White, and M. Davis, *Ap. J.* **271**, 417 (1983).
27. P. R. Shapiro, C. Struck-Marsell, and A. Melott, *Ap. J.* **275**, 413 (1983).
28. N. Kaiser, *Ap. J.* **273**, L17 (1983).
29. A. A. Starobinskii and V. Sahni, *Sov. Phys. JETP* (in press).
30. V. N. Lukash, I. D. Novikov, and A. A. Starobinskii, *Sov. Phys. JETP* **69**, 1484 (1975).
31. V. F. Shvartzman, *JETP Lett.* **9**, 315 (1969).
32. G. F. Smoot and P. M. Lubin, *Ap. J.* **234**, L83 (1979).
33. D. Fixen, E. Cheng, and D. Wilkinson, *Phys. Rev. Lett.* **50**, 620 (1983).
34. J. M. Bardeen, J. R. Bond, N. Kaiser and A. S. Szalay, Preprint Fermilab-Pub. 85/148-A (1985).
35. A. G. Doroshkevich and M. Yu. Khlopov, *Mon. Not. R. Astron. Soc.* **211**, 277 (1984).
36. A. G. Doroshkevich and M. Yu. Khlopov, *Pis'ma Astron. Zh.* **11**, 563 (1985).
37. L. A. Kofman, D. Yu. Pogosyan, and A. A. Starobinskii, *Pis'ma Astron. Zh.* **12**, 100 (1986).
38. P. J. E. Peebles, *Ap. J. Lett.* **263**, L1 (1982).
39. J. Einasto, A. A. Klypin, and E. Saar, *Mon. Not. R. Astron. Soc.* (in press).
40. L. A. Kofman and A. A. Starobinskii, *Pis'ma Astron. Zh.* **11**, 643 (1985).
41. E. I. Dorosheva and P. D. Naselskii, *Astron. Zh.* (in press).
42. M. N. Basko and A. G. Polnarev, *Mon. Not. R. Astron. Soc.* **191**, 207 (1980).
43. I. D. Novikov, *Astron. Zh.* **45**, 538 (1968).
44. V. N. Lukash, in *Contributed Papers, Eight International Conference on GRG* (Waterloo University, 1977), p. 237.
45. V. N. Lukash, in *Proceedings Symposium 104 IAU*, G. Abell and G. Chincarini, eds. (Reidel Publishing Co., 1982), p. 149.
46. G. S. Bisnovatyi-Kogan, V. N. Lukash, and I. D. Novikov, in *Proceedings of the Fifth Regular Meeting in Astronomy*, IAU/EPS, Liege, G.1.1 (1980).

47. V. N. Lukash, P. D. Naselskii, and I. D. Novikov, in *Proceedings of the Third Seminar on Quantum Gravity* (Moscow), M. A. Markov *et al.*, eds. (World Scientific, 1984), p. 675.
48. A. A. Starobinskii, *Pis'ma Astron. Zh.* **9**, 579 (1983).
49. J. R. Bond and G. Estathiou, *Ap. J.* **285**, L45 (1984).
50. I. A. Strukov, R. Z. Sagdeev, N. S. Kardashev, D. Skulachev, and N. Eysmont, in *Proceedings COSPAR Advances in Space Research* (Pergamon Press, New York, 1984).
51. J. M. Uson and D. T. Wilkinson, *Ap. J. Lett.* **277**, L1 (1984).
52. Yu. N. Parijskii, Z. N. Petrov, and A. N. Chernov, *Pis'ma Astron. Zh.* **3**, 483 (1977).
53. D. V. Korolkov and Yu. N. Parijskii, *Commun. Spec. Astrophys. Obs.* **41**, **42**, **43** (1985).
54. V. N. Lukash, in *Proceedings of Symposium 117, IAU*, "Dark Matter in the Universe" (Princeton), J. Kormendy, ed. (1985).
55. J. A. Wheeler, in *Astrofisica e Cosmologia, Gravitazione, Quanti e Relativita*, Centenario di Einstein, 1879–1979, negli sviluppi del pensiero scientifico di A. Einstein (Giunti Barbera, Firenze, 1979).

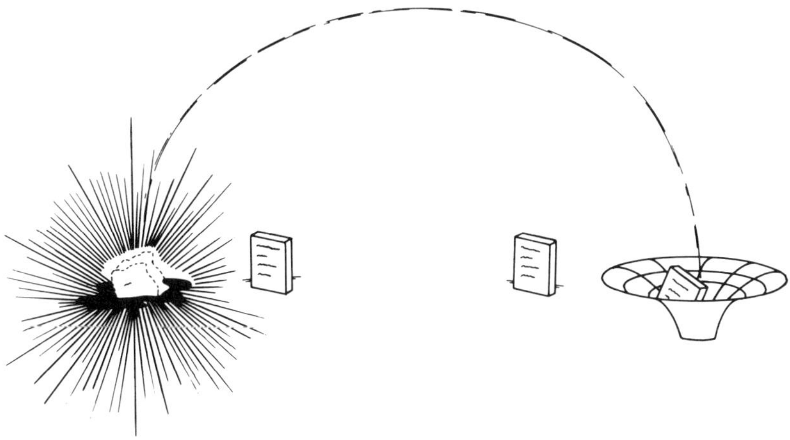

Plate 9

# Professor Wheeler and the Crack of Doom: Closed Cosmologies in the 5-d Kaluza–Klein Theory

**Richard A. Matzner**[1] **and Anthony Mezzacappa**[1]

*Received July 18, 1985*

*We study the classical and the quantum structures of certain 5-d Kaluza–Klein cosmologies. These models were chosen because their 4-d restriction is a closed, radiation-dominated, homogeneous, isotropic cosmology in the usual sense. The extra (field) dimension is taken to be a circle. In these models the solution starts from a 5-d curvature singularity with infinite circumference for the circle and zero volume for the 3-space. It evolves in finite proper time to a solution with zero dimension for the extra field direction. In the 5-vacuum case this is not a curvature singularity, but is a singularity of the congruence describing the physics, and in particular, the solution cannot causally be extended to the future of this point. In the 5-vacuum case this event coincides with the maximum of expansion of the 5-space. In the 5-dust cases, this point is a real 5-d curvature singularity. By adjustment it can be made to occur before or after the maximum of 3-expansion. The solution stops at that instant, but the 4-cosmology reveals* no *pathology up to the crack of doom. The quantum behavior is identical in these respects to the classical one.*

## DEDICATION

It is impossible to list all of John Wheeler's remarkable achievements. He has made contributions in nuclear physics, quantum mechanics, nuclear engineering, numerical physics, the interpretation of measurements, cosmology, etc. And to date this remarkable, energetic man pursues his love of science as avidly as ever.

[1] Center for Relativity and Physics Department, The University of Texas at Austin, Austin, Texas 78712.

John Wheeler's most substantial contribution must be as a teacher. Everyone who comes in contact with him comes away enlightened and enthused—layperson, Nobel prize winner, those of us in between. John Wheeler teaches us to use a problem to test the techniques we know. He tells us that if we have exhausted our techniques and ourselves then lay the problem aside, but do not discard it, for we *will* learn other skills that can be applied to it.

The behavior of closed cosmologies is one area that has been central to John Wheeler's thinking ever since he started to consider problems in relativity. The simple, metric 3-sphere cosmologies have the property that they recollapse after a finite lapse of proper time—the big bang is followed by a "big crunch." Do *all* closed cosmologies do this? One author (RM) has spent some time studying this question for anisotropic but homogeneous cosmologies. The answer is that most of these cosmologies do recollapse, but there are those that defy analysis, so we cannot say all do.

It was then natural when we began this study of Kaluza–Klein cosmology that we should include models like the closed cosmology. The surprising (we think) results are simply an example of the Wheeleresque serendipity that appears when we once again try a new technique on an old problem. Our Kaluza–Klein cosmologies do not live to the big crunch. Rather, they hear the crack of doom before they have had a chance to.

It is a pleasure to contribute this article to the birthday volume of our colleague, teacher and friend, John Wheeler.

## 1. INTRODUCTION

Following the discovery of the singularity theorems by Hawking and Penrose,[1] there was considerable interest in quantum cosmology motivated by the hope that quantum mechanics would provide an escape from the singularities inevitable in classical relativistic cosmology. While this still remains one of the main motivations for studying quantum cosmology, this branch of quantum gravity clearly has great utility as an approximation to the very difficult general quantum gravity problem. Further, as pointed out by Misner,[2] at times near the singularity where quantum gravity is important, it is physically plausible that the lowest modes of the gravitational field are dominant. One then expects that at those times the qualitative behavior of the full theory would be similar to that of the quantized cosmology.

In the past decade there has been a revival of interest in the Kaluza–Klein idea that space-time is higher dimensional.[3] This idea has been used extensively, and with some success, in formulating unified field theories of the fundamental interactions. If one takes the idea of extra dimensions seriously, a quantum cosmology would, of course, be based on a cosmology of the Kaluza–Klein type. In fact, at times when quantum cosmology would be relevant, the extra dimensions would not always be relatively small and, therefore, could not be ignored. Recently, several quantum Kaluza–Klein cosmologies have been investigated by Hu and Wu.(4)

In this paper we pick the topology of the solution and make use of the canonical formulation of Arnowitt, Deser, and Misner(5) with which some of the first quantum cosmological models were considered. Then, we obtain an exactly soluble equation for the cosmological wave function with boundary conditions emerging naturally from what is shown to be the analogous motion of a particle in an exponential potential. Our work will parallel Fishbone's treatment of the Kantowski–Sachs cosmology.(6,7)

## 2. OUR MODELS

Our models are generalizations of the Chodos–Detweiler(8) model. The Chodos–Detweiler model is a solution to the 5-dimensional vacuum Einstein equations $G_{AB}=0$ (A, B = 0,..., 4 with $x^4$ = the extra "field" dimension). The 4-dimensional space-time is a "flat" (i.e., $k=0$, where $k$ is the curvature parameter) homogeneous, isotropic cosmology, with a 3-space evolving from an initial zero-volume singularity at $t=0$ like a radiation-dominated cosmology. The field dimension begins with metrically infinite circumference at $t=0$, which decreases, approaching zero at $t=+\infty$ when the 3-space is infinite. It is the evolution of the field dimension which gives rise to the radiation source in the 3-space. Figure 1 schematizes this behavior.

*Our* models have closed 3-sphere (i.e., $k=+1$) spatial sections. As in the Chodos–Detweiler solution, we consider the vacuum case, and we also consider the 5-dimensional dust case. These solutions and some preliminary analysis were given by Lorentz-Petzold(9) and by Ishihara *et al.*,(10) respectively. We wish to obtain these solutions explicitly and examine them in detail, particularly focusing on the issue of the singularity in the fifth dimension and the recollapse of the closed three-space. We do so using a geometric construct fundamentally tied to the interpretation of Kaluza–Klein theories. We feel that these models are interesting in that, in both cases, the closed three-space does not recollapse because a singularity in the fifth dimension ends the evolution. Ishihara(11) has shown numerically that for $\mathbb{S}^3\times\mathbb{S}^n$ $(n>3)$, with $(4+n)$-dimensional

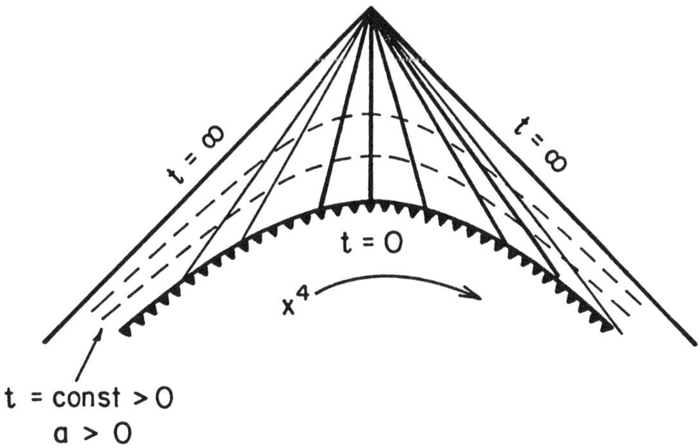


Fig. 1. The Chodos–Detweiler solution is a 5-dimensional Kaluza–Klein cosmology with the topology $\mathbb{R} \times \mathbb{R}^3 \times \mathbb{S}^1$. The metric of the spatial sections, $\mathbb{R}^3$, is a function only of time and describes the evolution of a "flat", radiation-filled Robertson–Walker cosmology. Although as a 5-space solution this is a vacuum model, there is an effective energy density in the 4-spaces which arises from the evolution of the field-dimension component of the metric. In this figure we have conformally compressed an infinite amount of evolution to a finite diagram. Each straight timelike line represents a 4-surface, the history of a 3-space from an initial zero-volume singularity to infinite expansion after an infinite amount of time. At the same time the metric circumference of the field dimension circle shrinks, going to zero after infinite proper time. The solution is homogeneous in the $x^4$ (field) direction, but with the $\mathbb{S}^1$ topology $x^4$ lies in the range $[0, 2\pi)$, indicated by the heavier drawn timelike lines. The downward-curving hyperbolae are surfaces of constant proper time and of constant value of the 3-space expansion factor, $a$.

dust, the three-space does not recollapse when certain initial conditions are imposed.

Our metric is

$$ds^2 = -dt^2 + h_{ij}(t)\, \omega^i \omega^j, \qquad i, j = 1, 2, 3, 4 \tag{1}$$

with the basis forms $\omega^1 = dr/\sqrt{1-r^2}$, $\omega^2 = r\, d\theta$, $\omega^3 = r \sin\theta\, d\phi$, and $\omega^4 = dx^4$, where the extra spatial dimension is coordinatized by the angle $x^4 \varepsilon [0, 2\pi)$. $h_{ij}(t)$ has the simple diagonal form

$$h_{ij}(t) = \text{diag}(a^2(t), a^2(t), a^2(t), \sigma^2(t)) \tag{2}$$

which shows that the physical 3-space remains metrically a 3-sphere of radius $a(t)$, while the extra dimension is a circle of circumference $2\pi\,|\sigma(t)|$.

## 3. CLASSICAL SOLUTIONS

We first consider solutions to the 5-dimensional vacuum Einstein equations, which in terms of $\sigma(t)$ and $a(t)$ are

$$G_{00}=0\colon 3\frac{\dot{\sigma}\dot{a}}{\sigma a}+3\left(\frac{\dot{a}}{a}\right)^2+\frac{3}{a^2}=0 \tag{3}$$

$$G_{ii}=0\colon \frac{\ddot{\sigma}}{\sigma}+2\frac{\ddot{a}}{a}+2\frac{\dot{\sigma}\dot{a}}{\sigma a}+\left(\frac{\dot{a}}{a}\right)^2+\frac{1}{a^2}=0, \qquad i=1,2,3 \text{ (no sum on } i) \tag{4}$$

$$G_{44}=0\colon 3\frac{\ddot{a}}{a}+3\left(\frac{\dot{a}}{a}\right)^2+\frac{3}{a^2}=0 \tag{5}$$

With a 5-dimensional dust source, (3) becomes

$$3\frac{\dot{\sigma}\dot{a}}{\sigma a}+3\left(\frac{\dot{a}}{a}\right)^2+\frac{3}{a^2}=8\pi\rho \qquad \text{(5-d dust)} \tag{6}$$

Recalling the Einstein equations for a closed, homogeneous, isotropic 4-space cosmology

$$3\left(\frac{\dot{a}}{a}\right)^2+\frac{3}{a^2}=8\pi\rho_{\text{eff}} \tag{7}$$

$$2\frac{\ddot{a}}{a}+\left(\frac{\dot{a}}{a}\right)^2+\frac{1}{a^2}=-8\pi p_{\text{eff}} \tag{8}$$

we see from Eqs. (3) and (4) that we may view

$$\rho_{\text{eff}}=-\frac{3}{8\pi}\frac{\dot{\sigma}\dot{a}}{\sigma a} \tag{9}$$

as an effective energy density in the 3-dimensional spatial sector due to the presence of the extra spatial dimension and

$$p_{\text{eff}}=\frac{1}{4\pi}\frac{\dot{\sigma}\dot{a}}{\sigma a}+\frac{1}{8\pi}\frac{\ddot{\sigma}}{\sigma} \tag{10}$$

as an effective pressure. From the $G_{44}=0$ equation (5), we obtain the equation of state

$$p_{\text{eff}}=\tfrac{1}{3}\rho_{\text{eff}} \tag{11}$$

so that, due to the presence of the extra dimension, there is an effective radiation content in the 3-dimensional spatial sector as in the case considered by Chodos and Detweiler.

The solution to (7) and (8) for a radiation-filled universe is well known and is given in parametrized form in Ref. 12. As a function of $t$

$$a(t)=\sqrt{a_{\max}^2-(t-a_{\max})^2} \tag{12}$$

where $a_{\max}$ is the value of $a(t)$ at the classical turning point $(t = a_{\max})$, corresponding to the maximum of expansion for the 3-volume. We also have

$$\sigma(t) = -C\frac{a_{\max} - t}{\sqrt{a_{\max}^2 - (t - a_{\max})^2}} \tag{13}$$

where $C$ is an arbitrary positive constant. The sign of $\sigma(t)$ in (13) is of no physical consequence since the circumference of the field dimension is $2\pi\,|\sigma(t)|$.

There is some difficulty in interpreting the higher-dimensional geometry of the solution. The usually unstated requirement on interpretation is that the field dimension is segregated from the other dimensions of the geometry. By this we mean there is a restriction on the coordinate transformations allowed when considering the solution. In some cases this restriction can be imposed with a degree of naturalness, as in the case considered by Chodos and Detweiler. In our case, however, we feel that it cannot be.

In the Chodos–Detweiler solution the 5-cosmology contains a congruence of preferred timelike 4-surfaces, namely the $\mathbb{R}^3 \times \mathbb{R}$ surfaces which represent the evolving 3-spaces. This congruence is parametrized by the value of $x^4$. This separation is essential for the interpretation: no interpretation is attempted, for instance, for observers who move across the $x^4 = \text{constant}$ line. Such an observer would have a time coordinate which is a combination of the coordinates $t$ and $x^4$. How does one age in the field direction? The congruence is an essential geometric construct tied to the interpretation.

In our case the construction has a congruence of $\mathbb{S}^3 \times \mathbb{R}$ 4-surfaces, the individual cosmologies parametrized by $x^4$. As in the Chodos–Detweiler solution, the 3-space evolves from an initial vanishing 3-volume singularity like a radiation-dominated, homogeneous, isotropic solution. The radiation source arises from the evolving 5th dimension, which again begins as a metrically infinite circle and decreases as the evolution proceeds. In our case, however, the field metric circumference reaches zero in a *finite* cosmic time, $t = a_{\max}$. While this is not a geometrical singularity of the 5-space, it is a singularity of the congruence upon which our interpretation is based. Continuing the model by simply applying the analytic forms (12) and (13) neglects the causal instability of posing data at the past vertex of a causal cone (in our case the vertex is a 3-sphere and a 4-dimensional Cauchy surface would be required) and the topological inconsistency of continuing the field dimension from a point to a circle. The only consistent conclusion is that the 4-space must end at that instant because the whole description has failed. The end of the model occurs at the instant of maximum expansion of the 3-spheres.

Figure 2 exhibits some features of our 5-dimensional cosmology. In this diagram the upward-directed straight lines represent the histories of 3-spheres, each with a particular value of $x^4$. Although the solution is homogeneous in $x^4$, only the specified range $[0, 2\pi)$ is physically realized. The hyperbolic curves represent 4-spaces of constant time (constant proper time from the initial singularity). At $t = a_{\max}$, these constant proper time

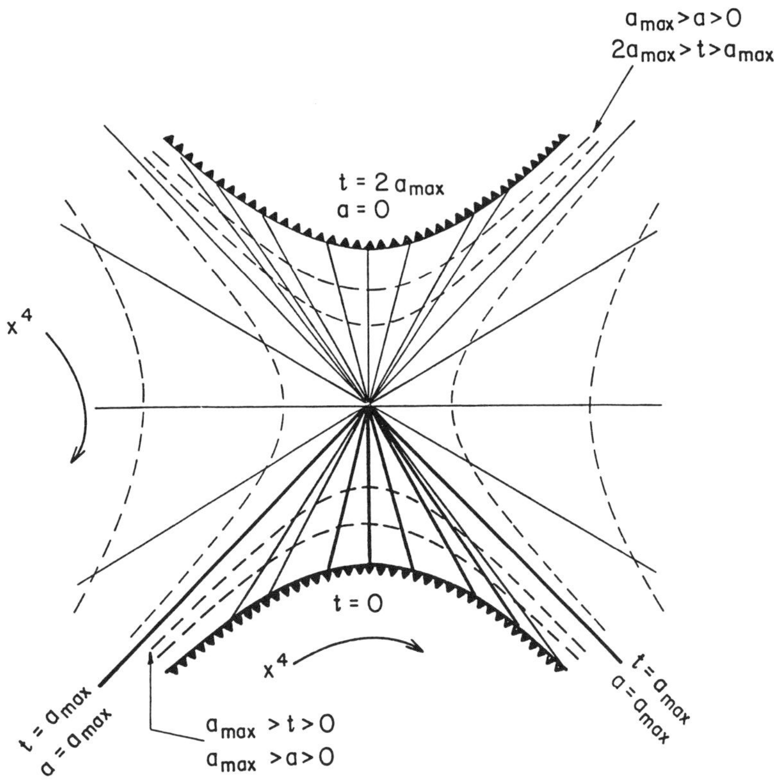


Fig. 2. A 5-vacuum $\mathbb{R} \times \mathbb{S}^3 \times \mathbb{S}^1$ cosmology. Because the model lives only a finite amount of time, we have a more faithful diagram of the dynamics than in the Chodos–Detweiler case. Again the range $x^4 \varepsilon [0, 2\pi)$ is drawn with heavier timelike lines and the downward-curving hyperbolae are surfaces of constant time and constant expansion factors. Notice that these surfaces approach a null limit as $t \to a_{\max}$. At that instant all the 3-spaces, which are up to then separated in the $x^4$ direction, pass through the same 3-sphere. The circle in the $x^4$ direction has shrunk and has degenerated into a point. This is a singularity in the congruence providing the interpretation of the model. We argue that this event is sufficiently disruptive that the cosmology cannot be continued into the upper (more lightly drawn) part of the diagram. Notice that complete data cannot be set at the lower central 3-sphere for *any* part of subsequent evolution in the upper part of the diagram. A 4-dimensional Cauchy surface would be required.

slices become null. The lines corresponding to different values of $x^4$ all pass through the same 3-sphere, represented as the central point. The structure attempting to describe the field label $x^4$ separately from the space-time variables has broken down.

What if one considers nonvacuum 5-space solutions? We generally expect that congruence type singularities such as those just described will become geometrically singular if matter is introduced because an infinite density (hence infinite curvature) arises where the world lines intersect. Hence, if 5-space "dust" is included, we would anticipate a real singularity at the 3-sphere of intersection.

In fact, we can solve the 5-dimensional Einstein equations with a (5-dimensional) dust source (4)–(6) to find

$$a(t)=\sqrt{a_{\max}^2-(t-a_{\max})^2} \tag{14}$$

$$\sigma(t)=-C\frac{t_0-t}{\sqrt{a_{\max}^2-(t-a_{\max})^2}} \tag{15}$$

where $C$ is an arbitrary positive constant, $a_{\max}$ is the time when the 3-sphere reaches its maximum volume, and $t_0$ is the time when the field dimension has zero circumference. In terms of $\sigma(t)$ and $a(t)$, the 5-space dust $\rho(t)$ is

$$\rho(t)=\frac{-3C\varepsilon}{8\pi\sigma a^3} \tag{16}$$

where $\varepsilon \equiv t_0-a_{\max}$. In the limit $\varepsilon \to 0$, we recover the vacuum solution. *From* (16), *it is evident that there is a real geometrical, 5-dimensional curvature singularity at the* 3-*sphere of intersection* $\sigma=0$ (see Fig. 3). For $\varepsilon>0$ $(\rho(t)>0)$, the singularity occurs after the 3-sphere has attained its maximum volume and has started to collapse. For $\varepsilon<0$ $(\rho(t)<0)$, the singularity occurs before the 3-sphere has attained its maximum volume (in 5 dimensions we know of no compelling reason to assume that $\rho(t)>0$). From (5) we still have

$$p_{\text{eff}}=\tfrac{1}{3}\rho_{\text{eff}} \tag{17}$$

but now

$$\rho_{\text{eff}}=\rho-\frac{3\dot{\sigma}\dot{a}}{8\pi\sigma a} \tag{18}$$

which is simply the effective energy density for the 5-dimensional vacuum case modified by the additional 5-dimensional dust source (of course $\sigma(t)$ and $a(t)$ are now slightly different than in the 5-vacuum model).

Notice that from the 4-space viewpoint there is no difference between

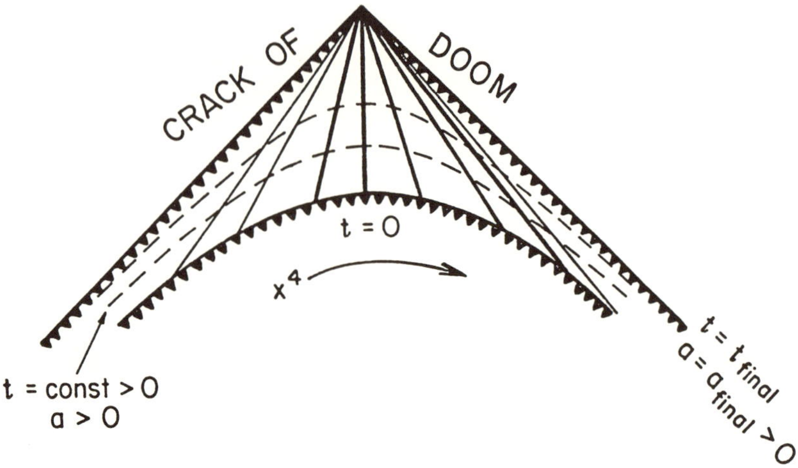


Fig. 3. When the 5-dimensional model contains a 5-dimensional *dust* source, the situation diagrammed in Fig. 2 is further modified. The 4-space still behaves like a radiation-filled Robertson–Walker cosmology, indistinguishable from the 5-vacuum model. As the field dimension shrinks, the density of the 5-dust increases, becoming singular as the fifth-dimension circumference attempts to go through zero. Hence, there is a 5-dimensional curvature singularity at some instant, which may be before or after the maximum of 3-space expansion. This instantaneous curvature singularity is completely unpredictable within the context of the radiation-filled Robertson–Walker cosmology and the universe ends at some finite 3-volume size, $a_{\text{final}}$.

the 5-vacuum and the 5-dust solutions. In both cases the 4-cosmology is a radiation-dominated, closed cosmology, which, however, at some instant in the history of the 4-cosmology, simply stops. In the 5-vacuum case one *could* argue (though we would disagree) that a reinterpretation allows continued 4-space evolution. But the real geometrical singularity of the 5-dust models is a literal crack of doom ending the world instantaneously. So on an ironic note this Kaluza–Klein model provides a bizarre answer to the

> Question: "Do 3-sphere cosmologies always recollapse?"
> Answer: "They should live so long!"

The singularity structure described here in the classical case survives when one quantizes the cosmologies. In this paper we quantize using the canonical ADM formulation, which we now describe.

## 4. THE ADM FORMALISM

In any canonical formalism we begin with a spacelike hypersurface in some manifold on which we specify a metric $h_{ij}(\mathbf{x}, t)$ and an extrinsic curvature $K_{ij}(\mathbf{x}, t)$ ($\mathbf{x}$ and $t$ are the coordinates on the manifold). The metric evolves according to Einstein's equations as we move the hypersurface forward in time. In the ADM formalism this is accomplished by following the lines in the manifold where all of the spatial coordinates are constant. One may consider these as integral curves generated by the "deformation" vector $N^\alpha(\mathbf{x}, t)$ (see Fig. 4).

In terms of $h_{ij}(\mathbf{x}, t)$, $N(\mathbf{x}, t)$ and $N^i(\mathbf{x}, t)$, the metric is

$$ds^2 = -N^2(\mathbf{x}, t)\,dt^2 + h_{ij}(\mathbf{x}, t)(dx^i + N^i(\mathbf{x}, t)\,dt)(dx^j + N^j(\mathbf{x}, t)\,dt) \quad (19)$$

We see that it has been decomposed into components ($N^\alpha(\mathbf{x}, t)$) which depend on the choice of coordinates in the manifold and components

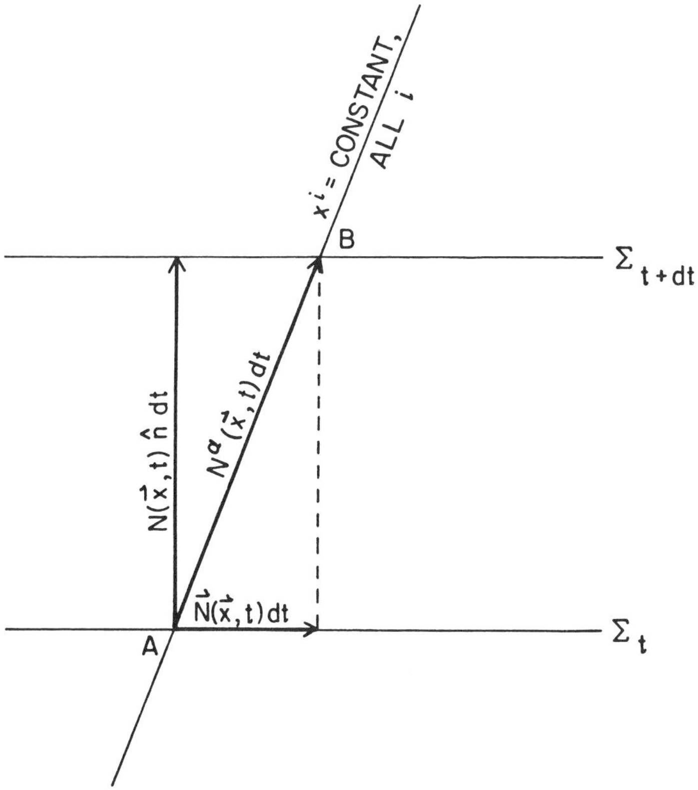


Fig. 4. Here $\Sigma_t$ is a spacelike hypersurface defined by $t$ = constant and $\Sigma_{t+dt}$ is a second spacelike hypersurface defined by $t + dt$ = constant. The point $A$ on $\Sigma_t$ is mapped into the point $B$ on $\Sigma_{t+dt}$ by $N^\alpha(\mathbf{x}, t)$. The components of $N^\alpha(\mathbf{x}, t)$ tangent and normal to the hypersurface, $\Sigma_t$, are $\mathbf{N}(\mathbf{x}, t)$ and $N(\mathbf{x}, t)$, respectively. They depend on the choice of coordinates $(\mathbf{x}, t)$ in the manifold.

$(h_{ij}(\mathbf{x}, t))$ which describe the spatial geometry. In a coordinate system where $N^i(\mathbf{x}, t) = 0$ and $N(\mathbf{x}, t) = N(t)$ (we shall restrict our considerations to this case below)

$$ds^2 = -N^2(t)\,dt^2 + h_{ij}(\mathbf{x}, t)\,dx^i\,dx^j \tag{20}$$

For our vacuum case we take as our action the 5-dimensional Einstein action

$$S = \frac{1}{16\pi}\int d^5x\,\sqrt{-g}\,{}^{(5)}R \tag{21}$$

with $G = c = 1$. In terms of $h_{ij}(t)$ and $N(t)$

$$S = \frac{1}{16\pi}\int dt\,d^4x\left(\pi^{ij}\frac{\partial h_{ij}}{\partial t} - N(t)\,\mathscr{H}\right) \tag{22}$$

where

$$\pi^{ij} = \sqrt{h}\,(K^{ij} - h^{ij}K) \tag{23}$$

$$\mathscr{H} = -\sqrt{h}\,[{}^{(4)}R - h^{-1}(\pi^{ij}\pi_{ij} - \tfrac{1}{3}\pi^2)] \tag{24}$$

are the momenta canonical to $h_{ij}(t)$ and the super Hamiltonian, respectively.[2] Because $N^i(\mathbf{x}, t) = 0$, the extrinsic curvature $K_{ij}(\mathbf{x}, t)$ has a simple expression in terms of $N(t)$ and the metric $h_{ij}(t)$

$$K_{ij}(t) = \frac{1}{2N}\frac{\partial h_{ij}}{\partial t} \tag{25}$$

${}^{(4)}R$ is the intrinsic curvature of the 3-sphere plus extra dimension. $h \equiv \det h_{ij}$ and $\pi \equiv h^{ij}\pi_{ij}$. Variation of the action with respect to $N(t)$ yields the Hamiltonian constraint $\mathscr{H} = 0$, which reflects the invariance of the theory under time reparametrization.

[2] For a $d$-dimensional space-time we would have

$$S = \frac{1}{8\pi\bar{G}}\int dt\,d^{d-1}x\left(\pi^{ij}\frac{\partial h_{ij}}{\partial t} - N(t)\,\mathscr{H}\right)$$

$$\mathscr{H} = -\sqrt{h}\left[{}^{(d-1)}R - h^{-1}\left(\pi^{ij}\pi_{ij} - \left(\frac{1}{d-2}\right)\pi^2\right)\right]$$

where $\bar{G} = GV_D$ ($V_D$ is the volume of the compact $D$-dimensional space).

## 5. THE CLASSICAL SOLUTIONS

We would like to reduce the action to canonical form where the independent degrees of freedom become explicit. In so doing, an intrinsic time coordinate will also emerge (the precise meaning of this will become clear later).

We begin with the action

$$S=\frac{1}{16\pi}\int dt\, d^4x\, \pi^{ij}\frac{\partial h_{ij}}{\partial t} \tag{26}$$

where now some of the $\pi^{ij}$ are not independent but are determined by the constraint $\mathscr{H}=0$. Since the metric is only a function of the time, $t$, the spatial integration can be carried out immediately by referring the metric components to the basis $\{\omega^i\}$. We define $\int \omega^1 \wedge \omega^2 \wedge \omega^3 \wedge \omega^4=1$, so that

$$S=\frac{1}{16\pi}\int \pi^{ij}\, dh_{ij} \tag{27}$$

Here the canonical momenta $\pi^{ij}$ are related to the original momenta by this coordinate transformation. If we now write $h_{ij}(t)$ as

$$h_{ij}=e^{2\Omega}(e^{-2\beta})_{ij} \tag{28}$$

with

$$\beta_{ij}=\operatorname{diag}(\beta,\beta,\beta,-3\beta) \tag{29}$$

and appropriately transform the momenta, (27) becomes

$$S=\frac{1}{8\pi}\int \pi^i{}_i\, d\Omega-\pi^i{}_k\, d\beta_{ki} \tag{30}$$

In terms of $\sigma$ and $a$, $\Omega$ and $\beta$ are $\frac{1}{4}\log \sigma a^3$ and $\frac{1}{4}\log \sigma/a$, respectively. It is evident that $\beta$ will emerge as the intrinsic time variable, and with this in mind we write the momenta as

$$\pi^i{}_k=8\pi p^i{}_k+\tfrac{1}{4}\delta^i{}_k\pi^l{}_l \tag{31}$$

where $\pi^l{}_l\equiv 8\pi p_\Omega$ and

$$p^i{}_k=\tfrac{1}{12}\operatorname{diag}(H,H,H,-3H) \tag{32}$$

(30) can now be written in canonical form

$$S=\int p_\Omega\, d\Omega-H\, d\beta \tag{33}$$

with $\beta$ the intrinsic time coordinate (intrinsic in the sense that it is part of the spatial metric ($h_{ij}$) determining the geometry of the $\beta$ = constant hypersurface), $H$ the Hamiltonian, $\Omega$ the coordinate, and $p_\Omega$ its canonical momentum. From the Hamiltonian constraint we then obtain

$$H^2 = p_\Omega^2 + \frac{3}{16\pi^2} h\,{}^{(4)}R \tag{34}$$

The spatial curvature, ${}^{(4)}R$, is easily determined and is the curvature of the 3-sphere, $6/a^2$; the flat, extra spatial dimension does not contribute. In terms of $\Omega$ an $\beta$

$$H^2 = p_\Omega^2 + \frac{9}{8\pi^2} e^{6\Omega + 2\beta} \tag{35}$$

We can simplify the Hamiltonian and thus the equations of motion by making the Hamiltonian cyclic in the time coordinate, $\beta$. This has the added advantage that the Hamiltonian then becomes a constant of the motion. Our choice of transformation will be dictated by the structure of the super Hamiltonian (24). With $p_\beta \equiv -H$

$$\sqrt{h}\,\mathcal{H} = \frac{1}{2}\left(-p_\beta^2 + p_\Omega^2 + \frac{9}{8\pi^2} e^{6\Omega + 2\beta}\right) \tag{36}$$

which can be written as

$$\sqrt{h}\,\mathcal{H} = \frac{1}{2}\left(G^{AB} p_A p_B + \frac{9}{8\pi^2} e^{6\Omega + 2\beta}\right) \tag{37}$$

with

$$G^{AB} = \mathrm{diag}(-1, 1) \tag{38}$$

Here $p_A = p_B$ and $p_B = p_\Omega$. Equation (38) is a Lorentz metric in the minisuperspace ($\beta$, $\Omega$); $G^{AB} p_A p_B$ is then invariant under the Lorentz transformations

$$\Omega' = \gamma(\Omega - v\beta) \tag{39}$$

$$\beta' = \gamma(\beta - v\Omega) \tag{40}$$

with $\gamma = 1/\sqrt{1 - v^2}$. For $v = -1/3$

$$\sqrt{h}\,\mathcal{H} = \frac{1}{2}\left(-p_\beta^2 + p_\Omega^2 + \frac{9}{8\pi^2} e^{2\sqrt{8}\,\Omega}\right) \tag{41}$$

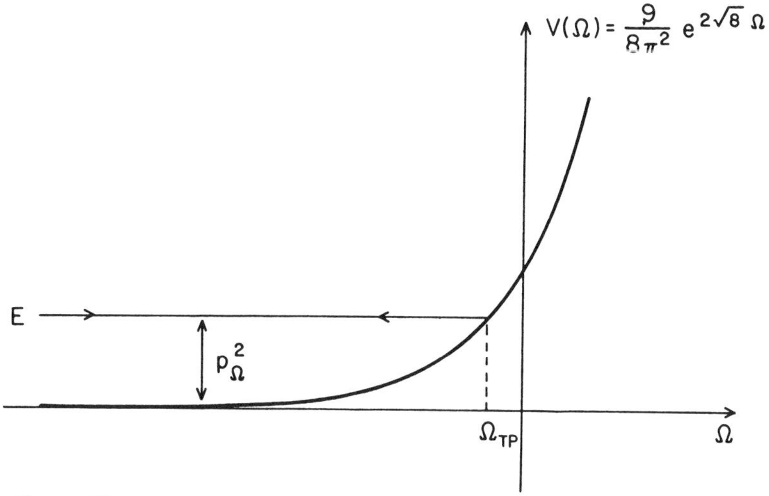


Fig. 5. The universe point will follow the same path as a particle of energy $E$ in the potential $V(\Omega)=(9/8\pi^2)\exp(2\sqrt{8}\,\Omega)$. It moves from the asymptotic region $\Omega \to -\infty$ toward the classical turning point corresponding to the maximum of expansion for the 4-volume. After the turning point it moves in the opposite manner. The momentum $p_\Omega$ for the motion is $\pm[E-(9/8\pi^2)\exp(2\sqrt{8}\,\Omega)]^{1/2}$, which changes sign at the turning point, the positive (negative) sign corresponding to the incoming (outgoing) branch. The value of $\Omega$ at the turning point is $\Omega_{\mathrm{TP}}=(1/2\sqrt{8})\log(8\pi^2 E/9)$.

where we have dropped primes. Then

$$H^2 = p_\Omega^2 + \frac{9}{8\pi^2} e^{2\sqrt{8}\,\Omega} \tag{42}$$

In terms of $\sigma(t)$ and $a(t)$, $\Omega$ and $\beta$ are now $(1/\sqrt{8})\log \sigma a^2$ and $(1/\sqrt{8})\log \sigma$, respectively. It is evident that (42) is the Hamiltonian of a particle in a 1-dimensional exponential potential, $V(\Omega)=(9/8\pi^2)\,e^{2\sqrt{8}\,\Omega}$ (see Fig. 5).

The classical solution is found by solving the canonical equations of motion (below, $\mathscr{H}$ is defined to be expression (41))

$$\begin{aligned}
&\frac{d\beta}{d\lambda}=\frac{\partial\mathscr{H}}{\partial p_\beta}, &&\frac{d\Omega}{d\lambda}=\frac{\partial\mathscr{H}}{\partial p_\Omega}\\
&\frac{dp_\beta}{d\lambda}=-\frac{\partial\mathscr{H}}{\partial\beta}, &&\frac{dp_\Omega}{d\lambda}=-\frac{\partial\mathscr{H}}{\partial\Omega}\\
&\frac{d\mathscr{H}}{d\lambda}=\frac{\partial\mathscr{H}}{\partial\lambda}
\end{aligned} \tag{43}$$

where we have introduced the time parameter, $\lambda$, in the following way:

$$\frac{d\beta}{d\lambda} \equiv H \tag{44}$$

The solution is

$$\begin{aligned} \Omega(\lambda) &= \frac{1}{\sqrt{8}} \log\left(2\left(\frac{8\pi^2 E}{9}\right)^{1/2} \frac{e^{\sqrt{8E}\,|\lambda|}}{1+e^{2\sqrt{8E}\,|\lambda|}}\right) \\ \beta(\lambda) &= H\lambda + \beta_0 \end{aligned} \tag{45}$$

In the asymptotic regions $\lambda \to \pm\infty$, we have

$$\begin{aligned} \Omega \pm \beta = 0 \qquad (H > 0) \\ \Omega \mp \beta = 0 \qquad (H < 0) \end{aligned} \tag{46}$$

The behavior of the solution in terms of $\sigma(t)$ and $a(t)$ is the following: for $H > 0$, $a(t)$ decreases monotonically from a maximum value, $a_{\max}(\beta_0, E) = (32\pi^2 E/9)^{1/4}\, e^{-\sqrt{2}\,\beta_0}$, to zero while $\sigma(t)$ increases monotonically from zero to positive infinity; for $H < 0$, we obtain the time-reversed motion. The $H < 0$ solution agrees with the solution we obtained previously by solving the Einstein equations directly. As a model for reality the latter case is more appropriate. We also note that during the evolution the 4-dimensional volume, $\sigma a^3$, has a maximum, which is attained at the classical turning point (see Fig. 5).

If we now include a 5-dimensional dust source, the super Hamiltonian (24) becomes[8]

$$\mathscr{H}' = \mathscr{H} - 16\pi\mu \tag{47}$$

where $\mu$ is an arbitrary constant. Then

$$H^2 = p_\Omega^2 + \frac{9}{8\pi^2}\, e^{6\Omega + 2\beta} + \frac{3\mu}{\pi}\, e^{4\Omega} \tag{48}$$

where we have dropped primes. We can utilize the solution (14) and (15) to determine which terms in (48) are dominant near the initial and final singularities where we are interested in quantizing the cosmology. One finds that

$$\frac{(9/8\pi^2)\, e^{6\Omega + 2\beta}}{(3\mu/\pi)\, e^{4\Omega}} = \frac{3C}{8\pi\mu}\,(t_0 - t) \tag{49}$$

As $t \to t_0$ the dust term is dominant, so

$$H^2 \simeq p_\Omega^2 + \frac{3\mu}{\pi} e^{4\Omega} \tag{50}$$

Furthermore, we restrict ourselves to the case where $Ct_0/\mu \gg 1$. Then as $t \to 0$ the curvature term dominates, so that

$$H^2 \simeq p_\Omega^2 + \frac{9}{8\pi^2} e^{6\Omega + 2\beta} \tag{51}$$

Near the initial singularity the Hamiltonian is the same as in the vacuum model and the solution in that region is therefore given by (45). Near the final singularity the canonical equations of motion (43) can be readily solved with the Hamiltonian (50), giving

$$\begin{aligned} \Omega(\lambda) &= \frac{1}{2} \log \left( 2 \sqrt{\frac{E}{3\mu}} \frac{e^{2\sqrt{E}|\lambda|}}{1 + e^{4\sqrt{E}|\lambda|}} \right) \\ \beta(\lambda) &= H\lambda + \beta_0 \end{aligned} \tag{52}$$

where $E$ has the same interpretation as in the vacuum case. Once again [see (46)], for $\lambda \to +\infty$ we have

$$\begin{aligned} \Omega + \beta = 0 \qquad (H > 0) \\ \Omega - \beta = 0 \qquad (H < 0) \end{aligned} \tag{53}$$

## 6. QUANTIZATION

### 6.1. The 5-Vacuum Case

We quantize our models using a Klein–Gordon quantization. We make the substitutions

$$H \to i \frac{\partial}{\partial \beta}, \qquad p_\Omega = -i \frac{\partial}{\partial \Omega} \tag{54}$$

in (42), which yields the wave equation

$$-\frac{\partial^2 \psi}{\partial \beta^2} = -\frac{\partial^2 \psi}{\partial \Omega^2} + \frac{9}{8\pi^2} e^{2\sqrt{8}\,\Omega} \psi \tag{55}$$

for $\psi = \psi(\beta, \Omega)$. Since the potential term does not depend on $\beta$, we can write the wave function as

$$\psi(\beta, \Omega) = e^{-ip_\beta \beta} f(\Omega) \tag{56}$$

If we substitute (56) into (55) we obtain an *ordinary* differential equation for $f(\Omega)$, which after a change of variable, $z = (9/8\pi^2)^{1/2} e^{\sqrt{8}\,\Omega}$, becomes a Bessel equation

$$z^2 \frac{d^2 g}{dz^2} + z \frac{dg}{dz} - \frac{1}{8}(z^2 - p_\beta^2)\, g = 0 \tag{57}$$

where $g(z) = f(\Omega(z))$. The exact solution to this equation is

$$g(z) = C_1 I_{ip_\beta/\sqrt{8}}\left(\frac{1}{\sqrt{8}} z\right) + C_2 K_{ip_\beta/\sqrt{8}}\left(\frac{1}{\sqrt{8}} z\right) \tag{58}$$

with $I_{i\nu}(z)$ and $K_{i\nu}(z)$ modified Bessel functions of imaginary order.

For $z \to +\infty$ ($\Omega \to +\infty$), we demand that the wave function be exponentially damped as this region is classically forbidden. For $z \to +\infty$, $I_\nu(z) \sim (1/\sqrt{z})\, e^z$ and $K_\nu(z) \sim (1/\sqrt{z})\, e^{-z}$; we therefore demand that $C_1 = 0$. Then

$$\psi(\beta, \Omega) = C e^{-ip_\beta \beta} K_{ip_\beta/\sqrt{8}}\left(\frac{3}{8\pi} e^{\sqrt{8}\,\Omega}\right) \tag{59}$$

To study the small argument behavior of the wave function ($\Omega \to -\infty$, which is the region near the initial and final singularities), we express $K_{i\nu}(z)$ as

$$K_{i\nu}(z) = \frac{1}{\sin i\nu}\left[I_{i\nu}(z) - I_{-i\nu}(z)\right] \tag{60}$$

For $z \to 0$, $I_\nu(z) \sim (1/\Gamma(1+\nu))(z/2)^\nu$ so that

$$K_{i\nu}(x) \sim \frac{1}{\sin i\nu}\left[\frac{1}{\Gamma(1+i\nu)}\left(\frac{z}{2}\right)^{i\nu} - \frac{1}{\Gamma(1-i\nu)}\left(\frac{z}{2}\right)^{-i\nu}\right] \tag{61}$$

In the asymptotic region $\Omega \to -\infty$, our wave function is then (for $H > 0$)

$$\psi(\beta, \Omega) = N(p_\beta)\left[e^{ip_\beta(\Omega - \beta)} - e^{2i\delta(p_\beta)} e^{-ip_\beta(\Omega + \beta)}\right] \tag{62}$$

with

$$N(p_\beta) = C\left(\frac{9}{32\pi^2}\right)^{ip_\beta/2\sqrt{8}} \csc\left(\frac{ip_\beta}{\sqrt{8}}\right) \Gamma^{-1}\left(1 + \frac{ip_\beta}{\sqrt{8}}\right) \tag{63}$$

$$\delta(p_\beta) = \frac{1}{2i}\log\left(\left(\frac{9}{32\pi^2}\right)^{-ip_\beta/\sqrt{8}} \frac{\Gamma(1 + ip_\beta/\sqrt{8})}{\Gamma(1 - ip_\beta/\sqrt{8})}\right) \tag{64}$$

The wave function corresponds to an incoming plane wave and an outgoing phase-shifted plane wave. This is precisely the form we would expect it to have in a region where the motion is analogous to the motion of a free particle.

The most general solution to (55) that obeys the boundary conditions is a superposition of wave functions of the type (59)

$$\psi(\beta, \Omega) = \int dp_\beta\, b(p_\beta)\, e^{-ip_\beta \beta} K_{ip_\beta/\sqrt{8}}\left(\frac{3}{8\pi} e^{\sqrt{8}\,\Omega}\right) \tag{65}$$

which has the form of a superposition of stationary states. We can construct a localized wave function moving on the average with the classical path by making an appropriate choice for $b(p_\beta)$. For example, with

$$b(p_\beta)\, N(p_\beta) = e^{-(p_\beta - \bar{p}_\beta)^2 \Delta^2/4} \tag{66}$$

where $\bar{p}_\beta = E$, the wave function in the asymptotic region corresponds, for $H > 0$, to an incoming wave packet centered about the classical path $\Omega - \beta = 0$ [see (46)] with width $\Delta$, moving on the average with momentum $\bar{p}_\beta$, and to an outgoing wave packet centered about the path $\Omega + \beta = 2\delta(\bar{p}_\beta)/\bar{p}_\beta$ with width $\Delta$, moving with momentum $-\bar{p}_\beta$. For $H < 0$, the incoming packet moves along the path $\Omega + \beta = 0$ [see (46)] and the outgoing packet along the path $\Omega - \beta = 2\delta(\bar{p}_\beta)/\bar{p}_\beta$. (For $H < 0$, $N(p_\beta)$ and $\delta(p_\beta)$ are given by (63) and (64) with $p_\beta \to -p_\beta$.) Furthermore, of most importance is the fact that the asymptotic region $\Omega \to -\infty$ is nondispersive, so the classical singularity structure discussed earlier is unchanged by quantization. The wave packet simply proceeds unaltered and unimpaired toward the final singularity, following the classical solution path.

### 6.2. The 5-Dust Case

To quantize the 5-dust model we use (51) near the initial singularity and (50) near the final singularity. Since these approximations are valid only in the asymptotic region $\Omega \to -\infty$, we cannot impose boundary conditions on the wave function at $\Omega \to +\infty$ as we did in obtaining (59). We can, however, demand that near the initial singularity the wave function correspond to an incoming plane wave and near the final singularity that it correspond to an outgoing plane wave. The wave function near the initial singularity is simply given by the 5-vacuum wave function (58), since in this region the dust term may be neglected. If the wave function here is to correspond to an incoming plane wave, we must demand (for $H > 0$) that $C_2 = 0$, since $K_{i\nu}(z)$ contains, for small arguments, both incoming and outgoing plane wave components. (For $H<0$, we must begin with a solution of the form (69) before applying this procedure.) Then

$$\psi(\beta, \Omega)=n(p_\beta)\, e^{ip_\beta(\Omega-\beta)} \tag{67}$$

where $n(p_\beta)$ is determined from the asymptotic form of $I_{i\nu}(z)$ for small arguments. To obtain the wave function near the final singularity we make the substitutions (54) in (50) to obtain the wave equation

$$-\frac{\partial^2\psi}{\partial\beta^2}=-\frac{\partial^2\psi}{\partial\Omega^2}+\frac{3\mu}{\pi}\,e^{4\Omega}\psi \tag{68}$$

Proceeding as we did before we obtain the solution

$$\psi(\beta, \Omega)=e^{-ip_\beta\beta}C_1 I_{-ip_\beta/2}\left(\sqrt{\frac{3\mu}{4\pi}}\,e^{2\Omega}\right)+C_2 K_{ip_\beta/2}\left(\sqrt{\frac{3\mu}{4\pi}}\,e^{2\Omega}\right) \tag{69}$$

In the asymptotic region near the final singularity ($\Omega\rightarrow-\infty$), we again demand that $C_2=0$. (For $H<0$, we must start with a solution of the form (58) before doing this.) Then

$$\psi(\beta, \Omega)=m(p_\beta)\, e^{-ip_\beta(\Omega+\beta)} \tag{70}$$

where $m(p_\beta)$ is determined as before. We proceed, as in the 5-vacuum case, to construct localized quantum states on the average following the classical solution path. *Once again the asymptotic region near the initial and final singularities is nondispersive, so the classical 5-dust singularity structure carries over into the quantized cosmology.*

## 7. THE KANTOWSKI–SACHS UNIVERSE

The evolution for the 5-vacuum cosmology is very similar to that of the vacuum Kantowski–Sachs 4-dimensional cosmology, which begins in a state with vanishing 2-sphere volume and ends up, after a finite cosmic time has elapsed, in a state with a finite 2-sphere volume. For the Kantowski–Sachs cosmology, the third spatial dimension has the line topology and evolves from an infinite length to zero length.

The quantized cosmologies are also very similar. In the ADM formalism the Kantowski–Sachs metric

$$ds^2=-dt^2+h_{ij}(t)\,dx^i\,dx^j \tag{71}$$

with $dx^1=d\theta$, $dx^2=\sin\theta\, d\phi$, $dx^3=dz$ and

$$h_{ij}(t)=\operatorname{diag}(a^2(t), a^2(t), \sigma^2(t)) \tag{72}$$

is written as

$$h_{ij} = e^{2\Omega}(e^{-2\beta})_{ij} \tag{73}$$

with

$$\beta_{ij} = \text{diag}(\beta, \beta, -2\beta) \tag{74}$$

Conjugate to the $\beta_{ij}$ are

$$p^i{}_k = \tfrac{1}{6}\text{diag}(H, H, -2H) \tag{75}$$

As in the 5-vacuum model the problem can be reduced to the motion of a particle in a 1-dimensional exponential potential

$$H^2 = p_\Omega^2 + \frac{3}{16\pi^2} e^{2\sqrt{3}\,\Omega} \tag{76}$$

with solutions

$$\begin{aligned} \Omega(\lambda) &= \frac{1}{\sqrt{3}} \log\left(2\left(\frac{16\pi^2 E}{3}\right)^{1/2} \frac{e^{\sqrt{3E}\,|\lambda|}}{1 + e^{2\sqrt{3E}\,|\lambda|}}\right) \\ \beta(\lambda) &= H\lambda + \beta_0 \end{aligned} \tag{77}$$

Once again, as $\lambda \to \pm\infty$, (46) hold. The wave equation

$$-\frac{\partial^2 \psi}{\partial \beta^2} = -\frac{\partial^2 \psi}{\partial \Omega^2} + \frac{3}{16\pi^2} e^{2\sqrt{3}\,\Omega} \psi \tag{78}$$

has the solution

$$\psi(\beta, \Omega) = Ce^{-ip_\beta \beta} K_{ip_\beta/\sqrt{3}}\left(\frac{1}{4\pi} e^{\sqrt{3}\,\Omega}\right) \tag{79}$$

which in the asymptotic region $\Omega \to -\infty$ behaves as an incoming plane wave and an outgoing phase-shifted plane wave. This asymptotic region is nondispersive, so the wave packet solutions are unaltered and the classical singularity structure is preserved.

So far we have seen that the quantized Kantowski–Sachs cosmology and the quantized 5-vacuum cosmology considered in this paper are very similar. But the two do differ in one important respect. Both quantum states behave most classically in the asymptotic region $\Omega \to -\infty$, and most quantum mechanically in the region near the maximum of volume (3-volume for Kantowski–Sachs, 4-volume, i.e., including the field dimension, here; this occurs at the classical turning point of the potential problem). This just reflects the usual behavior of a wave packet scattering off of a potential. For the Kantowski–Sachs universe the region near the maximum of expansion corresponds to the present, as this is where the universe is 3-dimensional. We are thus led to the odd conclusion that this universe would now behave most quantum mechanically. While the location of the present in the Kaluza–Klein case depends on the choice of $\beta_0$ and $E$, it is not *necessary* that it be in the region near the maximum of 4-space expansion (the classical turning point of the potential problem). In fact, we could choose $\beta_0$ and $E$ so that the present would correspond to the asymptotic region $\Omega \to -\infty$ of the outgoing branch where the wave function behaves most classically. Thus, in the context of Kaluza–Klein cosmologies, a very peculiar feature of some of the first quantum cosmological models has been removed.

## 8. CONCLUSION

We saw that for both the 5-vacuum and 5-dust cosmologies with the topology $\mathbb{S}^3 \times \mathbb{S}^1$, the evolution of the 3-space ended at some finite cosmic time due to a singularity in the 5th dimension. As a result, the 3-space did not recollapse in the normal manner. In quantizing our models, we saw that this singularity structure is preserved. Both problems were reducible to particle motion in an exponential potential, the region near the singularity corresponding to the "free" nondispersive region of the potential. Localized quantum states moving on the average with the classical solution path proceed unaltered and unimpaired toward the final singularity.

Some of the successes in the recent schemes for unification can be attributed to the application of the Kaluza–Klein idea to particle theory. From a particle-physics point of view, it is desirable to have a static internal space, since the gauge couplings are related to the circumferences of the internal compact dimensions.(13) Randjbar-Daemi, Salam and Strathdee(14) considered a universe with three-sphere spatial sections and a $k$-sphere as the internal compact space. With the stress-energy arising from a one-loop thermal term, corresponding to a gas of noninteracting scalar particles, they obtained a solution with a static internal space and a three-space evolving as a radiation-dominated Robertson–Walker universe. Since the singularities we presented in this paper were crucial to the issue of the recollapse of the closed three-spaces, in a more realistic model with a static internal space the closed three-space may evolve in the usual manner.

## ACKNOWLEDGMENTS

It is a pleasure to thank J. A. Wheeler, B. S. Dewitt, W. Fischler, L. C. Shepley, G. G. Nagao, and K. L. Thompson for discussions on this subject.

One of us (AM) would like to thank his wife, Mary Ellen, for proofreading the manuscript. This work was supported in part by NSF grant PHY 8404931.

## REFERENCES

1. S. W. Hawking, *Proc. R. Soc. London A* **300**, 182 (1967); R. Penrose, *Phys. Rev. Lett.* **14**, 57 (1965).
2. C. W. Misner, *Phys. Rev.* **186**, 1319 (1969).
3. Th. Kaluza, Sitzingsber, *Preuss. Akad. Wiss. Phys. Math. Kl.*, 966 (1921); O. Klein, *Z. Phys.* **37**, 895 (1926).
4. X. M. Hu and Z. C. Wu, *Phys. Lett. B* **149**, 87 (1984).
5. R. Arnowitt, S. Deser, and C. W. Misner, *Gravitation: An Introduction to Current Research*, L. Witten, ed. (Wiley, New York, 1962), p. 227.
6. R. Kantowski and R. K. Sachs, *J. Math. Phys.* **7**, 443 (1966).
7. M. Ryan, *Hamiltonian Cosmology* (Springer-Verlag, New York, 1972), p. 39.
8. A. Chodos and S. Detweiler, *Phys. Rev. D* **21**, 2167 (1980).
9. D. Lorenz-Petzold, *Phys. Lett. B* **149**, 79 (1984).
10. H. Ishihara, K. Tomita, and H. Nariai, *Prog. Theor. Phys.* **71**, 859 (1984).
11. H. Ishihara, *Prog. Theor. Phys.* **72**, 376 (1984).
12. C. W. Misner, K. S. Thorne, and J. A. Wheeler, *Gravitation* (Freeman, San Francisco, 1973), p. 734.
13. P. Candelas and S. Weinberg, *Nucl. Phys. B* **237**, 397 (1984).
14. S. Randjbar-Daemi, A. Salam, and J. Strathdee, *Phys. Lett.* **135B**, 388 (1984).

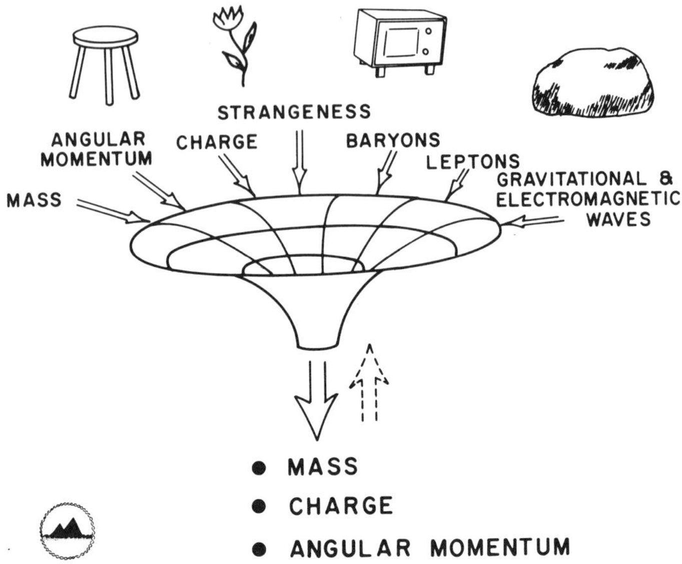
STRANGENESS
ANGULAR MOMENTUM
CHARGE
BARYONS
LEPTONS
MASS
GRAVITATIONAL & ELECTROMAGNETIC WAVES
MASS
CHARGE
ANGULAR MOMENTUM

*Plate 10*

# Gravitation and Spontaneous Symmetry Breaking[1]

**Jacob D. Bekenstein**[2]

*Received June 3, 1985*

*It is pointed out that the Higgs field may be supplanted by an ordinary Klein–Gordon field conformally coupled to the space-time curvature, and with very small, real, rest mass. Provided there is a bare cosmological constant of order of its square mass, this field can induce spontaneous symmetry breaking with a mass scale that can be as large as the Planck–Wheeler mass, but may be smaller. It can thus play a natural role in grand unified theories. In the theory presented here the physical cosmological constant is small, being of order of the squared mass, and can meet observational constraints without having to be cancelled accurately. The physical gravitational constant differs somewhat from the coupling constant in Einstein's equation, and is temperature dependent in the broken symmetry regime. Symmetry restoration occurs at high temperature.*

This paper is dedicated to Professor John Archibald Wheeler on the occasion of his 75th birthday.

John Wheeler is not only one of the luminaries in the fields of general relativity, quantum gravity, and foundations of the quantum theory. He is also a great teacher, not only in the classroom and as an impressive lecturer before wide audiences, but also as an inspiring guide to his graduate students throughout their careers. It is in this latter capacity that he had a deep and lasting influence on my way of thinking and choice of investigation motifs. From Wheeler I learned, among myriad other things, that the phycisist must be ready to delve into widely separate subjects if the problem requires it; no superspecialization for him. Once, upon my saying something about "relativists believe that...," he commented that there is no (should not be, he meant) such a thing as a relativist, only a physicist. It is in this spirit that I permit myself to write on quantum field theory, a sub-

[1] Research supported by the Arnow Chair in Astrophysics.

[2] Physics Department, Ben-Gurion University of the Negev, Beersheva 84105, Israel.

ject I am not normally associated with. I hope John is not sorry for that maxim when he reads this.

Many happy returns, John!

## 1. INTRODUCTION

There can be no question that spontaneous symmetry breaking is one of the key ideas in modern physics. It plays a leading role in modern views about the appearance of spatial order in ferromagnets,[1] convecting fluids,[2] and liquid crystals.[3] It enters into the theories of the transitions to superconductivity[4] and lasing.[5] And it is the principal actor in the theory of the electroweak interaction.[6] The mechanisms for spontaneous symmetry breaking in all these instances involve an order parameter which, by virtue of being subject to a certain "potential," can settle into a value connoting high order for some regime of an external "handle," i.e., temperature. In most instances the order parameter is phenomenological in origin, and we would normally regard it as a convenient way to summarize some more fundamental (or detailed) description of the interactions. For example, the Landau–Ginzburg phenomenological order parameter in superconductivity, the condensate wave function, is a substitute for the detailed description of Cooper-pair quantum dynamics. In relativistic field theory the situation may be different. At least if we are not too speculative, we can regard the order parameter, the Higgs field, as a fundamental entity. It may not stand for anything but itself. If so, relativistic field theory is a good arena (to use a favorite Wheeler image) in which to scrutinize the role of an order parameter in spontaneous symmetry breaking.

In the simplest version of a field theory with broken symmetry, spontaneous symmetry breaking is implemented by coupling gauge and fermion fields to a scalar field $\phi$ endowed with *negative* squared mass and a quartic self-coupling.[7] This is the Higgs field. Even at the classical (tree) level these properties suffice to give the effective potential for $\phi$ minima at nonvanishing values of $\phi$. Thus $\phi$ can acquire a ground state value which is responsible, via the said couplings, for generating fermion and vector boson masses. In this manner is gauge invariance reconciled with the presence of masses in the theory.

This said, one must admit that the Higgs field is a strange beast. One is reminded of a famous remark of Niels Bohr's which we might paraphrase as "if upon hearing about the Higgs field, it does not strike you as crazy, you have not understood the subject." Indeed there are odd features here. It is hard to swallow the reversed sign of the squared mass. That reversal is made to "save the phenomenon" and seems to have no deeper reason. It is true that the excitations of $\phi$ above its ground state value have positive squared mass.[8] But this seems poor consolation for the odd aspect of the fundamental wave equation if the Higgs field is indeed fundamental. It must, however, be mentioned that quantum effects allow an escape. In Coleman and E. Weinberg's model[9] one-loop corrections coming from the gauge couplings of $\phi$ contribute to the effective potential in such a way as to allow spontaneous symmetry breaking even for vanishing or small positive squared Higgs mass. This model is used in the electroweak theory of S. Weinberg, Salam, and Glashow.[6,10]

The Coleman–Weinberg model retains the quartic self-coupling which I find odd. It is true that quartic self-couplings are not unknown in field theory. For example, in non-Abelian gauge theories the gauge field Lagrangian density has terms quartic in the gauge potentials. But the strengths with which these enter are set by the same gauge coupling constants which govern the gauge couplings of fermions. By contrast the Higgs quartic term is governed by a new coupling constant $\lambda$. There seem to be no fundamental reasons for preferring one value of $\lambda$ over another.

A further problem with the Higgs field is that the vacuum energy density associated with its ground state value has dire consequences in cosmology: it is tens of orders of magnitude too large to be consistent with observational limits on the cosmological constant.[11] This large energy density has been exploited for the construction of inflationary cosmological models for the infant universe.[12] In these the small observed cosmological constant is explained by assuming the vacuum energy is essentially cancelled out by a "bare" cosmological constant. Thus the theory relies on fine tuning of the parameters (at the level of $10^{-120}$ or better), not a natural procedure. Although the literature recalls many attempts to ameliorate this problem,[13] the large vacuum energy engendered by the Higgs field is definitely a detracting feature.

John Wheeler has ever been a faithful promoter of Einstein's view that all interactions in nature are geometry at a deep level. Just as general relativity and its experimental success established this by the 1960's for gravitation, gauge theories have done the equivalent for the electroweak interaction, and apparently also for the strong interaction, in the 1970's. Gauge fields relate to the geometries of internal parameter spaces just as curvature relates to the geometry of spacetime. Could it not be that just as the coupling to gauge fields in Coleman and Weinberg's strategy succeeds in triggering spontaneous symmetry breaking without invoking "tachionic" masses, so may a coupling to curvature?

In what follows I describe one possible way to do this. It supplants the villainous Higgs field by an ordinary Klein–Gordon field, which by virtue of a coupling to curvature, can mimic Higgs behavior without involving "tachionic" mass or quartic self-interaction at the fundamental level. Provided the Klein–Gordon field is light, the observable cosmological constant is small in this approach; there is no cosmological problem. And because gravitation enters fundamentally, this approach leads naturally to a spontaneous symmetry breaking mass scale of the order of the Planck–Wheeler mass ($m_{\rm PW}$). This is not too far above the scale needed in grand unification ($10^{-4}m_{\rm PW}$).

In what follows we work with metric signature $(-+++)$. Our conventions for the curvature tensor are those of Misner, Thorne, and Wheeler.[14] We work in units with $\hbar = c = 1$; for these $G = m_{\rm PW}^{-2}$.

## 2. ACTION AND EQUATIONS

The action of the theory breaks up naturally into three parts: matter, Klein–Gordon field, and gravitation. The matter action takes the form

$$S_m = \int L_m(-g)^{1/2}\, d^4x \tag{1}$$

where $L_m$ is to be a scalar built out of matter fields in the usual way, but with fermion masses replaced by $g\phi$ where $g$ are appropriate coupling constants, and $\phi$ our Klein–Gordon field. Gauge field contributions are left in their customary form without mass terms. The gauge bosons are to acquire mass via the minimal coupling of $\phi$ to them, and this coupling we regard as part of the action for $\phi$. However, in order to keep things simple, we shall exclude scalar-gauge and fermion-gauge couplings in this paper.

A striking feature of fermion and gauge actions is that they are conformally invariant, namely, if the metric is rescaled as $g_{\mu\nu} \to g_{\mu\nu}\Omega^2$ with $\Omega$ an arbitrary function of space-time event, and the fields are rescaled by appropriate powers of $\Omega$ (3/2 for fermions and 0 for gauge fields), then the actions are unchanged.[15] The mentioned invariance is also respected by fermion mass terms of the type we postulated provided $\phi \to \phi\Omega^{-1}$, while minimal coupling of scalar and spinor fields to gauge fields also preserves conformal invariance.[15]

Several deep reasons exist for regarding conformal invariance as a desirable symmetry of fundamental physics.[16–18] Thus we are impelled to try to preserve this invariance in our choice of action for $\phi$. Ignoring gauge couplings we may take

$$S_\phi = -\tfrac{1}{2}\int (\phi_{,\mu}\phi^{\mu}_{;} + m^2\phi^2 + \tfrac{1}{6}R\phi^2)(-g)^{1/2}\, d^4x \tag{2}$$

where $R$ is the Ricci curvature associated with the metric. (In considering conformal transformations we cannot remain in the framework of flat space-time.) Were $m=0$, the action Eq. (2) would be exactly conformally invariant.[16] As we shall see, we cannot do without $m$, but in deference to conformal invariance we shall assume it is very small. In flat space-time Eq. (2) describes a standard massive scalar field with *no* quartic coupling.

We shall here regard gravitation as classical metric generated by quantum fields (semiclassical approach). As a gravitational action we select the standard Einstein–Hilbert action with cosmological constant

$$S_g = \frac{m_{\rm PW}^2}{16\pi} \int (R+2\Lambda)(-g)^{1/2}\, d^4x \tag{3}$$

It is evident that this action is *not* conformally invariant. Gravitation is thus the agent that breaks conformal invariance and introduces a scale into the theory.

Varying $S_\phi + S_m$ with respect to $\phi$ gives us the equation for $\phi$

$$\Box\phi - \tfrac{1}{6}R\phi - m^2\phi = -\delta S_m/\delta\phi \tag{4}$$

where $\Box$ denotes the d'Alembertian. Note that the equation is linear in $\phi$ (we assume $L_m$ contains at most second powers of masses). It is possible to eliminate $R$ from Eq. (4) by resorting to Einstein's equations. These are obtained by varying $S_g + S_\phi + S_m$ with respect to $g^{\mu\nu}$ and replacing the stress energy tensors comming from $S_\phi$ and $S_m$ by their expectation values (or more directly by using effective actions instead of $S_\phi$ and $S_m$). We thus get

$$G_{\mu\nu} = 8\pi m_{\rm PW}^{-2}[\langle T_{\mu\nu}\rangle + \langle\theta_{\mu\nu}(m,\phi)\rangle] + \Lambda g_{\mu\nu} \tag{5}$$

where $\langle\ \rangle$ denotes a quantum expectation value (already regularized), $T_{\mu\nu}$ the stress energy tensor of gauge and fermion fields, and $\theta_{\mu\nu}$ the stress energy tensor for $\phi$[17,19]:

$$\theta_{\mu\nu}(m,\phi) \equiv \phi_{,\mu}\phi_{,\nu} - \tfrac{1}{2}(\phi_{,\sigma}\phi^{\sigma}_{,} + m^2\phi^2)\, g_{\mu\nu} + \tfrac{1}{6}(\Box\phi^2 g_{\mu\nu} - \phi^2_{,\mu;\nu} + G_{\mu\nu}\phi^2) \tag{6}$$

## 3. SPONTANEOUS SYMMETRY BREAKING

For the moment we only need the trace of Eq. (5) which is

$$R = -8\pi m_{\rm PW}^{-2}[\langle T\rangle + \langle\phi\Box\phi - 2m^2\phi^2 - \tfrac{1}{6}R\phi^2\rangle] - 4\Lambda \tag{7}$$

where $\langle T\rangle = g^{\mu\nu}\langle T_{\mu\nu}\rangle$. We may eliminate several terms from Eq. (7) by calculating $\langle \phi\Box\phi\rangle$ with the help of Eq. (4):

$$\langle \phi\Box\phi\rangle = (m^2 + \tfrac{1}{6}R)\langle \phi^2\rangle - \langle \phi\, \delta S_m/\delta\phi\rangle \tag{8}$$

Classically, we may infer from substitution of Eq. (8) into Eq. (7) and comparison with Eq. (5) that in the limit $m \to 0$, and in the absence of matter, the trace of $\theta_{\mu\nu}$ vanishes. In quantum theory this is no longer true; one-loop effects contribute an "anomalous" trace to $\theta_{\mu\nu}$, a linear combination of $R^2$, $R_{\mu\nu}R^{\mu\nu}$, and $\Box R$.[20] Such terms must thus be included in the square brackets in Eq. (7); however, since similar terms will be contributed by $\langle T\rangle$ itself (matter also has a trace anomaly—see below), we do not exhibit them explicitly.

As shown in the Appendix, it follows from conformal invariance alone that, classically,[15]

$$T = \phi\, \delta S_m/\delta\phi \tag{9}$$

When fermion one-loop contributions to $\langle T\rangle$ are taken into account, their trace anomaly modifies this equation to (see Appendix)

$$\langle T\rangle = \langle \phi\, \delta S_m/\delta\phi\rangle + O(R^2) \tag{10}$$

where $O(R)$ is short for the just mentioned curvature invariants. When this $\langle T\rangle$ is substituted into Eq. (7), the curvature invariants are multiplied by $m_{\mathrm{PW}}^{-2}$.

So long as the curvature of space-time is small in Planck–Wheeler units, ($Rm_{\mathrm{PW}}^{-2} \ll 1$), the anomalies can evidently be neglected as compared with $R$ in the left-hand side of Eq. (7). (But evidently this assumption is not correct in early cosmology.) Replacing thus $\langle T\rangle$ in Eq. (7) by $\langle \phi\, \delta S_m/\delta\phi\rangle$ and substituting Eq. (8) into Eq. (7) we get, after some cancellations,

$$R \simeq 8\pi m_{\mathrm{PW}}^{-2} m^2 \langle \phi^2\rangle - 4\Lambda \tag{11}$$

If we now replace $R$ in Eq. (4) by this last expression we obtain the equation for the quantum field $\phi$ in the form

$$\Box\phi + \tilde{M}^2\phi = -\delta S_m/\delta\phi \tag{12}$$

with

$$\tilde{M}^2 \equiv -m^2 + \tfrac{2}{3}\Lambda - \tfrac{4}{3}\pi m_{\mathrm{PW}}^{-2} m^2 \langle \phi^2\rangle \tag{13}$$

Evidently the squared mass parameter, $-\tilde{M}^2$, can be negative provided $\Lambda$ is positive and not too small. Thus one of the criteria for Higgs field behavior has been met without $\phi$ being a Higgs field at the fundamental level. It is the conformal coupling of $\phi$ to gravity which is responsible.

There is more in Eq. (12) that meets the eye since $\tilde{M}^2$ is state dependent: it depends on $\langle\phi^2\rangle$. To disclose the full Higgs behavior, let us take the expectation value of Eq. (12). With the notation $\sigma = \langle\phi\rangle$ and $\eta = \phi - \sigma$ we have $\langle\phi^2\rangle = \sigma^2 + \langle\eta^2\rangle$ so that

$$\Box\sigma + M^2\sigma - \lambda\sigma^3 = -\langle\delta S_m/\delta\sigma\rangle \tag{14}$$

where

$$M^2 \equiv -m^2 + \tfrac{2}{3}\Lambda - \tfrac{4}{3}\pi m_{\rm PW}^{-2} m^2 \langle\eta^2\rangle \tag{15}$$

$$\lambda \equiv \tfrac{4}{3}\pi m_{\rm PW}^{-2} m^2 \tag{16}$$

We see that $\sigma$ obeys an equation with standard quartic self-coupling, even though no such coupling entered fundamentally. We also note that the coupling strength $\lambda$ is fully determined, and small by assumption.

From Eq. (14) we infer the effective potential for $\sigma$:

$$V(\sigma) = -\tfrac{1}{2}M^2\sigma^2 + \tfrac{1}{4}\lambda\sigma^4 \tag{17}$$

This is the usual Higgs potential. Actually there should be a radiative correction to it coming from scalar loops.[(10)] To one-loop this is proportional to $m^4$ and we shall thus neglect it since the terms in Eq. (17) are of order $m^2 m_{\rm PW}^2$ [see Eq. (17) and Sect. 4]. Having excluded by fiat gauge couplings of $\phi$, we need not consider here radiative corrections to $V$ due to gauge boson loops which are of different scale[(9,10)] (see Sect. 6). The effective potential has two equal minima at

$$\sigma_m = \pm M\lambda^{-1/2} = \pm m_{\rm PW}(\Lambda/(2\pi m^2) - 3/(4\pi) - m_{\rm PW}^{-2}\langle\eta^2\rangle)^{1/2} \tag{18}$$

and so spontaneous symmetry breaking can occur in the usual manner. Thus fermions (and also gauge bosons in a more complete theory) can acquire masses given by $g\sigma_m$.

## 4. TEMPERATURE EFFECTS IN THE GRAVITATIONAL CONSTANT

Equation (18) tells us that a necessary condition for $\sigma_m$ to be real (a must for spontaneous symmetry breaking) is

$$\Lambda > \tfrac{3}{2}m^2 \tag{19}$$

We shall henceforth assume this condition to be satisfied. At the present primitive level of the theory we can supply no deep reasons for this ordering of the parameters. It is also evident that unless the $\langle\eta^2\rangle$ term in Eq. (18) nearly cancels the other two, the mass scale generated (essentially $\sigma_m$) is of order $m_{\text{PW}}$. A mass scale more like that required by grand unified theories will follow if Eq. (19) is a near equality at the $10^{-4}$ level. Of course this requires fine tuning, but of a kind which is a far cry from the ultrahigh precision fine tuning required to get rid of the cosmological constant in standard broken symmetry theories.

We now inquire into the significance of the $\langle\eta^2\rangle$ term. The equation for the quantum field $\eta$ follows by substracting Eq. (14) from Eq. (12). A nice cancellation results from substitution of Eqs. (15), (16), and (18), after which we are left with

$$\Box\eta = -(\delta S_m/\delta\phi - \langle\delta S_m/\delta\sigma\rangle) \tag{20}$$

We see that $\eta$ is a *massless* field. Hence, in a thermal state $\langle\eta^2\rangle$ must take on the value appropriate for massless spinless quanta: $\langle\eta^2\rangle = t^2/12$, where $t$ is the temperature. With this substitution in Eq. (18) we find that

$$|\sigma_m| = \frac{1}{\sqrt{12}}(t_c^2 - t^2)^{1/2}, \qquad t_c \equiv \sqrt{\frac{6}{\pi}}\, m_{\text{PW}} \left(\frac{\Lambda}{m^2} - \frac{3}{2}\right)^{1/2} \tag{21}$$

We see that $t_c$ plays the role of a critical temperature at which the minima of $V$ merge at $\sigma_m = 0$. For $t > t_c$ there is no spontaneous symmetry breaking, namely, the symmetry is restored. As we shall soon see, consistency of the theory requires that $\Lambda$ not much exceed $m^2$. Therefore, $t_c$ is about $m_{\text{PW}}$ or lower. And if $\Lambda$ is tuned to $m^2$ in the manner described above to achieve the grand unification mass scale, then $t_c$ is itself at this scale. Further lowering of $t_c$ should result when finite temperature one-loop corrections to the effective potential are taken into account.[21]

Interesting temperature effects also occur in connection with the physical gravitational constant $G_p$ which is to be distinguished from the coupling constant $G$. To identify $G_p$ we substitute in Einstein's equations (5) $\phi = \sigma_m + \eta$ (assuming we are in the broken symmetry regime), and rewrite them in the form

$$G_{\mu\nu} - \Lambda_p g_{\mu\nu} = 8\pi G_p \langle T_{\mu\nu} + \theta_{\mu\nu}(0, \eta)\rangle \tag{22}$$

with

$$G_p = m_{\text{PW}}^{-2}(1 - 4\pi/3 \cdot m_{\text{PW}}^{-2}\sigma_m^2)^{-1} \tag{23}$$

$$\Lambda_p = [\Lambda - 4\pi m_{\text{PW}}^{-2} m^2(\sigma_m^2 + \langle\eta^2\rangle)]\, G_p m_{\text{PW}}^2 \tag{24}$$

and with $\theta_{\mu\nu}$ defined by Eq. (6).

Alongside the matter stress energy tensor $T_{\mu\nu}$, there enters as source of gravitation the corresponding tensor for a *massless* conformally coupled scalar field $\theta_{\mu\nu}(0, \eta)$. Evidently the quantities $G_p$ and $\Lambda_p$ are the physical gravitational and cosmological constants. The passage from Eq. (5) to Eqs. (22)–(24) is not entirely strightforward, and a pair of ambiguities have to be settled in order to make it. First, one must decide which fraction of $G_{\mu\nu}\phi^2$ in Eq. (6) is to become part of the stress energy tensor in Eq. (22), and which part goes into "renormalizing" $m_{\text{PW}}$. Our specific choice was predicated on the requirement that the stress energy tensor of $\eta$ be that of a *conformally* coupled field, and this dictates the appearance of only the $G_{\mu\nu}\eta^2$ in $\theta_{\mu\nu}$. Next, one has to decide how to split the $m^2\phi^2$ term appearing in Eq. (6) into contributions to stress energy tensor and to $\Lambda_p$. Since by Eq. (21) $\eta$ is massless, we move the entire term in question into $\Lambda_p$.

We see from Eq. (23) that $G_p$ in the broken symmetry regime is larger than $m_{\text{PW}}^{-2}$, and is temperature dependent since $\sigma_m$ is temperature dependent. The possibility of a temperature-dependent gravitational constant in theories with spontaneous symmetry breaking was noticed by Linde,[22] while Davies considered a temperature-dependent gravitational constant in a theory involving a conformally coupled scalar field.[23] Our procedure for identifying $G_p$ is essential to show that the temperature dependence is unavoidable. For example, had we included the full term $G_{\mu\nu}\phi^2$ in $G_p$, the temperature dependence would have cancelled out because in our theory $\sigma_m^2 + \langle\eta^2\rangle$ is strictly constant as shown by Eq. (18).

Substituting Eq. (18) into Eq. (23) and taking into account that $\langle\eta^2\rangle = t^2/12$, we see that $G_p$ starts at

$$G_p(t=0) = \tfrac{3}{2} m_{\text{PW}}^{-2}(3 - \Lambda/m^2)^{-1} \tag{25}$$

decreases monotonically with $t$, reaches a minimum equal to $m_{\text{PW}}^{-2}$ at $t = t_c$, and then stays constant with $t$ in the restored symmetry regime (where $\sigma_m = 0$). It is a matter of experience that $G_p$ is positive, at least at low temperature. In order for this to be true the condition

$$\Lambda < 3m^2 \tag{26}$$

must be satisfied. Comparing this with Eq. (19) we see that $\Lambda$ is limited to a narrow range of values whose scale is set by $m^2$. If $\Lambda$ is tuned to $\frac{3}{2}m^2$ in order to push $t_c$ well below the Planck–Wheeler mass, then $G_p(t=0) \approx m_{\text{PW}}^{-2}$ already.

## 5. THE COSMOLOGICAL CONSTANT

To exhibit the physical cosmological constant, let us substitute Eq. (18) with the thermal $\langle\eta^2\rangle$ into Eq. (24). After some rearrangement we get

$$\Lambda_p = \frac{3}{2}m^2 \frac{1-\Lambda/3m^2}{1-\Lambda/3m^2+\pi t^2/18m_{\mathrm{PW}}^2} \tag{27}$$

It follows from condition (26) that $\Lambda_p$ is itself positive. Evidently, $\Lambda_p$ is also temperature dependent. However, if $\Lambda$ has been tuned to $\frac{3}{2}m^2$, then the term $t^2m_{\mathrm{PW}}^{-2}$ in Eq. (27) is negligible ($t < t_c \ll m_{\mathrm{PW}}$) so that $\Lambda_p$ is essentially temperature independent in the broken symmetry regime. In this case

$$\Lambda_p \approx \tfrac{3}{2}m^2 \tag{28}$$

If no fine tuning is resorted to, we can again recover Eq. (28) at low temperature. In this case $\Lambda_p$ is insensitive to the precise value of $\Lambda$.

The important thing about Eq. (28) is that the scale of $\Lambda_p$ is set by the small scale $m$ and not by the large scale $\sigma_m$. Thus the physical cosmological constant can satisfy the cosmological constraints (this requires $m$/electron mass $< 10^{-27}$) while the mass scale generated by spontaneous symmetry breaking is large, possibly as large as the Planck–Wheeler mass. The advantage over the standard Higgs model is obvious: large elementary particle masses do not imply a gigantic Higgs vacuum energy with cosmological complications. Likewise, a mass scale appropriate to grand unification can be obtained with only a moderate (by the standards of the subject) amount of fine tuning.

In summary, the theory presented here avoids the darker features of the Higgs field, negative square mass and quartic self-coupling, at the expense of coupling the surrogate scalar field to the space-time curvature. This coupling provides a natural avenue for the entrance of the Planck–Wheeler mass into the physics of spontaneous symmetry breaking. This mass sets the natural scale for generated rest masses. Lower masses such as needed in grand unified theories, or the electroweak theory, can be had at the cost of fine tuning one parameter of the theory with accuracies $10^{-4}$ and $10^{-17}$, respectively. This is far short of the precision required in the standard theory. In the present theory, despite the large mass scale which enters, no large cosmological constant appears, and so it does not have to be cancelled with preposterous precision as in the standard theory.

But the present theory also has shortcomings. First, the all important scalar-gauge couplings that would be responsible for generating vector boson masses are not encompassed. Their introduction will raise new issues. The couplings should give rise to a new class of divergent Feynman diagrams containing one gauge boson loop. The counterterm in the renormalization scheme would seem to have to be a $\lambda\phi^4$ term. This means that after renormalization we should be left with a new contribution to the effective quartic coupling of our scalar field, a contribution over and above that in Eq. (16) coming from conformal coupling to gravity. Which contribution predominates? The renormalization contribution is not computable; it is an experimental parameter. If it does not turn out to be small compared to Eq. (16), it will change the discussion radically.

Another product of the neglected scalar-gauge couplings should be a Coleman–Weinberg type correction to the effective potential in Eq. (17). This has not yet been examined. Finite temperature corrections to the effective potential have yet to be calculated by the proven method of effective actions; the makeshift approach to temperature effects employed here can only be a stopgap measure. And the influence of the trace anomaly coming from fermion-gauge couplings[(24)] on the quantum dynamics of $\phi$ has yet to be taken into account.

Finally, it is well to remember that in any application to grand unification the scalar field will be multicomponent. As pointed out by L. Parker,[(25)] it is not clear that the number of distinct quartic coupling parameters generated by the extension of the theory sketched here will suffice to cause spontaneous breaking of the grand symmetry in just the right way—to $SU(3)\times SU(2)\times U(1)$. It may be that the $\lambda$'s associated with renormalization will play a useful role here.

## ACKNOWLEDGMENTS

I thank Leonard Parker for many clarifications, and for pointing out several errors in the previous version of the paper. My colleages David Eichler and Aharon Davidson also helped with their critical comments. Some of the ideas described here were conceived in the stimulating atmosphere of the Aspen Center for Physics in 1983.

## 6. APPENDIX

Here we derive Eq. (9) and describe where the trace anomaly correction in Eq. (10) comes from. We assume the action $S_m$ for fermions and gauge fields is conformally invariant, that is, that it is unchanged under

$$g_{\mu\nu} \to g_{\mu\nu}\Omega^2 \tag{29}$$

$$\psi \to \psi\Omega^{-3/2} \tag{30}$$

$$A \to A \tag{31}$$

where $A$ denotes collectively the gauge potentials while $\psi$ denotes collectively the fermion fields and adjoints. As shown in Ref. 15, $S_m$ is indeed conformally invariant if it is the sum of Dirac and Yang–Mills actions, and provided mass terms are of the form $g\bar{\psi}\psi\phi$, where $\phi$ is our Klein–Gordon field which transforms as $\phi \to \phi\Omega^{-1}$. However, the precise form of $S_m$ is unimportant here.

If we carry out an infinitesimal conformal transformation of the metric and all fields ($\Omega = 1 + \delta\Omega$), the invariance of $S_m$ gives

$$g^{\mu\nu}T_{\mu\nu} - \phi\, \delta S_m/\delta\phi = \tfrac{3}{2}\psi\, \delta S_m/\delta\psi \tag{32}$$

Here $T_{\mu\nu} = -2\delta S_m/\delta g^{\mu\nu}$ and we are taking into account that $\delta g^{\mu\nu} = -2g^{\mu\nu}\,\delta\Omega$, $\delta\phi = -\phi\,\delta\Omega$, and $\delta\psi = -\frac{3}{2}\psi\,\delta\Omega$. No $\delta S_m/\delta A$ appears since $\delta A = 0$ for a conformal transformation. We stress that Eq. (32) is valid classically as well as in the quantum domain.

If we now impose the classical equations of motion $\delta S_m/\delta\psi = 0$ and carry out the trace of $T_{\mu\nu}$, we get precisely Eq. (9). (Note that we cannot impose $\delta S_m/\delta\phi = 0$ since $\phi$ also appears in $S_\phi$.)

Instead we may regard $\phi$ and $\psi$ as quantum fields and take the expectation value of Eq. (32) to get

$$\langle T\rangle - \langle\phi\, \delta S_m/\delta\phi\rangle = \tfrac{3}{2}\langle\psi\, \delta S_m/\delta\psi\rangle \tag{33}$$

which contains a term not found in Eq. (32). Thinking in terms of functional integrals, it is easy to see what kind of contribution it will make. At tree-level the extra term must vanish because the extremal path in the functional integral is precisely that for which $\delta S_m/\delta\psi = 0$. Thus the dominant contribution to the last term in Eq. (33) comes from one-loop order which takes into account paths departing slightly from the classical path $\delta S_m/\delta\psi = 0$.

It is unnecessary to compute the one-loop contribution to the extra term to guess its form. It suffices to take note of one-loop computations of $\langle T\rangle$ for fermion fields.(24,26) This quantity does not vanish even if the fermion mass terms are excluded: there is a "trace anomaly." Part of it comes from gauge couplings,(24) and is thus beyond the scope of this paper. Coupling to metric results, in the massless fermion case, is an anomaly which is a linear combination of $R^2$, $R_{\mu\nu}R^{\mu\nu}$, and $\Box R$.(26) Since in the massless case the second term in the left-hand side of Eq. (33) is absent, we see that the extra term's one-loop piece is identical to that of $\langle T \rangle$, i.e., a linear combination of $R^2$, $R_{\mu\nu}R^{\mu\nu}$, and $\Box R$. This establishes Eq. (10) in the massless case.

In the massive fermion case there could, on dimensional grounds, appear an additional term of the form $m_{\mathrm{F}}^2 R$ in the right-hand side of Eq. (33) ($m_{\mathrm{F}}$ is the fermion's mass; we exclude the possibility of an $m_{\mathrm{F}}^4$ term because it would also appear in flat space-time). This anomaly will also make a negligible contribution compared to the pure $R$ term in Eq. (7) because $m_{\mathrm{F}}/m_{\mathrm{PW}} \ll 1$, and so we have dropped it from Eq. (10).

## REFERENCES


1. L. D. Landau, E. M. Lifshitz, and L. P. Pitaevskii, *Electrodynamics of Continuous Media* (Pergamon, Oxford, 1984), p. 135.
2. M. G. Velarde and C. Normand, *Sci. Am.* **243**, No. 1, 92 (1980).
3. L. D. Landau and E. M. Lifshitz, *Statistical Physics, Part 1* (Pergamon, Oxford, 1980), p. 440.
4. E. M. Lifshitz and L. P. Pitaevskii, *Statistical Physics, Part 2* (Pergamon, Oxford, 1980), p. 178.
5. M. G. Velarde, in *Change and Organization in Complex Systems*, M. Alonso, ed. (Paragon House, New York, 1986).
6. S. Weinberg, *Phys. Rev. Lett.* **19**, 1264 (1967); A. Salam, in *Elementary Particle Theory: Groups and Analyticity*, N. Svartholm, ed. (Almqvist and Wiksell, Stockholm, 1968); S. Glashow, *Nucl. Phys.* **22**, 579 (1961).
7. P. W. Higgs, *Phys. Rev. Lett.* **12**, 132 (1964); F. Englert and R. Brout, *Phys. Rev. Lett.* **13**, 321 (1964); G. S. Guralnik, C. R. Hagen, and T. W. Kibble, *Phys. Rev. Lett.* **13**, 585 (1964).
8. E. S. Abers and B. W. Lee, *Phys. Rep. C* **9**, 1 (1973); A. D. Linde, *Rep. Prog. Phys.* **42**, 389 (1979).
9. S. Coleman and E. Weinberg, *Phys. Rev. D* **7**, 1888 (1973).
10. K. Huang, *Quarks, Leptons and Gauge Fields* (World Scientific, Singapore, 1982).
11. A. D. Linde, *Pis. Red. Zh. Eksp. Teor. Fiz.* **19**, 320 (1974) [*JETP Lett.* **19**, 183 (1974)]; S. A. Bludman and M. A. Ruderman, *Phys. Rev. Lett.* **38**, 255 (1977).
12. A. Guth, *Phys. Rev. D* **23**, 347 (1971); A. Albrecht and P. J. Steinhardt, *Phys. Rev. Lett.* **48**, 120 (1982); A. D. Linde, *Phys. Lett. B* **108**, 389 (1982); R. H. Brandenberger, *Rev. Mod. Phys.* **57**, 1 (1985).
13. A. D. Dolgov, in *The Very Early Universe*, S. Hawking, G. Gibbons, and S. Siklos, eds. (Cambridge University Press, Cambridge, 1983); F. Wilczek, *ibid.*; L. Parker, *Phys. Rev. Lett.* **50**, 1009 (1983); S. W. Hawking, in *Unified Theory of Elementary Particles*, P. Breitenlohner and H. P. Dürr, eds. (Springer, Berlin, 1982).
14. C. W. Misner, K. S. Thorne, and J. A. Wheeler, *Gravitation* (Freeman, San Francisco, 1973).
15. J. D. Bekenstein and A. Meisels, *Phys. Rev. D* **22**, 1313 (1980).
16. R. Penrose, in *Relativity, Groups and Topology*, C. DeWitt and B. S. DeWitt, eds. (Gordon and Breach, New York, 1964); S. Deser, *Ann. Phys. (N.Y.)* **59**, 248 (1970).
17. C. G. Callan, S. Coleman, and R. Jackiw, *Ann. Phys. (N.Y.)* **59**, 42 (1970).

18. F. Hoyle and J. Narlikar, in *Cosmology, Fusion and Other Matters*, F. Reines, ed. (Colorado Associated Universities Press, Boulder, 1972); A. Zee, *Ann. Phys. (N.Y.)* **151**, 431 (1983); L. Smolin, *Nucl. Phys. B* **160**, 253 (1979); E. S. Fradkin and G. A. Vilkovisky, *Phys. Lett. B* **73**, 209 (1978).
19. L. Parker, *Phys. Rev. D* **7**, 976 (1973).
20. D. Capper and M. Duff, *Nuovo Cimento A* **23**, 173 (1974); J. S. Dowker and R. Critchley, *Phys. Rev. D* **13**, 3224 (1976); S. L. Adler, J. Lieberman, and Y. J. Ng, *Ann. Phys. (N.Y.)* **106**, 279 (1977); L. S. Brown and J. P. Cassidy, *Phys. Rev. D* **15**, 2810 (1977).
21. S. Weinberg, *Phys. Rev. D* **9**, 3357 (1974); L. Dolan and R. Jackiw, *Phys. Rev. D* **9**, 3320 (1974).
22. A. D. Linde, *Phys. Lett. B* **93**, 394 (1980).
23. P. C. W. Davies, *Phys. Lett. B* **110**, 111 (1982).
24. S. L. Adler, J. C. Collins, and A. Duncan, *Phys. Rev. D* **15**, 1712 (1977); J. C. Collins, A. Duncan, and S. D. Joglekar, *Phys. Rev. D* **16**, 438 (1977).
25. L. Parker, private communication.
26. S. M. Christensen, *Phys. Rev. D* **17**, 964 (1978); L. Parker, in *Recent Developments in Gravitation*, M. Levy and S. Deser, eds. (Plenum Press, New York, 1979).

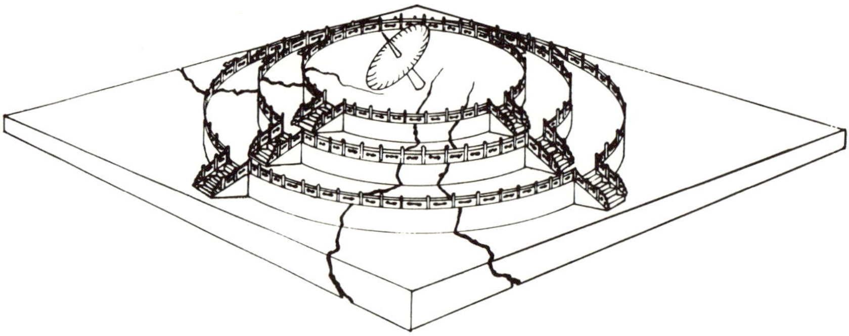

*Plate 11*

# The Birth of Time

**J. Géhéniau[1] and I. Prigogine[1]**

*Received November 8, 1985*

*The formulation of the second law of thermodynamics in the frame of general relativity is reconsidered in the case of an istotropic homogeneous universe. We show that there appears then a direct link between the cosmological state of the universe, as expressed in terms of conformal coordinates, and quantities such as energy density, pressure, and entropy associated with the description of nature. In the early universe there appears a kind of phase transition due to transfer of gravitational energy to matter associated with the cosmological expansion if the universe starts with a non-Euclidian (space) state. As a result, we may envisage the possibility of a "cold big-bang model," in which the universe would start at zero temperature and entropy. The temperature goes then through a maximum before entering the present area of cooling related to adiabatic expansion.*

## 1. INTRODUCTION

It is a great privilege to participate in this volume honoring John Archibald Wheeler. Over the years we had the opportunity to discuss often with him, and invariably we would end pondering the nature of time. This is a problem on which no consensus has been reached among physicists. Einstein's saying "time (as irreversibility) is an illusion" is well known, but on the other hand great progress has been achieved over the last decades in the description of irreversible processes as well as in linking irreversibility to instability of motion.[(1)] The role of irreversible processes in the early universe, specifically in terms of nonequilibrium phase transitions, has become a popular subject.[(2,3)] However, as has been noticed by Page and others,[(4)] this presupposes (and does not prove) the asymmetry of time and

[1] Service de Chimie Physique II, Campus Plaine, CP-231, ULB, 1050 Brussels, Belgium, and Center for Studies in Statistical Mechanics, The University of Texas, Austin, Texas 78712.

the validity of the second law. As is well known, the problem of cosmology and the problem of gravitational collapse are closely related. Wheeler has called the gravitational collapse "the greatest crisis in the physics of all time".[5] Misner, Thorne, and Wheeler describe in most vivid terms "the fate of a man who falls into a singularity at $r=0$".[6]

Wheeler's dramatic descriptions have been a powerful incentive for us to reconsider the problem of entropy and time in the context of cosmology.

## 2. NONEQUILIBRIUM PROCESSES IN THE EARLY UNIVERSE

A remarkable insight into the structure of the universe is given by the so-called "entropy" $S$, defined as the ratio of the number densities of photons to baryons. The usually accepted value is

$$S \sim 10^8\text{–}10^9$$

Today baryons are considered to be nonequilibrium objects which are either in a metastable state or decaying, while the photons are "waste products" which can no longer decay into other known forms of matter. Such a duality is typical for coupled nonequilibrium processes. Solar radiation gives rise to positive entropy production, most of which leads to dissipation in the form of photons but a small part of which is used to build biomolecules. In a schematic way,

$$\frac{d_i S}{dt} = \underset{>0}{\text{photons}} + \underset{<0}{\text{biomolecules}} \geqslant 0$$

An even simpler example is the following: A two-component system containing a mixture of hydrogen and nitrogen is maintained in a temperature gradient. As a result, one of the components, say $H_2$, accumulates in compartment I, the other in compartment II. Again, the entropy production is due to two processes—heat transfer and "antidiffusion."

$$\frac{d_i S}{dt} = \underset{>0}{\text{heat transfer}} + \underset{<0}{\text{"antidiffusion"}} \geqslant 0$$

If at some moment we discontinue the thermal gradient, the concentration of the components in the two compartments will slowly equalize. Such coupled phenomena are well known.

The important point is that the part of the system which is slowly evolving to equilibrium has been itself brought out of equilibrium by a *non-*

*equilibrium process.* May it be that matter is the nonequilibrium part of some cosmic nonequilibrium process which has also produced the black body radiation?

We believe that we see a possibility that it may have been so by reconsidering the status of the second law in general relativity. We intend to give arguments that a new formulation of the second law is necessary which links the evolution of entropy to the cosmological state of the universe. We shall then show that we have to expect a cosmic nonequilibrium process leading to entropy production through the transfer of gravitational energy to matter. This process occurs at the time when the early closed or open universe leads to a "quasi-Minkowskian" description, as is the case in conformal coordinates. But let us first comment on the status of the second law in general relativity.

## 3. SECOND LAW IN GENERAL RELATIVITY

The formulation of the second law in general relativity is based on the principle of general covariance (or equivalently on the principle of equivalence).[7,8] Using for the stress-energy tensor the usual perfect gas model, we then obtain for the evolution of the entropy density

$$T\, d(sR^3) = d(\rho R^3) + p\, dR^3 = 0 \tag{1}$$

where $R$ is the radius of the universe (as it appears in the Robertson–Walker metric) and where $\rho$ and $p$ are respectively the energy density and the pressure. The vanishing of (1) expresses that evolution is *adiabatic.*[7] However, the following facts should be kept in mind:

(a) The transition from a reversible dynamical description to an irreversible thermodynamic description corresponds to the transition from dynamical groups to semigroups. A rigorous theory now exists for a class of highly unstable dynamical systems called K-flows (for Kolmogorov).[9] This transition is based on a selection principle which has a nonlocal character in time (in other words, on the asymptotic behavior of the system for $t \to +\infty$ or $t \to -\infty$). Therefore the appeal to the equivalence principle is doubtful. In more picturesque terms, in Einstein's elevator thought-experiment all is local (in space-time) except the passenger, which cannot be dissociated from the irreversible process that sustains his life.

(b) The validity of the second law in arbitrary coordinate systems is also subject to doubt. The very concept of instability (i.e., as exhibited by positive Lyapounov exponents) may be destroyed by redefining the time scale.[10]

(c) The distinction between what is reversible and what is not is carried out in nonrelativistic physics by separating the entropy flow from the entropy production.[16] Because of the use of covariant derivatives this is not easy to perform in the frame of general relativity.

For these reasons we try a different approach which we shall describe now.

## 4. ALTERNATIVE FORMULATION OF THE SECOND LAW

To express most simply the idea of conservation (and therefor also of production or creation), it is useful to start with a "pseudo-tensor" $\tau^{\alpha\beta}$ such as[8,12]

$$\partial_\beta \tau^{\alpha\beta} = 0 \tag{2}$$

with

$$\tau^{\alpha\beta} = \tau^{\alpha\beta}_{(\mathrm{m})} + \tau^{\alpha\beta}_{(\mathrm{gr})} \tag{3}$$

The pseudotensor $\tau^{\alpha\beta}$ satisfies a Minkowski type of conservative equation and may be split into a matter part $\tau^{\alpha\beta}_{(\mathrm{m})}$ and a gravitational part $\tau^{\alpha\beta}_{(\mathrm{gr})}$ such that

$$u_\alpha \partial_\beta \tau^{\alpha\beta}_{(\mathrm{m})} = -u_\alpha \partial_\beta \tau^{\alpha\beta}_{(\mathrm{gr})} \tag{4}$$

In this sense we have now exchange of energy between gravitation and matter.

For reasons explained in Landau and Lifshitz[12] we use their definition of the pseudotensor which leads to

$$\tau^{\alpha\beta}_{(\mathrm{m})} = (-g)\, T^{\alpha\beta} \tag{5}$$

where $T^{\alpha\beta}$ is the stress-energy tensor of a perfect gas.

In addition, we prescribe that the metric is conformal (Minkowskian); it means

$$ds^2 = F^2 \eta_{\alpha\beta}\, dx^\alpha\, dx^\beta \tag{6}$$

and that $F$ tends to a finite constant when $t \to \infty$.

Let us consider in succession the case of an open universe ($k = -1$), a spatially flat universe ($k = 0$), and a closed universe ($k = +1$).

(a) For an open universe we may write $F = F(s)$, where the variable $s$ is a function of the cosmological time (see Tauber[14]). As a specific example for a universe filled by radiation or highly relativistic particles ($p - \rho/3$), thc conformal factor $F$ has the form

$$F \propto 1 - 1/s^2 \tag{7}$$

The origin of the universe in cosmological time corresponds to $s = 1$. The quantity tends asymptotically to a constant.

Taking into account the transfer of gravitational energy to matter [see (4)], it is easy to show[(15)] that the conventional formulation of the second law (1) is now replaced by

$$\begin{aligned} T\,d(sR^3) &= d(\rho R^3 F^3) + pF^3\,dR^3 \\ &= \rho R^3\,dF^3 \geqslant 0 \end{aligned} \tag{8}$$

The essential new element is the inclusion of the conformal factor $F$ [see (6)]. When $F$ is a constant, we recover Eq. (1). There are three remarkable features:

(1) We may write

$$T\,d(sR^3) = d(\tilde{\rho} R^3) + \tilde{p}\,dR^3 \tag{9}$$

which is formally identical to (1) but now the effective energy density $\tilde{\rho}$ and the effective pressure $\tilde{p}$ depend on the cosmological state of the universe through $F$.

(2) According to (9), the thermodynamic evolution is no longer adiabatic. There is a positive entropy production corresponding to an increase of $F$ which occurs in the expanding universe.

Thermodynamic theory cannot specify the microscopic nature of the mechanism of entropy production. It may be related to the formation of unstable particles ("mini black holes"?) which decay into baryons and photons.[(11,15)]

(3) The effective energy $\rho R^3 F^3$ reaches a maximum in time, and later decreases as the result of the adiabatic expansion, when $F$ is nearly constant.

The existence of the maximum is related to the "specific entropy" of the universe (ratio of the number densities of photons to baryons). It exists only when the entropy is larger than some threshold close or equal to the experimental order of magnitude $10^8$–$10^9$. The time at which it appears is shifted to earlier times when the "entropy" increases toward infinity.[2] A numerical discussion will be given elsewhere. It appears that there is a kind

[2] We thank Dr. E. Gunzig for these remarks.

of phase transition heating up the early universe and related to the transition from a non-Minkowskian open universe to an effective Minkowskian universe for which $F$ is approximated by a constant.

(b) In the case of a "flat" universe $k=0$ the theory cannot be applied as in no phase of the temporal evolution is $F$ approximated by a constant. The conditions for applying the pseudotensor are not satisfied and we do not expect any "phase transition" as the universe is born flat and remains so.

(c) In the case of a closed universe the situation is similar to that of the open universe. However, the conformal factor $F$ is now a product of two factors[(14)] of which only one depends on the cosmological age, while the other is a kinematic factor which is not related to entropy. Hence we have again to expect a phase transition in the early universe. A new feature is that in the contracting phase near the singularity $R=0$ we have to expect a *negative* entropy production which will make the singularity inaccessible. It is clear that while "creation" of ultraheavy unstable particles out of gravitation may be possible, the inverse process of building such particles from hadrons and photons may thermodynamically not be possible. In this way the singularity $R=0$ would be shielded. Similar conditions apply to collapsing stars.[(16)]

Perhaps they may prevent man, when he comes near to a singularity, from the sad fate described so vividly by Wheeler.

## 5. CONCLUDING REMARKS

In the picture presented above, the large-scale structure of the universe is traced back to irreversible processes occurring in the early universe at the juncture between the transition from a non-Minkowskian to a Minkowskian structure. In philosophical terms, of which we hope John Wheeler would not disapprove, "*time* precedes existence."

We do not consider the formulation of the second law in general relativity as a solved problem—far from that. There exists in classical mechanics a method which reduces through the use of a time-dependent transformation all Hamiltonians $H$ to $H=0$!. In this view the distinction between stable and unstable dynamical systems is erased and with it the possibility of a microscopic formulation of the second law. Inversely, if we consider the second law as a basic dynamical principle, we have to restrict the transformation theory. This is also true in general relativity. A quantitative formulation of this idea is, however, only a program for the future.

We feel, however, confident that we can apply the second law to the quasi-Lorentz regime as described in the conformal coordinates. This is in agreement with the phenomenological validity of the second law. Our method is then a way of "extrapolating" (or "interpolating") this result so that it covers the cosmological evolution. The fact that a simple link then appears between thermodynamics and the basic geometrical structure of the universe seems to us a most attractive feature.[15]

## ACKNOWLEDGMENTS

We would like to thank E. Gunzig, M. Henneaux, L. Shepley, and V. Szebehely for many helpful conversations. This work was supported in part by a grant from the Robert A. Welch Foundation.

## REFERENCES

1. B. Misra, *Proc. Natl. Acad. Sci. USA* **75**, 1627 (1978); B. Misra, I. Prigogine, and M. Courbage, *Physica A* **98**, 1 (1979); E. Tirapegui and S. Martinez, *Phys. Lett. A* **110**, 81 (1985).
2. A. D. Linde, *Rep. Prog. Phys.* **47**, 925 (1984).
3. R. H. Brandenberger, *Rev. Mod. Phys.* **57**, 1 (1985).
4. D. N. Page, *Nature* **304**, 39 (1983).
5. C. W. Misner, K. S. Thorne, and J. A. Wheeler, *Gravitation* (Freeman, San Francisco, 1973), p. 1196.
6. C. W. Misner, K. S. Thorne, and J. A. Wheeler, *Gravitation* (Freeman, San Francisco, 1973), p. 860.
7. R. C. Tolman, *Relativity Thermodynamics and Cosmology* (Clarendon Press, Oxford, 1950).
8. S. Weinberg, *Gravitation and Cosmology Principles and applications of the General Theory of Relativity* (Wiley, New York, 1972).
9. Cf. B. Misra and I. Prigogine, *Lett. Math. Phys.* **7**, 421 (1983).
10. L. Cesari, *Asymptotic Behavior and Stability Problems in Ordinary Differential Equations* (Springer, Berlin, 1963).
11. A. Casher and F. Englert, *Phys. Lett. B* **104,** 117 (1981); E. Gunzig, "Self-consistent cosmogenesis," in *Relativity, Supersymmetry and Cosmology* (Wiley Eastern, New Delhi, 1985); E. Gunzig and P. Nardone, *Gen. Relativ. Gravit.* **16**, 305 (1984).
12. L. Landau and E. Lifshitz, *Teoriia Polia* (Nauka, Moscow, 1973); *The Classical Theory of Fields*, tr. by. N. Hamevmesh (Pergamon, Oxford, 1975).
13. L. Infeld and A. Schild, *Phys. Rev.* **68**, 250 (1950).
14. G. E. Tauber, *J. Math. Phys.* **8**, 118 (1967).
15. I. Prigogine, J. Géhéniau, and E. Gunzig, to appear in *Proc. Natl. Acad. Sci. USA.*
16. S. R. De Groot and P. Mazur, *Non-Equilibrium Thermodynamics* (North-Holland, Amsterdam, 1962).

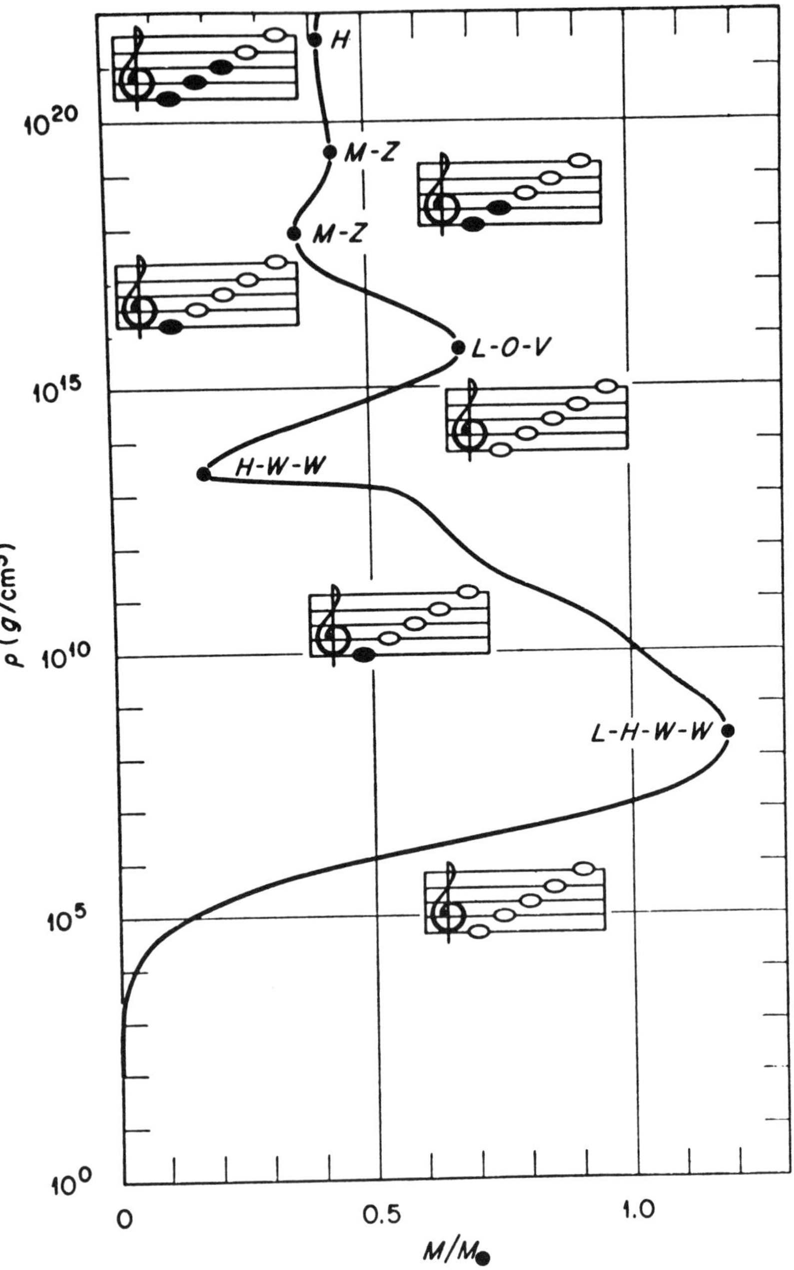


*Plate 12*

# Black Holes and Quantum Coherence

**Robert M. Wald**[1]

*Received July 16, 1985*

*We attempt to gain some insight into the issue of whether pure states evolve to density matrices in the black hole evaporation process by examining the mode functions of the particles entering the black hole which are correlated with the particles which escape to infinity. We show that these particles enter the black hole singularity at relatively early times. This tends to support the view that pure states evolve to density matrices, i.e., that in this process quantum coherence is lost.*

One of the most remarkable features of black holes in classical general relativity is their uniqueness. The two-parameter Kerr family comprises the only stationary, vacuum solutions of Einstein's equation which describe black holes; similar "no hair" theorems hold when the presence of matter fields is permitted (see Ref. 1 for a summary of results). Thus, one expects a black hole to quickly settle down to a final state where all information about the matter which formed the black hole is lost except for a few parameters, such as total mass and angular momentum. Although the proofs of the "no hair" theorems were due to others, John Wheeler played a central role in stimulating the development of this subject by recognizing the importance and physical significance of these issues at a very early stage.

Perhaps the most remarkable feature of the theory of black holes which has been discovered to date is the particle creation phenomenon. Semiclassical theory (where gravity is treated classically but matter fields are treated via quantum field theory) predicts that black holes emit thermal, black body radiation.[(2)] As a result of this radiation, a black hole loses mass and should "evaporate" completely within a finite amount of time.

[1] Enrico Fermi Institute and Department of Physics, University of Chicago, Chicago, Illinois 60637.

In Figs. 1 and 2 I have drawn standard space-time diagrams describing spherically symmetric collapse to a black hole in (1) classical general relativity and (2) semiclassical theory where particle creation and consequent black hole evaporation are taken into account. In both cases, there is information concerning the state of the system at the "time" corresponding to the slice $\Sigma$ which cannot be deduced by examining only the portion of $\Sigma$ lying outside the black hole; there are many possible "black hole internal states" that correspond to the same "external state." In both cases, if we consider a slice $\Sigma'$ which enters the singularity, there is information loss in that we cannot deduce the initial state of the system from a knowledge of the system on *all* of $\Sigma'$ (including the portion of $\Sigma'$ within the black hole); the domain of dependence of $\Sigma'$ does not equal the entire space-time. However, in the case of Fig. 1, we can foliate the entire space-time by slices like $\Sigma$ which are Cauchy surfaces, i.e., they contain all the information about the system. In the case of Fig. 2, this is impossible, since a foliation must include slices like $\Sigma''$ which are not Cauchy surfaces.

The space-time structure of Fig. 2 yields a new phenomenon for a quantum field: loss of quantum coherence. On a slice such as $\Sigma$, there unavoidably exist correlations between the field on the portion of $\Sigma$ outside the black hole and the field inside the black hole. Hence, even assum-

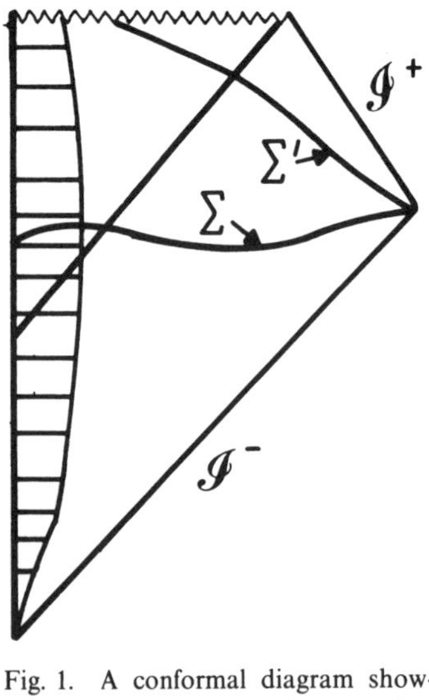


Fig. 1. A conformal diagram showing a spherically symmetric space-time where collapse to a black hole occurs in accord with classical general relativity.

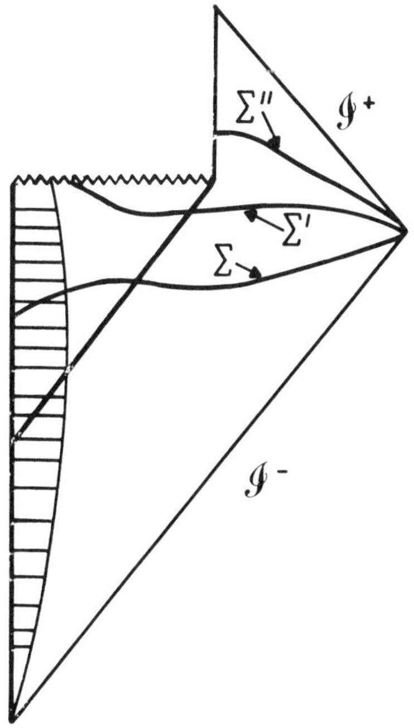


Fig. 2. A conformal diagram showing the space-time structure resulting from formation and subsequent evaporation of a black hole in accord with the standard semiclassical picture.

ing that the quantum field initially was in a pure state, the state of the field on the portion of $\Sigma$ outside the black hole can be described only by a density matrix. This same comment applies, of course, to Fig. 1. However, in the case of Fig. 1, since one can foliate the entire space-time with slices like $\Sigma$, if we include the state of the field inside the black hole in our description, the evolution of the field can be described in terms of an ordinary Hamiltonian evolution, with pure states evolving only to pure states. This is not possible in Fig. 2. There, the degrees of freedom associated with the states of the field inside the black hole propagate into the singularity and disappear. The *entire* state of the quantum field at time $\Sigma''$ can be described only by a density matrix.

Many physicists have expressed discomfort with this notion of evolution of pure states into density matrices (see, e.g., Ref. 3), since it is seen as a violation of Hamiltonian dynamics and unitarity. I do not share this feeling, since the *local* field dynamics is quite ordinary in the semiclassical picture (except, of course, at the singularity itself) and probabilities are conserved; my views in favor of the semiclassical picture are given in more detail elsewhere.[4] Nevertheless, there certainly remains room for doubt of the semiclassical picture, since its description of space-time structure cannot be expected to be valid near the black hole singularity.

One of the most valuable lessons that John Wheeler has taught most of his students is that a great deal often can be learned by asking blunt and, perhaps, even naive, physically motivated questions. I will attempt to apply this lesson here. In the semiclassical calculations of the quantum state vector describing the field prior to black hole evaporation (i.e., on a slice such as $\sum$ in Fig. 2), the particles which emerge to infinity are directly correlated with particles which enter the black hole.[5] The loss of quantum coherence at time $\sum''$ results from the disappearance of these particles into the black hole singularity. In the semiclassical picture, exactly when do these particles enter the singularity?

If the particles stay near the black hole horizon and enter the singularity only just prior to the disappearance of the black hole, then one plausibly could imagine that a small modification of the semiclassical picture could result in the escape of these particles and restoration of quantum coherence. On the other hand, if these particles enter the singularity at relatively early times, one would have to reject considerably more of the semiclassical picture to avoid the loss of quantum coherence. However, I will leave it to the reader to interpret the significance (if any) of the results and will now merely proceed with the calculation.

Perhaps the least trivial aspect of the calculation is to state in more precise detail the space-time structure given by the semiclassical picture of black hole formation and evaporation shown in Fig. 2. First, we remind the reader of the physical picture that when a spherical body of mass $M$ collapses [with $M \gg M_p \equiv (\hbar c/G)^{1/2}$] it initially forms a Schwarzschild black hole of mass $M$. Until the final moments of the evaporation (and away from the singularity), the quantum effects produce only a very tiny stress-energy. Hence, since the space-time is spherically symmetric, the local departures from a Schwarzschild geometry always are very small. However, because of the steady flux of positive energy to infinity and flux of negative energy into the black hole singularity, these quantum corrections produce the important long-term cumulative effect of decreasing the value of $M$, until the black hole finally evaporates completely.

Thus, consider the surface $S_0$, shown in Fig. 3, which is formed by joining the radially ingoing null surface which hits the singularity simultaneously with the surface of the star to the radially outgoing null surface which, if extended backward to the origin, would reach the origin at a dynamical time scale (i.e., at a proper time $\tau \sim M$) before the instant of

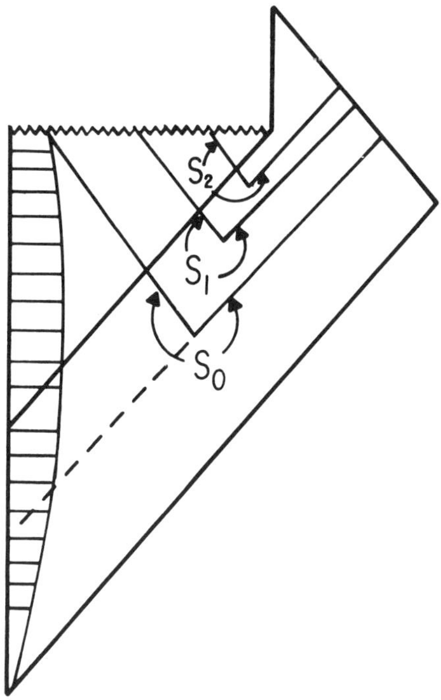


Fig. 3. A conformal diagram illustrating how a large portion of the space-time of Fig. 2 can be viewed as a sequence of "Schwarzschild segments."

black hole formation. Then near the surface $S_0$, the space-time geometry should be accurately approximated by the Schwarzschild geometry of mass $M$. Furthermore, if we move $S$ "upwards" by the (approximate) time translation isometry of this (approximate) Schwarzschild geometry, the space-time geometry should remain very accurately the Schwarzschild geometry of mass $M$ until the net quantum energy flux to infinity and into the singularity becomes a non-negligible fraction of $M$. This occurs at Schwarzschild coordinate time $t \sim M(M/M_p)^2$, and the surface obtained by translating $S_0$ by this amount is labeled $S_1$. Thus, the region between $S_0$ and $S_1$ in Fig. 3 should be very nearly isometric to a corresponding region of the Schwarzschild space-time of mass $M$. However, to obtain an accurate approximation to the space-time geometry above $S_1$, we must use a slightly smaller value of the Schwarzschild mass; the region between $S_1$ and its appropriate time translate, $S_2$, should be very nearly isometric to a portion of the Schwarzschild space-time with this new value of mass. Continuing this process, we see that a large portion of the space-time of Fig. 3 should be well approximated by a sequence of Schwarzschild segments of decreasing mass. Note, incidentally, that the manner in which the Schwarzschild segments are stacked together here is quite different near the horizon from the way one could similarly approximate the Vaidya metric by a sequence of Schwarzschild segments.

Let us focus attention, now, on one of these Schwarzschild segments; for definiteness, we choose the first such segment lying between $S_0$ and $S_1$, which has mass $M$. In this segment, we introduce outgoing and ingoing null coordinates by

$$u = t - r_* \tag{1}$$

$$v = t + r_* \tag{2}$$

where $t$ is the Schwarzschild time coordinate and $r_*$ is the "tortoise coordinate"

$$r_* = r + 2M \ln \left| \frac{r}{2M} - 1 \right| \tag{3}$$

first introduced by Wheeler and Regge in their pioneering paper on Schwarzschild perturbations.[6] For convenience, we choose the "origin" of $t$ so that the surface of the collapsing star hits the singularity at $t = 0$. It follows then that the ingoing null portion of $S_0$ is labeled by

$$v = 0 \tag{4}$$

The outgoing null portion of $S_0$ will be labeled by

$$u = C \tag{5}$$

where the magnitude of the constant $C$ will be

$$|C| \lesssim M \tag{6}$$

Consider now a wave packet of the type discussed in Refs. 2 and 5, representing a particle which reaches $\mathscr{I}^+$ at retarded time $u_0$, where $u_0 \ll M(M/M_p)^2$, so that it lies within the first "Schwarzschild segment" considered here. If we propagate this packet backward into the past, the portion of it which does not get scattered to $\mathscr{I}^-$ by the external Schwarzschild geometry will be described on the null surface $v = 0$ by a wave packet centered about the same retarded time $u_0$. The particle inside the black hole which is correlated with this particle will be described along $v = 0$ by a similar wave packet centered about the point inside the black hole which lies at the same affine parameter distance from the horizon as the particle wave packet outside the black hole[5] as shown in Fig. 4. Since near the horizon the affine parameter distance $U$ from the horizon is directly related to $u$ by $U \approx -e^{-u/4M}$, this implies that the wave packet representing the particle inside the black hole also is centered about the retarded time labeled $u_0$. [The coordinate $u$ defined by Eq. (1) is double-valued in the full Schwarzschild geometry, as it ranges from $-\infty$ to $\infty$ both outside and inside the black hole.] When does this particle enter the singularity?

The wave packets representing the particles both inside and outside the black hole are ultrahigh frequency at $v = 0$. Hence, we may appeal to geometrical optics and simply trace the path of an "outgoing" radial null geodesic inside the black hole starting at the event labeled by retarded time $u_0$ and advanced time $v = 0$ (see Fig. 4). The critical issue is whether this geodesic reaches the singularity while remaining within the original

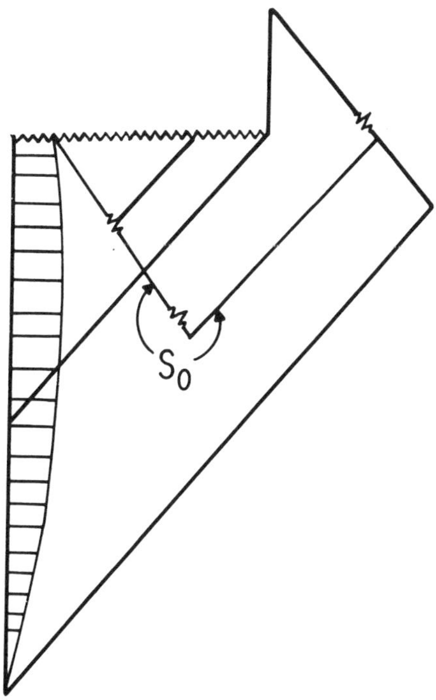


Fig. 4. A conformal diagram of the same space-time as Figs. 2 and 3, illustrating wave packets representing a particle escaping to infinity and its correlated partner inside the black hole.

"Schwarzschild segment." If it does not, this would tend to indicate the possibility of its avoiding the singularity for long enough to be able to escape during the final stage of black hole evaporation.

Let us calculate the trajectory of the geodesic during the interval it remains in the original Schwarzschild segment. This is very simple to do, since the outgoing radial geodesics satisfy $u = \text{const}$, so the trajectory is given by

$$t = r_* + u_0 \tag{7}$$

But would this reach all the way to the singularity, $r_* = 0$, at a time

$$t = u_0 \tag{8}$$

Since $u_0 \ll M(M/M_p)^2$, this means that the entire geodesic lies within the original Schwarzschild segment.

We have focused attention here only on the first "Schwarzschild segment," but similar analysis and conclusions apply to the latter segments (until the final moments of black hole evaporation when $M \sim M_p$). Thus, in the semiclassical picture, the particles which are correlated with the ones which escape to infinity enter the black hole singularity at relatively early times.

## ACKNOWLEDGMENT

This research was supported in part by NSF grant PHY 84-16691 to the University of Chicago.

## REFERENCES

1. R. M. Wald, *General Relativity* (University of Chicago Press, Chicago, 1984).
2. S. W. Hawking, *Commun. Math. Phys.* **43**, 199 (1975).
3. G. 't Hooft, "On the Quantum Structure of a Black Hole," to be published.
4. R. M. Wald, in *Quantum Gravity*, S. M. Christensen, ed. (Adam Hilger, Bristol, 1984).
5. R. M. Wald, *Commun. Math. Phys.* **45**, 9 (1975).
6. T. Regge and J. A. Wheeler, *Phys. Rev.* **108**, 1063 (1957).

| | Electromagnetic | Gravitational |
|---|---|---|
| Potential | $\Delta A \sim (\hbar c)^{1/2}/L$ | $\Delta g \sim L^*/L$ |
| Field | $\Delta F \sim (\hbar c)^{1/2}/L^2$ | $\Delta(\text{acceleration}) \sim c^2 L^*/L^2$ |
| Space curvature | $\cdots$ | $\Delta R \sim L^*/L^3$ |

*Plate 13*

# Heuristic Viewpoint Concerning the Thermal Ambience Relative to an Accelerated Frame

**Ulrich H. Gerlach**[1]

*Received January 31, 1986*

*A linear wave field in its Minkowski ground state is analyzed heuristically by two observers, one inertial, the other accelerating in a linear and uniform way. Relative to the accelerated observer, the zero-point oscillations of each Minkowski plane wave mode have an unusual Fourier spectrum. Its intensity is (i) the same for all plane wave modes, (ii) isotropic, and (iii) that of a thermal ambience (relative to the accelerated frame). The temperature of this ambience is given by the Davies–Unruh formula* $kT = (\hbar/2\pi)(g/c)$. *The existence of this ambience is incompatible with the requirement that the laws of nature be formulated in terms of tensors.*

## 1. MOTIVATION

If one wants to see how to focus on the essentials, concentrate on the central theme of a picture and its basic implications for physics, then it is difficult to point to someone who is doing it better than John A. Wheeler. This twentieth-century Leonardo da Vinci has inspired a whole generation of physicists working in the field. The *two really big principles* of physics, as he likes to remind us, are the principle of relativity and the quantum principle.

Everyone knows the principle of special relativity: "The laws of nature are the same in all inertial reference frames."

There also is the principle of relativity in its broadest sense: "The laws of physics are valid (i.e., the same) in every coordinate frame."(1) This has also become known as the principle of "general covariance": "The general laws of nature are to be expressed by equations that hold good for all

[1] Department of Mathematics, The Ohio State University, Columbus, Ohio 43210.

systems of coordinates, that is, are covariant with respect to any substitutions whatever."[2]

This manner of thinking about the laws of nature is expressed nowadays by saying that "... the coordinates that we use are arbitrary and have no physical meaning..." or "in general relativity the usually introduced coordinates are entirely arbitrary and have no physical meaning."[3]

This assessment is extended to the mathematical formulation of nature's laws by saying: The "comma-goes-to-semicolon" rule yields "no change at all physically or geometrically; change is due only to switch in reference frame from Lorentz to non-Lorentz!"[4]

Do these assessments also prevail in the context of quantum mechanics? To find out, consider one of nature's basic manifestations of mass-energy, a free wavefield. In such considerations it is difficult to point to properties more typical of quantum mechanics than those of the ground state of the system. Its zero-point or vacuum fluctuations constitute a crude but nearly universal way of identifying the ground state unambiguously. In fact, each normal mode of the free wavefield is a simple harmonic oscillator. Consequently, the vacuum fluctuation description is exact, and constitutes a practical and very picturesque way of dealing with quite a number of experimentally significant processes involving a quantized wave field.[5] There is another more universal and powerful way of dealing with a quantum process. This way describes the process in terms of a sum over very simple processes.[6] This way is also very picturesque, but it requires a correspondingly greater investment than the rudimentary technology associated with vacuum fluctuations.

In dealing with questions of principle it is precisely the roughness, i.e., the paucity and simplicity of the assumptions, that can illuminate the essentials underlying the principle.

## 2. ACCELERATED REFERENCE FRAME

Our immediate aim is to compare and contrast the vacuum fluctuations relative to an inertial frame with those relative to an "accelerated frame." Recall that an "inertial frame" consists of a global lattice of meter rods and clocks.[7] An "inertial observer" would be one who is located at any one of these lattice sites. He can collect and analyze experimental data from his assistants located at the other lattice sites. Thus one can say that there is a one-to-one relationship between inertial frames and inertial observers.

An accelerated frame and its accelerated observer are similarly defined.

The frame consists of a lattice work of accelerated robust clocks whose world lines are given by[8]

$$t = \xi \sinh g\tau, \qquad x = \xi \cosh g\tau, \qquad y = Y, \qquad z = Z \tag{1}$$

where $\xi$ ($>0$), $Y$, and $Z$ held constant characterize a lattice point. The flat space-time metric relative to these coordinates is

$$ds^2 = -\xi^2 g^2\, d\tau^2 + d\xi^2 + dY^2 + dZ^2$$

The time parameter $\tau$ is not proper time, but it can (and should!) be used to instanteneously synchronize clocks at different lattice sites.

An "accelerated observer" would be one who traces out the hyperbolic world line ($\xi > 0$, $Y$, $Z$ held constant) of one of the lattice sites and who thereby can collect experimental data from any of his robust assistants at the other lattice sites in the limited "right-hand Rindler sector" $|t| < x$, i.e., $\xi > 0$.

## 3. VACUUM FLUCTUATIONS

The vacuum fluctuations can be described in classical terms as an incoherent superposition of normal modes in their ground states. The normal modes we consider are those of a zero-rest-mass scalar field in a large cavity of length $a$ and cross section area $L^2$. They are

$$\Psi_n = A_n \sin(\omega_n t + \beta_n) \sin k_x x \sin k_y y \sin k_z z \tag{2}$$

where

$$\omega_n = \sqrt{k_x^2 + k_y^2 + k_z^2} \tag{3}$$

with

$$k_x = \frac{n_1 \pi}{a}, \quad k_y = \frac{n_2 \pi}{L}, \quad k_z = \frac{n_3 \pi}{L} \qquad (n_1, n_2, n_3 = 1, 2 \cdots) \tag{4}$$

is the Minkowski frequency, and

$$\beta_n = \beta(n_1, n_2, n_3)$$

is the phase, a random function of the modes. The modes are quantum mechanical oscillators in their ground states whose energies are $\frac{1}{2}\hbar\omega_n$. Consequently, the amplitude $A_n$ of each has the value

$$A_n = \left(\frac{2\hbar}{a\omega_n}\right)^{1/2} \frac{1}{2L}$$

These vacuum fluctuations affect an inertial detector, for example the electron of a hydrogen atom. There the response of that electron to the fluctuating field causes most of the Lamb shift.[5] The oscillatory behavior of three typical modes as seen by an inertial observer is given in Fig. 1.

The important question now is this: How do these vacuum fluctuations look when measured by an accelerated observer? In other words, how does the electron of an accelerated hydrogen atom respond to the vacuum fluctuations? (How can one accelerate a neutral hydrogen atom? John Wheeler has a ready answer: blast it with a steady beam of neutrinos!)

It is easy to transform the fluctuating wave field from an inertial to an accelerated reference frame: one simply evaluates the field, Eq. (2), along the world line, Eq. (1), of the accelerated hydrogen atom. The result is shown in Fig. 2. The time-dependent Doppler shift is due to the observer's motion. It imparts a "frequency modulation" to each normal mode $\Psi_{n-1}$, $\Psi_n$, $\Psi_{n+1}$, etc.

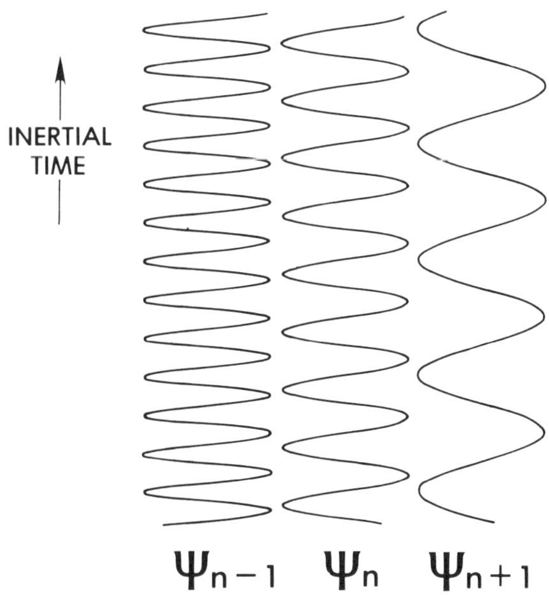


Fig. 1. Time evolution of the amplitude of three plane wave modes at a fixed point in an inertial frame. Their amplitudes are given by Eq. (1). Their frequency and propagation vector components relative to this frame are given by Eqs. (3) and (4).

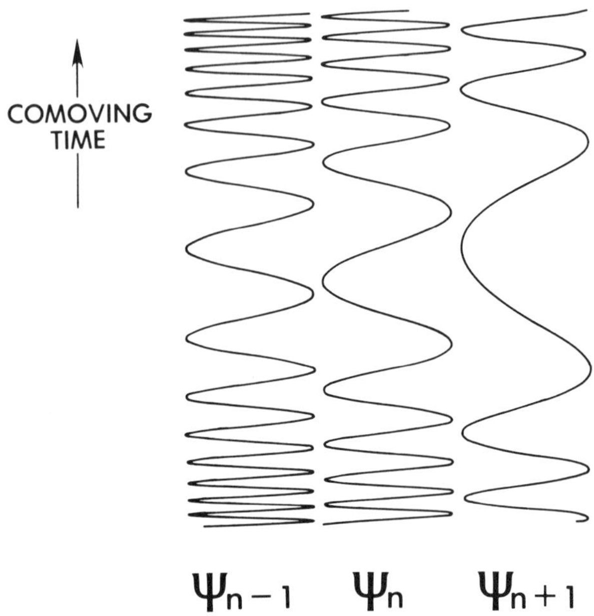


Fig. 2. Evolution of the same three modes as in Fig. 1, but their amplitude evolution is depicted at a fixed point relative to a linearly uniformly accelerated frame. The comoving time is the time parameter $\tau$ in Eq. (1). The amplitudes $\Psi_{n-1}$, $\Psi_n$, and $\Psi_{n+1}$, obtained by inserting Eq. (1) into Eq. (2), suffer a dynamically changing Doppler shift. The remarkable feature about these frequency-modulated waves is that their respective Fourier spectral powers, Eq. (11), are essentially the same and have the form "zero-point fluctuations" plus "thermal power" plus "thermal fluctuations."

As viewed by the accelerated observer, these modes have a most striking property: the intensities of their Fourier spectra are identical and are equal to the Planckian blackbody spectrum. But perhaps the most stunning feature is that each of these spectra also includes the spectrum of the fluctuation away from the mean, thereby mandating the existence of a thermal phenomenon.

It is difficult to get away from the thought that the appearance of the thermal spectrum (mean, as well as fluctuation away from the mean) among the special functions of mathematical physics is nature's attempt to tell us something important.

## 4. THE SPECTRUM

What are the vacuum fluctuations, in particular their intensity spectra, at the location of the uniformly accelerated observer with his detector, say a hydrogen atom?

The answer is obtained by having the accelerated observer Fourier analyze each of the normal modes, Eq. (2). He must use, of course, the Fourier basis,

$$e^{i\omega\tau}/\sqrt{2\pi}$$

associated with his own time (=Killing) axis. The procedure is straightforward. He evaluates Eq. (2) along his own world line (thereby obtaining Fig. 2) and then Fourier analyzes the result.

The typical standing plane wave mode,

$$\Psi_n(t, x, y, z) = \Phi_{n,+} + \Phi_{n,-} + \Phi^*_{n,+} + \Phi^*_{n,-} \tag{5}$$

is composed of the complex exponential traveling plane waves

$$\begin{aligned} \Phi_{n,\pm} &= \pm\frac{A_n}{2}\exp[-i\omega_n t \pm ik_x x - i\beta_n]\sin k_y y \sin k_z z \\ &= \pm\frac{A_n}{2}\exp[-ik\xi\sinh(\tau \mp \theta_n) - i\beta_n]\sin k_y y \sin k_z z \end{aligned} \tag{6}$$

and their complex conjugates. Here we have introduced with the help of

$$\omega_n = k\cosh\theta_n, \qquad k_x = k\sinh\theta_n, \qquad k = \sqrt{k_y^2 + k_z^2}$$

the (pseudo) polar coordinates $k$ and $\theta_n$ on the positive mass shell, Eq. (3). Furthermore, we have introduced the dimensionless time parameter by letting

$$g\tau = \bar{\tau}$$

and then dropping the bar. Recall that the accelerated frame coordinates $\xi > 0$ and $\tau$ can be viewed as (pseudo) polar coordinates on space-time. But they cover only the right-hand Rindler sector, $|t| < x$.

With the help of Eqs. (5) and (6), the Fourier amplitude of $\Psi_n$ in Fig. 2 becomes

$$\Psi_n(\omega, \xi, y, z) = \frac{1}{\sqrt{2\pi}} \int_{-\infty}^{\infty} \Psi_n(t, x, y, z)\Big|_{\text{worldline}} e^{i\omega\tau}\, d\tau$$

$$= \frac{iA_n}{\sqrt{2\pi}} [e^{\pi\omega/2} e^{-i\beta n} + e^{-\pi\omega/2} e^{i\beta n}]$$

$$\times \sin \omega\theta_n\, K_{i\omega}(k\xi) \sin k_y y \sin k_z z \qquad (7a)$$

where

$$\int_{-\infty}^{\infty} e^{\pm ik\xi \sinh(\tau - \theta_n)} e^{i\omega\tau}\, d\tau = 2e^{\mp \pi\omega} e^{i\omega\theta_n} K_{i\omega}(k\xi) \qquad (7b)$$

defines $K_{i\omega}(k\xi)$, the Bessel function of imaginary argument and imaginary order $i\omega$ (the "McDonald function"). The spectral intensity of the mode $\Psi_n$ is

$$|\Psi_n(\omega, \xi, y, z)|^2 = g_n(\omega, y, z)\,|K_{i\omega}(k\xi)|^2 \times 2[\cosh \pi\omega + \cos 2\beta_n] \qquad (8)$$

where

$$g_n(\omega, y, z) = |iA_n \sin \omega\theta_n \sin k_y y \sin k_z z|^2 \qquad (9)$$

## 5. THE THERMAL AMBIENCE

What is so remarkable about this spectral intensity? The answer is threefold: (i) thermal nature; (ii) isotropy, and (iii) uniformity.

(i) *Thermal Nature.* In the accelerated frame the normal mode corresponding to the Fourier amplitude, Eq. (7a), is

$$e^{-i\omega\tau} \Psi_n(\omega, \xi, y, z)$$

For $(k_y^2 + k_z^2)\, \xi^2 > \omega^2$ it represents an evanescent wave, and for $(k_y^2 + k_z^2)\, \xi^2 < \omega^2$ it represents a propagating wave. This follows from the wave equation

$$\left\{ \xi \frac{d}{d\xi} \xi \frac{d}{d\xi} + \omega^2 - \xi^2 (k_y^2 + k_z^2) \right\} \Psi_n(\omega, \xi, y, z) = 0 \qquad (10)$$

it satisfies. Semiclassically speaking, a particle with conserved energy $\hbar\omega$

and transverse momentum $\hbar\sqrt{k_y^2+k_z^2}$ has an instantaneous squared momentum in the $\xi$ direction given by

$$\left[\frac{\omega^2}{\xi^2}-(k_y^2+k_z^2)\right]\hbar^2=(p_\xi)^2$$

In fact, the semiclassical approximation is

$$\omega^2 \gg (k_y^2+k_z^2)\,\xi^2$$

In this approximation, the Rindler spectrum, Eq. (7), associated with the Minkowski normal mode, Eq. (2), in its ground state is

$$\begin{aligned}|\Psi_n(\omega, \xi, y, z)|^2 \\ &= g_n(\omega, y, z)\,h(\omega, k\xi)\left\{\tfrac{1}{2}\coth\frac{\hbar\omega}{2kT}+\tfrac{1}{2}\cos 2\beta_n \operatorname{csch}\frac{\hbar\omega}{2kT}\right\} \\ &= g_n(\omega, y, z)\,h(\omega, k\xi)\left\{\tfrac{1}{2}+N+\sqrt{N(N+1)}\cos 2\beta_n\right\} \qquad (11)\end{aligned}$$

Where

$$\begin{aligned}N &= \frac{1}{\exp \hbar\omega/kT-1} \\ h(\omega, k\xi) &= \frac{4\pi}{\omega}\sin^2\left[\omega \ln\frac{k\xi}{2}+\arg\Gamma(i\omega)\right]\end{aligned} \qquad (12)$$

and $\Gamma$ is the gamma function.

The importance of Eq. (11) lies in the quantity in the curly brackets. It consists of three terms.

The first two have the form of the zero-point fluctuation spectrum augmented by a thermal blackbody spectrum whose temperature is

$$T=\frac{1}{k}\frac{\hbar}{2\pi}\frac{g}{c}$$

(The conventional units have been recovered by the replacement $\omega \to \omega c/g$.)

That the second term must be given a thermal interpretation follows from the third term. Without it such a thermal interpretation would be impossible. This third term constitutes the thermodynamically required r.m.s. deviation[(9)] away from the thermal mean $N$.

(ii) *Isotropy.* Recall that each Minkowski normal mode, Eq. (2), and its Rindler spectral amplitude, Eq. (7), is identified by the set of integers

$$n = (n_1, n_2, n_3)$$

in Eq. (4). An accelerated detector (such as (i) the aforementioned hydrogen atom or (ii) a linearly accelerated but otherwise free electron) responds to each of these modes in an appropriately weighted way. This response consists therefore of a linear combination of the Rindler spectral intensities, Eq. (11). Their coefficients $g_n$ and $h$ are very rapidly oscillating functions of direction. Their mean values are [see Eqs. (9) and (12)]

$$\text{Mean value } g_n(\omega, y, z) = A_n^2/8$$

$$\text{Mean value } h(\omega, k\xi) = 2\pi/\omega$$

which are direction-independent. Thus a detector will register an *isotropic* thermal ambience. This spatial isotropy extends itself to the stress-energy tensor as well as to the various polarization features in the case of the electromagnetic field.[10,11] It also agrees with a careful quantum mechanical analysis.[12]

(iii) *Uniformity*. As viewed by an accelerated observer, there is a dramatic difference between different Minkowski standing wave modes. This difference can be seen in Fig. 2. Nevertheless their Rindler spectra, Eq. (11), or rather their mean values [obtained by combining Eqs. (13), (14) with Eq. (11)]

$$\text{Mean value } |\Psi(\omega, \xi, y, z)|^2 = (A_n^2\pi/4\omega)\left\{\frac{1}{2} + \frac{1}{\exp \hbar\omega/kT - 1}\right\}$$

are identical. They consist of the zero-point noise and the thermal noise. That nature chooses to hide these two quantum features among the special functions of mathematical physics makes us wonder whether she perhaps is not trying to tell us something important. Is it relevant that Bessel functions, Eq. (7b), form a representation of the translation group in Minkowski space-time? Vilenkin[13] confines his attention to functions restricted to the respective four Rindler coordinate neighborhoods. However, Gerlach[14] considers the globally defined "Minkowski Bessel modes."

## 6. CONCLUSION

The heuristic viewpoint confronts us with the following fact: a linearly uniformly accelerated detector registers a thermal ambience whereas an inertial detector registers nothing. This fact may be difficult to accept because it goes counter to one of Einstein's ideas which has been implemented with great success in classical (as compared to quantum) physics.

This idea is the principle of "general covariance" already mentioned in the introduction. Its modern day formulation holds that "Physics is simple only when viewed locally: that is Einstein's great lesson."[15] This simplicity is expressed mathematically by casting the laws of physics in terms of tensors. They are local, eventwise defined quantities which give one coordinate system the same physical status as any other, i.e., the coordinates are "entirely arbitrary and have no physical meaning."

It is clear that Einstein's principle of "general covariance" does not apply in a quantum mechanical setting. If a tensor is zero relative to one coordinate frame, then it is zero in all coordinate frames. The fact that a detector in an accelerated frame registers a thermal ambience while a detector in an inertial frame registers nothing violates this tensorial rule. In other words, quantum phenomena cannot be expressed in terms of pointwise locally defined tensorial laws. The reason is simple: the phenomenon of the accelerated observer registering the thermal ambience is a highly nonlocal one. The phenomenon depends on the detector tracing out a "substantial" portion of its hyperbolic world line—substantial enough for the detector to change its velocity (relative to an inertial observer) by a substantial fraction of the velocity of light.

In fact, as one can see from the limits in Eq. (7a), to obtain the exact thermal spectral amplitude requires that the detector trace out the hyperbolic worldline from its distant past all the way to its distant future.

## REFERENCES

1. A. Einstein, *The Meaning of Relativity* (Princeton University Press, Princeton, New Jersey, 1955), 5th edn., p. 61.
2. A. Einstein, *Ann. Phys.* **49**, 1916, translated in H. A. Lorentz *et al.*, *The Principle of Relativity* (Dover, New York, 1923), p. 117.
3. E. Wigner, in open Discussion following papers by S. S. Chern and I. Regge, in *Some Strangeness in the Proportion*, H. Woolf, ed. (Addison-Wesley, Reading, Massachusetts, 1980), pp. 285–286.
4. C. W. Misner, K. S. Thorne, and J. A. Wheeler, *Gravitation* (Freeman, San Francisco, 1973), Chap. 16.
5. T. A. Welton, *Phys. Rev.* **74**, 1157 (1948).
6. R. P. Feynman and A. R. Hibbs, *Quantum Mechanics and Path Integrals* (McGraw-Hill, New York, 1965).
7. E. F. Taylor and J. A. Wheeler, *Space-time Physics* (Freeman, San Francisco, 1966).
8. W. Rindler, *Am. J. Phys.* **34**, 1174 (1966).

9. An analogous result also prevails if the accelerated observer has (according to his own clock) (a) a uniform drift velocity or (b) a uniform orbiting motion in addition to what otherwise would be uniform linear acceleration. U. H. Gerlach, *Phys. Rev. D* **27**, 2310 (1983).

10. U. H. Gerlach, *Phys. Rev. D* **27**, 2310 (1983).
11. T. Boyer, *Phys. Rev. D* **21**, 2137 (1980); P. Candelas and D. Deutsch, *Proc. R. Soc. London A* **354**, 79 (1977).
12. P. G. Grove and A. C. Ottewill, *Class. Quantum Gravit.* **2**, 373 (1985).
13. N. J. Vilenkin, in *Special Functions and the Theory of Group Representations* (American Mathematical Society, Providence, Rhode Island, 1968).
14. U. H. Gerlach, "The Rindler Condensate," preprint, The Ohio State University, 1985.
15. C. W. Misner, K. S. Thorne, and J. A. Wheeler, *Gravitation* (Freeman, San Francisco, 1973), Chap. 1.

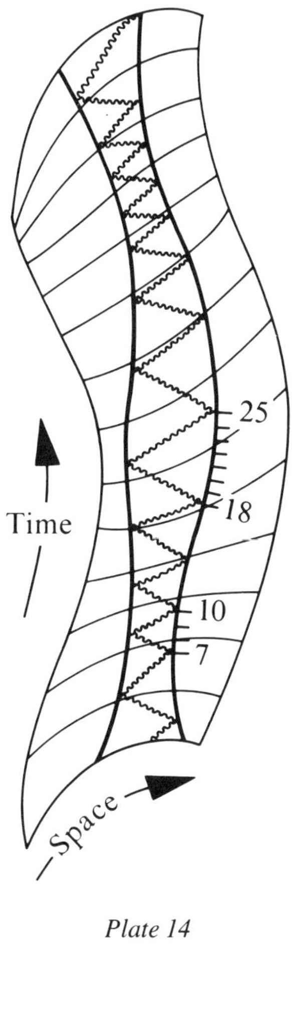


*Plate 14*

# Test of the Gravitomagnetic Field via Laser-Ranged Satellites

**Ignazio Ciufolini**[1]

*Received September 18, 1985*

*We describe a new experiment to measure the gravitomagnetic field of the Earth. This field, a consequence of the general relativistic formulation of Mach's principle (WEM—Wheeler–Einstein–Mach principle), has never been detected. The idea is to measure the Lense–Thirring precession of the nodal lines of two laser-ranged satellites with supplementary inclinations. In this way it is possible to separate the relativistic nodal precession from the classical nodal precession due to the multipole moments of the Earth.*

I know of no one—apart from Einstein himself—who did more than John Archibald Wheeler to clarify Mach's principle in general relativity.[1,2]

The existence of the gravitomagnetic field (see below) can be thought of as a consequence of the general relativistic formulation of Mach's principle or the "WEM—Wheeler–Einstein–Mach principle"[2]:

"Inertia here is given by mass-energy there."

In other words, the orientation of the axes—gyroscopes—of the local inertial frames is influenced by the motion and the distribution of mass and energy in the universe. In electrodynamics electric currents produce the magnetic field **B**. Similarly, in general relativity mass currents produce a field called, by analogy, the *gravitomagnetic* field. In the same way as the electromagnetic tensor $F^{\alpha\beta}$ can be written in terms of the components of the electric field **E** and the magnetic field **B**, in general relativity the metric tensor $g_{\alpha\beta}$ can be decomposed into an "electric" part $g_{00}$, a "magnetic" part

[1] Center for Theoretical Physics, Center for Relativity, and Physics Department, The University of Texas, Austin, Texas 78712.

$g_{0i}$, and a spatial part $g_{ik}$.[1] The curl of $g_{0i}$, in the weak-field and slow-motion limit, is the gravitomagnetic field.

For a mass $M$ rotating with angular momentum $J$ (Kerr metric[3]), in the weak-field and slow-motion limit, the gravitomagnetic "potential" is given, in isotropic coordinates, by

$$g_{03} \equiv g_{0\phi} = \frac{2GJ}{c^3 r} \sin^2 \theta \tag{1}$$

This term was discovered in 1918 by Lense and Thirring[4,5]; see also De Sitter.[6]

There are two principal manifestations for near-Earth orbits of the "dragging of inertial frames," due to the gravitomagnetic potential (1):

(I) The Lense–Thirring precession of the nodal lines (intersection of the orbital plane with the equatorial plane of the central body) of a test particle orbiting a spinning central body. This is the precession of the "enormous gyroscope" constituted by the particle orbital plane and was discovered by Lense and Thirring in 1918.[4,5]

(II) The Lense–Thirring–Schiff precession of the spin axis of a small gyroscope orbiting a spinning central body. This effect was discovered by Schiff in 1960[7] using the Lense–Thirring potential $g_{0\phi}$.

Oersted proved in 1820 that electric charge in motion produces magnetism; so far no one has proved that mass in motion produces the gravitomagnetic field. Several experiments have been proposed for this purpose.[8–11]

In particular, papers by Everitt[12] and by Lipa, Everitt, and Fairbank[13] describe in detail the Stanford experiment to measure the Lense–Thirring–Schiff precession of a small gyroscope orbiting the Earth.

In 1975 Everitt and Van Pattern[14,15] proposed to measure the Lense–Thirring nodal precession using two drag-free satellites, counter-rotating in the same polar plane. Another possible experiment[16,17] would be to detect the gravitomagnetic field measuring the orbital perturbations of Laser-ranged satellites.[2]

In particular, LAGEOS[20–22] (laser geodynamic satellite) is a high-altitude, small cross sectional area to mass ratio, spherical, laser-ranged satellite.

[2] A theoretical method to measure the Riemann curvature tensor using a generalized version of the geodesic deviation equation[18] and using four laser-ranged test particles is described in Ref. (19).

We know the LAGEOS position better than that of any other object in the sky. The accuracy in tracking its orbit is of the order of $10^{-8}$–$10^{-9}$, a few cm over $\simeq 5900$ km of altitude.

The LAGEOS semimajor axis is $a = 12270$ km, the period $P = 3.758$ h, the eccentricity $e = 0.004$, and the inclination (angle between the orbital plane and the Earth's equatorial plane) $I = 109.94°$.

The problem of measuring general relativistic perturbations on LAGEOS was first treated by Ashby and Bertotti in 1984[23] and by Matzner, Tapley, Schutz, Eanes, Vincent, and Ries (unpublished).

They studied the precession of the LAGEOS perigee. However, this relativistic perturbation is difficult to detect on LAGEOS because the measurable quantity is $e \times \delta\omega$, and not the precession $\delta\omega$ itself, while $e$ is very small for LAGEOS which has an almost circular orbit. It follows that the relativistic perturbation best measurable on satellites like LAGEOS with $e \ll 1$ is the precession of the nodal lines.[16,17,24]

For LAGEOS the total nodal precession is[20,21]

$$\dot{\Omega}_{\rm exp}^{\rm LAGEOS} \simeq 126°/\text{year}$$

and the Lense–Thirring precession is[16,24]

$$\dot{\Omega}_{\rm LT}^{\rm LAGEOS} = \frac{2GJ_\oplus}{c^2a^3(1-e^2)^{3/2}} \simeq 31 \text{ milli-arcsec/year} \tag{2}$$

However, the total nodal precession can be measured on LAGEOS with an accuracy of 1 or 2 milli-arcsec/year.[25,26]

Unfortunately the Lense–Thirring precession cannot be extracted from the experimental value of $\dot{\Omega}_{\rm exp}^{\rm LAGEOS}$ because of the uncertainty in the value of the classical precession[27]:

$$\begin{aligned}\dot{\Omega}_{\rm Class}^{\rm LAGEOS} = &-\frac{3}{2}n\left(\frac{R_\oplus}{a}\right)^2 \frac{\cos I}{(1-e^2)^2} \\ &\times \left\{J_2 + J_4\left[\frac{5}{8}\left(\frac{R_\oplus}{a}\right)^2 (7\sin^2 I - 4)\frac{(1+\frac{3}{2}e^2)}{(1-e^2)^2}\right] + \cdots\right\}\end{aligned} \tag{3}$$

Here $n = 2\pi/P$ is the orbital mean motion, $R_\oplus$ is the Earth's equatorial radius, and $J_{2n}$ are the even zonal harmonic coefficients.

This classical precession is due to the quadrupole and higher multipole mass moments of the Earth, measured by the coefficients $J_{2n}$. The orbital parameters $n$, $a$, and $e$ in formula (3) are measured with sufficient accuracy via the LAGEOS laser ranging,[21,26] and the inclination angle $I$ can be measured with sufficient accuracy over a long enough period of time[26] [the average inclination error over one year is of a few tenths of a milliarcsec, implying from Eq. (3) a nodal precession of a few milliarcsec/year]. However, the uncertainty in the classical precession $\dot{\Omega}^{\text{LAGEOS}}_{\text{Class}}$ arises from the uncertainty in the coefficients $J_{2n}$.

This uncertainty, relative to $J_2$, is of the order[28]: $\delta J_{2n}/J_2 \simeq 10^{-6}$.

For $J_2$ this corresponds, from formula (3), to an uncertainty in the nodal precession of $\simeq \pm 450$ milli-arcsec/year, and similarly for higher $J_{2n}$ coefficients. Therefore the uncertainty in $\dot{\Omega}^{\text{LAGEOS}}_{\text{Class}}$ is more than ten times larger than the Lense–Thirring precession.

One solution would be to orbit several high-altitude, laser-ranged satellites, similar to LAGEOS, to measure $J_2$, $J_4$, $J_6$, etc. and one satellite measure $\dot{\Omega}_{\text{LT}}$.

A second solution would be to orbit polar satellites; in fact, from formula (3) for polar satellites, being $I = 90°$, $\dot{\Omega}_{\text{class}}$ is equal to zero. As mentioned before, Everitt and Van Patten proposed in 1975 an experiment of this type with two drag-free, counter-rotating polar satellites.[14,15]

A third solution[16,24,29] would be to orbit a second satellite, of LAGEOS type, with the same semimajor axis, the same eccentricity, but the inclination supplementary to that of LAGEOS. Therefore, LAGEOS II should have the following orbital parameters:

$$I^{\text{II}} \simeq \pi - I^{\text{I}} \simeq 70°, \qquad a^{\text{II}} \simeq a^{\text{I}}, \qquad e^{\text{II}} \simeq e^{\text{I}}$$

With this choice, since the classical precession $\dot{\Omega}_{\text{class}}$ is linearly proportional to cos $I$, $\dot{\Omega}_{\text{class}}$ would be equal and opposite for the two satellites:

$$\dot{\Omega}^{\text{II}}_{\text{class}} = -\dot{\Omega}^{\text{I}}_{\text{class}}$$

By contrast, since the Lense–Thirring precession $\dot{\Omega}_{\text{LT}}$ is independent of the inclination [Eq. (2)], $\dot{\Omega}_{\text{LT}}$ will be the same in magnitude and sign for both satellites:

$$\dot{\Omega}^{\text{II}}_{\text{LT}} = \dot{\Omega}^{\text{I}}_{\text{LT}}$$

Adding the measured nodal precessions $\dot{\Omega}_{\text{exp}}$, we get

$$\dot{\Omega}^{\text{II}}_{\text{exp}} + \dot{\Omega}^{\text{I}}_{\text{exp}} - \dot{\Omega}^{\text{I+II}}_{\text{other forces}} = 2\dot{\Omega}^{\text{II}}_{\text{LT}} = 2\dot{\Omega}^{\text{I}}_{\text{LT}}$$

where $\dot{\Omega}^{\text{I+II}}_{\text{other forces}}$ is the sum of the nodal precessions of LAGEOS and LAGEOS II due to all the other forces. In particular, odd zonal harmonics, with $m=0$ and $l$ odd, and nonzonal harmonics, with $m\neq 0$, give periodic contributions to the nodal motion but a null secular nodal precession.(27) The contribution from non-linear even zonal harmonics(27) is mainly given by $(J_2)^2$. This contribution, being[3] only $\simeq 2\times 10^3$ times $\dot{\Omega}_{\rm LT}$, can be calculated with sufficient accuracy to extract the Lense–Thirring precession and anyhow is equal and opposite for the two satellites.

Lunar, solar, and planetary perturbations are also known with sufficient accuracy.(26,30) Concerning the atmospheric drag and the radiation pressure, we observe that LAGEOS-type satellites are built to minimize these effects.(20,21) The cross-sectional area to mass ratio is in fact for LAGEOS very small: $A/M\simeq 0.0069\ \text{cm}^2/\text{g}$. Furthermore, LAGEOS is a high-altitude satellite $[(a-R_{\oplus})\simeq 5900\ \text{km}]$, orbiting well above the dense atmosphere, with eccentricity very small ($e=0.004$); it is essentially spherically symmetric. For these reasons, atmospheric drag and radiation pressure are minimized, and it is possible—better than with nonspherical satellites—to calculate these perturbations with enough accuracy—up to a few tenths of a milli-arcsec/year(26,30)— to measure the Lense–Thirring precession.

Finally we mention the problem of calculating the influence of Earth tides on the LAGEOS nodal precession. Since the Earth tides are a periodical phenomenon, if the nodal precession will be averaged over a period of a few years, the uncertainty will be reduced to a few milli-arcsec/year.(26,30)

We observe that, in order to have less than 10% individual contributions to the experimental uncertainty, assuming the uncertainty of the $J_{2n}$ relative to $J_2$ to be(29) $\delta J_{2n}/J_2\simeq 10^{-6}$—a conservative hypothesis—it follows from (3) that the orbital parameters of LAGEOS II should be

$$a^{\rm II}\simeq a^{\rm I}\pm 16\ \text{km},\qquad I^{\rm II}\simeq\pi-I^{\rm I}\pm 0.13^\circ,\qquad e^{\rm II}\lesssim 0.04$$

However, with the improvement, in the next few years, of the knowledge of the $J_{2n}$, these limits will be relaxed.

For example, assuming $\delta J_{2n}/J_2\simeq 10^{-7}$, we would need[4]:

$$a^{\rm II}_{\rm II}\simeq a^{\rm I}\pm 160\ \text{km},\qquad I^{\rm II}\simeq\pi-I^{\rm I}\pm 1.3^\circ,\qquad e^{\rm II}\lesssim 0.4$$

A second satellite LAGEOS II—a joint mission between Italy and the United States—is to be launched in 1988 with the same semimajor axis and the same eccentricity as LAGEOS.(31)

[3] The nodal precession of LAGEOS due to $(J_2)^2$ is(26) $\dot{\Omega}_{(J_2)^2}\simeq 60$ arcsec/year.

[4] A value of the eccentricity $e^{\rm II}$ of LAGEOS II of the order of $10^{-2}$—$10^{-3}$ would be convenient to average to $0+O(e^{\rm II})$ the secular effect of the radiation pressure on the nodal precession.

If the inclination of LAGEOS II were chosen to be $I^{\text{II}} \simeq 70°$, after a few years of data, we could at last have a test of Mach's principle (in the general relativistic formulation) and a measurement of the gravitomagnetic field.

## ACKNOWLEDGMENTS

I thank Professors M. Demianski, B. DeWitt, L. Shepley, and S. Weinberg for helpful discussions and particularly Professors R. Matzner, B. E. Schutz, B. D. Tapley, J. A. Wheeler, and R. J. Eanes. This work was supported by the Center for Theoretical Physics, Center for Relativity, and Physics Department, the University of Texas, Austin, Texas, and by National Science Foundation Grants PHY 82-05717 and PHY84-04931.

## REFERENCES

1. C. W. Misner, K. S. Thorne, and J. A. Wheeler, *Gravitation* (Freeman, San Francisco, 1973). p. 1066.
2. J. Isenberg and J. A. Wheeler, "*Relativity, Quanta, and Cosmology in the Development of the Scientific Thought of Albert Einstein*," M. Pantaleo and F. deFinis, eds. (Johnson Reprint Corp., New York, 1979), Vol. I, p. 267.
3. R. P. Kerr, *Phys. Rev. Lett.* **11**, 237 (1963).
4. J. Lense and H. Thirring, *Phys. Z.* **19**, 156 (1918).
5. J. Lense and H. Thirring, English translation by B. Mashhoon *et al.*, *Gen. Relativ. Gravit.* **16**, 711 (1984).
6. W. DeSitter, *Mon. Not. R. Astron. Soc.* **76**, 699 (1916).
7. L. I. Schiff, *Proc. Natl. Acad. Sci.* **46**, 871 (1960).
8. V. B. Braginsky, C. M. Caves, and K. S. Thorne, *Phys. Rev. D* **15**, 2047 (1977).
9. V. B. Braginsky and A. G. Polnarev, *Sov. Phys. JETP Lett.* **31**, 415 (1980).
10. M. O. Scully in *Laser Spectroscopy IV*, H. Walther and K. W. Rothe, eds. (Springer-Verlag, New York, 1979), p. 27.
11. V. B. Braginsky, A. G. Polnarev, and K. S. Thorne, Caltech Goldenrod preprint in theoretical Astrophysics, Gravitational Physics, or Quantum Optics (Caltech, 1983).
12. C. W. F. Everitt, in *Experimental Gravitation*, B. Bertotti, ed. (Academic Press, New York, 1974), p. 331.
13. J. A. Lipa, W. M. Fairbank, and C. W. F. Everitt, in: *Experimental Gravitation*, B. Bertotti, ed. (Academic Press, New York, 1974), p. 361.
14. R. A. Van Patten and C. W. F. Everitt, *Phys. Rev. Lett.* **36**, 629 (1976).
15. R. A. Van Patten and C. W. F. Everitt, *Celest. Mech.* **13**, 429 (1976).
16. I. Ciufolini, Ph. D. thesis, University of Texas at Austin, August 1984.
17. I. Ciufolini, Bull. Am. Phys. Soc., Cleveland, Ohio, October 1984, **6**, 1169 (1985).
18. I. Ciufolini, to be published in *Phys. Rev. D.*
19. I. Ciufolini and M. Demianski, to be published in *Phys. Rev. D.*
20. D. E. Smith and P. J. Dunn, *J. Geophys. Res. Lett.* **7**, 437 (1980).

21. C. F. Yoder, J. G. Williams, J. O. Dickey, B. E. Schutz, R. J. Eanes, and B. D. Tapley, *Nature* **303**, 757 (1983).
22. R. J. Eanes, B. E. Schutz, and B. D. Tapley, paper in preparation.
23. N. Ashby and B. Bertotti, *Phys. Rev. Lett.* **52**, 485 (1984).
24. I. Ciufolini, *Phys. Rev. Lett.* **56**, 278 (1986).
25. W. E. Carter, D. S. Robertson, J. E. Pettey, B. D. Tapley, B. E. Schutz, R. J. Eanes, and M. Lufeng, *Science* **224**, 957 (1984).
26. B. D. Tapley, B. E. Schutz, and R. J. Eanes, private communication (1985).
27. W. M. Kaula, *Theory of Satellite Geodesy* (Blaisdell, Waltham, 1966).
28. F. J. Lerch, S. M. Klosko, R. E. Laubscher, and C. A. Wagner, *J. Geophys. Res.* **84**, 3897 (1979).
29. I. Ciufolini, *Bull. Am. Phys. Soc.*, Crystal City, Virginia, April 1985.
30. I. Ciufolini, submitted to *Celest. Mech.*
31. NASA/PSN LAGEOS II study group, final report, Goddard Space Flight Center, Greenbelt, Maryland 20771, July 1983.

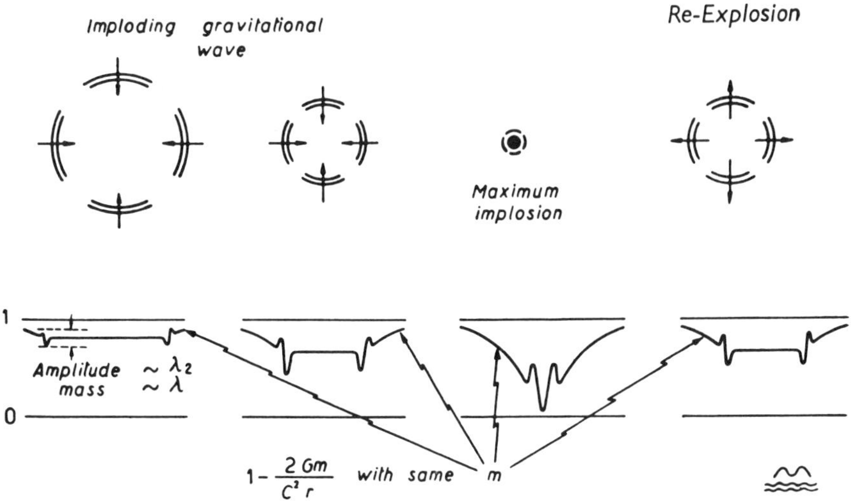
Imploding gravitational wave
Re-Explosion
Maximum implosion
1
0
Amplitude ~ $\lambda_2$
mass ~ $\lambda$
$1-\frac{2Gm}{c^2 r}$ with same m

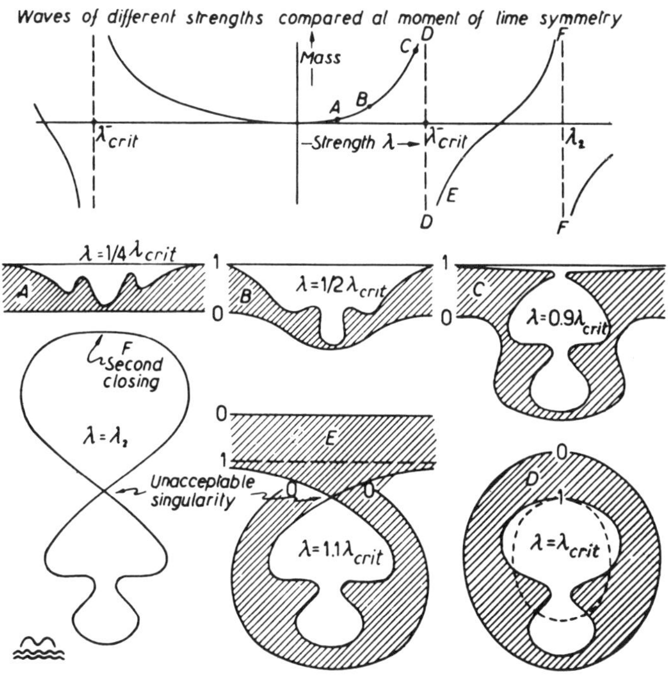
Waves of different strengths compared at moment of time symmetry
Mass
A
B
C
D
E
F
$\lambda^-_{crit}$
Strength $\lambda$
$\lambda^-_{crit}$
$\lambda_2$
$\lambda = 1/4\, \lambda_{crit}$
$\lambda = 1/2\, \lambda_{crit}$
$\lambda = 0.9 \lambda_{crit}$
F
Second closing
$\lambda = \lambda_2$
Unacceptable singularity
$\lambda = 1.1 \lambda_{crit}$
$\lambda = \lambda_{crit}$
1
0

Plate 15

# Static Electromagnetic Geon

**Marek Demianski**[1]

*Received August 6, 1985*

*A static, spherically symmetric, and asymptotically flat solution of coupled Einstein–Born–Infeld equations is presented. When the internal mass of the system is zero the resulting space-time is regular and describes static electromagnetic geon.*

Early in the fifties John Wheeler introduced the notion of an electromagnetic geon[(1)]: In the Einstein–Maxwell theory an electromagnetic geon is formed of electromagnetic radiation held together by mutual gravitational attraction. The gravitational field and the electromagnetic field of a geon is regular everywhere.

Here we would like to show that there is a spherically symmetric static solution of the Einstein field equation coupled with the nonlinear Born–Infeld electrodynamics, which is regular everywhere.

The space-time is described by a spherically symmetric line element

$$ds^2 = e^{\nu}\, dt^2 - e^{\lambda}\, dr^2 - r^2(d\theta^2 + \sin^2\theta\, d\phi^2) \tag{1}$$

where $\nu$ and $\lambda$ are functions of $r$ only.

The electromagnetic field is determined by an antisymmetric tensor $F_{\mu\nu}$, which is related in the usual way to the covariant potential $A_\mu$, $F_{\mu\nu} = A_{\nu,\mu} - A_{\mu,\nu}$. The electric and magnetic fields are defined by the relations

$$E_a = F_{\mu\nu} u^\mu v^\nu_a, \qquad B^a = -F^{\mu\nu} u_\mu v^a_\nu \tag{2}$$

[1] Institute for Theoretical Physics, University of Warsaw, Warsaw, Poland, and Center for Theoretical Physics, University of Texas Austin, Texas 78712.

where $\tilde{F}^{\mu\nu}$ is the dual tensor to $F^{\mu\nu}$, $u^{\mu}$ is the four velocity of an observer, and $v^{\mu}_{a}$ are three orthonormal spacelike vectors. The two invariants of the electromagnetic field tensor $S = -\frac{1}{4}F_{\mu\nu}F^{\mu\nu}$ and $P = -\frac{1}{4}F_{\mu\nu}\tilde{F}^{\mu\nu}$ are used to construct the Lagrangian. The Born–Infeld electrodynamics is derived from the Lagrangian of the form[2]

$$L = b^2[1 - \sqrt{1 - 2b^{-2}S - b^{-4}P^2}] \tag{3}$$

where $b$ is a constant which has the physical interpretation of a critical field strength.

The auxiliary fields $D$ and $H$ are defined with the help of an auxiliary tensor $H^{\mu\nu} = \partial L/\partial F_{\mu\nu}$

$$D^a = H^{\mu\nu}u_{\mu}v^a_{\nu}, \qquad H_a = \tilde{H}_{\mu\nu}u^{\mu}v^{\nu}_{a} \tag{4}$$

The field equations of the Born–Infeld electrodynamics assume the form

$$\nabla_{\mu}\tilde{F}^{\mu\nu} = 0, \qquad \nabla_{\mu}H^{\mu\nu} = 4\Pi j^{\nu} \tag{5}$$

The field equations could be easily generalized to include contribution from a magnetic monopole.

The spherically symmetric solution of the Born–Infeld equations describing a field produced by a point charge $q$ is

$$\begin{aligned} D_r &= \frac{q}{r^2}\, e^{-(1/2)(\nu+\lambda)} \\ E_r &= \frac{q}{\sqrt{r^4 + q^2/b^2}}\, e^{(1/2)(\nu+\lambda)} \end{aligned} \tag{6}$$

The energy momentum tensor of the Born–Infeld electrodynamics is

$$\begin{aligned} 4\Pi T^{\mu}_{\nu} = {} & \frac{1}{\sqrt{1 - 2b^{-2}S - b^{-4}P^2}}\, F^{\mu\lambda}F_{\lambda\nu} \\ & + g^{\mu}_{\nu}b^2\left[\frac{1 - 2b^{-2}S}{\sqrt{1 - 2b^{-2}S - b^{-4}P^2}} - 1\right] \end{aligned} \tag{7}$$

For the field of a point charge we have

$$\begin{aligned} 4\Pi T^0_0 = 4\Pi T^1_1 = {} & \frac{E_r^2\, e^{-(\nu+\lambda)}}{\sqrt{1 - E_r^2 b^{-2}\, e^{-(\nu+\lambda)}}} \\ & + b^2[\sqrt{1 - E_r^2 b^{-2}\, e^{-(\nu+\lambda)}} - 1] \\ 4\Pi T^2_2 = 4\Pi T^3_3 = {} & b^2[\sqrt{1 - E_r^2 b^{-2}\, e^{-(\nu+\lambda)}} - 1] \end{aligned} \tag{8}$$

The Einstein field equations for the stationary spherically symmetric space-time assume the form

$$-8\Pi T_1^1 = e^{-\lambda}\left(\frac{\nu'}{r} + \frac{1}{r^2}\right) - \frac{1}{r^2} \tag{9}$$

$$-8\Pi T_0^0 = e^{-\lambda}\left(\frac{1}{r^2} - \frac{\lambda'}{r}\right) - \frac{1}{r^2} \tag{10}$$

$$-8\Pi T_2^2 = -8\Pi T_3^3 = \frac{1}{2} e^{-\lambda}\left(\nu'' + \frac{\nu'2}{2} + \frac{\nu' - \lambda'}{r} - \frac{\nu'\lambda'}{2}\right) \tag{11}$$

Since the energy-momentum tensor satisfies $T_0^0 = T_1^1$, Eqs. (9) and (10) imply $\nu' + \lambda' = 0$, and without losing generality we can assume that $\nu + \lambda = 0$.

Equation (10) can be now written in the form

$$-2b^2\left(\frac{\sqrt{r^4 + a^4}}{r^2} - 1\right) = \frac{1}{r^2}(r\, e^{-\lambda})' - \frac{1}{r^2} \tag{12}$$

where $a^2 = q/b$. The general solution of this equation is

$$e^{-\lambda} = e^{\nu} = 1 - \frac{2m}{r} - \frac{2q^2}{3}\frac{1}{\sqrt{r^4 + a^4} + r^2} - \frac{4q^2}{3}\frac{1}{r}\int \frac{dr}{\sqrt{r^4 + a^4}} \tag{13}$$

here $m$ is an integration constant, which can be interpreted as an intrinsic mass. One can easily generalize this solution to incorporate a magnetic monopole. The gravitational field produced by a magnetic monopole $g$ and electric charge $q$ is described by

$$e^{-\lambda} = e^{\nu} = 1 - \frac{2m}{r} - \frac{2(q^2 + g^2)}{3\sqrt{r^4 + a^4} + r^2} - \frac{4(q^2 + g^2)}{3}\frac{1}{r}\int \frac{dr}{\sqrt{r^4 + a^4}} \tag{14}$$

where $a^4 = (q^2 + g^2/b^2)$, and $m$ is an intrinsic mass.

This solution can be explicitly given in terms of elliptic functions

$$e^{-\lambda} = e^{\nu} = 1 - \frac{2m}{r} - \frac{2(q^2 + g^2)}{3\sqrt{r^4 + a^4} + r^2} - \frac{2(q^2 + g^2)}{3}\frac{1}{r} F\left(\frac{1}{2}, \arccos\frac{a^2 - r^2}{a^2 + r^2}\right) \tag{15}$$

where $F(k, \phi)$ is the elliptic function of the first kind.

For large $r$ we have the following asymptotic expansion:

$$e^{\nu}=1-\frac{2m}{r}-\frac{4(q^2+g^2)\,K(1/2)}{3ar}+\frac{(q^2+g^2)}{r^2}+\cdots \tag{16}$$

A distant observer will associate with this solution a total mass $M=m+[2(q^2+g^2)\,K(1/2)]/3a$, total charge $q$, and total magnetic charge $g$.

The spherically symmetric, static, static solution of the coupled Einstein–Born–Infeld equtions possesses interesting properties. When the intrinsic mass $m$ is zero the line element is regular everywhere. It is not very difficult to check that the Riemann tensor is also regular everywhere and hence the space-time is singularity free. It is therefore an example of a static electromagnetic geon.

The purely electromagnetic mass of the geon is $m_{\rm el}=2q^{3/2}b^{1/2}K(1/2)/3$. Let us compute the electromagnetic mass of a geon produced by an elementary charge assuming that the critical field strength $b=e/r_e^2$, where $r_e$ is the classical electromagnetic radius. In this case $m_{\rm el}=(2K(1/2)/3)\,m_e=1.236m_e$, where $m_e$ is the electron mass. It is more reasonable, however, to use for the critical field strength the value arising from quantum considerations. Assuming that the potential difference across the Compton wavelength of an electron is equal to $2m_ec^2$ for the critical field strength, we obtain $b=2m_e^2c^3/eh$. This value of the critical field strength leads to $m_{\rm el}=0.15m_e$. This means that in a realistic situation the electromagnetic mass of an electron is smaller than its total. The static geon cannot therefore surve as a classical model of an electron.

I would like to thank David Park and Carl Wolf for discussions and comments, and John Wheeler for continuous interest and hospitality. This work was supported in part by the Center for Theoretical Physics of the University of Texas at Austin.

## REFERENCES

1. J. A. Wheeler, *Phys. Rev.* **97**, 511 (1955).
2. M. Born and L. Infeld, *Proc. R. Soc. London A* **144**, 425 (1934).

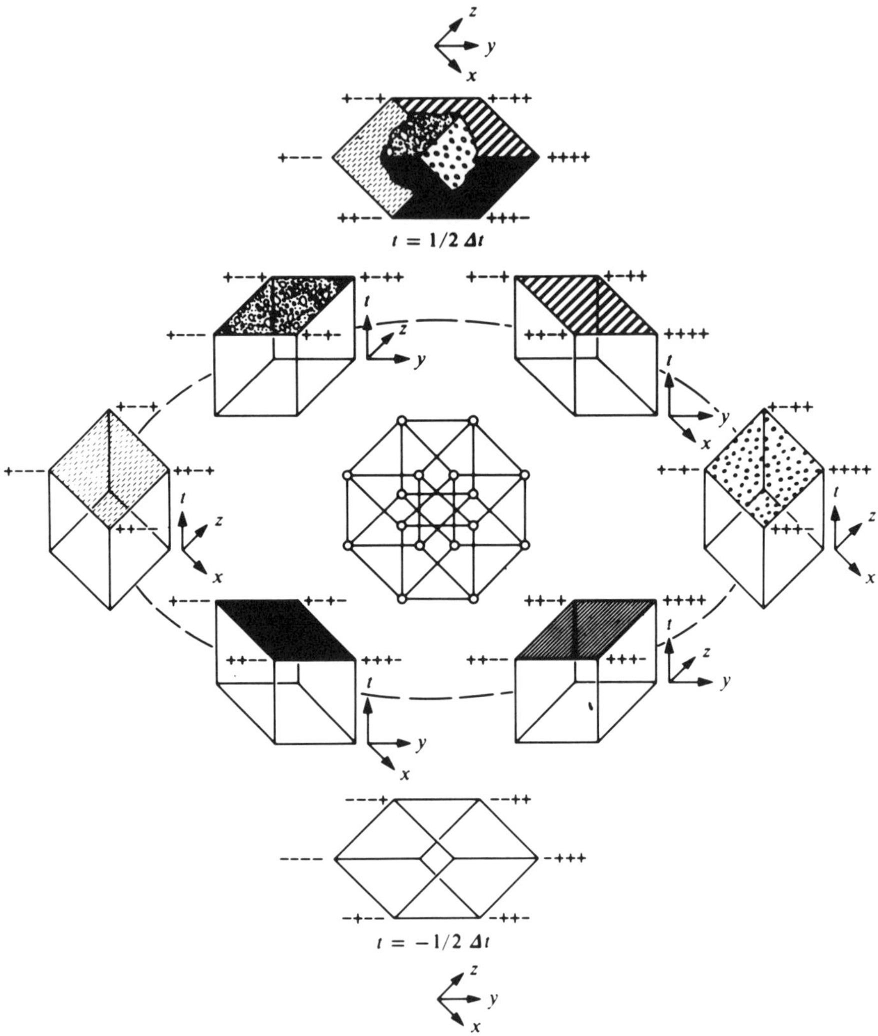


*Plate 16*

# The Geometrodynamic Content of the Regge Equations as Illuminated by the Boundary of a Boundary Principle[1]

**Warner Allen Miller**[2]

*Received September 16, 1985*

*In this paper the principle that the boundary of a boundary is identically zero ($\partial \circ \partial \equiv 0$) is applied to a skeleton geometry. It is shown that the left-hand side of the Regge equation may be interpreted geometrically as the sum of the moments of rotation associated with the faces of a polyhedral domain. Here the polyhedron, warped though it may be, is located in a lattice dual to the original skeleton manifold. This sum is related to the amount of energy-momentum (E-p) associated to the edge in question. In the establishment of this equation the ordinary Bianchi identity is rederived by applying the principle that the ($\partial \circ \partial \equiv 0$) in its (1–2–3)-dimensional formulation to polyhedral domain. Steps toward the derivation of the contracted Bianchi identity using this principle in its (2–3–4)-dimensional form are discussed. Preliminary results in this direction indicate that there should be one vector identity per vertex of the skeleton geometry.*

"Space acts on matter, telling it how to move. In turn, matter reacts back on space, telling it how to curve."—Ref. 3, Chap. 1.

## 1. INTRODUCTION: A WHEELER SUGGESTION

I know of no approach that offers itself for the derivation of the Einstein equations that gives one a clearer and deeper understanding of their geometrodynamical content than the approach that I have learned from John A. Wheeler—the approach utilizing the principle that the boundary of

[1] Submitted in honor of John Archibald Wheeler's 75th birthday. Original version submitted to Prof. Wheeler's Spring 1985 PHY393T course on Foundations of Physics taught at The University of Texas at Austin. Publication assisted by this Center and by NSF Grants PHY 8503890 and PHY 8205717.

[2] Center for Theoretical Physics, University of Texas at Austin, Austin, Texas 78712.

a boundary ($\partial \circ \partial \equiv 0$) is zero (hereafter termed the BB principle).[1–6] The central issue for the use of this principle in geometrodynamics is its ability to illustrate geometrically how the field (space-time curvature) is to be "wired up" to the source. This principle, when applied to general relativity, makes its appearance felt twice. It makes its appearance once in its (2–3–4)-dimensional formulation giving us the contracted Bianchi identity

$$\nabla \cdot \mathbf{G} = 0 \tag{1}$$

which makes the conservation of energy-momentum ($E$-$p$) an automatic consequence of Einstein's equations. To prove that this result is an identity, the BB principle has first to be employed in its (1–2–3)-dimensional form, giving us the ordinary Bianchi identity

$$R^{\alpha}_{\beta[\mu\nu;\lambda]} = 0 \tag{2}$$

Both uses of this principle lend themselves to geometric description in terms of standard curvature analysis on polyhedral domains (three- and four-dimensional polytopes). These domains have the property that their boundary is boundaryless. Needed for the establishment of the "linkage" between the source and the field is the moment of rotation trivector—a geometrical quantity introduced by Élie Cartan.[3,4] In this language, Einstein's equations read

$$\left(\begin{array}{l}\text{The dual of the moment of} \\ \text{rotation associated to} \\ \text{a three-dimensional region of} \\ \text{space-time}\end{array}\right) = 8\pi \left(\begin{array}{l}\text{Content of} \\ \text{energy-momentum} \\ \text{in that region of} \\ \text{space-time}\end{array}\right) \tag{3}$$

*Wheeler suggested that this principle may equally well be applied to Regge calculus (the approximation of space-time by flat Minkowskian building blocks) to give one a deeper geometrical insight into the Regge equations.* This paper reviews our progress in applying this principle to a skeleton geometry and presents the steps we have made toward the transcription of the content of the above three equations into the language of Regge calculus.[7] Although Tullio Regge introduced a skeleton version of the ordinary Bianchi identity, the author is not aware of a skeleton version of the contracted Bianchi identity or of a geometrical description, analogous to Eq. (3), of the Regge equations.

This paper is divided into three distinct parts. Part 1 (Sec. 2) applies the BB principle to a simplicial domain in the continuum. Here we recall the basic elements in the application of the BB principle to Einstein's theory. Part 2 of this paper, which includes Secs. 3–4, reviews the standard results of Regge calculus that are needed in the later sections. In Sec. 3 the notion of the "entourage" is introduced. In Sec. 4 we review the derivation of Regge equations from the Regge action principle. The third and final part of this paper (Secs. 5–10) describes the application of the principle that the ($\partial \circ \partial \equiv 0$) to a skeleton geometry. Section 6 shows that the dual lattice is a natural domain for the application of the BB principle in Regge calculus. In Sec. 7 the standard Regge equation (the statement that the Regge action must be an extremum with respect to the variation of the typical edge length) yields at last its geometrical interpretation—the left-hand side is the dual of the sum of the moments of rotation of a polyhedral domain. Section 8 shows that the Regge equation derived in the last section is independent of the placement of the fulcrum. In Sec. 9 the right-hand side of the Regge equation is interpreted as representing the amount of *E-p* associated with the corresponding edge dual to the polyhedral domain. Finally, Sec. 10 raises some questions concerning a conservation law for *E-p* in Regge calculus.

## 2. THE BOUNDARY OF A BOUNDARY IN THE CONTINUUM AS APPLIED TO A 4-SIMPLEX

To understand geometrically what the Riemann curvature tensor is and does is the first step towards a deeper insight into Einstein's theory of general relativity. Three vehicles lead one to such an understanding. These methods are (1) geodesic deviation,[8] (2) area deficit, and (3) parallel transport of a test vector around a closed loop. All methods lend themselves to the measurement of the curvature by experiments performed within the manifold, that is, with no need whatsoever to speak of an embedding of the manifold in a higher-dimensional space. All are expressible in the language of Regge calculus.[3] But one, namely the third, the method of parallel transport around a closed loop, will be used in the boundary of a boundary argument to arrive at a deeper understanding of Einstein's equations.

Qualitatively speaking, the curvature of an infinitesimal region of space-time as obtained by the parallel transport of a vector around a closed loop is given by

$$\text{Curvature} = \frac{\text{Angle of rotation of vector}}{\text{Area circumnavigated}} \tag{4}$$

The amount of rotation of the test vector when brought around the closed loop is analogous to the deficit angle in Regge calculus. There, all infor-

[3] Regarding geodesic motion on a skeleton geometry, see Ref. 9.

mation about the local curvature of space-time contained in the Riemann curvature tensor can be obtained by a finite collection of such experiments. Quantitatively speaking, the rotation can be represented by the evaluation of the curvature operator (a bivector-valued 2-form on space-time) at a point $p$ located at "the center" of a small oriented loop. In a coordinate system $\{x^\mu\}_{\mu=1,4}$ this operator can be decomposed as follows:

$$\mathscr{R} \equiv -(e_\alpha \wedge e_\beta)\, R^{|\alpha\beta|}{}_{|\mu\nu|}(dx^\mu \wedge dx^\nu) \qquad (5)$$

where $\{e_\alpha\}_{\alpha=1,4} = \{\partial/\partial x^\alpha\}_{\alpha=1,4}$ are the coordinate basis vectors dual to the basis 1-forms $\{dx^\mu\}_{\mu=1,4}$, and $R^{|\alpha\beta|}{}_{|\mu\nu|}$ is the Riemann tensor with vertical lines indicating summation over indices with $\alpha<\beta$ and $\mu<\nu$.

Consider a small loop encircling an oriented area; furthermore, assume that "the center" of this area is located at a point $p$. The change in a test vector $\mathbf{A}$ when parallel transported around this closed loop is

$$(\text{Change in } \mathbf{A}) = -\mathscr{R}(\mathbf{U} \wedge \mathbf{V})^{(p)}\, \mathbf{A} \qquad (6)$$

The bivector $(\mathbf{U} \wedge \mathbf{V})$ represents the area (its orientation as well as its magnitude) encircled by the loop, and the letter $p$ in parentheses indicates that the coefficients of $\mathscr{R}$ in (5) are calculated at the point $p$. This formula, expressing the usual law of parallel transport, will be used (1) to calculate the net rotation of $\mathbf{A}$ for the four faces of the tetrahedron in Fig. 1, and (2) to construct the trivector-valued 3-form of É. Cartan's moment of rotation.

In the spirit of Regge calculus, where the simplex plays a fundamental role, the remainder of this section, parts 2.1 and 2.2, makes use of a simplicial domain in the continuum to aid in the presentation of the principle here to a simplicial domain parallels the beautiful treatment by Wheeler of the principle applied to a cubical domain.[1,2] The following analysis can be, in principle, applied to any orientable polyhedron, for example a truncated octahedron.

### 2.1. The Application of the Principle in Its (1–2–3)-Dimensional Formulation

In the establishment of Einstein's equations as the dual of the sum of the moments of rotations, it will be useful to initially introduce the principle in its (1–2–3)-dimensional formulation by applying it to one of the tetrahedron faces of the simplicial domain of Fig. 2.

What should the net rotation of a test vector be after being parallel transported, with due regard to orientation, around the four faces of a tetrahedron? Zero! This is guaranteed by the topological fact that the one-

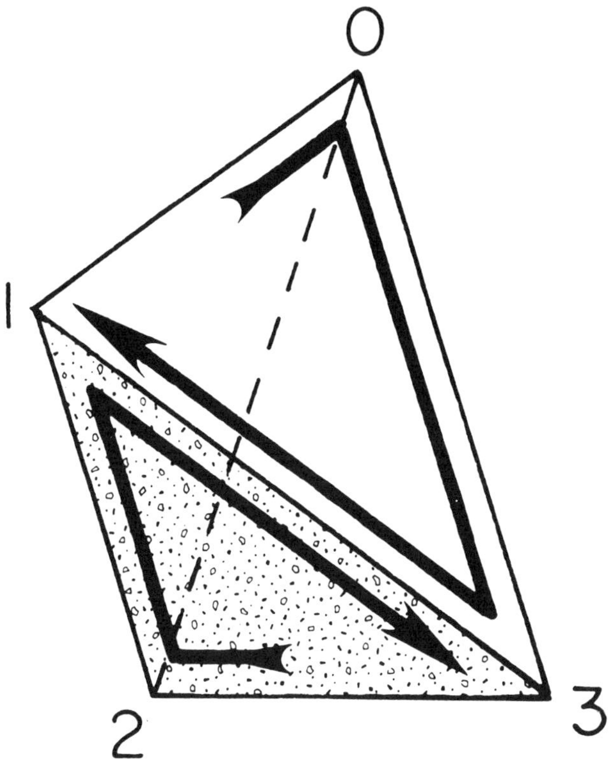


Fig. 1. The BB principle in its (1–2–3)-dimensional form illustrated for a tetrahedral domain. Shown above is an oriented tetrahedron. The orientation of the tetrahedron can be specified, up to an even permutation, by an ordering of its four vertices. The orientation of the tetrahedron shown above is (0123). The orientation of the tetrahedron induces an orientation on its two-dimensional boundary. The two-dimensional boundary of the tetrahedron consists of four oriented triangles and is given by $\partial(0123)=(012)-(013)+(023)-(123)$. The orientation of each triangle in turn induces an orientation on its one-dimensional boundary. The one-dimensional boundary of one of the two-dimensional triangles, say (012), consists of three oriented edges and is given by $\partial(012)=(01)-(02)+(12)$. If we sum the one-dimensional boundaries for each of the four triangles of the tetrahedron, the edges will cancel in pairs $[(01)-(01)=0]$. *We thus say that the tetrahedron is edgeless* or that *the one-dimensional boundary of the two-dimensional boundary of the three-dimensional region is zero.*

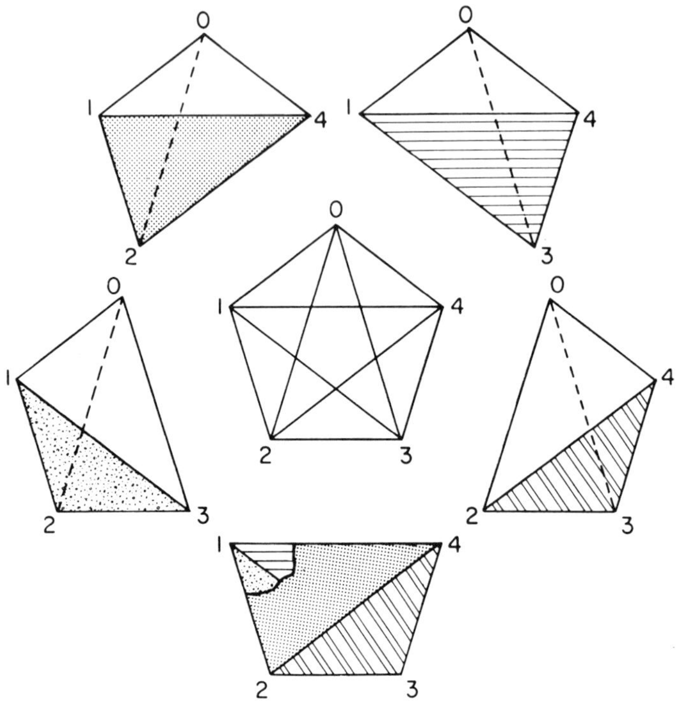


Fig. 2. Exploded view of a 4-simplex. In the central region of the drawling above is a two-dimensional projection of a 4-simplex (pentagon with a star inscribed). A simplex has five vertices, ten edges, ten triangles, and five tetrahedrons. Its three-dimensional boundary consisting of five tetrahedrons is shown "exploded off" into the space-time around the simplex. The fact that there is no creation of $E$–$p$ within the space-time region bounded by the simplex is an automatic consequence of the closure of this region. The net flow of $E$–$p$, "into" or "out of" the bottom tetrahedron (1234), through its four triangle faces is exactly compensated by a flow of the same amount "out of" or "into" the other four tetrahedrons. This point is illustrated in this diagram by the shading of the four triangle faces of (1234). Here a section of the tetrahedron (1234) is "cut out" revealing the shading of its back two faces. Each of these triangles, as well as the remaining six triangles, occur twice as faces of adjacent tetrahedrons and therefore cancel one another. In this sense we may say that the simplex is faceless or that the two-dimensional boundary of the three-dimensional boundary of the four-dimensional simplex is zero.

dimensional boundary of the two-dimensional boundary of a three-dimensional polyhedron is identically zero.[10] In other words, the tetrahedron is edgeless (Fig. 1). The vector in its travel around the perimeter of the tetrahedron will go an equal number of times "down" any given edge as "up" that edge as long as it encircled all the faces and was returned to its starting position. Thus, as long as the tetrahedron is small enough to ensure that the rotations associated with each of the four faces commute, the net rotation of the test vector will vanish. The "stage" is set for the derivation of the Bianchi identity using (1) this topological principle together with (2) the curvature operator [Eq. (5)].

Consider the oriented tetrahedron shown in Fig. 1. We (1) evaluate the rotation operator on each of its four triangular faces, (2) add each of the four expressions to obtain the "net" rotation, and (3) annul it to obtain the ordinary Bianchi identity. For this purpose, define the set of normalized tetrad basis vectors, $\{e_i\}$, by $(0i) = \Delta l^i e_i$ for $i = 1, 2, 3$ (summation over index $i$ is not intended). These vectors, although not orthonormal, span the tetrahedron in Fig. 1. Here $\Delta l^i$ is the length of the edge of the tetrahedron labeled $(0i)$. The vertex labeled "0" we take to be the origin. The orientation of the tetrahedron, given by an ordering of its vertices up to even permutations, is (0123). The rotation operator evaluated for triangle face (012) is given by

$$-\tfrac{1}{2}(e_\alpha \wedge e_\beta)\, R^{|\alpha\beta|}{}_{12}(0 + \tfrac{1}{3}(\Delta l^1 e_1 + \Delta l^2 e_2))\, \Delta l^1\, \Delta l^2 \tag{7}$$

Here, $\Delta l^1\, \Delta l^2$ is obtained by the evaluation of the rotation operator on the bivector $\tfrac{1}{2}(\Delta l^1 e_1 \wedge \Delta l^2 e_2)$ and is, in this coordinate system, an area associated with the triangle face (012). The vector $\tfrac{1}{3}(\Delta l^1 e_1 + \Delta l^2 e_2)$ gives the location where the Riemann tensor is to be evaluated. It is the vector that reaches from the origin, 0, to the barycenter of the triangle. When expanded to first order about point 0, this expression becomes

$$-\tfrac{1}{2}(e_\alpha \wedge e_\beta)(R^{|\alpha\beta|}{}_{12} + \tfrac{1}{3}R^{|\alpha\beta|}{}_{12,1}\, \Delta l^1 + \tfrac{1}{3}R^{|\alpha\beta|}{}_{12,2}\, \Delta l^2)\, \Delta l^1\, \Delta l^2 \tag{8}$$

Similar expressions present themselves when the curvature operator is evaluated for the faces (023) and $-(013)$. They will be identical up to reordering of the indices. The rotation associated with face (023) is

$$-\tfrac{1}{2}(e_\alpha \wedge e_\beta)(R^{|\alpha\beta|}{}_{23} + \tfrac{1}{3}R^{|\alpha\beta|}{}_{23,2}\, \Delta l^2 + \tfrac{1}{3}R^{|\alpha\beta|}{}_{23,3}\, \Delta l^3)\, \Delta l^2\, \Delta l^3 \tag{9}$$

The rotation associated with face $-(013)$ is

$$+\tfrac{1}{2}(e_\alpha \wedge e_\beta)(R^{|\alpha\beta|}{}_{13} + \tfrac{1}{3}R^{|\alpha\beta|}{}_{13,1}\, \Delta l^1 + \tfrac{1}{3}R^{|\alpha\beta|}{}_{13,3}\, \Delta l^3)\, \Delta l^1\, \Delta l^3 \tag{10}$$

The evaluation of the rotation about the remaining triangle face, $-(123)$, the face opposite point 0, will be just as straightforward but slightly more cumbersome. The orientation of this triangle face is given by the bivector

$$((\Delta l^3 e_3 - \Delta l^1 e_1) \wedge (\Delta l^2 e_2 - \Delta l^1 e_1))$$
$$= \Delta l^2\, \Delta l^3 (e_2 \wedge e_3) + \Delta l^1\, \Delta l^2 (e_1 \wedge e_2) - \Delta l^1\, \Delta l^3 (e_1 \wedge e_3) \qquad (11)$$

The Riemann tensor associated with this face is evaluated at the triangles barycenter. The vector reaching from the origin, point 0, to the barycenter is

$$\tfrac{1}{3}(\Delta l^1 e_1 + \Delta l^2 e_2 + \Delta l^3 e_3) \qquad (12)$$

where the factor $\frac{1}{3}$ once again appears. The rotation operator evaluated at the barycenter of triangle face $-(123)$ is given by the following expression:

$$+\tfrac{1}{2}(e_\alpha \wedge e_\beta)[R^{|\alpha\beta|}{}_{23}(0 + \tfrac{1}{3}(\Delta l^1 e_1 + \Delta l^2 e_2 + \Delta l^3 e_3))\, \Delta l^2\, \Delta l^3$$
$$+ R^{|\alpha\beta|}{}_{12}(0 + \tfrac{1}{3}(\Delta l^1 e_1 + \Delta l^2 e_2 + \Delta l^3 e_3))\, \Delta l^1\, \Delta l^2$$
$$- R^{|\alpha\beta|}{}_{13}(0 + \tfrac{1}{3}(\Delta l^1 e_1 + \Delta l^2 e_2 + \Delta l^3 e_3))\, \Delta l^1\, \Delta l^3] \qquad (13)$$

After we expand each of these three terms to first order in $\Delta l$ about the origin, the following terms appear:

$$+\tfrac{1}{2}(e_\alpha \wedge e_\beta)\, R^{|\alpha\beta|}{}_{23,1}\, \Delta l^1\, \Delta l^2\, \Delta l^3 - \text{(Expression (9))} \qquad (14)$$
$$+\tfrac{1}{2}(e_\alpha \wedge e_\beta)\, R^{|\alpha\beta|}{}_{12,3}\, \Delta l^1\, \Delta l^2\, \Delta l^3 - \text{(Expression (8))} \qquad (15)$$
$$-\tfrac{1}{2}(e_\alpha \wedge e_\beta)\, R^{|\alpha\beta|}{}_{13,2}\, \Delta l^1\, \Delta l^2\, \Delta l^3 - \text{(Expression (10))} \qquad (16)$$

Therefore, the net rotation matrix for this tetrahedron, evaluated to $(\text{length})^3$ and given by the summation of expressions (8)–(10) and (14)–(16), contains three terms. All other terms cancel. We know that the sum of these terms must vanish identically because this tetrahedron is "edgeless." The resulting equation is precisely the ordinary Bianchi identity, namely,

$$R^\alpha{}_{\beta 12,3} + R^\alpha{}_{\beta 23,1} + R^\alpha{}_{\beta 31,2} = 0 \qquad (17)$$

Here, the volume term, $\Delta l^1\, \Delta l^2\, \Delta l^3$, common to each of the terms has been factored out and the indices, 3 and 1, of the last term have been interchanged. The BB principle when linked to the geometric notion of curvature evaluated over the boundary of a closed three-dimensional region provides us with the Bianchi identities.

## 2.2. The Application of the BB Principle in Its (2–3–4)-Dimensional Formulation

We ask, how are we to "wire up" the field (space-time curvature) to the source to ensure an automatic conservation of the source? We accomplish this by applying the BB principle in its (2–3–4)-dimensional form to 4-simplex shown in Fig. 2.

Topologically the 4-simplex shown in Fig. 2 is closed. In other words, the five tetrahedrons that constitute it are interlocked so that none of their triangular faces are left exposed. Each of the ten triangular faces of the 4-simplex is shared by two tetrahedrons. The orientations induced on this triangle from these two tetrahedrons exactly cancel. This is the sense in which we say that the 4-simplex is faceless or that its boundary is boundaryless.

We want zero creation of source within the region of space-time bounded by this simplex, furthermore we want this as a consequence of the closure of the simplex. This implies that the $E$–$p$ contained within any one tetrahedron must be a sum of contributions from each of its four triangle faces. These contributions made by each triangle must cancel themselves. Why? Because the closure of the boundary of the simplex—$\partial \circ \partial \equiv 0$—automatically guarantees exactly that this is so. In other words, any flow of energy-momentum ($E$–$p$) out one of the tetrahedrons of the simplex through a triangular face gives rise to an inflow of ($E$–$p$) into the adjacent tetrahedron—the tetrahedron sharing the triangular face.

How are we to represent geometrically the content of $E$–$p$ within one of these tetrahedrons? How are we to guarantee its automatic conservation? As mentioned above, the conservation of source employs the notion of cancellation, in pairs, of the contributions of the triangular faces. For this reason the geometrical object sought must be associated with the triangle faces of the simplex. Wheeler[2] points out that the rotation bivectors associated with the faces are the wrong geometric quantities because their sum always gives zero. He points out that the proper geometric object is É. Cartan's "moment of rotation" trivector. Wheeler motivates this by analogy with a statics problem in mechanics.

The moment of the rotation operator[3] associated with the triangle face (012) of the tetrahedron (0123) is given by

$$\tau_{012} = \text{Moment of rotation operator for } (012) = (d\mathscr{P} \wedge \mathscr{R}) \qquad (18)$$

Here $\mathscr{R}$ is the rotation operator to be evaluated on the face (012) and $d\mathscr{P} = e_\mu \, dx^\mu$ is the mathematical representation of the moment arm of that rotation. The moment arm reaches from the fulcrum to the barycenter of the face. This three-indexed geometrical object, resembling a parallelepiped, is dual to a one-index object, a vector in the four-dimensional space-time, the 4-vector of $E$–$p$. Equation (3) connects this geometrical object to the integral of the Einstein tensor over a three-dimensional domain, one of the tetrahedrons.

Consider the calculation of the dual of the sum of the moments of rotation for the tetrahedron (0123), assumed here for convenience to lie in a spacelike hypersurface. The Bianchi identity allows complete freedom in the placement of the fulcrum (point where the moment is to be evaluated) in a local region about the tetrahedron. We choose the fulcrum at one vertex labeled "0."

We further restrict our attention to the time component of the $E$–$p$ 4-vector of this tetrahedron. Therefore the only contribution will come from the time component of the dual of the moment of rotation trivector. We can therefore disregard any of the timelike components of the four rotations; hence, the only face that will contribute to the lowest order will be the face opposite 0, namely triangle $-(123)$. The moment arm for this face, given by

$$\text{Moment arm} = \tfrac{1}{3}(\Delta l^1 e_1 + \Delta l^2 e_2 + \Delta l^3 e_3) \tag{19}$$

is the vector pointing from 0 to the barycenter of the triangle. The bivector giving the orientation of the face is given by (11). Therefore the moment of rotation operator given in Eq. (18) must be evaluated on the trivector (parallelepiped)

$$(\text{Moment arm}) \cdot (\text{Orientation of face } 123) = \tfrac{1}{6}\Delta l^1\, \Delta l^2\, \Delta l^3 (e_1 \wedge e_2 \wedge e_3) \tag{20}$$

By our *ansatz*, this tetrahedral domain lies on a spacelike hypersurface; therefore, not a single time component will appear. Hence the sum of the moments of rotation over the four faces of (0123), which is the rotation operator evaluated on the parallelepiped of Eq. (20), is

$$-(R^{12}{}_{12} + R^{23}{}_{23} + R^{31}{}_{31})\, \Delta l^1\, \Delta l^2\, \Delta l^3 \tag{21}$$

The term is simply the integral of the $G^0{}_0$ component of the Einstein tensor over the tetrahedron. The dual of this sum of moments of rotation over the faces of (0123) is a vector quantity. It represents $8\pi$ times the 0-component of the $E$–$p$ 4-vector.

So far we considered energy. What about the other components of $E$–$p$, for example the $x$-component of the momentum? The program to be followed is the same. Here the timelike components of the rotation will appear but the $x$ component of the net rotation will be disregarded.

## 3. CLOSED LOOPS, DEFICIT ANGLES, HINGES, AND THE "ENTOURAGE"

So far we have dealt with rotations and parallel transport around closed loops that is effectively the continuum limit. What are the corresponding ideas in Regge calculus?

Consider a four-dimensional simplicial space-time skeleton geometry $S$. The four-dimensional "building blocks" are *a priori* assumed to be flat Minkowskian simplexes with signature $(-+++)$. Simplexes have the advantage over the use of other polytopes because they are self-supporting: the specification of the ten edge lengths of a simplex completely fixes its geometry. The geometry of $S$ is therefore a function of its edge lengths and only its edge lengths.

The elements of $S$ are the vertices, edges, triangles, tetrahedrons, and simplexes that make up the skeleton. In the curvature analysis of $S$ and its application to Einstein's theory of general relativity (Regge calculus) the author has found the notion of the "entourage" of an element of $S$ fundamental in that it isolates the relevant geometrical objects associated to the element in question (Fig. 3). The "entourage" of an $m$-dimensional simplex $s_m \in S_n$, where $S_n$ is an $n$-dimensional simplicial skeleton geometry, is defined to be the set of all $p$-dimensional simplexes $(m < p \leqslant n)$ that share $s_m$.

$$\begin{pmatrix}\text{Entourage of} \\ s_m \in S_n\end{pmatrix} \equiv E(s_m \in S_n) = \{s_p \in S_n \mid s_m \text{ an element of } s_p\} \tag{22}$$

For example, in four dimensions the entourage of a triangle $h$ consists of all the tetrahedrons that share triangle $h$ and all the 4-simplexes that share or "hinge on" it.

When we examine a generic triangle hinge in the skeleton geometry, we find that it has a number of simplexes (for example, seven) in its entourage. The interior of each of these blocks is flat. Therefore we may, if we wish, choose a Minkowski coordinate system covering the interior of any one of the blocks. We may even extend this coordinate system into one ore more neighboring simplexes. In the example there is no trouble extending this coordinate patch to the left of the original simplex by four blocks and to the right through three blocks. However, a fault occurs at the tetrahedron where the left and right extensions meet. This discontinuity in the two coordinate patches across the common tetrahedron face of the two adjacent simplexes is a manifestation of the nonzero or $\delta$-function curvature concentrated at the triangle hinge. This discrepancy is the Regge-calculus manifestation of curvature.

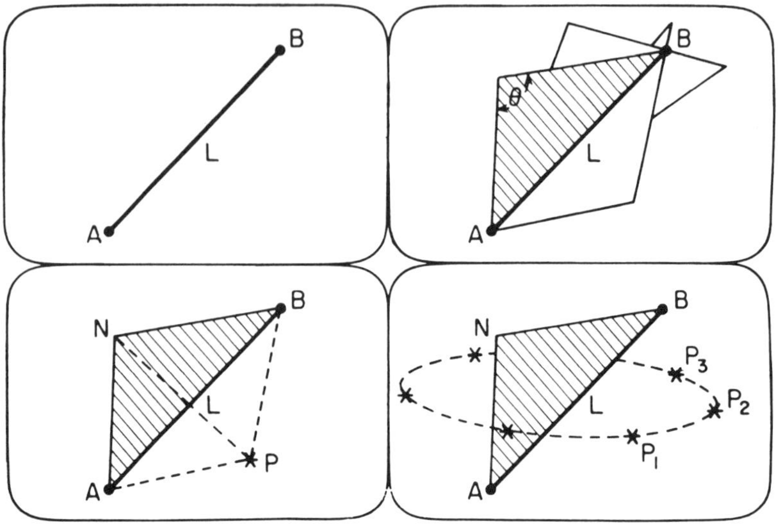


Fig. 3. Entourage of edge $L = AB$ in skeleton geometry. The entourage of edge $L$ consists of all triangles, tetrahedrons, and 4-simplexes that share $L$. Upper left: that edge in isolation. Upper right: set of all triangles in entourage of $L$. Also shown is the interior angle $\theta$ opposite to edge $L$ of the "singled-out" shaded triangle hinge $ABN$. This angle occurs in the Regge equation as $\cot\theta$ and is a function of the three edgelengths of the triangle. Lower left: typical tetrahedron face in entourage of $L$ and hinging on triangle $ABN$. Lower right: collection of all such tetrahedron faces hinging on triangle hinge $ABN$. The angle between any two adjacent tetrahedrons, say $ABNP_1$ and $ABNP_2$, is a hyperdihedral angle, $\Theta_{ABNP_1P_2}$, of a simplex $ABNP_1P_2$. The deficit angle $\varepsilon_{ABN}$ associated with the shaded triangle is given by $2\pi$ minus the sum of such hyperdihedral angles. Edges such as $P_1P_2$ are needed to define the hyperdihedral angles and are said to be "implicated" by (a function of) the deficit angle.

The coordinate systems at the fault line are discrepant by a Lorentz transformation. The parameter or "space-time angle" of this transformation needed to join the two coordinate systems is precisely the deficit angle associated with the hinge. In other words, no single coordinate system can cover the interior of the seven simplexes unless of course there is zero curvature concentrated at the hinge.

A vector, after being parallel-transported around a loop contained within the seven simplexes and encircling the triangle hinge, returns to the starting point rotated. More precisely, the component of the vector lying in the plane perpendicular to the hinge will be rotated by an amount equaling the deficit angle. The component of the vector lying in the plane of the triangle hinge will be unaltered. The Lorentz transformation describing this rotation, which is also the transformation that joined the two coordinate patches across the "fault line," leaves invariant the plane defined by the triangle.

Any Lorentz transformation may be expressed in the form

$$\mathscr{L} = \exp(iA) \tag{23}$$

where $A$ is a complex antisymmetric tensor.[11] We know that, in general, $A$ is expressible as the sum of two bivectors; however, the transformations considered here will either represent pure boosts or spatial rotations and never a combination. As a result $A$ may be expressed as a single bivector. This bivector gives the orientation and the magnitude of the rotation and will therefore be referred to as the "rotation bivector."

Consider the shaded hinge ($ABN$) shown in Fig. 3. Associated to this hinge is a deficit angle $\varepsilon_{ABN}$. The "rotation bivector" associated to this triangle is

$$\mathscr{R}_{ABN} = \text{Rotation bivector associated to hinge } ABN = \frac{*(\mathbf{NB} \wedge \mathbf{NA})}{2 \text{ Area } ABN} \varepsilon_{ABN} \tag{24}$$

Here $(\mathbf{NB} \wedge \mathbf{NA})$ is the bivector representing the plane of the triangle and $*(\mathbf{NB} \wedge \mathbf{NA})$ is its dual tensor

$$*(\mathbf{NB} \wedge \mathbf{NA})_{\mu\nu} = \tfrac{1}{2}\varepsilon_{\mu\nu\rho\sigma}(\mathbf{NB} \wedge \mathbf{NA})^{\rho\sigma} \tag{25}$$

This is orthogonal to the triangle hinge. The dual object, which strictly speaking is a 2-form, can be thought of here without any difficulty as the corresponding bivector. The Lorentz transformation associated to this hinge is therefore given by

$$\mathscr{L}_{ABN} = \exp\left[ i \frac{*(\mathbf{NB} \wedge \mathbf{NA})}{2 \text{ Area } ABN} \varepsilon_{ABN} \right] \tag{26}$$

Such a transformation leaves invariant any vector in the plane defined by the triangle hinge $(\mathbf{NB} \wedge \mathbf{NA})$ and may be referred to as a "plane Lorentz transformation."[11]

Three cases present themselves depending on the nature of the hinge. First, when the triangle hinge is timelike, the plane of rotation will be spacelike and the rotation bivector will be real, representing a spatial rotation. Second, if the hinge is spacelike, then the plane of rotation will be timelike and the rotation bivector will be pure imaginary, representing a boost. Finally, we do not consider the case where the hinge is null.

The deficit angle $\varepsilon_{ABN}$ occurring in the above expression is assumed to be small. This can be assured if one makes the skeleton mesh fine compared to the scale of the ripples in the continuum geometry that is idealized as being approximated. Naturally, for any skeleton to make sense, all such ripples in that background geometry have to be assumed—and are assumed here—to die away at high wave numbers. According to Eq. (4) the deficit angle (angle of rotation of the test vector) is of second order in the scale $L$ of the mesh.

$$\varepsilon_{ABN} \propto CL^2 \tag{27}$$

Here $C$ is the upper bound of the components of the curvature tensor of the skeleton associated to the triangles in the entourage of the edge $L = AB$. Therefore for relatively fine meshes (fine with respect to the scale of the ripples) the transformation representing the space-time rotation of the test vector around the triangle hinge $ABN$ is

$$\mathscr{L}_{ABN} = 1 + \frac{1}{2} \frac{{}^*(\mathbf{NB} \wedge \mathbf{NA})}{\text{Area } ABN} \varepsilon_{ABN} + \cdots \tag{28}$$

The sense of the rotation, whether clockwise or counterclockwise, can be specified by assigning an orientation to the triangle hinge. This order reflects the order of the factors in wedge product above. The orientation adopted here is "$A$ before $B$ before $N$" (Levi–Civita tensor with convention $\varepsilon_{0123} = +1$ is assumed).

Before using this rotation matrix to construct the skeletal moment of rotation trivector it may be useful to review the variational approach to the Regge equations.

## 4. REGGE'S EQUATIONS VIA VARIATIONAL PRINCIPLE

In 1961 Tullio Regge proposed an action principle for a skeleton manifold which resembles the familiar Einstein–Hilbert action principle for geometrodynamics.[9] It reads

$$I = (1/8\pi) \sum_{\text{hinges } h} A_h \varepsilon_h \tag{29}$$

where $A_h$ is the area of the two-dimensional hinge of the skeleton and $\varepsilon_h$ is the "angle of rattle" or "deficit angle" associated with the hinge, $h$. With respect to what do we vary this action? We know that in the continuum theory we vary the corresponding Einstein–Hilbert action with respect to the metric. Therefore, the natural variables with respect to which to vary the Regge action are the edge lengths, $L$, that define the skeleton geometry,

$$L^2 \rightarrow L^2 + \delta(L^2) \tag{30}$$

The equations that arise from extremizing (29) with respect to such variations are known as the Regge equations. There is one Regge equation for each edge length $L$:

$$\sum_{\substack{\text{hinges } h \\ \text{in entourage} \\ \text{of edge } L}} \cot \theta_{Lh} \varepsilon_h = 0 \tag{31}$$

Here, $\theta_{Lh}$ is the interior angle of triangle hinge $h$, opposite edge $L$. This angle is a function solely of the three edge lengths of the triangle. On the other hand, the deficit angle "implicates" (is a function of) many more edge lengths. For example, when there are $N$ simplexes in the "entourage" (Fig. 3) of hinge $h$, then $\varepsilon_h$ implicates $(4N+3)$ edge lengths.

There is one Regge equation for each edge length; therefore, it might seem that there is no freedom at all in the choice of the edge lengths. Yet, exactly such freedom is an essential part of the idea of approximating a smooth geometry by a skeleton geometry.

This apparent a disappearance of all degrees of freedom is misleading. Only a fraction of the $N$ Regge equations for an $N$-edge skeleton really put any condition on edge lengths. The remaining and more numerous Regge equations are effectively nonoperative. The analysis required to demonstrate this point—on which we acknowledge valuable discussions with Tsvi Piran—is deep and sophisticated and will be taken up at another place.[19]

Here we want to understand how the Regge equations relate to the concept of the "moment of rotation" arising from the application of the BB principle. Three features mentioned above are relevant to this inquiry. First, there is one equation per edge length in the skeleton. Second, the number of terms in each equation equals the number of triangles in the entourage of the relevant edge. Finally, the entourage of each equation (i.e., the entourage of the relevant edge $L$) consists of the triangles, tetrahedrons, and simplexes that share $L$.

## 5. THE BOUNDARY OF A BOUNDARY PRINCIPLE FOR A SKELETON GEOMETRY—THE MOTIVATION

The author is unaware of any derivation of Regge's equations—other than the one to be given here—that does not start with his action principle. We ask, what is the geometrodynamic content of the Regge equations? The Regge action principle, however, does not reveal the geometric content of the equations. This is not surprising when we recall the analogous continuum treatment of gravitation. There Hilbert's 1915 variational approach to Einstein's field equations gives great insight into the initial-value formulation—telling us what to fix at the spacelike boundaries—but does not immediately illuminate the geometrodynamic content of the equations. We therefore turn to Élie Cartan's application of the principle that the $(\partial \circ \partial \equiv 0)$ to the skeleton geometry, recalling how useful this approach was in the continuum. In this spirit this paper applies this principle to a skeleton geometry and concludes: Just as the BB principle provides us with the true content of Einstein's equations, so does it provide us with the true content of the corresponding Regge equations.

Four conclusions follow from applying the BB principle to a skeleton geometry. First (Sec. 6) the natural domain for applying the BB principle to a skeleton geometry is the lattice dual to the original lattice, warped though the faces of the cells of this dual lattice may be. Second (Secs. 7 and 9) we recover the left-hand side of the standard Regge equation (31), associated to an edge in the skeleton, when we evaluate the sum of the moments of rotation of the faces of the polyhedron dual to that edge. This moment of rotation appears to be related to the amount of $E$–$p$ associated with the edge in question. Third (Sec. 8) the freedom in choice of the location of the fulcrum (the point about which moment of rotation is taken), as guaranteed by the vanishing of the net rotation of the faces of the polyhedral domain, is only an approximate freedom.

The reason is simple. Finite rotations do not in general commute. Nevertheless, the resulting discrepancy in the moment of rotation is of fourth order in the scale of the lattice compared to the third-order expression for the moment of rotation itself. Therefore the discrepancy can be made percentagewise as small as we wish by making the skeleton fine enough. Finally (Sec. 10) we discuss the link of the BB principle, or curvature, in its (2–3–4)-dimensional form to conservation of $E$–$p$.

## 6. THE DUAL LATTICE AS A NATURAL DOMAIN FOR THE APPLICATION OF THE BB PRINCIPLE IN REGGE CALCULUS

We say that a lattice $S^*$ is dual to the original skeleton geometry, $S$, if their respective "face-lattices" (vertices, polygons, polyhedra, and cells) are antiisomorphic.[12–14] In other words, the lattice $S^*$ is dual to the original skeleton geometry if there is a 1–1 correspondence between (1) vertices in $S$ and 4-cells in $S^*$, (2) edges in $S$ and polyhedra in $S^*$, (3) triangles in $S$ and polygons in $S^*$, (4) tetrahedrons in $S$ and edges in $S^*$, and (5) 4-simplexes in $S$ and vertices in $S^*$. The lattice $S^*$ will be referred to as a dual lattice," and $S$ as the "original skeleton geometry."

The author has not found a more natural domain for the application of the BB principle to a skeleton geometry $S$ than the dual lattice, $S^*$. Specifically, the antiisomorphic map between $S$ and $S^*$ establishes for each edge in $S$ an entourage in $S$ identical to that implicit in the corresponding Regge equations. First, $S^*$ provides us with one polyhedral domain $P_L$ for each edge $L$ in $S$. Second, the faces of this domain in $S^*$ are in 1–1 correspondence with those triangles of $S$ that lie in the entourage of $L$. The perimeter of one such face of $P_L$ can be thought of as a closed loop encircling the corresponding $L$-containing hinge in $S$; therefore, this dual lattice provides us with that one loop per triangle hinge which is needed in order to realize the components of the curvature of the skeleton geometry (Fig. 4). Third, these polyhedral domains, warped though they are in general, have the topological property that their boundary is boundaryless. Each of these regions will be used to calculate a "net" moment of rotation trivector. This "net" rotation trivector (one for each polyhedron $P_L$) is the sum of the moments associated to each of the faces of $P_L$. It contains the same number of terms as the number of triangles in the entourage of $L$. Therefore it contains the same number of terms as the corresponding Regge equation.

The Levi–Civita tensor [Eq. (25)] relates the orientation of an element in $S$ to the orientation of its counterpart in $S^*$. For example, the

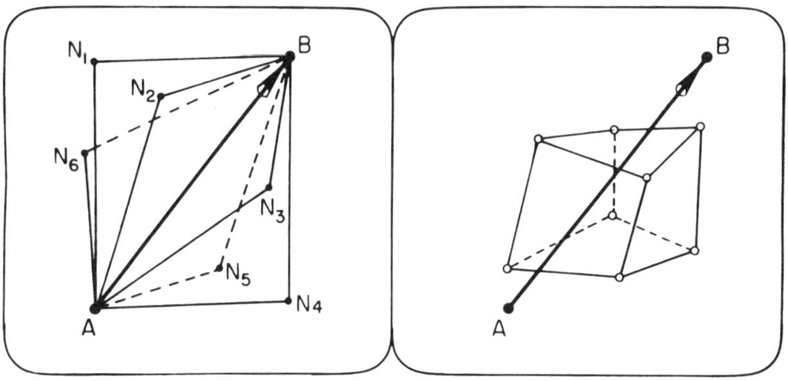


Fig. 4. Polyhedral cells in the dual lattice as a natural domain for the application of the BB principle. Left: Typical edge $L = AB$ in the skeleton geometry $S$ with its entourage of six triangles surrounding it. The orientation of $L$ is indicated by the arrow reaching from point $A$ to point $B$. The orientation of $L$ ($A$ before $B$) induces an orientation on the six triangles ($A$ before $B$ before $N_i$). Right: warped cube $P_L$ of dual lattice $S^*$ associated to edge $L$. There is a 1–1 correspondence between any given square face on the cube and the corresponding triangle $ABN_i$. The perimeter of the square is to be thought of as a closed loop encircling the corresponding triangle hinge. When $S^*$ is the "circumcentric" dual lattice, then the vertices of the cube are located at the "circumcenter" of the simplexes in the entourage of $L$.

specification of the orientation of an edge $L$ in $S$ induces an orientation on the polyhedron $P_L$ in $S^*$ dual to $L$. This linkage is exploited in Sec. 7.

There are many lattices $S^*$ that are dual to the original lattice $S$. If we impose more restrictions on the dual lattice, we can single out unique $S^*$'s. In this paper we make use of two uniquely defined but completely distinct dual lattices each in its appropriate context. One is the "barycentric" dual lattice. It is defined to be the dual lattice $S^*$ whose vertices are the barycenters of the simplexes in $S$. The other is the "circumcentric" dual lattice. The "circumcentric" dual lattice is defined to be the dual lattice whose vertices are the "circumcenters" of the simplexes in $S$. Thus, each 4-simplex is inscribable within a 4-sphere (four-dimensional hyperboloid if timelike) of radius determined uniquely by the ten edge lengths of the 4-simplex. The center of this sphere is the "circumcenter" of the simplex. The choice of one dual lattice over another dual lattice should not influence the physics (the skeletal approximation of a solution of Einstein's equations); nevertheless, it may influence the exact form of the equations obtained by skeletonization. In order to recover the typical Regge equation in its standard form as a sum of moments of rotation associated with the faces of a polyhedral domain in $S^*$, we find (Sec. 9) we have to give a favored status to the "circumcentric" dual lattice.

Associated with each vertex in $S$ is a four-dimensional cell in the dual lattice. The faces of these cells are the polyhedrons discussed above. The boundary of a boundary principle in its (2–3–4)-dimensional formulation will be applied to these blocks in Sec. 10.

## 7. THE LEFT-HAND SIDE OF THE REGGE EQUATION AS A SUM OF MOMENTS OF ROTATION OVER A POLYHEDRAL DOMAIN

We calculate here the dual of the sum of the moments of rotation associated with the faces of the polyhedral domain $P_L$ dual to a given edge $L$ and to show that it is identical to the left-hand side of the Regge equation (31) associated with that edge. Three questions present themselves. First, how shall the orientation of the polyhedron $P_L$ be specified? Second, where shall the fulcrum be placed? Third, where do we evaluate the rotation operator associated with the faces of $P_L$? Here it is shown that if the fulcrum is placed at the center of the edge $L$ and if the rotation bivector is evaluated at the "circumcenter" of the corresponding triangle hinge, then the Regge equation associated to $L$ is recovered in its standard form.

The orientation of an edge $L = AB$ in a simplicial skeleton geometry $S$ can be specified by giving it a direction. In the following argument $L$ is directed from vertex $A$ to vertex $B$, i.e., $\mathbf{L} = \mathbf{AB}$. Associated to this edge is a polyhedron $P_L$ in the dual lattice $S^*$. By construction there is a 1–1 correspondence between the faces of $P_L$ and the triangle hinges $ABN_i \in E(L \in S)$. The orientation of $L$ induces an orientation on $P_L$. This can be seen as follows: (1) the orientation given to $L$ ($A$ before $B$) induces an orientation on the triangles $ABN_i$ in the entourage of $L$ ($A$ before $B$ before $N_i$) and (2) the orientation of each of these triangles fixes an orientation on its corresponding dual face of $P_L$ in accordance with the convention chosen for the Levi–Civita tensor (in our case $\varepsilon_{0123} = +1$).

Consider, once again, the triangle $ABN_i$ in the entourage of edge $L = AB$. This triangle, if spacelike, can be inscribed within a circle of radius $(AB\ BN\ BA)/(4 \cdot \text{Area } ABN)$. The center of this circle, also called here the "circumcenter" of the triangle, is the space-time point where the rotation bivector will be evaluated. But where do we place the fulcrum? Anywhere in a local neighborhood that we find convenient. The Bianchi identity, derived and discussed in the next section, is the vehicle that provides us this latitude. Since the polyhedral region is dual to the edge $L = AB$, and this edge in a sense "pierces" this region, it seems perfectly natural, and proves to be convenient, to choose as the fulcrum the center of this edge. This center we label "$O$." The moment arm, which is the vector reaching from point $O$ to the "circumcenter" $C$, of the triangle, (1) lies in the plane of the triangle, and (2) is orthogonal to the edge $L = AB$, as seen in Fig. 5. The length of the moment arm is,

$$|d\mathscr{P}| = \text{Length of moment associated to triangle } ABN_i = \tfrac{1}{2}L \cot\theta \quad (32)$$

Here, $\theta$ is the interior angle of triangle hinge $ABN_i$ opposite to edge $L = AB$. The moment arm is a function solely of the three edge lengths of the triangle hinge.

When the triangle $ABN_i$ is timelike there is a unique hyperboloid of eccentricity $\sqrt{2}$ and directrices $\pm(AB \cdot BN \cdot NA)/(4\sqrt{2} \cdot \text{Area } ABN)$ that "encloses" the triangle. Four cases present themselves. First, when its three edges are spacelike it is inscribable on a timelike hyperbola. Second, when its three edges are timelike it is inscribable on a spacelike hyperbola. Third, when two of its edges are spacelike and the third timelike, a spacelike hyperbola is required. Finally, when two of its three edges are timelike, it is inscribable on a timelike hyperbola. Nevertheless, Eq. (32) would remain unchanged except that the hyperbolic cotangent would be used; in other words, the equation, modified in this fashion, would allow for boosts and timelike or null moment arms.

We make use of this moment arm to determine the moment of rotation trivector associated with the circuit around the face of the polyhedron $P_L$ dual to triangle $ABN_i$. The moment of rotation trivector for this circuit, given qualitatively by

$$\tau_{ABN_i} = \begin{pmatrix}\text{Distance from}\\ O \text{ to } C\end{pmatrix} \cdot \begin{pmatrix}\text{Area of parallelogram spanned by}\\ \text{the two unit vectors } CI \text{ and } CF\end{pmatrix}$$

$$= \begin{pmatrix}\sim(1/2)\text{ Altitude or "circumcentric"}\\ \text{radius to face of polyhedron } AB\end{pmatrix}$$

$$\cdot\,(\text{Curvature})\cdot\begin{pmatrix}\text{Area}\\ \text{circumnavigated}\end{pmatrix}$$

$$= \begin{pmatrix}\text{Curvature}\\ \text{of space-time}\end{pmatrix} \cdot \begin{pmatrix}\sim\text{Volume of polyhedron}\\ \text{associated to face}\end{pmatrix} \tag{33}$$

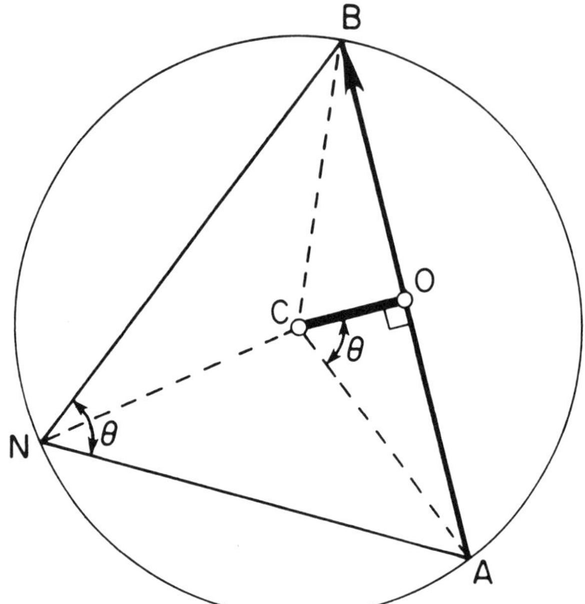


Fig. 5. Construction of moment arm. The moment arm $OC$ associated with face (dual to triangle $ABN$) of polyhedron (dual to edge $AB$) in lattice dual to original skeleton geometry is shown above. The triangle hinge $ABN$ of skeleton geometry is inscribed within its circle. Point labeled $C$ represents the center of the circle and point $O$ the center of edge $AB = L$. Length of moment arm is given by $OC = \frac{1}{2}L \cot\theta$. To see this, let $\angle NBC$ be the angle between $NB$ and $NC$. Then $\angle NBC + \angle NOA = \theta$. Furthermore, by construction, triangles $NBC$, $NAC$, and $ABC$ are isosceles. Therefore $\angle CBN = \pi - 2\angle NBC$ and $\angle CNA = \pi - 2\angle NOA$. But $\angle CAB = 2\pi - \angle CBN - \angle CNA = 2(\angle NBC + \angle NOA) = 2\theta$. Hence $\angle COA = \frac{1}{2}\angle CAB = \theta$. By construction, $OAC$ is a right triangle with edge $OA = L/2$. Therefore $OC = \frac{1}{2}L \cot\theta$.

is the exterior product of the moment arm $(d\mathscr{P})$ and the rotation bivector [Eq. (24)].

$$(d\mathscr{P}_{OC} \wedge \mathscr{R}_{ABN_i})^* = \left(\frac{i}{4}\right) L \cot\theta \left[\frac{\mathbf{OC} \wedge (\mathbf{BN}_i \wedge \mathbf{AN}_i)^*}{|OC| \text{ Area } ABN_i}\right]^* \tag{34}$$

The dual of this quantity is a 4-vector and points along edge $L = AB$. Why? Because, (1) the moment arm $OC$ lies in the plane of the triangle hinge, and (2) the rotation bivector is orthogonal to the plane of the triangle. Therefore, the dual of the moment of rotation associated with triangle $ABN_i$ is given by

$$\left(\begin{array}{l}\text{Dual of the moment of rotation}\\ \text{trivector associated with the}\\ \text{circuit around the face (dual}\\ \text{to triangle } ABN_i\text{) of } P_L\end{array}\right) = \frac{i}{4} \cot\theta \, \varepsilon_{ABN_i} \widehat{AB} \tag{35}$$

Here $\widehat{AB}$ represents the unit vector oriented along edge $\mathbf{L} = \mathbf{AB}$.

The moment of rotation vector associated with the circuit (path of transport of test vector) around the face dual to triangle $ABN_i$ depends on the location of the fulcrum. However, when we take the fulcrum at the center of the edge $L = AB$ and evaluate the rotation operator at the circumcenter of triangle $ABN_i$ then—and only then—will the moment of rotation vector point along edge $L$. In this case all of the trivectors of $P_L$ share a common orientation, that of $L$. Only when we add all the moments of rotation associated with the circuit around the faces of $P_L$ will the resulting expression be independent of the choice of the fulcrum (Sec. 8). The skeletal analogue of the left-hand side of Eq. (3) corresponds to such a sum

$$\left(\begin{array}{l}\text{Dual of moment of rotation}\\ \text{associated with the circuit}\\ \text{of one face summed over all}\\ \text{faces of polyhedron } P_L\end{array}\right) = \frac{i}{4} \sum \cot\theta \, \varepsilon_{ABN_i} \widehat{AB} \tag{36}$$

To annul this sum (in the case where there is no source) is the final step that gives the standard Regge equation associated with edge $L = AB$ (Fig. 6).

The warping of the polyhedral domain $P_L$ did not enter into the calculation presented above. Why? Because the rotation bivector associated with the circuit around a face of $P_L$ is independent of the bumps and wiggles in the circuit. It only requires that the circuit or loop encircle the corresponding hinge.

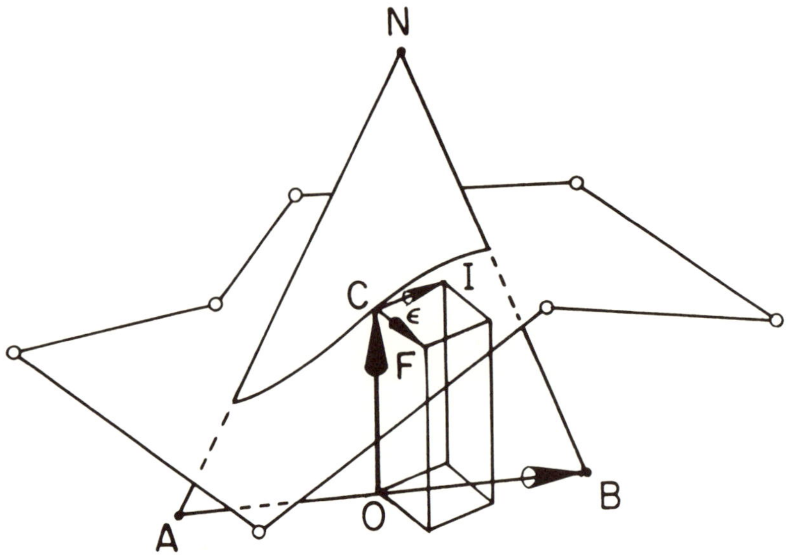


Fig. 6. Moment of rotation associated with one of the triangles that radiates out from a given edge. Shown here is the moment of rotation trivector $(OC \wedge CI \wedge CF)$ associated to triangle hinge $ABN$ based on edge $AB$. The top face of the parallelepiped $(CI \wedge CF)$ represents the rotation bivector giving the orientation of the plane of rotation and the magnitude of the rotation. Here, as in every other triangle in the skeleton geometry, the plane of rotation is orthogonal to the plane of the triangle. The angle between the unit test vector $CI$ (vector before rotation around the perimeter of the perimeter of the dual face) and the same vector $CF$ after parallel transport is equal to the deficit angle $\varepsilon_{ABN}$. The moment arm $OC$ (also shown in Fig. 5) reaches from the center of $AB$ to the "circumcenter" of the triangle $ABN$. By construction $OC$ lies in the plane of the triangle and is perpendicular to edge $AB$; therefore, in the four-dimensional space the vector $AB$ is orthogonal to the parallelepiped.

The interpretation of the moment arm $OC$ as roughly half the altitude or "circumcentric" radius to the face in question is not ciear to the author. If the triangle $ABN_i$ were severely elongated the corresponding moment arm $OC$ would be large. In other words the circumcenter of the triangle would lie at a point far removed from the triangle. How can this be reconciled? We raise the following two questions:

1. When a triangle hinge is elongated, not just one but all three of its edge lengths will give rise to moment arms that are excessively long; moreover, two of the moment arms will be long and point towards the interior of the triangle and the other will be long and point away from the triangle region. When the contributions of these three moment arms are added, will their unphysical parts cancel?

2. The "circumcentric" dual lattice appears fundamental and needed for the recovery of the standard form of the Regge equations. At some places in a "circumcentric" dual lattice, a given polyhedral domain may be far removed from the edge dual to it in $S$. The four-dimensional cells in such a lattice may even overlap, requiring the incorporation of the concept of "negative volumes." Furthermore there is a "transitivity" of a kind present in this dual lattice that makes it appealing. For example, if a tetrahedron is inscribed within a 2-sphere, then automatically its four triangle faces are inscribed within circles that are cross sections of the 2-sphere. In particular, if one of these triangles is elongated, then it will necessarily be inscribed in a large circle and therefore the 2-sphere will have a radius at least as large as the radius of this circle. Can this "transitivity" provide a geometrical interpretation for the moment arms?

The nature of this sum of moments of rotation as related to the right-hand side of the Regge equation is discussed below.

## 8. FREEDOM IN CHOICE OF THE FULCRUM AND THE SKELETAL BIANCHI IDENTITY

That the sum of the moments of rotation of the faces of polyhedron $P_L$ in $S^*$ is independent of the placement of the fulcrum is a consequence of the BB principle in its (1–2–3)-dimensional form,

$$\sum \tau = \sum (d\mathscr{P} + \text{const}) \wedge \mathscr{R} = \sum d\mathscr{P} \wedge \mathscr{R} + \underbrace{\text{const} \wedge \sum \mathscr{R}}_{\text{vanishes}} \tag{37}$$

We are assured, if the $P_L$ is small enough, that the net rotation of a test vector when parallel transported around its faces will be zero. The last term in the above equation is therefore zero and the sum of moments of rotation is independent of the displacement of the fulcrum by the constant. The rotation of a test vector around the face of $P_L$ dual to the triangle $ABN$ is given by Eq. (26) in Sec. 2. The product of these space-time rotations—one for each face of the polyhedron—will be the identity transformation.

$$\prod \mathscr{L}_{ABN_i} = 1 = \text{Identity matrix} \tag{38}$$

This is the skeletal form of the ordinary Bianchi identity. Since the deficit angles are assumed to be small, the left-hand side of the above equation may equally well be expressed as the sum of the rotation bivectors [Eq. (24)] associated with the faces of $P_L$

$$\frac{i}{2}\cdot \sum_{\substack{\Delta\, ABN_i \\ \text{in entourage} \\ \text{of edge } L}} \left[\frac{{}^*(\mathbf{N}_i\mathbf{B}\wedge \mathbf{N}_i\mathbf{A})}{\text{Area } ABN_i}\right]\cdot \varepsilon_{ABN_i}=0 \tag{39}$$

Although the rotations are small, they are nevertheless finite; therefore, the space-time rotations in Eq. (38) will not in general commute. Consequently the Bianchi identity will only be an approximate identity and the corresponding Regge equation derived in the last section will be only approximate. How close an approximation? We saw in Sec. 2 that the deficit angles were of second order ($L^2$) in the local scale of the lattice; hence, the equations ($\sum$ moment arm times deficit angle) are third-order equations in $L$. The failure of the rotations to commute is of fourth order, but this is only part of the story. The reader is referred to Ref. 15. It is clear though that the identity can be made as exact as one wishes if a suitable refinement of the mesh is made.

## 9. THE RIGHT-HAND SIDE OF THE REGGE EQUATION

We ask, "What is to be interpreted as being on the right-hand side of the Regge equation?" The zero vector, if there is no source. But what if there is a flow of $E$–$p$ in the space-time skeleton? We refer to the continuum once again for guidance. In Eq. (3) we saw that the right-hand side was a vector quantity giving the content of $E$–$p$ within the three-dimensional region and that it was equal, up to a factor of $8\pi$, to the dual of the "net" moment of rotation for that region. We therefore suggest that the right-hand side of the corresponding skeletal equation stands for the amount of $E$–$p$ associated within the edge $L$ dual to polyhedral domain $P_L$. The full (source + geometry) Regge equation analogous to Eq. (3) appears to be

$$\begin{pmatrix} \text{Dual of the moment of rotation} \\ \text{associated with the circuit of} \\ \text{one face summed over all faces} \\ \text{of polyhedron } P_L \text{ in } S^* \text{ dual} \\ \text{to edge } L \text{ in skeleton geometry} \end{pmatrix} = 8\pi \begin{pmatrix} \text{Content of energy-momentum} \\ \text{associated with edge } L \text{ in } S \end{pmatrix} \tag{40}$$

It is not clear what is meant by the "amount of $E$–$p$ contained in" the warped polyhedral domain $P_L$. This is why we refer to the right-hand side as "the amount of $E$–$p$ associated with" edge $L$.

We recall that each moment of rotation trivector associated with a face of $P_L$ has the same orientation—each is directed along the edge $L$ in $S$ dual to $P_L$ in $S^*$. Therefore, if this interpretation of the $E$–$p$ 4-vector is corroborated, then the skeletal structure may be thought of as a "circuit" carrying $E$–$p$.

In general the polyhedral domain $P_L$ will be warped. In other words, such a polyhedron cannot ordinarily be embedded in a flat three-dimensional space. Therefore, the volume enclosed by such a polyhedral region does not have a clear or precise meaning. Moreover, the amount of warping of the dual lattice does not depend on the scale of the lattice. The warping is not a higher-order effect in the local scale $L$ of the lattice and therefore cannot be neglected. This difficulty brings us face to face with the true nature of the skeletal approximation. We ask, "Where does the approximative character of Eq. (40) reveal itself?" An averaging process may be the key in answering this question. For example, if we examine a region of the skeleton geometry containing $N$ edges, then the volume of the dual domain associated to these $N$ edges becomes more precise. Here the dual domain is the aggregate sum of all the polyhedral domains associated to the $N$ edges. This collection of polyhedral domains may have a nice "surface" volume when viewed at a "suitable" distance. This entire issue of "what locality means" in Regge calculus is deep. The author hopes that the approach to the skeletal equations presented here may be helpful in resolving a few of these difficulties. This will be discussed in another place.

## 10. TOWARDS A CONSERVATION LAW IN REGGE CALCULUS

Although a skeletal version of the contracted Bianchi identity does not at this time exist, it is our belief that the application of the principle that the $(\partial \circ \partial \equiv 0)$ in its (2–3–4)-dimensional formulation to a four-dimensional cell in the dual lattice will produce such an identity. In this section we present a few preliminary results of qualitative character that have been obtained. We point out here that the nonlocal nature of the curvature analysis that is needed in order to obtain the identity appears to be the key issue involved.

Consider a vertex $A$ in the original skeleton geometry $S$. Emanating from this vertex will be a collection of edges, each of which is labeled for convenience as $L_i = AB_i$. These edges are precisely the edges found in the entourage of $A$. Associated to this vertex is a four-dimensional cell $C_A$ in the dual lattice $S^*$; furthermore, there is a 1–1 correspondence between the edges emanating from $A$ and the three-dimensional polyhedral faces $P_{L_i}$ of $C_A$. We apply the BB principle to this 4-cell $C_A$. Our hope lies in the realization of a skeletal version of the contracted Bianchi identity as a result of this application.

An orientation chosen for the vertex $A$ (source or sink, + or −, etc.) will induce an orientation on $C_A$. Once oriented, $C_A$ has the topological property that its boundary is boundaryless. This fact assures us that the sum, over the polyhedral domains $P_{L_i}$ of the "net" moment of rotation vectors will be identically zero.

We recall that the "net" moment of rotation vector for any given $P_L$ was introduced in Sec. 7. To obtain the contracted Bianchi we need only sum these vector quantities; however, each moment vector is evaluated at the center of their corresponding edge $L_i$. The evaluation of this sum involves the parallel transportation of the vectors to a common point, say $A$. This summation at point $A$, we believe, is a skeletal analogue of Eq. (1). Once obtained, it may be used in conjunction with the full Regge equation [Eq. (40)] to obtain a skeletal conservation law. These issues, which we believe are of utmost importance, will be discussed in another place.[(19)]

## ACKNOWLEDGMENTS

For helpful discussions I thank T. Banchoff, L. Brewin, C. Fuchs, C. McCarter, T. Regge, W. Schleich, R. Sorkin, T. Piran, R. Williams, and very especially A. Kheyfets. Above all others, I thank Prof. John Archibald Wheeler for the guidance he has given me in my research and for the many hours of discussions we shared on this and related topics. I also thank Z. Davis for her help in preparing this manuscript.

## REFERENCES

1. J. A. Wheeler, "Physics and Austerity: Law Without Law," A Working Paper, The University of Texas, (1982), pp. 61–83; published in Chinese as "Physics and Austerity," Anhui, Science and Technology Publication, China August 1982.
2. J. A. Wheeler, "Particles and Geometry," in *Unified Theories of Elementary Particles*, P. Breithenlohner and H. Durr, eds. (Springer-Verlag, New York, 1982), pp. 189–217.
3. C. W. Misner, K. S. Thorne, and J. A. Wheeler, "Bianchi Identities and the Boundary of a Boundary Is Zero," in *Gravitation* (W. H. Freeman, San Francisco, 1973), Chap. 16.
4. É. Cartan, 1928 and 1946, *Leçons sur la Géométrie des Espaces de Riemann* (Gauthier-Villars, Paris, France); English translation by Robert Hermann (Mathematical Sciences Press, Brookline, Massachusetts, 1983), Chap. 8.
5. W. A. Miller and J. A. Wheeler, "4-Geodesy," paper presented at the Congress "Galactic and Extragalactic Dark Matter," Rome, 28–30 June 1983; to appear in *Nuovo Cimento* (1985).

6. A. Kheyfets, "The Boundary of a Boundary Principle: A Unified Approach," to appear, *Found. Phys.* (1986).
7. T. Regge, "General Relativity without Coordinates," *Nuovo Cimento* **19**, 558–571 (1961).
8. I. Ciufolini, "On a Generalized Geodesic Deviation Equation," Chap. 2 of dissertation, "*Theory and Experiments in General Relativity and Other Metric Theories of Gravity*," The University of Texas (1984).
9. R. M. Williams and G. F. R. Ellis, "Regge Calculus and Observations. I. Formalism and Applications to Radial Motion and Circular Orbits," *Gen. Relativ. Gravit.* **13**, 361 (1981).
10. J. R. Munkres, *Elements of Algebraic Topology* (Addison-Wesley, Menlo Park, California, 1984), pp. 26–32.
11. H. Bacry, *Lectures on Group Theory and Particle Physics* (Gordon and Breach, New York, 1977), Chap. 7.
12. A. Brøndsted, *An Introduction to Convex Polytopes* (Springer-Verlag, New York, 1983), p. 64.
13. B. Grünbaum, *Convex Polytopes* (Wiley, New York, 1967), pp. 46, 137, 243, and 285.
14. H. S. M. Coxeter, *Regular Polytopes* (Dover, New York, 1973), 3rd. edn., pp. 6 and 60.
15. J. Cheeger, W. Müller, and R. Schrader, "On the Curvature of Piecewise Flat Spaces," *Commun. Math. Phys.* **92**, 405 (1984).
16. T. D. Lee, "Time as a Dynamical Variable," a talk given at the Shelter Island II Conference, June 2, 1983, Columbia University Preprint CU-TP-266 (1983), p. 18.
17. C. W. F. Everitt, *James Clark Maxwell* (Charles Scribners, New York, 1975), Chap. 10.
18. John A. Wheeler, private communication.
19. W. A. Miller, "Null-Strut Geometrodynamics and the Inchworm Algorithm," in *Dynamical Spacetimes and Numerical Relativity*, J. Centrella, ed. (Cambridge Univ. Press, 1986), in press.

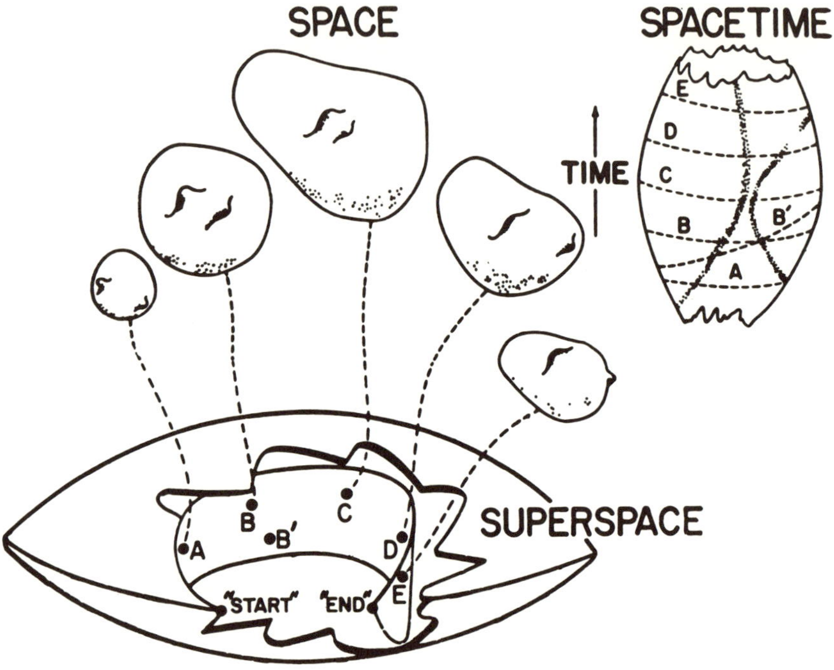


*Plate 17*

# Canonical Geometrodynamics and General Covariance

**Karel V. Kuchař**[1]

*Received July 2, 1985*

*By extending geometrodynamical phase space by embeddings and their conjugate momenta, one can homomorphically map the Lie algebra of space-time diffeomorphisms into the Poisson algebra of dynamical variables on the extended phase space.*

As a relativist, I take physical laws to be independent of time, place, orientation, and the state of motion: The credo recited under the name of general covariance. As a person, I am painfully aware that particular events, unlike the laws which govern them, depend on all of these circumstances. There was a time when my place, orientation, and my state of motion were severely restricted by accidents of initial conditions. This is why I felt it to be a great event when, in 1963, through a "Window on the West" (Prague in the rewritten role of St. Petersburg of Count Francesco Algarotti[(1)]), I went to the Les Houches Summer School, not to read about relativity as it was done in the past, but to feel its presence. It was there I first met John Wheeler. Five years later, just as the proverbial window was being shut again, I had the good luck to be on my way to Princeton as his assistant. Translating myself back to 1963, I realize with astonishment that he was then about as old as I am now. Recollecting the span of years between then and now and the span of subjects and problems which I learned from John Wheeler in that span of years, I find it not out of place to celebrate his anniversary with a brief summary of one such problem—*pars pro toto*—which, when appropriately worded, sounds an emotional echo of this introduction: the problem of the intrinsic tension between the idea of

[1] Department of Physics, The University of Utah, Salt Lake City, Utah 84112.

general covariance of physical laws and Hamiltonian dynamics, according to which physical events are evolved in time.

It was from John Wheeler's lectures in Les Houches that I first learned about geometrodynamics, that utmost version of relativistic dynamics in which space-time itself is envisaged as a dynamical history of a Riemannian geometry evolving along a foliation of spacelike hypersurfaces. I watched with fascination sequences of embedding diagrams which John Wheeler drew step by step in colored chalk on the blackboard. The dynamical process by which this was done mirrored the spirit of geometrodynamics itself. I watched the expansion and recontraction of the Friedmann universe, the mouthful of a star swallowed down the narrowing throat during gravitational collapse, or the strangling of the double horn of the Schwarzschild geometry leading inevitably to the pinch-off which separated those on the left from those on the right, the bugles sounding the coming of an apocalyptic singularity. Not all of these found their way into the written version of the notes[(2)] which, like other great undertakings, remained a torso. People talking about space-times as finished objects may use transparencies and write papers, but watching John Wheeler drawing his diagrams was enough to persuade me that the proper medium for geometrodynamics is chalk on blackboard.

I also gathered from John Wheeler's lectures that his geometric *theatrum mundi* was not being shown to us for its own sake but as a prolegomenon to any (or at least some) future quantization. The geometrodynamical histories were supposed to interfere with each other and lead to the propagator of the probability amplitude for geometry. One could introduce momentum conjugate to the geometry and these together were expected to obey uncertainty relations. Geometrodynamics could be formulated in the phase space of these variables and quantized the canonical way. However, a concurrent sequence of lectures by Bryce DeWitt[(3)] started with a stern warning: "A chief goal of these lectures is to develop a framework within which the quantization of fields possessing infinite dimensional invariance groups may be carried out in a manifestly covariant fashion. The requirement of manifest covariance means that we shall adopt an over-all spacetime view from the outset and ignore canonical formulations." The disparity between canonical and covariant was visible from my very first encounter with geometrodynamics and quantum gravity.

I should explain now this disparity arises in the formalism of canonical geometrodynamics. A geometry on a given hypersurface may be viewed as a point in an infinitely dimensional space for which John Wheeler coined the term superspace. But what about the hypersurfaces which carry these geometries? When Dirac,[(4)] Bergmann,[(5)] and Arnowitt, Deser, and

Misner[6] developed the geometrodynamical formalism, they used to identify the hypersurfaces of the foliation with constant time hypersurfaces in an ad hoc chosen space-time coordinate system. This is fine as long as one does not want to compare the evolution along two different foliations,[7] to study the change of geometry when a hypersurface is bent or tilted about a fixed space-time point,[8] or to change the space-time coordinate system at will without changing the foliation.[9] As my references show, from time to time it is important to do all of these things. I thus found it convenient to introduce another infinitely dimensional space, which I called hyperspace,[9] encompassing all the spacelike hypersurfaces. A spacelike hypersurface may be considered as an instant of time; maybe it would have been better to call this space supertime. Terminology apart—and this is the point which I want to stress—the proper language to describe geometrodynamics is that of variational differential equations in hyperspace. Or, because it is formally easier to work with embeddings rather than with hypersurfaces and with metrics rather than with geometries, factorizing the spatial diffeomorphisms only at the end of the day, one can work with variational differential equations in the space of embeddings. What I have are thus the embeddings from the space manifold $\Sigma$ (which for simplicity I take to be compact) to the space-time manifold $\mathscr{M}$,

$$X \in \mathrm{Emb}(\Sigma, \mathscr{M}) : x \in \Sigma \mapsto \mathrm{X} = X(x) \in \mathscr{M} \tag{1}$$

and the metrics $\gamma$ induced by the space-time metric $g$ on these embeddings,

$$\gamma = X * g \tag{2}$$

After introducing coordinates $x^a(x)$ in $\Sigma$ and $\mathrm{X}^\alpha(\mathrm{X})$ in $\mathscr{M}$, the embeddings $X$ may be parametrized by four functions $X^\alpha(x^a)$ of three coordinates $x^a$, and the metrics $\gamma$ written in the component form

$$\gamma_{ab}(x; X] = g_{\alpha\beta}(X(x))\, X^\alpha{}_{,a}(x)\, X^\beta{}_{,b}(x) \tag{3}$$

The square bracket in $\gamma_{ab}(x; X]$ emphasizes that $\gamma_{ab}$ is a function of $x$, but a functional of the embedding. This is what the round and square brackets mean throughout this paper.

Of course it is not only the evolution of geometry which may be viewed in this fashion. Ordinary fields, like the scalar or the electromagnetic field, may also be restricted to the embeddings and, if needed, projected parallel and perpendicular to them. The restriction of the scalar field $\psi(\mathrm{X})$ thus yields

$$\psi(x) = \psi(X(x)) \tag{4}$$

and the projections of the electromagnetic four-potential $A_\alpha(\mathrm{X})$ lead to the expressions

$$\begin{aligned} A_\perp(x) &= -n^\alpha(x)\, A_\alpha(X(x)) \\ A_a(x) &= X^\alpha{}_{,a}(x)\, A_\alpha(X(x)) \end{aligned} \tag{5}$$

The unit normal $n^\alpha(x; X]$ to the embedding $X(x)$ is characterized by the conditions

$$\begin{aligned} g_{\perp\perp} &:= g_{\alpha\beta}(X(x))\, n^\alpha(x)\, n^\beta(x) = -1 \\ g_{\perp b} &:= -g_{\alpha\beta}(X(x))\, n^\alpha(x)\, X^\beta{}_{,b}(x) = 0 \end{aligned} \tag{6}$$

which explains why in geometrodynamics the only dynamically relevant variable is the spatial metric (3).

The space Riem($\Sigma$) of all such Riemannian metrics is thus a configuration space of geometrodynamics and the space $C^\infty(\Sigma, R)$ of the scalar fields (4) a configuration space for any dynamics of such a field. The configuration space of the scalar field coupled to gravity is $C^\infty(\Sigma, R) \times \mathrm{Riem}(\Sigma)$. In Hamiltonian dynamics, the fields are evolved in the phase space which is the cotangent bundle over the appropriate configuration space.

The actual dynamics follows from the space-time action: the Hilbert action

$$S[g] = \int_{\mathscr{M}} d^4\mathrm{X}\, |\det g(\mathrm{X})|^{1/2}\, R[g] \tag{7}$$

in source-free gravity, and the scalar field action

$$S[\psi] = -\frac{1}{2} \int_{\mathscr{M}} d^4\mathrm{X}\, |\det g(\mathrm{X})|^{1/2}\, g^{\alpha\beta}(\mathrm{X})\, \psi_{,\alpha}(\mathrm{X})\, \psi_{,\beta}(\mathrm{X}) \tag{8}$$

in the simplest case of the massless scalar field. The process by which the space-time action is cast into the canonical form is well known[6,10] and I shall not go through its twists and turns. The canonical action

$$S[\gamma, \pi; N, \vec{N}] = \int_{\mathbb{R}} dt \left( \left( \int_\Sigma d^3x\, \pi^{ab}(x)\, \dot{\gamma}_{ab}(x) - H(N, \vec{N}) \right) \right) \tag{9}$$

which governs the evolution of the metric $\gamma_{ab}(x)$ and its conjugate momentum $\pi^{ab}(x)$ in source-free geometrodynamics is a functional of these variables as well as of the lapse $N(x)$ and the shift $N^a(x)$ multipliers. The Hamiltonian $H(N, \vec{N})$ is constructed by smearing the expressions

$$\begin{aligned} H(x; \gamma, \pi] = \tfrac{1}{2}(\det \gamma)^{-1/2}(&\gamma_{ac}\gamma_{bd} + \gamma_{ad}\gamma_{bc} \\ &- \gamma_{ab}\gamma_{cd})\, \pi^{ab}\pi^{cd} - (\det \gamma)^{1/2}\, R[\gamma] \end{aligned} \tag{10}$$

and

$$H_a(x; \gamma, \pi] = -2\pi^b_{a|b} \tag{11}$$

called super-Hamiltonian and supermomentum, by the lapse function and the shift vector:

$$\begin{aligned} H(N, \vec{N}) &= H(N) + H(\vec{N}) \\ H(N)[\gamma, \pi] &= \int_\Sigma d^3x\, N(x)\, H(x; \gamma, \pi] \\ H(\vec{N})[\gamma, \pi] &= \int_\Sigma d^3x\, N^a(x)\, H_a(x; \gamma, \pi] \end{aligned} \tag{12}$$

The multipliers $N(x)$ and $N_a(x)$ specify the stacking of the embeddings $X(x)$ within a foliation $X(x, t)$. They are the projections of the deformation four-vector $\dot{X}^\alpha(x, t)$ into directions $n^\alpha(x; X]$ and $X^\alpha{}_{,a}(x; X]$ normal and parallel to the embedding in question,

$$\dot{X}^\alpha(x) = N(X)\, n^\alpha(x) + N^a(x)\, X^\alpha{}_{,a}(x) \tag{13}$$

In geometrodynamics, however, they are not varied as functionals of the embedding, but rather as some arbitrarily prescribed functions of $x^a$ and $t$. This is an important point in our discussion. The arbitrariness of the multipliers is a direct consequence of the arbitrariness of all the components of the space-time metric $g_{\alpha\beta}(\mathrm{X})$ in the Hilbert action (7).

The variation of the multipliers leads to the constraints

$$H(N) = 0 = H(\vec{N}) \quad \forall N, \vec{N}, \qquad \text{or} \quad H(x) = 0 = H_a(x) \tag{14}$$

while the variation of $\gamma_{ab}(x)$ and $\pi^{ab}(x)$ yields the Hamilton equations

$$\begin{aligned} \dot{\gamma}_{ab}(x) &= \{\gamma_{ab}(x), H(N, \vec{N})\} \\ \dot{\pi}^{ab}(x) &= \{\pi^{ab}(x), H(N, \vec{N})\} \end{aligned} \tag{15}$$

When a source field, like $\psi(\mathrm{X})$, is nonderivatively coupled to gravity, as in (8), the scheme stays essentially unchanged. The gravitational super-Hamiltonian (10) and supermomentum (11) are merely complemented by the ordinary energy and momentum densities of the field, viz., by

$$H^\psi(x; \psi, \pi, \gamma] = \tfrac{1}{2}(\det \gamma)^{-1/2}\, \pi^2 + \tfrac{1}{2}(\det \gamma)^{1/2}\, \gamma^{ab}\psi_{,a}\psi_{,b} \tag{16}$$

and

$$H^\psi{}_a(x; \psi, \pi] = \pi\psi_{,a} \tag{17}$$

Equations (9)–(12) and (14)–(15) summarize the basic scheme of canonical geometrodynamics. Let me now restate the main problem: to what extent does this scheme reflect space-time covariance which is such a prominent feature of general relativity? The Hilbert action (7) from which the theory started is invariant under the action of the diffeomorphism group Diff $\mathscr{M}$ of the space-time manifold. The Lie algebra LDiff $\mathscr{M}$ of this group can be identified with the set of all (complete) vector fields $\mathbf{U}(\mathrm{X})$, $\mathbf{V}(\mathrm{X})$,... on $\mathscr{M}$ closed under the Lie bracket

$$[\mathbf{UV}] = -(U^\beta V^\alpha{}_{,\beta} - V^\beta U^\alpha{}_{,\beta}) \frac{\partial}{\partial \mathrm{X}^\alpha} \tag{18}$$

In an ordinary Poincaré invariant field theory on flat Minkowskian background, each generator of the Poincaré group is represented by a dynamical variable on the phase space of the field, so that the Lie algebra of the Poincaré group is homomorphically mapped into the Poisson bracket algebra of these dynamical variables. Moreover, each dynamical variable (the total energy, momentum, angular momentum, and the generalized angular momentum) generates the evolution of the field carried by a flat hypersurface when this hypersurface is displaced by the corresponding generator (time translation, space shift, rotation, or the Lorentz boost) of the Poincaré group. Is a similar construction valid in canonical geometrodynamics? Can one find a homomorphic mapping from the Lie algebra LDiff $\mathscr{M}$ of space-time vector fields into the Poisson algebra of dynamical variables on the geometrodynamical phase space $T^*$ Riem$(\Sigma)$?

Because the Hamiltonian (12) generates the evolution of the metric under deformation of the embedding, it is natural to study the algebra of the Poisson brackets among the $H(\vec{N})$'s and $H(N)$'s as a possible answer to this question. These Poisson brackets were evaluated a long time ago starting with the work of Dirac[2] and called (somewhat misleadingly, as we shall see) the Dirac algebra:

$$\{H(\vec{M}), H(\vec{N})\} = H(\vec{J}) \tag{19}$$

$$\{H(\vec{M}), H(N)\} = H(J) \tag{20}$$

$$\{H(M), H(N)\} = H(\vec{K}) \tag{21}$$

Here, the smearing functions $\vec{J}$, $J$, and $\vec{K}$ on the right are constructed from the smearing functions $M$, $\vec{M}$ and $N$, $\vec{N}$ on the left by the prescriptions

$$\vec{J} = [\vec{M}\vec{N}] = -L_{\vec{M}}\vec{N} \tag{19$'$}$$

[2] Dirac derived these relations in unsmeared form for parametrized field theories.[11] For the evaluation of the corresponding quantum commutators in canonical geometrodynamics, see Ref. 12. For the smeared form, see Ref. 13.

$$J = L_{\vec{M}} N \tag{20'}$$

$$K^a = \gamma^{ab}(MN_{,b} - NM_{,b}) \tag{21'}$$

Equation (19), (19′) is a promising beginning: it tells us that the mapping $\vec{N} \rightarrow H(\vec{N})$ is an antihomomorphism[3] from the Lie algebra LDiff $\Sigma$ of the *spatial* diffeomorphism group into our Poisson algebra. The second equation (20), (20′) is equally promising by telling us that $H(x)$ behaves as a scalar density under spatial diffeomorphisms. The third equation (21), (21′) is a total disaster. Because the composition rule (21′) refers not only to the lapse functions $M$ and $N$ which describe the deformations but also to the element $\gamma_{ab}$ of the phase space on which these deformations act, the collection of all composition rules (19′)–(21′) does not form any algebra at all. Thus the full group of space-time diffeomorphisms has somehow gotten lost in the transition from the Hilbert action (7) to the canonical action (9).

Various aspects of this problem have worried me for years. During the past academic year which I spent in London I found that they were worrying Chris Isham even more. He has been enthusiastically developing a group theoretical approach to quantum gravity[(14)] which requires the canonical group of the theory (describing the "kinematics" of gravity like the Heisenberg group describes the "kinematics" of ordinary quantum mechanics) to be unitarily implemented together with the dynamical group yielding the evolution. He seemed to hold the right canonical group but the space-time diffeomorphism group which was its natural dynamical partner has, as I have just explained, disappeared from the canonical formalism. The situation was exasperatingly funny as if taken from a farce in which the groom fails to appear in the church while the bride and all the relatives are waiting. We thus set out together to fetch the groom to the altar; what follows is mostly the result of our joint effort.[(15–16)] I admit that even if one finds the groom there is no guarantee (in the best 19th century English tradition[(17)]) that the marriage will ever be consummated but, we felt, it is worth a try.

Our search can be described as a pursuit of two related leads:

(1) Space-time diffeomorphisms Diff $\mathscr{M}$ act on space-time points $\mathrm{X} \in \mathscr{M}$ and the generators of LDiff $\mathscr{M}$ are space-time vector fields $\mathbf{U}(\mathrm{X})$. By the left action, Diff $\mathscr{M}$ can also be considered as acting on the embeddings $X \in \mathrm{Emb}(\Sigma, \mathscr{M})$, each generator $\mathbf{U}(\mathrm{X})$ displacing the embedding $X(x)$ by the amount $\mathbf{U}(X(x))$. On the other hand, any explicit reference to the space-time points X and the embeddings $X(x)$ is conspicuously absent in the canonical action (9). How can one ever succeed in representing space-

[3] Why antihomomorphism rather than homomorphism is explained in Ref. 16.

time diffeomorphisms in a canonical formalism in which embeddings are lost and the only remaining variables are *spatial* fields $\gamma_{ab}(x)$, $\pi^{ab}(x)$?

(2) On the way from the Hilbert action (7) to the canonical action (9), the Lie algebra (18) of space-time diffeomorphisms got somehow distorted into the Dirac "algebra" of the lapses and shifts. What is the geometric meaning of this "algebra," and can the distortion be undone?

Let me pursue the first lead. There is a class of canonical theories, called parametrized theories, which contain the embedding variables $X(x)$ among the configuration variables of the phase space. Their ancestors among finite-dimensional systems are really ancient,(18) while the field relatives themselves were considered by Dirac,(11) albeit only in a flat Minkowski space-time. They have so many similarities to canonical geometrodynamics that whenever I meet a problem there I invariably try to clarify it by inventing and solving a similar problem in a parametrized field theory. Following this poor man's way of looking at things (borrowing another phrase from John Wheeler's vocabulary) I spent a lot of time studying various features of such theories.(19–20) Here are some which are relevant to our problem:

Consider a field, like $\psi(X)$, which does not interact with gravity, but propagates on a given metric background $g_{\alpha\beta}(X)$. Because the metric $g_{\alpha\beta}(X)$ enters the spacetime action (8) as an external potential rather than as a field variable, the action is no longer invariant under space-time diffeomorphisms. To make it invariant, one must "parametrize" it by introducing new integration coordinates $t, x \in \mathbb{R}x\Sigma$ and adjoining the mappings $\Sigma \rightarrow \mathcal{M}$, i.e., the embedding variables $X^\alpha(x^a)$, to the configuration space. Here, $X^\alpha(x^a)$ are considered as four scalar fields, the coordinates $X^\alpha$ being just those coordinates in which the background metric $g_{\alpha\beta}(X^\gamma)$ has the given form (e.g., the Minkowskian form $\eta_{\alpha\beta}$ when $X^\alpha$ are pseudo-Cartesian coordinates on flat background). The action (8) is then conceived as a functional of the embedding fields $X(x, t)$ as well as of the scalar field $\psi(x, t) = \psi(X(x, t))$.

As before, this action can be cast into the canonical form

$$S[\psi, \pi; X] = \int_{\mathbb{R}} dt \int_{\Sigma} d^3x \, (\pi(x)\, \dot{\psi}(x) - N(x)\, H^{\psi}(x) - N^a(x)\, H^{\psi}{}_a(x)) \tag{22}$$

in the field variables $\psi(x)$, $\pi(x)$. However, the lapse and the shift are no longer some externally prescribed functions of $x$ and $t$, but are the functionals

$$N(x; X] = -n_\alpha(x; X]\, \dot{X}^\alpha \tag{23}$$

and

$$N^a(x; X] = X^a_\alpha(x; X]\, \dot{X}^\alpha$$
$$X^a_\alpha(x; X] := g_{\alpha\beta}(X(x))\, X^\beta{}_{,b}(x)\, \gamma^{ba}(x; X] \tag{24}$$

of a foliation of embeddings. Their functional form is obtained from the given function $g_{\alpha\beta}(\mathrm{X})$ by defining the induced metric $\gamma_{ab}(x; X]$ and the normal $n_\alpha(x; X]$ through Eqs. (3) and (6). Moreover, they are *linear* in the deformation vector $\dot{X}^\alpha(x)$. The momenta $P_\alpha(x)$ conjugate to the embedding variables thus do not depend on $X^\alpha$; their introduction leads to the constraints $\Pi_\alpha(x) = 0$, with

$$\Pi_\alpha(x) := P_\alpha(x) - n_\alpha(x; X]\, H^\psi(x; X, \psi, \pi] + X^a_\alpha(x, X]\, H^\psi(x; \psi, \pi] \tag{25}$$

under which the complete Hamiltonian form of the action (22),

$$S[\psi, \pi; X, P] = \int_{\mathbb{R}} dt \int_\Sigma d^3x\, (\pi\dot{\psi} + P_\alpha \dot{X}^\alpha) \tag{26}$$

must be varied. Alternatively, I can adjoin the constraints to the action by a Lagrange multiplier $N^\alpha(x, t)$ and vary all the variables freely,

$$S[\psi, \pi; X^\alpha, P_\alpha, N^\alpha] = \int_{\mathbb{R}} dt \int_\Sigma d^3x\, (\pi\dot{\psi} + P_\alpha \dot{X}^\alpha - N^\alpha \Pi_\alpha) \tag{27}$$

The variation of the multiplier leads back to the constraints $\Pi_\alpha = 0$. In a parametrized theory both the field variables and the embedding variables are subject to the Hamilton equations following from the Hamiltonian

$$\Pi(\mathbf{N}) = \int_\Sigma d^3x\, N^\alpha(x)\, \Pi_\alpha(x) \tag{28}$$

The Hamilton equation for $X^\alpha(x)$ ensures that the multiplier $N^\alpha(x)$ has the meaning of the deformation vector: $\dot{X}^\alpha(x) = N^\alpha(x)$. The Hamilton equations for $\psi(x)$ and $\pi(x)$ are equivalent to the wave equation for $\psi(\mathrm{X})$, and the Hamilton equation for $P_\alpha(x)$ works so that the constraint is preserved in time.

Our new candidate for representing space-time diffeomorphisms is the Poisson algebra of the Hamltonians $\Pi(\mathbf{N})$. One can learn by cumbersome though enlightening calculation[15] (for a really poor man, cumbersome calculations can often be enlightening) that this algebra is abelian,

$$\{\Pi(\mathbf{M}), \Pi(\mathbf{N})\} = 0 \qquad \forall \mathbf{M}, \mathbf{N} \tag{29}$$

This result is so surprising that a good fundamental reason must lie behind it. There is one. Consider the variation of the canonical action (25)–(26)

with respect to the foliation $X(x)$. Since the space-time action (8) is manifestly independent of any such choice, it follows that the *value* of the canonical action (25)–(26) must be likewise independent if evaluated on canonical fields $\psi(x)$, $\pi(x)$ and $X^\alpha(x)$, $P_\alpha(x)$ appropriately related to the space-time fields. I have shown[21] that this foliation independence of the parametrized action leads directly to the vanishing Poisson bracket (29).

Superficially, the abelian algebra (29) seems even less related to the Lie algebra (18) of Diff $\mathscr{M}$ than the Dirac "algebra" (19)–(21), (19′)–(21′). However, one can displace the embedding not by the action of externally prescribed four-vectors $\mathbf{M}(x)$ and $\mathbf{N}(x)$, but by the four-vectors $\mathbf{U}(x, X] = \mathbf{U}(X(x))$ and $\mathbf{V}(x, X] = \mathbf{V}(X(x))$ which correspond to the generators $\mathbf{U}(\mathrm{X})$ and $\mathbf{V}(\mathrm{X})$ of LDiff $\mathscr{M}$. Such four-vectors now depend on the embedding, and the Poisson bracket $\{\Pi(\mathbf{U}), \Pi(\mathbf{V})\}$ acquires additional terms which arise through the action of the embedding momenta $P_\alpha(x)$ in the constraints (25) on such embedding variables. As a result,

$$\{\Pi(\mathbf{U}), \Pi(\mathbf{V})\} = \Pi([\mathbf{U}\mathbf{V}]) \tag{30}$$

which shows that the mapping

$$\mathbf{U} \mapsto \Pi(\mathbf{U})[X, P, \psi, \pi] = \int_\Sigma d^3x\, U^\alpha(X(x))\, \Pi_\alpha(x)[X, P, \psi, \pi] \tag{31}$$

is a homomorphism from the Lie algebra of Diff $\mathscr{M}$ into the Poisson algebra of dynamical variables of the parametrized field theory.

To summarize, a parametrized field theory is a generally covariant theory by virtue of the introduction of embeddings as supplementary canonical variables, and the embeddings make it possible to represent LDiff $\mathscr{M}$ within the canonical formalism. However, what possible bearing does all of this have on gravity which is covariant even without the explicit introduction of embeddings and which does not offer us constraints which satisfy (29), but rather the Dirac "algebra" (19)–(21), (19′)–(21′)?

To regain a sense of direction, I shall pursue the second lead. What is the geometric meaning of the Dirac "algebra" and can one mimic it in a parametrized field theory? The answer to this problem is implicit in Dirac's work[11]; it was spelled out in detail by Teitelboim,[22] and expressed in terms of geometry of the embedding space by myself.[9] Instead of describing the deformation of an embedding by a space-time vector field $N^\alpha(x)$ along that embedding, one describes it by the projections of that vector field into directions normal and parallel to the embedding, i.e., by the lapse function and the shift vector. In other words, instead of parametrizing the embedding phase space $T^*\,\mathrm{Emb}(\Sigma, \mathscr{M})$ by $X^\alpha(x) \in \mathrm{Emb}(\Sigma, \mathscr{M})$ and by the momentum $P_\alpha(x)$ corresponding to the holonomic basis $\delta/\delta X^\alpha(x)$ in $T_X\,\mathrm{Emb}(\Sigma, \mathscr{M})$, parametrize it by the anholonomic momenta

$$P(x) = n^\alpha(x; X]\, P_\alpha(x) \qquad \text{and} \qquad P_a(x) = X^\alpha{}_{,a}(x; X]\, P_a(x) \tag{32}$$

corresponding to the anholonomic basis

$$n^\alpha(x; X]\, \delta/\delta X^\alpha(x) \qquad \text{and} \qquad X^\alpha{}_{,a}(x; X]\, \delta/\delta X^\alpha(x)$$

in $T_x\,\mathrm{Emb}(\Sigma, \mathscr{M})$. In this anholonomic basis, the deformation is characterized by the anholonomic components $N(x)$ and $N^\alpha(x)$,

$$P(N, \vec{N}) = P(N) + P(\vec{N}) = \int_\Sigma d^3x\, N(x)\, P(x) + \int_\Sigma d^3x\, N^a(x)\, P_a(x) \tag{33}$$

The normal in the expression (32) refers back to the background metric $g_{\alpha\beta}(\mathrm{X})$ through (6) and it is thus no surprise that the projection $\gamma_{ab}(x; X]$ of that metric finds its way into the Poisson brackets among the dynamical variables (33). In fact, such Poisson brackets durifully reproduce the Dirac "algebra." Though $\gamma_{ab}(x; X]$ are not canonical coordinates in our phase space, they depend on the canonical coordinates $X(x)$, and the old trouble that one is not dealing with a true algebra reappears in a new guise.

So far, there is no field dynamics involved in this argument. However, one can express the full constraint (25) in the same anholonomic basis, arriving at the expressions

$$\Pi(N)[X, P, \psi, \pi] = \int_\Sigma d^3x\, N(x)\, n^\alpha(x; X]\, \Pi_\alpha(x; X, P, \psi, \pi] \tag{34}$$

and

$$\Pi(\vec{N})[X, P, \psi, \pi] = \int_\Sigma d^3x\, N^a(x)\, X^\alpha{}_{,a}(x; X]\, \Pi_\alpha(x; X, P, \psi, \pi] \tag{35}$$

which again obey the Dirac "algebra." Expressed in the anholonomic basis, parametrized field theories present the same problems as canonical geometrodynamics. By returning to the holonomic basis, the problems disappear and representation of space-time diffeomorphisms becomes possible.

This suggests that canonical geometrodynamics should likewise be formulated in a holonomic basis in $T^*\,\mathrm{Emb}(\Sigma, \mathscr{M})$. So far, however, the embedding variables are not even an explicit part of the geometrodynamical phase space. Because the Hilbert action is already invariant under space-time diffeomorphisms and its canonical form already involves the constraints, it is often maintained that the embedding variables are already hidden among the standard phase space variables $\gamma_{ab}$, $\pi^{ab}$.[7]

This does not help much, however, unless they can be explicitly exhibited. This is a difficult task which has never been quite satisfactorily realized. It is thus best to forget that the embedding variables may be already there and adjoin them once more explicitly to the standard phase space. If this procedure is redundant, the formalism takes care of it automatically by imposing additional constraints on the extended space variables.

In the anholonomic basis, the embedding variables and the metric variables are clearly decoupled. The embeddings are evolved by the dynamical variables $P(N)$ and $P(\vec{N})$, which do not contain the geometric variables $\gamma_{ab}$, $\pi^{ab}$, and the geometric variables are evolved by the dynamical variables $H(N)$ and $H(\vec{N})$, which do not contain the embedding variables $X^\alpha$, $P_\alpha$. All variables together are evolved by the total generators

$$\Pi(N) = P(N) + H(N) \qquad \text{and} \qquad \Pi(\vec{N}) = P(\vec{N}) + H(\vec{N}) \tag{36}$$

which again satisfy the Dirac "algebra." Constraints can be imposed separately on $P(N)$, $P(\vec{N})$ and $H(N)$, $H(\vec{N})$ or, which is the same thing, imposed in the form (14) and

$$\Pi(N) = 0 = \Pi(\vec{N}) \qquad \forall N, \vec{N} \tag{37}$$

To return to the holonomic basis, we must recombine the variables $\Pi(N)$ and $\Pi(\vec{N})$ into

$$\Pi(\mathbf{N}) = \int_\Sigma d^3x\, N^\alpha(x)(P_\alpha(x) - n_\alpha(x)\, H(x) + X^a_\alpha(x)\, H_a(x)) \tag{38}$$

The trouble here is that $n_\alpha(x)$ and $X^a_\alpha(x)$ cannot be considered as known functionals of embeddings because the space-time metric $g_{\alpha\beta}(\mathrm{X})$ is no longer given, its construction being the very aim of geometrodynamics. What is the way out of this dilemma?

In a parametrized theory, the coordinates $\mathrm{X}^\alpha$ in space-time were selected so as to give $g_{\alpha\beta}(\mathrm{X})$ the assigned form. In geometrodynamics, $g_{\alpha\beta}(\mathrm{X})$ can never be completely fixed, but at least it can be limited by "coordinate conditions."[4] In the simplest case, one can require that the space-time coordinates $\mathrm{X}^\alpha$ be such that they cast $g_{\alpha\beta}(\mathrm{X})$ into the Gaussian form:

$$g_{00}(\mathrm{X}) = 0 = g_{0i}(\mathrm{X}) \tag{39}$$

Then, given an embedding $X^\alpha(x)$ and the Riemannian metric $\gamma_{ab}(x)$ induced on it, one can reconstruct the remaining components $g_{ij}(\mathrm{X})$ of the

[4] I am using this term and approach to simplify the discussion. A more elegant and "covariant" way to introduce such conditions is through an auxiliary geometric structure.(16)

space-time metric in the coordinate cobasis $d\mathrm{X}^{\alpha}$ as some (local) functionals of $\gamma_{ab}(x)$ and $X^{\alpha}(x)$. With the mixed components (39) fixed by the Gaussian conditions, $g_{\alpha\beta}(X(x))$ along the embedding is determined as a (local) functional of $\gamma_{ab}(x)$ and $X^{\alpha}(x)$:

$$g_{\alpha\beta} = g_{\alpha\beta}(x; X, \gamma] \tag{40}$$

The coefficients $n_{\alpha}(x)$ and $X^{a}_{\alpha}(x)$ in the generator (38) are thereby also turned into some specific local functionals of $\gamma_{ab}(x)$ and $X^{\alpha}(x)$ through Eq. (6).

The coordinate conditions are necessary to make the evolution generated by the dynamical variable (38) in the holonomic basis unambiguous. Without them, the space-time metric which is built by this evolution would itself be determined only up to a space-time diffeomorphism.

By this procedure, the generator (38) becomes a well-defined functional on the extended phase space $T^*(\mathrm{Emb}(\Sigma, \mathscr{M}) \times \mathrm{Riem}(\Sigma))$ and one can calculate the Poisson bracket $\{\Pi(\mathbf{M}), \Pi(\mathbf{N})\}$. The charms stilll work and one finds that the algebra is abelian:

$$\{\Pi(\mathbf{M}), \Pi(\mathbf{N})\} = 0 \tag{41}$$

From what I have done in parametrized theories it is obvious what must be done to represent space-time diffeomorphisms in canonical geometrodynamics. Take the generators $\mathbf{U}(\mathrm{X})$, $\mathbf{V}(\mathrm{X})$ of LDiff $\mathscr{M}$, restrict them to the embeddings, and use such functionals of the embeddings $\mathbf{U}(x; X] = \mathbf{U}(X(x))$ and $\mathbf{V}(x; X] = \mathbf{V}(X(x))$, instead of the externally prescribed functions $\mathbf{M}(x)$ and $\mathbf{N}(x)$ in the Poisson bracket (41). The constructed mapping

$$\mathbf{U}(\mathrm{X}) \mapsto \Pi(\mathbf{U})[X, \Pi, \gamma, \pi] \tag{42}$$

is a homomorphism from the Lie algebra LDiff $\mathscr{M}$ into the Poisson algebra on the extended phase space:

$$\{\Pi(\mathbf{U}), \Pi(\mathbf{V})\} = \Pi([\mathbf{UV}]) \tag{43}$$

The generators $\Pi(\mathbf{U})$ which represent space-time diffeomorphisms automatically preserve both sets of constraints, (14) and (37). Therefore, if the system starts at the intersection of these constraint surfaces, it stays on it during the dynamical evolution generated by $\Pi(\mathbf{U})$. The generators consistently evolve the embeddings and the physical gravitational data on the embeddings, building a Ricci-flat space-time by this process.

After a long seminar in Princeton in which the significance of the results got all but lost in the maze of technical details, I remember John Wheeler asking the speaker (his inimitable style unfortunately getting blurred by my paraphrasing from memory): "Suppose you leave this building and get killed in an accident. What would be your last words to us which we could remember and from which we could reconstruct your theory?" I fear I am liable to be asked the same question at this moment.

These are the words I would mumble. First: to represent space-time diffeomorphisms in geometrodynamics, one needs not only superspace, but also hyperspace. One needs not only metrics, but also embeddings. Second: I learned that often I must overcome my instincts in order to achieve my aims. As a skier I learned that I must turn my body right to make a left turn and as a relativist I learned that I must squeeze a physical system by stress to slow down its gravitational collapse. Now I learned I must impose "coordinate conditions" to make canonical geometrodynamics covariant. Physics is full of paradoxes.

Covariant canonical field theory is one of them. Twenty years after the 1963 Les Houches Summer School at which I met John Wheeler and heard from him about canonical geometrodynamics, Bryce DeWitt[(23)] repeated his stern warning: "Twenty years ago, at the Les Houches session bearing the same title as the present one, I began my lecture with the words "A chief goal of these lectures is to develop a framework within which the quantization of fields possessing infinite-dimensional invariance groups may be carried out in a manifestly covariant fashion. The requirement of manifest covariance means that we shall adopt an over-all spacetime view from the outset and ignore canonical formulations." The goal today remains the same. I have been astonished to see over the intervening years how physicists, when expounding the fundamentals of field theory, have failed to apply the lessons that relativity theory taught them early in this century."

As a participant in that 1963 Les Houches session of twenty years ago, brought in from Eastern Europe by a grant from its organizers, I feel obliged to answer the implicit question why I have not learned my lesson there properly. Like a medieval heretic, I can put at least part of the blame on the Early Fathers whose statements are often in subtle disagreement with the dogmas agreed on by later councils. Five years before the 1963 Les Houches school, Dirac summarized his work on canonical relativity with the words[(4)]: "The exact Hamiltonian for the theory of gravitation... turns out to be rather simpler than one might have expected. ...I am inclined to believe from this that four-dimensional symmetry is not a fundamental property of the physical world... From the mathematical point of view the loss of four-dimensional symmetry is to be regretted merely because it means a loss of transforming possibilities in the equations. It is amply compensated for by the increase in transforming possibilities arising from one's being able to make contact transformations in the Hamiltonian equations."

Dirac's and DeWitt's statements, taken at their face value, are clear opposites. They should not, however, be taken as statements of facts, but rather as statements of programs. I feel that both are misleading in identifying the canonical approach with an automatic loss of space-time "covariance." After all, the Peierls bracket which DeWitt advocates in his notes is just one of the ways of introducing a symplectic structure into a "covariant" formalism, and the use of arbitrary spacelike hypersurfaces unconnected with any choice of coordinates which Dirac discusses (and rejects) in his paper is just one of the ways of making canonical geometrodynamics "covariant." In one way or another, one can hope to make the "covariant" formalism "canonical" or the "canonical" formalism "covariant"; unfortunately, it is not easy to see at the beginning which way is the best. Therefrom stems the difference in emphasis of the two programs.

The fact that Dirac's and DeWitt's statements sound contradictory does not thus necessarily mean that one of them is false. Rather, it very well may be that, in the sense of Thomas Mann's and Niels Bohr's aphorism,[(24)] both are deep truths.

## ACKNOWLEDGMENT

This work was partially supported by NSF Grant PHY85-03653. I want to thank Julian Barbour, Richard Price, and Christopher Stone who helped me to polish the manuscript of this paper.

## REFERENCES

1. L. Réau, *L'Art russe*, II, 18n, Paris, 1921.
2. J. A. Wheeler, "Geometrodynamics and the Issue of the Final State," in *Relativity, Groups and Topology*, C. DeWitt and B. DeWitt, eds. (Gordon and Breach, New York, 1964).
3. B. S. DeWitt, "Dynamical Theory of Groups and Fields," in *Relativity, Groups and Topology*, C. DeWitt and B. DeWitt, eds. (Gordon and Breach, New York, 1964).
4. P. A. M. Dirac, *Proc. Roy. Soc. London A* **246**, 333 (1958).
5. P. G. Bergmann *et al., Phys. Rev.* **80**, 81 (1950).
6. R. Arnowitt, S. Deser, and C. W. Misner, "The Dynamics of General Relativity," in *Gravitation: An Introduction to Current Research*, L. Witten, ed. (Wiley, New York, 1962), and references cited therein.
7. K. Kuchař, *J. Math. Phys.* **13**, 768 (1972).
8. K. Kuchař, *J. Math. Phys.* **17**, 792 (1976).

9. K. Kuchař, *J. Math. Phys.* **17**, 777 (1976).
10. K. Kuchař, *J. Math. Phys.* **18**, 1589 (1977).
11. P. A. M. Dirac, *Lectures on Quantum Mechanics*, Belfer Graduate School of Science, Yeshiva, New York (1964).
12. B. S. DeWitt, *Phys. Rev.* **160**, 1113 (1967).
13. K. Kuchař, *J. Math. Phys.* **22**, 2640 (1981).
14. C. J. Isham, "Global and Topological Aspects of Quantum Gravity," in *Relativity, Groups and Topology II*, B. S. DeWitt and R. Stora, eds. (North-Holland, Amsterdam, 1984).
15. C. J. Isham and K. V. Kuchař, *Ann. Phys. (N.Y.)*, **164**, 288 (1985).
16. C. J. Isham and K. V. Kuchař, *Ann. Phys. (N.Y.)*, **164**, 316 (1985).
17. J. Evans, *John Ruskin* (Oxford University Press, New York, 1954).
18. C. Lanczos, *The Variational Principles of Mechanics* (University of Toronto Press, Toronto, 1970), 4th edn., Chap. V, §6.
19. K. Kuchař, "Canonical Methods of Quantization," in *Quantum Gravity II: A Second Oxford Symposium*, C. J. Isham, R. Penrose, and D. Sciama, eds. (Clarendon Press, Oxford, 1981).
20. K. Kuchař, *J. Math. Phys.* **17**, 801 (1976); **19**, 390 (1978); **25**, 1647 (1982); J. B. Hartle and K. V. Kuchař, *J. Math. Phys.* **25**, 57 (1984).
21. K. Kuchař, *J. Math. Phys.* **17**, 801 (1976), §12.
22. C. Teitelboim, *Ann. Phys. (N.Y.)* **79**, 542 (1973).
23. B. S. DeWitt, "The Spacetime Approach to Quantum Field Theory," in *Relativity, Groups and Topology II*, B. S. DeWitt and R. Stora, eds. (North-Holland, Amsterdam, 1984).
24. P. A. M. Dirac, "Versatility of Niels Bohr," in *Niels Bohr*, S. Rozental, ed. (North-Holland, Amsterdam, 1967); *The Reader's Digest Treasury of Modern Quotations* (Crowell, New York, 1975), p. 510.

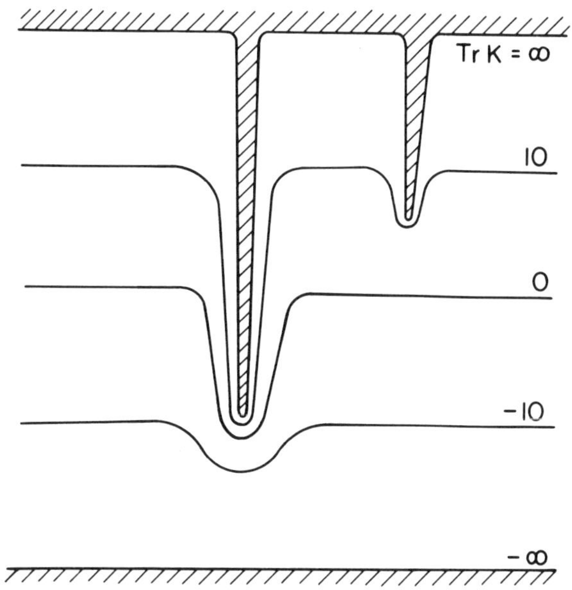


*Plate 18*

# Boundary Terms in the Action Principles of General Relativity

**James W. York, Jr.**[1]

*Received October 31, 1985*

*I address the question: "What is fixed on the boundary in the action principles of general relativity?" Four forms of the action are considered: the Einstein action, the Hilbert action, the first order action, and what may be called the cosmological action. The relationships and boundary data of these actions are described geometrically. Formal passage to the "Euclidean" forms of these actions is effected in detail.*

## DEDICATION

Among the many qualities that make John Wheeler a great physicist is his ability to ask the right questions. The first such question that struck me forcibly, when I had the good fortune in 1968 to become one of his assistants, was, "What is fixed on the boundary in the action principle for general relativity?" I found out years later that he had already seen the potential importance of this issue at the 1957 conference on "The Role of Gravitation in Physics" held at Chapel Hill. The question had been partially answered, implicity, in a number of papers that appeared in the 1960's. However, it still seemed that the matter could be put into a more explicit and geometrically clear form. Moreover, as there are several interesting forms of the action principle, their relationships seemed worth investigating. That the results of the investigation proved fruitful in several areas of general relativity bears witness to Professor Wheeler's penetrating insight.

---

[1] Institute of Field Physics, Department of Physics and Astronomy, University of North Carolina, Chapel Hill, North Carolina 27514.

It is perhaps amusing that this paper can be considered as an elaboration of the solution of Exercise 21.24 of the well-known text[1] by Misner, Thorne, and Wheeler. I respectfully delicate it to Professor Wheeler in the year of his seventy-fifth birthday.

## 1. INTRODUCTION

Since the early years of the general theory of relativity, it has been recognized that there are several versions of Hamilton's principle ("the action principle") that can be used to obtain the Einstein field equation for matter-free space-times. One version is based on the "Hilbert action"[2] $S_{\mathrm{H}}$, where

$$16\pi S_{\mathrm{H}} = \int_M \sqrt{g}\, R\, d^4x \tag{1}$$

Here $R$ is the scalar curvature of the space-time metric $g_{ab}$ and $g$ denotes the absolute value $|\det g_{ab}|$ of the determinant of the metric. In (1) we vary $g_{ab}$ and obtain second-order (rather than fourth-order) field equations because the second derivatives of the metric that appear in $R$ occur only in a divergence form. We obtain

$$16\pi \delta S_{\mathrm{H}} = -\int_M \sqrt{g}\, G^{ab} \delta g_{ab}\, d^4x + \int_{\partial M} \sqrt{g}\, (g^{ab}\delta^c_d - g^{ac}\delta^b_d)\, \delta\Gamma^d_{ab} (d^3x)_c \tag{2}$$

where $\delta\Gamma^d_{ab}$ denotes

$$\delta\Gamma^d_{ab} = \tfrac{1}{2} g^{de} [\nabla_a(\delta g_{eb}) + \nabla_b(\delta g_{ea}) - \nabla_e(\delta g_{ab})] \tag{3}$$

(I shall not consider the Palatini method[3] in which the metric and connection are varied independently.) We see that the Hilbert action is stationary ($\delta S_{\mathrm{H}} = 0$) if the Einstein equation $G^{ab} = 0$ is satisfied *and* if certain normal derivatives of $g_{ab}$ are fixed in a particular way on the boundary $\partial M$. Precisely how the normal derivatives are fixed will be displayed below.

The usual point of view is to regard the Einstein equations as second-order equations in $g_{ab}$. Hence, for the dynamical theory, it is more natural to regard $g_{ab}$, rather than its normal derivatives, as fixed on the boundary of the region of integration in the action. This leads us to consider the first-order action used by Einstein.[4] This is also closely related to the action $S_1$ that leads directly to the standard canonical form of the theory.[5] The Einstein $S_E$ involves only the metric and its first derivatives:

$$16\pi S_E = \int_M \sqrt{g}\, g^{ab} (\Gamma^c_{ad}\Gamma^d_{cb} - \Gamma^c_{ab}\Gamma^d_{dc})\, d^4x \tag{4}$$

Variation of the metric yields the field equations if the metric $g_{ab}$ is fixed on the boundary. The difference between the Einstein and Hilbert Lagrangian densities is a divergence term.

It is clear that $\mathscr{L}_H$ has much simpler behavior with respect to coordinate transformations than does $\mathscr{L}_E$. The former is a scalar density of weight one, while the latter does not transform homogeneously. To sharpen the specification of what must be fixed on the boundary, I will describe a first order action $S_1$, closely related to $S_E$, that can be written in manifestly covariant form. It is appropriate to derive geometrically the relation between $S_H$ and $S_1$. I have given such a derivation before,[(6)] but here those results will be extended somewhat to allow, among other things, an easy transition between the action for Lorentzian signature space-times and that for "real Euclidean space-times," which in recent years have been the focus of many investigations in quantum gravity. Also, it enables a precise statement of the "cosmological action" in which the conformal three-geometry and the extrinsic curvature scalar are fixed on the space-time boundary.

## 2. GEOMETRY OF THE ACTION

Consider a smooth family of nonintersecting slices of space-time of which one is to be regarded as a piece of the space-time boundary $\partial M$. Let $n^a$ be the unit normal vector field of this family, with $n_a n^a = \varepsilon$, $\varepsilon = +1$ if $n^a$ is spacelike, $\varepsilon = -1$ if $n^a$ is timelike. The metric induced on a slice is given by $\gamma_{ab} = g_{ab} - \varepsilon n_a n_b$ with covariant derivative $D_a$. The second fundamental tensor $K_{ab}$ of a slice will be *defined* by

$$K_{ab} = -\tfrac{1}{2}\pounds_n \gamma_{ab} \tag{5}$$

where £ denotes the Lie derivative. The *orientation* is defined by the requirement that $n^a$ be the unit outward-pointing *vector* field at the boundary. Hence the surface element is $d^3\Sigma_a = \varepsilon n_a \sqrt{\gamma}\, d^3x$, where $\gamma$ is the absolute value of the three-dimensional determinant of $\gamma_{ab}$. One has $g = \alpha^2 \gamma$, where $\alpha$ is defined below. The definition of $K_{ab}$ as a "velocity" is convenient when one considers the formal relation of Lorentzian and Euclidean actions for gravity (Sect. 5).

The Gauss–Codazzi equations for the slices involve projection using $\perp^a_b = \delta^a_b - \varepsilon n^a n_b$ of the space-time curvature tensor $R_{abcd}$ to obtain the induced curved tensor $\mathscr{R}_{abcd}$ of $\gamma_{ab}$. One finds upon projection of all free indices to the right of $\perp$

$$\perp R_{abcd} = \mathscr{R}_{abcd} + \varepsilon(K_{ad}K_{bc} - K_{ac}K_{bd}) \tag{6}$$

$$\perp R_{abcd} n^d = D_b K_{ac} - D_a K_{bc} \tag{7}$$

Suppose the slices are defined locally by $t = \text{const}$, where $t$ is a scalar. Then a "lapse function" $\alpha > 0$ is given by $\alpha^{-2} = \varepsilon g^{ab}(\nabla_a t)(\nabla_b t) > 0$ and a "shift vector" is any $\beta^a$ such that $\beta^a n_a = 0$. From these we obtain a vector field $t^a = \alpha n^a + \beta^a = (\partial/\partial t)^a$ threading the slices and leading up to the boundary. Then, with $a^a = n^b \nabla_b n^a (n_a a^a = 0)$,

$$R_{acbd} n^c n^d = \alpha^{-1}(\pounds_t - \pounds_\beta) K_{ab} + K_{ac} K^c_b + D_{(a} a_{b)} - \varepsilon a_a a_b \tag{8}$$

From (6)–(8) and $\sqrt{g} = \alpha \sqrt{\gamma}$ we calculate

$$\sqrt{g}\, R = \sqrt{g}[\mathscr{R} - \varepsilon(K_{ab}K^{ab} + K^2)] + 2\varepsilon \sqrt{\gamma}(\pounds_t - \pounds_\beta) K + \sqrt{g}\, \nabla_a a^a \tag{9}$$

with $K = \gamma^{ab} K_{ab}$. From (5) we have

$$(\pounds_t - \pounds_\beta) \gamma_{ab} = -2\alpha K_{ab} \tag{10}$$

from which it follows that

$$2\varepsilon \sqrt{\gamma}(\pounds_t - \pounds_\beta) K = 2\varepsilon(\pounds_t - \pounds_\beta)(\sqrt{\gamma}\, K) + 2\varepsilon \sqrt{g}\, K^2 \tag{11}$$

where

$$(\pounds_t - \pounds_\beta)(\sqrt{\gamma}\, K) = \nabla_a(\sqrt{g}\, K n^a) \tag{12}$$

Substitution of these formulas into (9) yields

$$\sqrt{g}\, R = \sqrt{g}[\mathscr{R} - \varepsilon(K_{ab}K^{ab} - K^2)] + 2\sqrt{g}\, \nabla_a[\varepsilon K n^a + a^a] \tag{13}$$

Though the first term on the right of (13) contains further second derivatives of the metric (in $\mathscr{R}$), we can repeat the above argument for each non-null element $B_A$ of the boundary, with $\partial M$ being the sum of the $B_A$. Then, upon integration of (13), we obtain the relation between the first order and Hilbert actions:

$$S_1 = \frac{1}{16\pi} \int_M \sqrt{g}\, R\, d^4x - \sum_A \frac{\varepsilon_A}{8\pi} \int_{B_A} \sqrt{\gamma}\, K\, d^3x \tag{14}$$

## 3. WHAT IS FIXED ON THE BOUNDARY?

By computing the variations of (14) and (2), we can readily obtain the boundary conditions appropriate for the Hilbert and Einstein actions. One finds

$$(\delta S_1)_{\partial M} = \sum_A \frac{\varepsilon_A}{16\pi} \int_{B_A} \sqrt{\gamma}(K^{ab} - K\gamma^{ab})\, \delta\gamma_{ab}\, d^3x \tag{15}$$

Hence, in Hamilton's principle using $S_1$, one fixes the intrinsic three-metric on the boundary rather than all the $g_{ab}$'s.

In a discussion of the boundary variation term of Hilbert's action principle, it is convenient to introduce the quantity

$$N^{ab} \equiv \sqrt{\gamma}(K\gamma^{ab} - K^{ab}) \tag{16}$$

which will be recognized as the momentum $\pi^{ab}$ canonically conjugate to the three-metric $\gamma_{ab}$[5] in the case where $B_A$ is spacelike and $n^a$ is future-oriented, i.e., a hypersurface of constant "time." Using (16), we see that (15) is equivalent to

$$(\delta S_1)_{\partial M} = \sum_A -\frac{\varepsilon_A}{16\pi}\int_{B_A} N^{ab}\delta\gamma_{ab}\, d^3x \tag{17}$$

Similarly, for the Hilbert principle we find

$$(\delta S_{\mathrm{H}})_{\partial M} = \sum_A \frac{\varepsilon_A}{16\pi}\int_{B_A} \gamma_{ab}\delta N^{ab}\, d^3x \tag{18}$$

Hence, $N^{ab}$ must be fixed on the boundary in this case. Therefore, the normal derivatives of the metric that are fixed in the Hilbert action principle are given by recognizing that

$$N^{ab} = -\tfrac{1}{2}g\gamma^{ac}\gamma^{bd}\pounds_n(g^{-1}\gamma_{ab}) \tag{19}$$

with $\pounds_n$ being the appropriate normal derivative operator. Notice that in four-dimensional space-time, the boundary conditions for the stationarity of both $S_{\mathrm{E}}$ and $S_{\mathrm{H}}$ involve six, rather than 10, conditions. It is well known that this is because the lapse $\alpha$ and the shift $\beta^a$ are "dynamically" irrelevant.

## 4. THE COSMOLOGICAL ACTION

Consider space-times of topology $R \times \Sigma$, where $\Sigma$ is spacelike and closed ($\partial\Sigma = 0$). The boundary $\partial M$ of $M$ may now be regarded as the sum of two closed spacelike slices $\Sigma_{t_2}$ and $\Sigma_{t_1}$. The boundary variation in Hilbert's action for a slice $\Sigma_t$ has the form

$$\delta S_{\mathrm{H}} = \frac{-1}{16\pi}\int_{\Sigma_t} \gamma_{ij}\delta\pi^{ij}\, d^3x \tag{20}$$

where Latin indices refer to a basis adapted to the spacelike slice $\Sigma_t$, whose normal vector is future-oriented, and $\pi^{ij}$ is the ADM momentum conjugate to $\gamma_{ij}$. Actually, because ADM used units in which $16\pi G = c = 1$ (rather than $G = c = 1$), we should use the momentum

$$p^{ij} = (16\pi)^{-1}\,\pi^{ij} \tag{21}$$

with

$$\delta S_{\rm H} = -\int_{\Sigma_t} \gamma_{ij}\,\delta p^{ij}\, d^3x \tag{22}$$

Expression (22) can be written in an interesting form if we introduce alternative variables

$$\tilde{\gamma}_{ij} = \gamma^{-1/3}\gamma_{ij} = \text{conformal three-metric} \tag{23}$$

$$\tilde{p}^{ij} = \gamma^{1/3}(p^{ij} - \tfrac{1}{3}p\gamma^{ij}) \tag{24}$$

$$p = \gamma_{ij}\,p^{ij} \tag{25}$$

$$\tilde{p} = \gamma^{-1/3}p = 2\gamma^{1/6}K(16\pi)^{-1} \tag{26}$$

Then

$$\delta S_{\rm H} = \int_{\Sigma_t} (\tilde{p}^{ij}\delta\tilde{\gamma}_{ij} - \gamma^{1/3}\delta\tilde{p})\, d^3x \tag{27}$$

This breakup is meaningful because both $\tilde{\gamma}_{ij}$ and $\tilde{p}^{ij}$ commute with both $\gamma^{1/3}$ and $\tilde{p}$. This can be verified[7] by using the Poisson brackets

$$\{\gamma_{ij}(x),\, p^{kl}(x')\} = \tfrac{1}{2}(\delta_i^k\,\delta_j^l + \delta_j^k\,\delta_i^l)\,\delta(x, x') \tag{28}$$

Thus, the alternative view arises that the Hilbert action is stationary if $G^{ab} = 0$ and if we fix on the boundary the conformal three-metric $\tilde{\gamma}_{ij}$ and the "extrinsic time" variable $\tilde{p} = (8\pi)^{-1}\,\gamma^{1/6}K$. However, $\tilde{p}$ is not a scalar with respect to spatial coordinate transformations; rather, it is a scalar density of weight one-third. Hence, its being constant ($\partial_i\,\tilde{p} = 0$) on a slice is not independent of the choice of spatial coordinates on the slice. This means that $\tilde{p}$ is not a natural choice for an extrinsic time variable. This situation led to the suggestion[6] that a spatial scalar such as $K$, or some monotonic function of $K$, would make a natural extrinsic time variable in cosmology.

A simple choice of an extrinsic time scalar is $K$ itself. We observe that setting

$$\{K(x),\, p_K(x')\} = \delta(x, x') \tag{29}$$

implies

$$p_K = -(12\pi)^{-1}\sqrt{\gamma} \tag{30}$$

When $G^{ab}=0$, the action variation analogous to (27) is

$$\delta S_K=\int_{\Sigma_t}(\tilde{p}^{ij}\delta\tilde{\gamma}_{ij}+p_K\delta K)\,d^3x \tag{31}$$

where the "cosmological action" $S_K$ is given by

$$S_K=\frac{1}{16\pi}\int_M\sqrt{g}\,R\,d^4x+\frac{1}{24\pi}\int_{\partial M}\sqrt{\gamma}\,K\,d^3x \tag{32}$$

The actions $S_1$, $S_{\mathrm{H}}$, and $S_K$ differ, respectively, by boundary integrals.

Suppose that the slice $\Sigma_t$ is such that $K=\text{const}$. Then among the Einstein equations is the "momentum constraint" $D_j p^{ij}=0$, which, when $K=\text{const}$, can be written as[7]

$$\tilde{D}_j\tilde{p}^{ij}=0 \tag{33}$$

where $\tilde{D}_j$ is the covariant derivative operator associated with $\tilde{\gamma}_{ij}$. Hence, a spatial diffeomorphism generated by $\xi^i$ induces (with $\delta\lambda$ an infinitesimal parameter)

$$\tilde{\gamma}_{ij}\rightarrow\tilde{\gamma}_{ij}+\delta\lambda(\tilde{D}_i\tilde{\xi}_j+\tilde{D}_j\tilde{\xi}_i-\tfrac{2}{3}\tilde{\gamma}_{ij}\tilde{D}_k\tilde{\xi}^k) \tag{34}$$

where $\tilde{\xi}^i=\xi^i$ and $\tilde{\xi}_i=\tilde{\gamma}_{ij}\tilde{\xi}^j$. In this case

$$\tilde{p}^{ij}\delta\tilde{\gamma}_{ij}=[2\tilde{D}_j(\tilde{p}^{ij}\tilde{\xi}_i)-2\tilde{\xi}_i\tilde{D}_j\tilde{p}^{ij}]\,\delta\lambda \tag{35}$$

We then see from (33) that spatial diffeomorphisms will not affect $\delta S_K$. We conclude that $S_K$ is stationary when $G^{ab}=0$ and when a constant $K$ and the conformal three-*geometry* (conformal three-metric *modulo* spatial diffeomorphisms) are fixed on closed boundary slices $\Sigma_{t_2}$ and $\Sigma_{t_1}$. This way of viewing geometrodynamics has recently found interesting applications in the work of Hartle and Hawking,[8] and Hawking,[9] concerning properties of the "wave function of the universe." We can easily recognize from its definition that $K$ is dimensionally an inverse time; one can think of it as a local mean inverse "Hubble time." Thus, one could also view

$$T\equiv -cK^{-1},\qquad c=\text{const}>0 \tag{36}$$

as a natural extrinsic time. One finds that

$$\{T(x),\,p_T(x')\}=\delta(x,x') \tag{37}$$

$$p_T=-\frac{16\pi}{3c}\gamma^{-1/2}(\text{tr }p)^2 \tag{38}$$

The corresponding action is again the cosmological action $S_K$ given by (32). The slicing $T = t$ will define the same family of hypersurfaces as $K = t$, but they will be labeled differently.

## 5. THE EUCLIDEAN ACTION

A number of investigations concerning quantum gravity in cosmology[8,9] and the Hawking effect[10–12] involve the use of an action appropriate to a metric with Euclidean signature $(++++)$. The first-order Euclidean action $I_1$ is easily obtained from (14) by observing that all of the $\varepsilon_A = +1$ in this case and then changing the overall sign to make the "kinetic" term $K_{ab}K^{ab}$ in (13) positive.

Alternatively, it is instructive to consider a formal "rotation" of the time by the substitution

$$t \to -it \tag{39}$$

In the volume term of (14), this induces

$$\sqrt{g}\, R\, d^4x \to -i\sqrt{g}\, R\, d^4x \tag{40}$$

If the boundary surface element $B_A$ in (16) is *spacelike* ($\varepsilon_A = -1$), we have

$$\sqrt{\gamma}\, d^3x \to \sqrt{\gamma}\, d^3x \tag{41}$$

and, because $K$ is dimensionally an *inverse* time (alternatively, $K = -\pounds_n \ln\sqrt{\gamma}$ is a velocity), we have

$$K \to iK \tag{42}$$

Hence the surface terms in (14), when $B_A$ is *spacelike*, transform as

$$+\sqrt{\gamma}\, K\, d^3x \to +i\sqrt{\gamma}\, K\, d^3x \tag{43}$$

Similarly, when $B_A$ is *timelike* ($\varepsilon_A = +1$), one has

$$\sqrt{\gamma}\, d^3x \to -i\sqrt{\gamma}\, d^3x \tag{44}$$

$$K \to K \tag{45}$$

and overall for a timelike boundary surface

$$-\sqrt{\gamma}\, K\, d^3x \to +i\sqrt{\gamma}\, K\, d^3x \tag{46}$$

We desire that $S_E \to iI_1$, so that $\exp(iS_E) \to \exp(-I_1)$ in the functional integral. Therefore, combining the above results,

$$I_1 = -\frac{1}{16\pi}\int_M \sqrt{g}\, R\, d^4x + \frac{1}{8\pi}\oint_{\partial M} \sqrt{\gamma}\, K\, d^3x \tag{47}$$

To $I_1$ it is sometimes convenient also to add a constant, that is, a quantity depending only on some fixed "background" geometry.[10–12] Recall that with the present conventions, $K = -\nabla_a n^a$ and $n^a$ is the outward-pointing unit normal vector of the boundary hypersurface. Note that from (36) and (42), one has

$$T \to -iT \tag{48}$$

just as in the defining transformation of "Euclidean time" (39).

A similar analysis applied to the cosmological action $S_K$ shows that its Euclidean form is

$$I_K = -\frac{1}{16\pi}\int_M \sqrt{g}\, R\, d^4x + \frac{1}{24\pi}\int_{\partial M} \sqrt{\gamma}\, K\, d^3x \tag{49}$$

In conclusion, I hope that this simple geometrical discussion of the action will be useful to researchers in gravity and that it answers fully Professor Wheeler's question: "What is fixed on the boundary in the action principle for general relativity?"

## REFERENCES

1. C. W. Misner, K. S. Thorne, and J. A. Wheeler, *Gravitation* (Freeman, San Francisco, 1973).
2. D. Hilbert, *Königl. Gesell. Wiss. Göttingen, Nachr., Math.-Phys. Kl.*, 395 (1915).
3. A. Palatini, *Rend. Circ. Mat. Palermo* **43**, 203 (1919).
4. A. Einstein, *Preuss. Akad. Wiss. Berlin, Sitzber.*, 1111 (1916).
5. R. Arnowitt, S. Deser, and C. W. Misner, in *Gravitation*, L. Witten, ed. (Wiley, New York, 1962).
6. J. W. York, *Phys. Rev. Lett.* **28**, 1082 (1972).
7. Y. Choquet-Bruhat and J. W. York, in *General Relativity and Gravitation I*, A. Held, ed. (Plenum, New York, 1980).
8. J. B. Hartle and S. W. Hawking, *Phys. Rev. D* **28**, 2960 (1983).
9. S. W. Hawking, "The Arrow of Time in Cosmology," preprint, April 1985.
10. G. W. Gibbons and S. W. Hawking, *Phys. Rev. D* **15**, 2752 (1977).
11. S. W. Hawking, in *General Relativity*, S. W. Hawking and W. Israel, eds. (Cambridge University Press, Cambridge, 1979).
12. J. W. York, "Black-Hole Thermodynamics and the Euclidean Einstein Action," to appear in *Phys. Rev.* (1986).

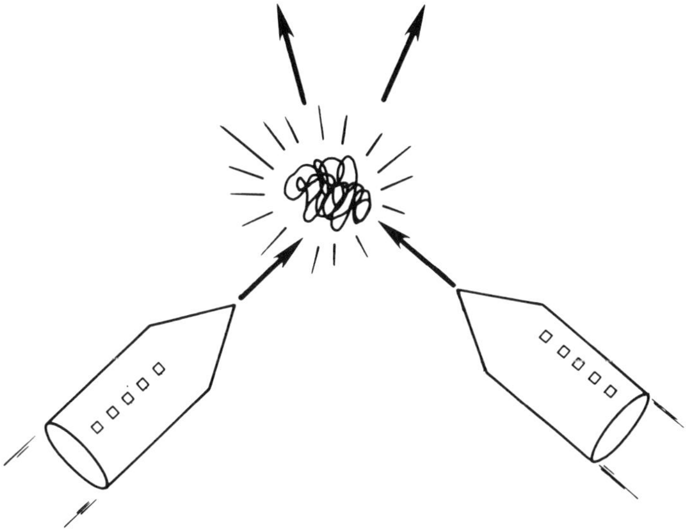

*Plate 19*

# Bianchi Identities and the Automatic Conservation of Energy-Momentum and Angular Momentum in General-Relativistic Field Theories

**Friedrich W. Hehl**[1,2] **and J. Dermott McCrea**[1]

*Received September 23, 1985*

*Automatic conservation of energy–momentum and angular momentum is guaranteed in a gravitational theory if, via the field equations, the conservation laws for the material currents are reduced to the contracted Bianchi identities. We first execute an irreducible decomposition of the Bianchi identities in a Riemann–Cartan space-time. Then, starting from a Riemannian space-time with or without torsion, we determine those gravitational theories which have automatic conservation: general relativity and the Einstein–Cartan–Sciama–Kibble theory, both with cosmological constant, and the nonviable pseudoscalar model. The Poincaré gauge theory of gravity, like gauge theories of internal groups, has no automatic conservation in the sense defined above. This does not lead to any difficulties in principle. Analogies to 3-dimensional continuum mechanics are stressed throughout the article.*

"... the conservation of momentum–energy is to be made geometric in character and automatic in action by the following prescription: *Identify the stress–energy with the moment of rotation* ... ."—C. W. Misner, K. S. Thorne, and J. A. Wheeler (1973).

## 1. INTRODUCTION

Experience tells us that all forms of energy gravitate. If we investigate then the gravitational properties of all physical agents contained in a certain (3-dimensional) spatial volume element $dV$, the net energy $dE$, or, in a field-

[1] Department of Mathematical Physics, University College, Dublin 4, and Dublin Institute for Advanced Studies, 10 Burlington Road, Dublin 4, Ireland.

[2] Permanent address: Institute for Theoretical Physics, University of Cologne, D-5000 Köln 41, W. Germany.

theoretical context, the energy density $dE/dV$, acts and should be acted upon gravitationally. The energy $dE$ is the time component of the energy-momentum vector $dP_\alpha$ ($\alpha, \beta \ldots = 0, 1, 2, 3$ are anholonomic or Lorentz indices). Appealing to the relativity principle, we find the *vector-valued energy-momentum 3-form*

$$\mathbf{\Sigma} = \mathbf{e}^\alpha \Sigma_\alpha = \frac{1}{3!} \mathbf{e}^\alpha \Sigma_{\alpha\mu\nu\rho} \vartheta^\mu \wedge \vartheta^\nu \wedge \vartheta^\rho = \mathbf{e}^\alpha \Sigma_\alpha{}^\gamma \eta_\gamma \tag{1.1}$$

where $\{\vartheta^\mu\}$ is a basis of 1-forms and $\eta_\gamma := (1/3!) \eta_{\gamma\mu\nu\sigma} \vartheta^\mu \wedge \vartheta^\nu \wedge \vartheta^\sigma$. If $\Sigma$ acts on a volume element trivector, it produces the energy-momentum vector $dP_\alpha$. The vector-valued 3-form $\Sigma$ is one source of gravity.

Let us view space-time as a (pseudo-)Riemannian manifold $V_4$ with the curvature 2-forms $\Omega^{\alpha\beta} = (1/2) F_{\mu\nu}{}^{\alpha\beta} \vartheta^\mu \wedge \vartheta^\nu$, satisfying the Bianchi identities $D\Omega^{\alpha\beta} \equiv 0$. If we carry a vector, by parallel displacement, around an infinitesimal closed contour, its *rotation* with respect to the vector we started with is characterized by the bivector valued 2-form

$$\mathbf{\Omega} = \mathbf{e}_\alpha \wedge \mathbf{e}_\beta \, \Omega^{\alpha\beta} \tag{1.2}$$

($\mathbf{e}_\alpha$ = basis vectors). If we want an automatic local conservation of energy, ensuing from the Bianchi identities, we can follow the Cartan prescription[(7–10)] as spelled out above by MTW[(48)] (see also Refs. 71 and 72). The *moment* of rotation

$$(\vartheta^\gamma \mathbf{e}_\gamma) \wedge \mathbf{\Omega} = (\mathbf{e}_\gamma \wedge \mathbf{e}_\alpha \wedge \mathbf{e}_\beta) \, \Omega^{\alpha\beta} \wedge \vartheta^\gamma \tag{1.3}$$

is a conserved *trivector valued 3-form.* Therefore, apart from factors and from taking the $^\star$ duals of the trivectors, the rotation, after building up that mysterious moment, has now been tailored and is ready to be identified with the energy-momentum 3-form:

$$(\mathbf{e}_\gamma \wedge \mathbf{e}_\alpha \wedge \mathbf{e}_\beta)^\star \Omega^{\alpha\beta} \wedge \vartheta^\gamma = l^2 \mathbf{e}_\delta \Sigma^\delta \tag{1.4}$$

($l$ = Planck length). This is Einstein's field equation.

Why a moment of rotation? Formally this is quite clear: The rotation (1.2) is a 2-form and we have to do something in order to upgrade it to a 3-form. MTW argue that "... the rotation associated with a given face is the analog of the force ... in mechanics." Would not an unbiased student opt for the moment of a force as an analog of the rotation? Moreover, it does not fit in with our usual intuition to link a rotation, or a moment of a rotation, to energy-momentum, which itself is linked to translations within the Noether machinery.

The main thrust of our article will be, taking the deeplying and beautiful derivation of the Einstein equation as given in MTW, to try to better understand the reasons which force us, in the context of general relativity, to take the moment of the rotation as an equivalent for material energy-momentum. On our way to this understanding, we need a few technicalities, like the irreducible decompositions of the Bianchi identities, which may be worth knowing, quite apart from their applications in the present work.

Bearing in mind the Euclidean group or, in 4 dimensions, the Poincaré group, one would like to link a translation with the rotation of (1.2). Indeed, as had been shown by Cartan, if we carry a vector around a closed infinitesimal contour by means of a Cartan displacement (see Schouten[54])[3], this yields, besides the rotation $\mathbf{e}_\alpha \wedge \mathbf{e}_\beta\, \Omega^{\alpha\beta}$ of (1.2), a *translation*

$$\mathbf{\Theta} = \mathbf{e}_\alpha \Theta^\alpha \tag{1.5}$$

provided the space-time is enriched by a torsion 2-form $\Theta^\alpha = (1/2) F_{\beta\gamma}{}^\alpha \vartheta^\beta \wedge \vartheta^\gamma$. Thus the space-time is assumed to be a Riemann–Cartan one, or a $U_4$, i.e., a manifold with a metric of Minkowskian signature and a metric-compatible nonsymmetric connection. Readers no interested in the torsion aspect of our work can at first put torsion safely to zero; up to the discussion of the Cartan map in Sect. 5, its existence is not vital for our considerations.

We would like to dedicate this article to John Wheeler on the occasion of his 75th birthday. May he have many more years of creative thinking and productive work to look forward to.

## Notation

In this subsection, a brief description is given of the terminology and notation that we shall use. For greater detail the reader is referred to Thirring[61] and Trautman.[65] Apart from Sect. 3, where 3-dimensional Euclidean space is considered, we shall be dealing with a 4-dimensional space-time, i.e., a 4-dimensional differentiable manifold with normal hyperbolic metric $g$ of signature $(-1, 1, 1, 1)$ and a metric-compatible connection which, in general, will have nonzero torsion. Such a space-time will be called a Riemann–Cartan space-time, or a $U_4$, while, if the torsion is zero, it is pseudo-Riemannian, or a $V_4$. Anholonomic vector bases (tetrads) are denoted by $\mathbf{e}_\alpha$, $(\alpha = 0, 1, 2, 3)$ and the corresponding dual bases of 1-forms by $\vartheta^\alpha$. We take the bases $\mathbf{e}_\alpha$ to be *orthonormal*, although the formalism described here is valid for any basis for which $g_{\alpha\beta} \equiv g(\mathbf{e}_\alpha, \mathbf{e}_\beta)$ are constants.

[3] Identifying the Cartan displacement around a contour with the Frank–Burgers circuit in crystal physics leads to the description of crystal dislocations by means of the torsion tensor (see Kondo,[30] Bilby,[6] and Kröner[34]).

In terms of a holonomic basis $\partial_i \equiv \partial/\partial x^i$ associated with a local coordinate system $\{x^i\}$ $(i=0, 1, 2, 3)$, we have $\mathbf{e}_\alpha = e^i{}_\alpha \partial_i$ and $\vartheta^\alpha = e_i{}^\alpha dx^i$, with $e_i{}^\alpha e^i{}_\beta = \delta^\alpha_\beta$. Anholonomic indices will be denoted by Greek letters and holonomic ones by Roman letters.

The torsion 2-forms $\Theta^\alpha$ and the curvature 2-forms $\Omega_\beta{}^\alpha$ are defined by

$$\Theta^\alpha = d\vartheta^\alpha + \omega_\beta{}^\alpha \wedge \vartheta^\beta \tag{1.6}$$

and

$$\Omega_\beta{}^\alpha = d\omega_\beta{}^\alpha + \omega_\mu{}^\alpha \wedge \omega_\beta{}^\mu \tag{1.7}$$

respectively, where $\omega_\beta{}^\alpha$ are the connection 1-forms. In terms of the basis of 1-forms, we have

$$\omega_\beta{}^\alpha = \Gamma_{\mu\beta}{}^\alpha \vartheta^\mu, \qquad \Theta^\alpha = \tfrac{1}{2} F_{\mu\nu}{}^\alpha \vartheta^\mu \wedge \vartheta^\nu, \qquad \Omega_\beta{}^\alpha = \tfrac{1}{2} F_{\mu\nu\beta}{}^\alpha \vartheta^\mu \wedge \vartheta^\nu \tag{1.8}$$

where $\Gamma_{\mu\beta}{}^\alpha$, $F_{\mu\nu}{}^\alpha$, and $F_{\mu\nu\beta}{}^\alpha$ are the tetrad components of the respective objects (see Ref. 40). Since the connection is metric-compatible and the $g_{\alpha\beta}$ are constants, the connection 1-forms are antisymmetric:

$$\omega_{\alpha\beta} = -\omega_{\beta\alpha} \tag{1.9}$$

The Bianchi identities,

$$D\Theta^\alpha \equiv \Omega_\beta{}^\alpha \wedge \vartheta^\beta, \qquad D\Omega_\alpha{}^\beta \equiv 0 \tag{1.10}$$

follow immediately from (1.6) and (1.7). Here $D$ is the exterior *covariant* derivative.

We denote the volume element 4-form by $\eta$. Thus, for a general oriented basis $\{\vartheta^0, \vartheta^1, \vartheta^2, \vartheta^3\}$,

$$\eta = \sqrt{-g}\, \vartheta^0 \wedge \vartheta^1 \wedge \vartheta^2 \wedge \vartheta^3 \tag{1.11}$$

where $g = \det(g_{\alpha\beta})$. For an orthonormal tetrad $\sqrt{-g} = 1$. Alternatively, we may write

$$\eta = \frac{1}{4!} \eta_{\alpha\beta\gamma\delta}\, \vartheta^\alpha \wedge \vartheta^\beta \wedge \vartheta^\gamma \wedge \vartheta^\delta \tag{1.12}$$

where $\eta_{\alpha\beta\gamma\delta} = \sqrt{-g}\, \varepsilon_{\alpha\beta\gamma\delta}$ and $\varepsilon_{\alpha\beta\gamma\delta}$ is the Levi–Civita permutation symbol with $\varepsilon_{0123} = +1$. Associated with $\eta$, we have the following forms which, together with $\eta$, span the algebra of exterior forms on the space-time:

$$\eta_\alpha = \mathbf{e}_\alpha \rfloor \eta = {}^*\vartheta_\alpha = \frac{1}{3!} \eta_{\alpha\mu\nu\rho} \vartheta^\mu \wedge \vartheta^\nu \wedge \vartheta^\rho \tag{1.13}$$

$$\eta_{\alpha\beta} = \mathbf{e}_\beta \lrcorner \eta_\alpha = *(\vartheta_\alpha \wedge \vartheta_\beta) = \frac{1}{2}\eta_{\alpha\beta\mu\nu}\vartheta^\mu \wedge \vartheta^\nu \tag{1.14}$$

$$\eta_{\alpha\beta\gamma} = \mathbf{e}_\gamma \lrcorner \eta_{\alpha\beta} = *(\vartheta_\alpha \wedge \vartheta_\beta \wedge \vartheta_\gamma) = \eta_{\alpha\beta\gamma\mu}\vartheta^\mu \tag{1.15}$$

$$\eta_{\alpha\beta\gamma\delta} = \mathbf{e}_\delta \lrcorner \eta_{\alpha\beta\gamma} = *(\vartheta_\alpha \wedge \vartheta_\beta \wedge \vartheta_\gamma \wedge \vartheta_\delta) \tag{1.16}$$

Here, the symbol $\lrcorner$ denotes the interior product of a vector with a form and $*$ denotes the Hodge dual operator. In addition to the formulas given, for instance, in Ref. 65 for the exterior products of the basis 1-forms $\vartheta^\alpha$ with the forms defined in (1.13)–(1.16), we note the following useful results:

(a) for $\Phi$ and $\Psi$ of the same degree,

$$*\Phi \wedge \Psi = *\Psi \wedge \Phi \tag{1.17}$$

(b) for a $p$-form $\Phi$,

$$\vartheta^\alpha \wedge (\mathbf{e}_\alpha \lrcorner \Phi) = p\Phi \tag{1.18}$$

$$*(\Phi \wedge \vartheta_\alpha) = \mathbf{e}_\alpha \lrcorner *\Phi \tag{1.19}$$

An operator analogous to the Hodge dual for forms may be defined for skew products of the basis vectors. We denote this operator by $\star$ and write it on the right:

$$\mathbf{e}_\alpha^\star = \frac{1}{3!}\eta_\alpha{}^{\mu\nu\rho}(\mathbf{e}_\mu \wedge \mathbf{e}_\nu \wedge \mathbf{e}_\rho) \tag{1.20}$$

$$(\mathbf{e}_\alpha \wedge \mathbf{e}_\beta)^\star = \frac{1}{2}\eta_{\alpha\beta}{}^{\mu\nu}(\mathbf{e}_\mu \wedge \mathbf{e}_\nu) \tag{1.21}$$

$$(\mathbf{e}_\alpha \wedge \mathbf{e}_\beta \wedge \mathbf{e}_\gamma)^\star = \eta_{\alpha\beta\gamma}{}^\mu \mathbf{e}_\mu \tag{1.22}$$

Thus, for the bivector-valued 2-form

$$\mathbf{\Omega} = \mathbf{e}_\alpha \wedge \mathbf{e}_\beta\, \Omega^{\alpha\beta} = \frac{1}{2}\mathbf{e}_\alpha \wedge \mathbf{e}_\beta F_{\mu\nu}{}^{\alpha\beta}\vartheta^\mu \wedge \vartheta^\nu \tag{1.23}$$

we have

$$*\Omega^{\alpha\beta} = \mathbf{e}_\alpha \wedge \mathbf{e}_\beta\, *\Omega^{\alpha\beta} \tag{1.24}$$

and

$$\begin{aligned}\Omega^\star &= (\mathbf{e}_\alpha \wedge \mathbf{e}_\beta)^\star\, \Omega^{\alpha\beta} \\ &= \frac{1}{2}\eta_{\alpha\beta}{}^{\mu\nu}(\mathbf{e}_\mu \wedge \mathbf{e}_\nu)\, \Omega^{\alpha\beta} \\ &= \mathbf{e}_\mu \wedge \mathbf{e}_\nu\, \Omega^{\star\mu\nu}\end{aligned} \tag{1.25}$$

## 2. IRREDUCIBLE DECOMPOSITION OF THE BIANCHI IDENTITIES

Let us first of all turn our attention to the irreducible decomposition of the torsion and curvature 2-forms. In a 4-dimensional Riemann–Cartan space-time we have[4]

$$\begin{aligned} \Theta^\alpha = {}^{(1)}\Theta^\alpha \quad &+ \quad {}^{(2)}\Theta^2 \quad + \quad {}^{(3)}\Theta^\alpha \\ \sim \text{TENTOR} \; &+ \; \text{TRATOR} + \text{AXITOR} \\ \sim 16 \quad &\oplus \quad 4 \quad \oplus \quad 4 \end{aligned} \tag{2.1}$$

and

$$\begin{aligned} \Omega^{\alpha\beta} &= {}^{(1)}\Omega^{\alpha\beta} + {}^{(2)}\Omega^{\alpha\beta} + {}^{(3)}\Omega^{\alpha\beta} + {}^{(4)}\Omega^{\alpha\beta} + {}^{(5)}\Omega^{\alpha\beta} + {}^{(6)}\Omega^{\alpha\beta} \\ &\sim \text{WEYL} + \text{PAIRCOM} + \text{PSCALAR} + \text{RICSYMF} + \text{RICANTI} + \text{SCALAR} \\ &\sim 10 \quad \oplus \quad 9 \quad \oplus \quad 1 \quad \oplus \quad 9 \quad \oplus \quad 6 \quad \oplus \quad 1 \end{aligned} \tag{2.2}$$

For more detail the reader is referred to Appendix A1. The tensor piece of the torsion is ${}^{(1)}\Theta^\alpha$, its trace ${}^{(2)}\Theta^\alpha$, while ${}^{(3)}\Theta^\alpha$ is the axial piece. In the curvature decomposition (2.2), ${}^{(1)}\Omega^{\alpha\beta}$ represents the Weyl piece with its usual symmetries $\mathbf{e}_\alpha \rfloor {}^{(1)}\Omega^{\alpha\beta} = 0$, ${}^{(1)}\Omega^{\alpha\beta} \wedge \vartheta_\alpha = 0$ and the "pair commutator" ${}^{(2)}\Omega^{\alpha\beta}$ has the symmetries ${}^{(2)}F_{\mu\nu}{}^{\alpha\beta} = -{}^{(2)}F^{\alpha\beta}{}_{\mu\nu}$, in terms of its tensor components. Thc tcnsor components of ${}^{(3)}\Omega^{\alpha\beta}$ are totally antisymmetric in all four indices and the last three pieces ${}^{(4)}\Omega^{\alpha\beta}$, ${}^{(5)}\Omega^{\alpha\beta}$, and ${}^{(6)}\Omega^{\alpha\beta}$ represent the symmetric tracefree, the antisymmetric, and the scalar piece of the Ricci 1-form $\mathbf{e}_\alpha \rfloor \Omega^{\alpha\beta}$, respectively. Clearly, in a $V_4$, only WEYL, RICSYMF, and SCALAR survive.

In 3 dimensions, we find

$$\text{Torsion}_3 = \text{TENTOR}_3 + \text{TRATOR}_3 + \text{AXITOR}_3 \sim 5 \oplus 3 \oplus 1 \tag{2.3}$$

$$\text{Curvature}_3 = \text{RICSYMF}_3 + \text{RICANTI}_3 + \text{SCALAR}_3 \sim 5 \oplus 3 \oplus 1 \tag{2.4}$$

Observe that the first three curvature pieces of (2.2) disappear in 3 dimensions. As a result, for 3-dimensional "gravity" the vacuum field does not exist, so that such models have exclusively topological interest.

[4] Cartan[(8,9)] has already given the irreducible decomposition in his notation. In tensor component notation it was redone by, for instance, Hayashi and Shirafuji,[(18)] Lenzen,[(36)] and Baekler *et al.*[(3)]. For the corresponding exterior form calculations, see Wallner.[(69,70)] The names for the various irreducible parts, as given above, feature in our REDUCE computer programs in the corresponding decomposition routines (see Ref. 42).

We shall also need the irreducible decomposition of the Bianchi identities. For arbitrary dimensions, the Bianchi identities read

$$D\Theta^{\alpha} \equiv \Omega_{\beta}{}^{\alpha} \wedge \vartheta^{\beta} \tag{2.5}$$

$$D\Omega^{\alpha\beta} \equiv 0 \tag{2.6}$$

In Appendices A2 and A3 we have provided formulas for the decomposition of a vector-valued and a bivector-valued 3-form in 4 dimensions. Since ${}^{(\alpha)}\Omega^{\alpha\beta} \wedge \vartheta_{\beta} = 0$ for $a = 1, 4, 6$, Eq. (2.5) can be put into the form

$$D\Theta^{\alpha} \equiv ({}^{(2)}\Omega_{\beta}{}^{\alpha} + {}^{(3)}\Omega_{\beta}{}^{\alpha} + {}^{(5)}\Omega_{\beta}{}^{\alpha}) \wedge \vartheta^{\beta} \tag{2.7}$$

The results of Appendix A2 then yield the decomposition

$$*D\Theta^{\alpha} - \frac{1}{4} *(D\Theta^{\beta} \wedge \vartheta_{\beta})\vartheta^{\alpha} + \frac{1}{2} \mathbf{e}^{\alpha} \rfloor (*D\Theta^{\beta} \wedge \vartheta_{\beta}) \equiv *({}^{(2)}\Omega_{\beta}{}^{\alpha} \wedge \vartheta^{\beta}) \tag{B1.1}$$

$$*D\Theta^{\alpha} \wedge \vartheta_{\alpha} \equiv *({}^{(5)}\Omega^{\beta\alpha} \wedge \vartheta_{\beta}) \wedge \vartheta_{\alpha} \tag{B1.2}$$

$$*(D\Theta^{\alpha} \wedge \vartheta_{\alpha}) \equiv *({}^{(3)}\Omega^{\beta\alpha} \wedge \vartheta_{\beta} \wedge \vartheta_{\alpha}) \tag{B1.3}$$

Note that in these identities the PAIRCOM in (B1.1), the RICANTI in (B1.2), and the PSCALAR in (B1.3) show up separately and are expressed in terms of the torsion tensor. In conformity with this, [see (2.2)], the identities (B1.1), (B1.2), (B1.3) correspond to a $(9 \oplus 6 \oplus 1)$-decomposition. Obviously, for vanishing torsion, these pieces of the curvature 2-form vanish also. The identity (B1.2) is the contracted first Bianchi identity.

The second identity (2.6) will be decomposed in a similar way. Using Appendix A3, we find

$$*D\Omega^{\alpha\beta} - \frac{2}{3} *(D\Omega^{[\alpha|\gamma} \wedge \vartheta_{\gamma})\vartheta^{|\beta]} + \frac{1}{6} \mathbf{e}^{\alpha} \rfloor \{\mathbf{e}^{\beta} \rfloor (*D\Omega^{\gamma\delta} \wedge \vartheta_{\gamma} \wedge \vartheta_{\delta})\} \equiv 0 \tag{B2.1}$$

$$*D\Omega^{\alpha\beta} \wedge \vartheta_{\alpha} \wedge \vartheta_{\beta} \equiv 0 \tag{B2.2}$$

$$*(D\Omega^{\alpha\beta} \wedge \vartheta_{\beta}) \equiv 0 \tag{B2.3}$$

Componentwise, the decomposition (B2.1), (B2.2), (B2.3) corresponds to $(16 \oplus 4 \oplus 4)$ independent equations. For a $V_4$, (B2.3) and the second term on the left-hand side of (B2.1) drop out. Equation (B2.2) is the contracted second Bianchi identity.

Depending on the purpose, alternative forms of the identities may be useful. Simple algebra yields the following formulas to be used later:

$$\eta_{\alpha\mu\nu} \wedge D\Theta^{\alpha} \equiv \eta_{\alpha\mu\nu} \wedge \Omega^{\beta\alpha} \wedge \vartheta_{\beta} \tag{B1.2$'$}$$

$$\vartheta_{[\nu} \wedge D\Theta_{\mu]} \equiv \vartheta_{[\nu} \wedge (\Omega_{\mu]}{}^{\beta} \wedge \vartheta_{\beta}) \tag{B1.2)$''$ (}$$

$$\eta_{\alpha\beta\mu} \wedge D\Omega^{\alpha\beta} \equiv 0 \tag{B2.2)$'$ (}$$

$$D\Omega^{\alpha\beta} \wedge \vartheta_{\beta} \equiv 0 \tag{B2.3)$'$ (}$$

In analyzing general relativity, E. Cartan[5] [7, 8] took 3-dimensional continuum mechanics in order to relate the equilibrium conditions of continuum statics to the Bianchi identities of a $U_3$. What we need then, in order to follow up his train of thought, is the specific form of the Bianchi identities in 3 dimensions. Since $^*D\Omega^{\alpha}$ and $^*D\Omega^{\alpha\beta}$ are, respectively, vector-valued and bivector-valued 0-forms in 3 dimensions, each of the identities (2.5) and (2.6) has only 3 components and cannot be decomposed any further. Indeed, in 3 dimensions, we lose no generality by rewriting (2.5) and (2.6) as

$$(^*D\Theta^{\alpha})\vartheta_{\alpha} \equiv {}^*(\Omega^{\alpha\beta} \wedge \vartheta_{\beta})\vartheta_{\alpha} \tag{2.8}$$

$$(^*D\Omega^{\alpha\beta})\vartheta_{\alpha} \wedge \vartheta_{\beta} \equiv 0 \tag{2.9}$$

[5] A translation of Cartan's article[(7)] into English has been provided by Kerlick[(27)] in his Princeton thesis of 1975 and is reprinted in Bergmann and DeSabbata.[(5)] Trautman[(63)] has given a very interesting analysis of Cartan's article using the exterior form calculus.

**Table I.** Bianchi Identities in 4-Dimensional Riemann–Cartan Geometry[a]

| | | |
|---|---|---|
| | (Bianchi 1) $D\Theta^{\alpha} \equiv \Omega_{\beta}{}^{\alpha} \wedge \vartheta^{\beta}$ | 16 |
| (Bianchi 1.1) | $^*D\Theta^{\alpha} - \frac{1}{4}{}^*(D\Theta^{\beta} \wedge \vartheta_{\beta})\vartheta^{\alpha} + \frac{1}{2}\mathbf{e}^{\alpha} \rfloor (^*D\Theta^{\beta} \wedge \vartheta_{\beta}) \equiv {}^*({}^{(2)}\Omega_{\beta}{}^{\alpha} \wedge \vartheta^{\beta})$ | 9 |
| (Bianchi 1.2) | $^*D\Theta^{\alpha} \wedge \vartheta_{\alpha} \equiv {}^*({}^{(5)}\Omega^{\beta\alpha} \wedge \vartheta_{\beta}) \wedge \vartheta_{\alpha}$ | 6[b] |
| (Bianchi 1.3) | $^*(D\Theta^{\alpha} \wedge \vartheta_{\alpha}) \equiv {}^*({}^{(3)}\Omega^{\beta\alpha} \wedge \vartheta_{\beta} \wedge \vartheta_{\alpha})$ | 1 |
| | (Bianchi 2) $D\Omega^{\alpha\beta} \equiv 0$ | 24 |
| (Bianchi 2.1) | $^*D\Omega^{\alpha\beta} - \frac{2}{3}{}^*(D\Omega^{[\alpha\|\gamma} \wedge \vartheta_{\gamma})\vartheta^{\|\beta]} + \frac{1}{6}\mathbf{e}^{\alpha} \rfloor \{\mathbf{e}^{\beta} \rfloor (^*D\Omega^{\gamma\delta} \wedge \vartheta_{\gamma} \wedge \vartheta_{\delta})\} \equiv 0$ | 16 |
| (Bianchi 2.2) | $^*D\Omega^{\alpha\beta} \wedge \vartheta_{\alpha} \wedge \vartheta_{\beta} \equiv 0$ | 4[b] |
| (Bianchi 2.3) | $^*(D\Omega^{\alpha\beta} \wedge \vartheta_{\beta}) \equiv 0$ | 4 |

[a] Irreducible decompositions of the Bianchi identities of a $U_4$ under the local Lorentz group. Torsion $\Theta^{\alpha}$, curvature $\Omega^{\alpha\beta}$, basis 1-forms $\vartheta^{\alpha}$, basis vectors $\mathbf{e}_{\beta}$, PAIRCOM = ${}^{(2)}\Omega$, PSCALAR = ${}^{(3)}\Omega$, RICANTI = ${}^{(5)}\Omega$. Alternative ways for presenting the different pieces of the Bianchi identities are given in the text. The numbers in the right-hand column indicate the number of independent components in each line.

[b] These pieces survive in 3 dimensions.

**Table II.** Bianchi Identities in General Relativity[a]

| Exterior calculus | | Ricci calculus | |
|---|---|---|---|
| (Bianchi 1) $^{(2)}\Omega^{\alpha\beta} = {}^{(3)}\Omega^{\alpha\beta} = {}^{(5)}\Omega^{\alpha\beta} \equiv 0$ | $\Longleftrightarrow$ | $R_{[\alpha\beta\gamma]}{}^{\delta} \equiv 0$ | 16 |
| (Bianchi 2.1) $^*D\Omega^{\alpha\beta} + \frac{1}{6}\mathbf{e}^{\alpha} \rfloor \{\mathbf{e}^{\beta} \rfloor (^*D\Omega^{\gamma\delta} \wedge \vartheta_{\gamma} \wedge \vartheta_{\delta}) \equiv 0$ | $\Longleftrightarrow$ | $\nabla_{\delta} C_{\alpha\beta\gamma}{}^{\delta} - \nabla_{[\alpha} R_{\beta]\gamma} + \frac{1}{6}\partial_{[\alpha} R g_{\beta]\gamma} \equiv 0$ | 16 |
| (Bianchi 2.2) $^*D\Omega^{\alpha\beta} \wedge \vartheta_{\alpha} \wedge \vartheta_{\beta} \equiv 0$ | $\Longleftrightarrow$ | $\nabla_{\beta}(R_{\alpha}{}^{\beta} - \frac{1}{2}\delta_{\alpha}^{\beta} R) \equiv 0$ | 4 |

[a] Irreducible decompositions of the Bianchi identities under the local Lorentz group in general relativity. The curvature components are defined as follows: $\Omega^{\alpha\beta} = (1/2) R_{\mu\nu}{}^{\alpha\beta} \vartheta^{\mu} \wedge \vartheta^{\nu}$, $R_{\alpha\beta} = R_{\gamma\alpha\beta}{}^{\gamma}$ = Ricci tensor, $C_{\mu\nu}{}^{\alpha\beta}$ = Weyl tensor. Left: Exterior calculus. Right: Ricci-calculus, $\nabla$ = covariant derivative. PAIRCOM = $^{(2)}\Omega$, PSCALAR = $^{(3)}\Omega$, RICANTI = $^{(5)}\Omega$. $V_4$-curvature = WEYL $\oplus$ RICSYMF $\oplus$ SCALAR = $^{(1)}\Omega + {}^{(4)}\Omega + {}^{(6)}\Omega \sim 10 \oplus 9 \oplus 1$.

and since $*(^{(\alpha)}\Omega^{\alpha\beta} \wedge \vartheta_{\beta}) \wedge \vartheta_{\alpha} = 0$ for $a \neq 5$, these two equations are simply the contracted Bianchi identities (B1.2) and (B2.2). Therefore, the transition from 4 to 3 dimensions eliminates the Bianchi identities (B1.1), (B1.3) as well as (B2.1), (B2.3). Consequently, the contracted identities (B1.2) and (B2.2) have a preferred position among their fellow-identities, since only they survive in 3 dimensions (compare also Table I). The Bianchi identities remaining, in the special case of vanishing torsion, are displayed in Table II.

For physical reasons, we want Bianchi identities (expressing the conservation laws) which have a "smooth" transition from $n=4$ to $n=3$, because only then can the Maxwell stresses of an electromagnetic field, for instance, constitute one part of the 4-dimensional energy-momentum 3-form. Already this simple argument singles out the *contracted Bianchi identities* (B1.2), (B2.2) as the ones relevant for our future discussion. Note that these (6 plus 4) identities, and the corresponding (3 plus 3) identities, (2.8) and (2.9), in 3 dimensions, display the number of parameters characteristic of the Poincaré and Euclidean groups, respectively.

## 3. EQUILIBRIUM CONDITIONS IN CONTINUUM STATICS

Having displayed the relevant results that we shall need concerning the Bianchi identities, we must now consider the other set involved in the isomorphism which one would like to establish, namely, the laws of balance of momentum and angular momentum in 3 and 4 dimensions.

Consider a time-independent state of stress in a 3-dimensional continuous medium. We suppose that the medium carries not only force stress but also moment stress. A so-called Cosserat continuum[33] is of such a nature. The force stress of the medium is characterized by a vector-valued 2-form $\mathbf{\Sigma}$ and the moment stress by a bivector-valued 2-form $\boldsymbol{\tau}$.

To see the meaning of these 2-forms, let us firstly consider the force stress. In terms of the basis $\mathbf{e}_\alpha$ and the dual basis $\vartheta^\alpha$, it may be expressed in the form

$$\mathbf{\Sigma}=\frac{1}{2}\mathbf{e}_\alpha \Sigma^\alpha{}_{\beta\gamma}\vartheta^\beta \wedge \vartheta^\gamma \tag{3.1}$$

In this section, all indices run from 1 to 3 only. Since $\mathbf{\Sigma}$ is a vector-valued 2-form, it maps pairs of vectors into vectors. Consider therefore an infinitesimal area element spanned by the two vectors

$$\delta\mathbf{u}=\delta u^\alpha \mathbf{e}_\alpha, \qquad \delta\mathbf{v}=\delta v^\alpha \mathbf{e}_\alpha \tag{3.2}$$

The force acting on the area element due to the stress $\mathbf{\Sigma}$ is

$$\begin{aligned}\delta\mathbf{f}&=\mathbf{\Sigma}(\delta\mathbf{u},\delta\mathbf{v})\\ &=\frac{1}{2}\mathbf{e}_\alpha \Sigma^\alpha{}_{\beta\gamma}\vartheta^\beta \wedge \vartheta^\gamma(\delta u^\mu \mathbf{e}_\mu, \delta v^\nu \mathbf{e}_\nu)\\ &=\mathbf{e}_\alpha \Sigma^\alpha{}_{\beta\gamma}\,\delta u^\beta\,\delta v^\gamma\end{aligned} \tag{3.3}$$

One may also express $\mathbf{\Sigma}$ in terms of the basis $\eta_\alpha$,

$$\mathbf{\Sigma}=\mathbf{e}_\alpha \Sigma^{\alpha\beta}\eta_\beta \tag{3.4}$$

where $\eta_\beta=\frac{1}{2}\eta_{\beta\mu\nu}\vartheta^\mu \wedge \vartheta^\nu$ and consequently

$$\Sigma^{\alpha\beta}=\frac{1}{2}\eta^{\beta\mu\nu}\Sigma^\alpha{}_{\mu\nu} \tag{3.5}$$

In a similar way, the bivector-valued moment stress 2-form $\boldsymbol{\tau}$ is expressible as

$$\boldsymbol{\tau}=\frac{1}{2}\mathbf{e}_\alpha \wedge \mathbf{e}_\beta\, \tau^{\alpha\beta}{}_{\gamma\delta}\,\vartheta^\gamma \wedge \vartheta^\delta \tag{3.6}$$

and the net moment acting on the area element is

$$\begin{aligned}\delta\mathbf{m}&=\boldsymbol{\tau}(\delta\mathbf{u},\delta\mathbf{v})\\ &=\frac{1}{2}\mathbf{e}_\alpha \wedge \mathbf{e}_\beta\, \tau^{\alpha\beta}{}_{\gamma\delta}\,\vartheta^\gamma \wedge \vartheta^\delta(\delta u^\mu \mathbf{e}_\mu, \delta v^\nu \mathbf{e}_\nu)\\ &=\mathbf{e}_\alpha \wedge \mathbf{e}_\beta\, \tau^{\alpha\beta}{}_{\gamma\delta}\,\delta u^\gamma\,\delta v^\delta\end{aligned} \tag{3.7}$$

As in the case of the force stress, we also have, for the moment stress, the dual representation

$$\boldsymbol{\tau} = \mathbf{e}_\alpha \wedge \mathbf{e}_\beta \, \tau^{\alpha\beta\gamma} \, \eta_\gamma \tag{3.8}$$

with

$$\tau^{\alpha\beta\mu} = \frac{1}{2} \eta^{\mu\gamma\delta} \tau^{\alpha\beta}{}_{\gamma\delta} \tag{3.9}$$

The conditions of equilibrium may be determined by considering the forces and moments acting on the faces of an infinitesimal cube of the material. In a state of equilibrium, the total force and moment have to vanish and, if there are no volume forces or moments present, the standard procedure of continuum mechanics (see Refs. 14, 26, and 66) yields for the static case the equilibrium conditions

$$D\Sigma_\alpha = 0 \tag{3.10}$$

$$D\tau_{\alpha\beta} = \vartheta_{[\beta} \wedge \Sigma_{\alpha]} \tag{3.11}$$

where $D$ is the exterior covariant derivative.

The components $\tau_{\alpha\beta\gamma\delta}$ of the moment stress 2-form are skew symmetric in both the first and the second pair of indices, a property shared by the curvature of a Riemann–Cartan manifold. If we express the force stress 2-form $\boldsymbol{\Sigma}$ in terms of the dual basis of vectors

$$(\mathbf{e}_\alpha \wedge \mathbf{e}_\beta)^\star = \eta_{\alpha\beta}{}^\gamma \mathbf{e}_\gamma \tag{3.12}$$

we obtain

$$\boldsymbol{\Sigma} = \frac{1}{2} (\mathbf{e}_\alpha \wedge \mathbf{e}_\beta)^\star \Sigma^{\alpha\beta}{}_{\gamma\delta} \vartheta^\gamma \wedge \vartheta^\delta \tag{3.13}$$

where

$$\Sigma^{\alpha\beta}{}_{\gamma\delta} = \eta^{\alpha\beta}{}_\mu \Sigma^\mu{}_{\gamma\delta} \tag{3.14}$$

Then the components $\Sigma_{\alpha\beta\gamma\delta}$ also have the same symmetries as the Riemann–Cartan curvature tensor. One might ask whether these correspondences go even further. Several authors have dealt with this in the case of the force stress tensor $\boldsymbol{\Sigma}$ (see Refs. 16 and 32). Before attempting to answer this question, we shall first of all generalize from our considerations of stresses in a continuous medium in 3 dimensions to energy and angular momentum of classical relativistic matter fields in a 4-dimensional space-time.

## 4. ENERGY-MOMENTUM AND ANGULAR MOMENTUM LAWS OF CLASSICAL MATTER FIELDS AS NOETHER IDENTITIES

> "Another theory of electricity which I prefer denies action at a distance and attributes electric action to tensions and pressures in an all pervading medium, these stresses being the same kind as those familiar to engineers, and the medium being identitcal with that in which light is supposed to be propagated."—J. C. Maxwell (1870).

In space-time, the inertial properties of the matter fields are represented by the energy-momentum 3-form $\boldsymbol{\Sigma}$ and the spin angular momentum 3-form $\boldsymbol{\tau}$. These are interpreted in a manner analogous to the 2-forms of force and moment stress in 3-dimensional continuum mechanics. In terms of an orthonormal tetrad field $\{\mathbf{e}_\alpha\}$, $\alpha = 0, 1, 2, 3$, and the dual basis of 1-forms $\{\vartheta^\alpha\}$, we have

$$\begin{aligned}\boldsymbol{\Sigma} &= \frac{1}{3!}\mathbf{e}^\alpha \Sigma_{\alpha\beta\gamma\delta}\, \vartheta^\beta \wedge \vartheta^\gamma \wedge \vartheta^\delta \\ &= \mathbf{e}^\alpha \Sigma_\alpha{}^\beta \eta_\beta \end{aligned} \tag{4.1}$$

where $\eta_\beta = (1/3!)\eta_{\beta\mu\nu\sigma}\vartheta^\mu \wedge \vartheta^\nu \wedge \vartheta^\sigma$ and

$$\Sigma_\alpha{}^\beta = \frac{1}{3!}\eta^{\mu\nu\sigma\beta}\Sigma_{\alpha\mu\nu\sigma} \tag{4.2}$$

are the components of the second-rank energy-momentum tensor. In the same way $\boldsymbol{\tau}$ may be expressed in the form

$$\begin{aligned}\boldsymbol{\tau} &= \frac{1}{3!}\mathbf{e}^\alpha \wedge \mathbf{e}^\beta \tau_{\alpha\beta\mu\nu\sigma}\, \vartheta^\mu \wedge \vartheta^\nu \wedge \vartheta^\sigma \\ &= \mathbf{e}^\alpha \wedge \mathbf{e}^\beta \tau_{\alpha\beta}{}^\gamma \eta_\gamma \end{aligned} \tag{4.3}$$

where

$$\tau_{\alpha\beta}{}^\gamma = \frac{1}{3!}\eta^{\mu\nu\sigma\gamma}\tau_{\alpha\beta\mu\nu\sigma} \tag{4.4}$$

are the components of the third-rank spin angular momentum tensor. The flux of momentum across a 3-element spanned by the infinitesimal 4-vectors $\delta\mathbf{u}$, $\delta\mathbf{v}$, $\delta\mathbf{w}$ is

$$\delta\mathbf{p} = \boldsymbol{\Sigma}(\delta\mathbf{u}, \delta\mathbf{v}, \delta\mathbf{w}) \tag{4.5}$$

and similarly for the flux of spin angular momentum $\delta\mathbf{s}$.

We must now determine the equations of balance for $\boldsymbol{\Sigma}$ and $\boldsymbol{\tau}$ analogous to Eqs. (3.10) and (3.11) of the last section. To do this, we assume that the dynamical behavior of the matter fields is determined by a Lagrangian. However, we also wish to include the contribution of the gravitational field to the balance of energy and angular momentum. Accordingly we take a material Lagrangian 4-form

$$L = L(\Psi, d\Psi, \vartheta^{\alpha}, \omega_{\alpha}{}^{\beta}) \tag{4.6}$$

where the $p$-forms $\Psi$ represent the matter field while the basis of 1-forms $\vartheta^{\alpha}$ and the connection 1-forms $\omega_{\alpha}{}^{\beta}$ are the gravitational potentials. In the absence of any reason to the contrary at this stage, we take the translational potentials $\vartheta^{\alpha}$ and the rotational potentials $\omega_{\alpha}{}^{\beta}$ to be independent, so that the background space-time is a Riemann–Cartan manifold. The material momentum current $\boldsymbol{\Sigma} = \mathbf{e}^{\alpha}\Sigma_{\alpha}$ and the spin current $\boldsymbol{\tau} = \mathbf{e}^{\alpha} \wedge \mathbf{e}^{\beta}\tau_{\alpha\beta}$ are defined by

$$\Sigma_{\alpha} = \frac{\delta L}{\delta \vartheta^{\alpha}} \quad \text{and} \quad \tau_{\alpha\beta} = \frac{\delta L}{\delta \omega^{\alpha\beta}} \tag{4.7}$$

respectively, where $\delta$ denotes the variational derivative. We assume that the matter field equations $\delta L/\delta\Psi = 0$ are fulfilled.

We furthermore require local Poincaré invariance of the Lagrangian (4.6) or, equivalently, that $L$ should be invariant under diffeomorphisms and tetrad rotations. This leads, via Noether's theorem, to a number of identities.[20,31,60,64] First of all, we find that

$$L(\Psi, d\Psi, \vartheta^{\alpha}, \omega_{\alpha}{}^{\beta}) = L(\Psi, D\Psi, \vartheta^{\alpha}) \tag{4.8}$$

In other words, the exterior derivative $d$ and the connection 1-forms occur in $L$ only in the combination

$$D\Psi \equiv d\Psi + \sigma_{\alpha\beta}\omega^{\alpha\beta} \wedge \Psi \tag{4.9}$$

the exterior *covariant* derivative of $\Psi$. Here, $\sigma_{\alpha\beta}$ are the spin generating operators. Secondly we find that

$$\Sigma_{\alpha} = \mathbf{e}_{\alpha} \rfloor L - (\mathbf{e}_{\alpha} \rfloor D\Psi) \wedge \frac{\partial L}{\partial D\Psi} - (\mathbf{e}_{\alpha} \rfloor \Psi) \wedge \frac{\partial L}{\partial \Psi} \tag{4.10}$$

and

$$\tau_{\alpha\beta} = \sigma_{\alpha\beta}\Psi \wedge \frac{\partial L}{\partial D\Psi} \tag{4.11}$$

i.e., $\Sigma_\alpha$ and $\tau_{\alpha\beta}$ turn out to be the canonical energy-momentum and spin tensors, respectively. Finally, there are the two identities that are of particular interest to us in the context of the present work, because they represent the balance of energy-momentum and angular momentum, viz.

$$D\Sigma_\alpha = (\mathbf{e}_\alpha \rfloor \Theta^\beta) \wedge \Sigma_\beta + (\mathbf{e}_\alpha \rfloor \Omega^{\beta\gamma}) \wedge \tau_{\beta\gamma} \tag{4.12}$$

and

$$D\tau_{\alpha\beta} = \vartheta_{[\beta} \wedge \Sigma_{\alpha]} \tag{4.13}$$

For vanishing spin, (4.13) yields the symmetry of the energy-momentum tensor and (4.12) can be rewritten as $\tilde{D}\Sigma_\alpha = 0$, where $\tilde{D}$ denotes the exterior covariant derivative with respect to the Levi–Civita connection, i.e., the torsion drops out and we recover the energy-momentum law of general relativity. For zero torsion *and* curvature, we recover the well-established conservation laws for energy-momentum and angular momentum in a Minkowskian space-time, equations that are identical in form with those of (3.10) and (3.11) for continuous media in three dimensions.

The Lagrangian formalism of this section was set up as a means to derive the equations of balance in 4-dimensional space-time. Lorentz-type forces show up on the right-hand side of (4.12)—unexpectedly perhaps—as soon as one assumes a Riemann–Cartan space-time and a material Lagrangian coupled minimally to it. Cartan,[8,9] starting from Eqs. (3.10) and (3.11), postulated them as energy-momentum and angular momentum laws, valid also in a 4-dimensional Riemann–Cartan space-time. As we can see from (4.13) and (4.12), Cartan's postulate for angular momentum was justified, but the overlooked the necessity of the forces on the right-hand side of the energy-momentum law.

Both types of forces are not unprecedented. The "curvature $\wedge$ spin"-force is known in general relativity (for vanishing torsion)[6] as the *Mathisson–Papapetrou force*[39,52] acting on spinning particles. The "torsion $\wedge$ energy-momentum"-force has a direct 3-dimensional analog in dislocation theory, where the dislocation density is identified with the torsion of a suitably defined deformation space (see footnote 3). Consequently we recover the Peach–Koehler formula "dislocation density $\wedge$ force stress" which, for a singular dislocation line, coincides exactly with the *Peach–Koehler force* acting on a dislocation in an external stress field (see Ref. 35).

[6] For vanishing torsion, (4.12) reads $\tilde{D}\Sigma_\alpha = (\mathbf{e}_\alpha \rfloor \Omega^{\beta\gamma}) \wedge \tau_{\beta\gamma}$. If one introduces the metric energy-momentum tensor $\sigma \sim \delta L/\delta g$, the Mathisson–Papapetrou force gets absorbed[19]: $\tilde{D}\sigma = 0$.

## 5. THE CARTAN MAP AND THE PSEUDOSCALAR MODEL

We want automatic conservation of energy and angular momentum. In other words, we wish to translate the Noether identities (4.12), (4.13),

$$D\Sigma_{\alpha} = (\mathbf{e}_{\alpha} \rfloor \Theta^{\beta}) \wedge \Sigma_{\beta} + (\mathbf{e}_{\alpha} \rfloor \Omega^{\beta\gamma}) \wedge \tau_{\beta\gamma} \tag{5.1}$$

and

$$D\tau_{\alpha\beta} = \vartheta_{[\beta} \wedge \Sigma_{\alpha]} \tag{5.2}$$

into Bianchi identities.

Equation (5.1) represents the four independent energy-momentum laws. On looking at the Bianchi identities, only (B2.2)′ and (B2.3)′,

$$\eta_{\alpha\beta\mu} \wedge D\Omega^{\alpha\beta} \equiv 0 \tag{5.3a}$$

$$D\Omega^{\alpha\beta} \wedge \vartheta_{\beta} \equiv 0 \tag{5.3b}$$

present themselves as possible candidates, because only they likewise have four independent components. Since (5.3b) vanishes in 3 dimensions, its use is restricted. For (5.2), however, there is only one possible choice, namely (B1.2)′ with its six independent components:

$$D\Theta^{\mu} \wedge \eta_{\mu\alpha\beta} \equiv \vartheta^{\gamma} \wedge \Omega_{\gamma}{}^{\mu} \wedge \eta_{\mu\alpha\beta} \tag{5.4}$$

compare also Table III.

**Table III.** Riemann–Cartan Space-Time and Conservation Laws[a]

| | | |
|---|---|---|
| (Bianchi 1.2)′ <br> $D\Theta^{\mu} \wedge \eta_{\mu\alpha\beta} \equiv \vartheta^{\nu} \wedge \eta_{\mu\alpha\beta}$ | | (Noether 1) <br> $D\Sigma_{\alpha} = (\mathbf{e}_{\alpha} \rfloor \Theta^{\beta}) \wedge \Sigma_{\beta} + (\mathbf{e}_{\alpha} \rfloor \Omega^{\beta\gamma}) \wedge \tau_{\beta\gamma}$ |
| | ⤡ ⤢ | |
| $D\Omega^{\mu\nu} \wedge \eta_{\mu\nu\alpha} \equiv 0$ <br> (Bianchi 2.2)′ | | $D\tau_{\alpha\beta} = \vartheta_{[\beta} \wedge \Sigma_{\alpha]}$ <br> (Noether 2) |
| Cartan | $\begin{cases} \Omega^{\mu\nu} \wedge \eta_{\mu\nu\alpha} + \Lambda\eta_{\alpha} = \kappa\Sigma_{\alpha} \\ \Theta^{\mu} \wedge \eta_{\mu\alpha\beta} = \kappa\tau_{\alpha\beta} \end{cases}$ | mass <br> spin |

[a] Automatic conservation of energy-momentum and angular momentum. Torsion $\Theta^{\alpha}$, curvature $\Omega^{\alpha\beta}$, basis 1-forms $\vartheta^{\alpha}$, basis vectors $\mathbf{e}_{\beta}$, material energy-momentum $\Sigma_{\alpha}$, material spin angular momentum $\tau_{\alpha\beta}$. By means of the Cartan map (→ Einstein–Cartan–Sciama–Kibble field equations) the Noether identities are equivalent to the contracted Bianchi identities (B1.2)′ and (B2.2)′. General relativity results for vanishing spin. The pseudoscalar model incorporates also automatic conservation; see (5.19), (5.20); it has, however, no decent 3-dimensional limit and must be rejected as unphysical.

### 5.1. The Case of General Relativity

In general relativity as Einstein "derived" it,[12,13] there is no need for torsion since the Newtonian point particle and its motion in non-inertial reference frames yields the Christoffel symbols as gravitational field strength and no additional geometrical structure is required. Moreover, in the continuum approximation, this concept of matter leads to a material fluid without intrinsic spin. Consequently, in general relativity we have

$$\Theta^{\alpha}=0, \qquad \tau^{\alpha\beta}=0 \tag{5.5}$$

All that is left over from Eqs. (5.1) to (5.4) of this section is

$$D\Sigma_{\alpha}=0, \qquad \vartheta_{[\beta}\wedge\Sigma_{\alpha]}=0 \tag{5.6}$$

and

$$\eta_{\alpha\beta\mu}\wedge D\Omega^{\alpha\beta}\equiv 0, \qquad \vartheta_{\beta}\wedge\eta_{\alpha\mu\nu}\wedge\Omega^{\beta\alpha}\equiv 0 \tag{5.7}$$

Note that (5.3b) is eliminated because of (2.5) and $D\vartheta_{\alpha}=0$. The correct identification is now obvious (remember that vanishing torsion implies $D\eta_{\alpha\beta\gamma}=0$), namely

$$\Omega^{\mu\nu}\wedge\eta_{\mu\nu\alpha}=\kappa\Sigma_{\alpha} \tag{5.8}$$

The constant $\kappa$ has to have the dimensions of 1/force. The moment of rotation ($\eta_{\mu\nu\alpha}\wedge\Omega^{\mu\nu}$), apart from a factor, is identified with the energy-momentum 3-form. The energy-momentum and angular momentum laws (5.6) are then identical with the contracted Bianchi identities (5.7). Incidentally, the identification

$$\Omega^{\mu\nu}\wedge\eta_{\mu\nu\alpha}+\Lambda\eta_{\alpha}=\kappa\Sigma_{\alpha} \tag{5.9}$$

with a cosmological constant $\Lambda$ is also consistent with the arguments given. This is simply Einstein's field equations with cosmological constant. The identification (5.9) has a 3-dimensional analogue in continuum statics (see Refs. 29 and 34 and Table IV) since in this case $\eta_{\mu\nu\alpha}$ is a 0-form and coincides with the totally antisymmetric unit tensor.

### 5.2. The Cartan Map and the Einstein–Cartan–Sciama–Kibble Theory

If, in a gauge-theoretical setting, one starts with a classical Dirac wave function instead of a point particle, torsion arises naturally as a translational field strength and the fluid representing such matter fields is one with intrinsic spin:

$$\Theta^{\alpha}\neq 0, \qquad \tau^{\alpha\beta}\neq 0 \tag{5.10}$$

**Table IV.** A Riemann–Cartan Stress Space $U_3$ and Continuum Statics[a]

| | |
|---|---|
| (Bianchi 1) $D\Theta^\alpha \equiv \vartheta^\beta \wedge \Omega_\beta{}^\alpha$ | $D\Sigma_\alpha = 0$ (equil. 1) |
| (Bianchi 2) $D\Omega_\alpha{}^\beta \equiv 0$ | $D\tau_{\alpha\beta} = \vartheta_{[\beta} \wedge \Sigma_{\alpha]}$ (equil. 2) |
| Cartan–Kröner map $\begin{cases} \Omega^{\alpha\beta} \wedge \eta_{\alpha\beta\mu} = \Sigma_\mu \\ \Theta^\alpha \wedge \eta_{\alpha\mu\nu} = \tau_{\mu\nu} \end{cases}$ | force stress<br>moment stress |

[a] Automatic equilibrium for force and moment stresses in 3-dimensional continuum statics (compare Cartan[(7,8)] and Kröner[(32)]): The metric of the stress space is represented by the stress functions, its torsion by the moment stress. Torsion $\Theta^\alpha$, curvature $\Omega^{\alpha\beta}$, basis 1-forms $\vartheta^\alpha$, basis vectors $\mathbf{e}_\beta$, force stress $\Sigma_\alpha$, moment stress $\tau_{\alpha\beta}$. Classical continuum statics results for vanishing moment stress. The map presented here is certainly correct in linear continuum statics. It is not clear to us whether it is also correct in its full nonlinearity.

The map (5.9) from geometry to matter is generalized in a straightforward manner:

$$\Omega^{\mu\nu} \wedge \eta_{\mu\nu\alpha} + \Lambda\eta_\alpha = \kappa\Sigma_\alpha \tag{5.11}$$

$$\Theta^\mu \wedge \eta_{\mu\alpha\beta} = \kappa\tau_{\alpha\beta} \tag{5.12}$$

To see this, we must prove that with the identification (5.11), (5.12),

(a) (5.3a) is equivalent to (5.1),

(b) (5.4) is equivalent to (5.2).

Let us begin with (a). Equation (5.3a) may be written in the form

$$D(\Omega^{\mu\nu} \wedge \eta_{\mu\nu\alpha}) = \Omega^{\mu\nu} \wedge D\eta_{\mu\nu\alpha} \tag{5.13}$$

Hence, on substituting (5.11),

$$\begin{aligned}\kappa D\Sigma_\alpha &= \Omega^{\mu\nu} \wedge D\eta_{\mu\nu\alpha} + \Lambda D\eta_\alpha \\ &= \Omega^{\mu\nu} \wedge \eta_{\mu\nu\alpha\gamma}\Theta^\gamma + \Lambda\eta_{\alpha\gamma} \wedge \Theta^\gamma \\ &= -\Omega^{\mu\nu} \wedge (\mathbf{e}_\alpha \rfloor \eta_{\mu\nu\gamma})\Theta^\gamma + \Lambda\eta_{\alpha\gamma} \wedge \Theta^\gamma \\ &= \{-\mathbf{e}_\alpha \rfloor (\Omega^{\mu\nu} \wedge \eta_{\mu\nu\gamma}) + (\mathbf{e}_\alpha \rfloor \Omega^{\mu\nu}) \wedge \eta_{\mu\nu\gamma}\} \wedge \Theta^\gamma + \Lambda\eta_{\alpha\gamma} \wedge \Theta^\gamma \\ &= -\{\mathbf{e}_\alpha \rfloor (\kappa\Sigma_\gamma - \Lambda\eta_\gamma)\} \wedge \Theta^\gamma + (\mathbf{e}_\alpha \rfloor \Omega^{\mu\nu}) \wedge (\Theta^\gamma \wedge \eta_{\mu\nu\gamma}) + \Lambda\eta_{\alpha\gamma} \wedge \Theta^\gamma\end{aligned} \tag{5.14}$$

Therefore,

$$D\Sigma_\alpha = (\mathbf{e}_\alpha \rfloor \Theta^\gamma) \wedge \Sigma_\gamma + (\mathbf{e}_\alpha \rfloor \Omega^{\mu\nu}) \wedge \tau_{\mu\nu} \tag{5.15}$$

since $(\Sigma_\gamma \wedge \Theta^\gamma)$ is a 5-form.

To prove (b) we note that Eq. (5.4) may be written in the form

$$D(\Theta^{\mu} \wedge \eta_{\mu\alpha\beta}) = \Omega^{\gamma\mu} \wedge \vartheta_{[\gamma} \wedge \eta_{\mu]\alpha\beta} \tag{5.16}$$

By (5.12), the left-hand side of (5.16) is $\kappa D\tau_{\alpha\beta}$ and by using the easily verifiable identities

$$\vartheta_{[\gamma} \wedge \eta_{\mu]\alpha\beta} = \vartheta_{[\beta} \wedge \eta_{\alpha]\gamma\mu}, \qquad \vartheta_{[\beta} \wedge \eta_{\alpha]} = 0 \tag{5.17}$$

we obtain Eq. (5.2).

The equations (5.11), (5.12) are the field equations of the Einstein–Cartan–Sciama–Kibble (EC) theory of gravitation.[22,28,57,62] Equation (5.11) can be rewritten so as to display the irreducible parts of the curvature relevant to it:

$$[{}^{(4)}\Omega^{\mu\nu} + {}^{(5)}\Omega^{\mu\nu} + {}^{(6)}\Omega^{\mu\nu}] \wedge \eta_{\mu\nu\alpha} + \Lambda\eta_{\alpha} = \kappa\Sigma_{\alpha} \tag{5.18}$$

Here ${}^{(5)}\Omega^{\mu\nu}$ is the only purely non-Riemannian piece. For vanishing spin (or torsion), (5.11), (5.12) lead back to the Einstein equation (5.9). Since ${}^{(a)}\Omega^{\mu\nu} \wedge \eta_{\mu\nu} = 0$ for $a = 1,\ldots, 5$, the Lagrangian 4-form $\Omega^{\mu\nu} \wedge \eta_{\nu\mu}$ of the EC theory may be rewritten as ${}^{(6)}\Omega^{\mu\nu} \wedge \eta_{\nu\mu}$, so that it only involves SCALAR.

## 5.3. The Pseudoscalar Model

So far we have not been concerned with (5.3b), since it vanishes in a $V_4$ (and in a $U_3$). Nevertheless, we should be conscious of the fact that (5.3b) represents four equations and could therefore be mapped onto (5.1). Indeed, the identification

$$\Omega_{\alpha\beta} \wedge \vartheta^{\beta} + \Lambda'\eta_{\alpha} = \kappa'\Sigma_{\alpha} \tag{5.19}$$

$$\Theta_{[\alpha} \wedge \vartheta_{\beta]} = \kappa'\tau_{\alpha\beta} \tag{5.20}$$

is successful in the sense of the program of automatic conservation. Again a "cosmological" constant is allowed and a constant $\kappa'$ with the dimension of 1/force is needed. The proof of the correctness of this ansatz runs along the same lines as the one given above and is left to the reader.

As we have seen in deriving (2.7), we may rewrite (5.19) in the form

$$[{}^{(2)}\Omega_{\alpha\beta} + {}^{(3)}\Omega_{\alpha\beta} + {}^{(5)}\Omega_{\alpha\beta}] \wedge \vartheta^{\beta} + \Lambda'\eta_{\alpha} = \kappa'\Sigma_{\alpha} \tag{5.21}$$

Hence (5.21) carries only the non-Riemannian pieces of the curvature and the left-hand side of (5.21) vanishes in the case of vanishing torsion (or spin). Therefore (5.19), (5.20) collapse in a $V_4$ and would appear to have no decent limit in the spinless case. Moreover, in contrast to the EC equations (5.11), (5.12), Eqs. (5.19), (5.20) make no sense in a $U_3$. They are exclusively confined to 4 dimensions, as can be seen from the fact that $\Sigma_\alpha$ is an $(n-1)$-form and $\Omega_{\alpha\beta} \wedge \vartheta^\beta$ is a 3-form, so $n$ is necessarily equal to 4. Consequently, (5.19), (5.20) do not represent the field equation of a viable gravitational theory.

The irreducible decomposition (2.2) suggests why the set (5.19), (5.20) should have been expected to arise in this context. The parity-violating gravitational Lagrangian $\Omega^{\alpha\beta} \wedge \vartheta_\alpha \wedge \vartheta_\beta$ is exactly the one that yields these as field equations and, since ${}^{(a)}\Omega^{\alpha\beta} \wedge \vartheta_\alpha \wedge \vartheta_\beta = 0$ for $a \neq 3$, involves only PSCALAR. For physical reasons, one would expect such a distinctly 4-dimensional and non-Riemannian Lagrangian to arise at most as a supplementary term in another Lagrangian. An interesting model, in which the PSCALAR term is added to the EC Lagrangian and in which parity-violating terms appear, has first been discussed by R. Hojman *et al.*[25]

## 6. ON POINCARÉ GAUGE THEORY

Does the argument concerning automatic conservation exclude all other gravitational model theories from a serious discussion? In order to approach the answer to this question, we shall first take a look at Maxwell's theory.

Let $\mathscr{H}$ be the induction 2-form unifying the electromgnetic quantities $(\mathbf{H}, \mathbf{D})$ and $\mathscr{J}$ the electric current 3-form. Then the inhomogeneous Maxwell equation reads

$$d\mathscr{H} = \mathscr{J} \tag{6.1}$$

The electromagnetic field strength 2-form $F$, uniting $(\mathbf{B}, \mathbf{E})$, obeys the homogeneous Maxwell equation

$$dF = 0 \tag{6.2}$$

In vacuo we have $\mathscr{H} = {}^*F$. The conservation of charge can be derived by the Noether procedure, but it is also guaranteed by the form of (6.1):

$$d\mathscr{J} = dd\mathscr{H} = 0 \tag{6.3}$$

Consequently (6.2), which is the analog of a Bianchi identity, has no direct bearing on the conservation of charge.

For a Yang–Mills field (see Refs. 51 and 67) the structure of these equations remains basically the same, except that the exterior derivative is replaced by the (gauge-) *covariant* exterior derivative. This yields a covariant conservation law in much the same sense as energy-momentum is only covariantly conserved in general relativity.

In the Poincaré gauge theory[7] we have the same general pattern. The basis 1-forms $\vartheta^\alpha$ and the connection 1-forms $\omega^{\alpha\beta}$ of a Riemann–Cartan space-time are considered to be the translational and rotational (or Lorentz) potentials, respectively, with $\Theta^\alpha$ and $\Omega^{\alpha\beta}$ as the corresponding field strengths. Let $L_{\text{tot}}$ be the total and $V$ the gravitational Lagrangian 4-forms. Then, in view of (4.8), we have

$$L_{\text{tot}} = V(\vartheta^\alpha, \Theta^\alpha, \Omega^{\alpha\beta}) + L(\Psi, D\Psi, \vartheta^\alpha) \tag{6.4}$$

The gravitational field momentum 2-forms

$$\mathscr{H}_\alpha = -\frac{\partial V}{\partial \Theta^\alpha}, \qquad \mathscr{H}_{\alpha\beta} = -\frac{\partial V}{\partial \Omega^{\alpha\beta}} \tag{6.5}$$

fulfill the (inhomogeneous) field equations

$$D\mathscr{H}_\alpha - \mathscr{E}_\alpha = \Sigma_\alpha \tag{6.6}$$

$$D\mathscr{H}_{\alpha\beta} - \mathscr{E}_{\alpha\beta} = \tau_{\alpha\beta} \tag{6.7}$$

where $\Sigma_\alpha$ and $\tau_{\alpha\beta}$ are the currents defined in (4.7). The *gravitational energy-momentum* 3-form,

$$\mathscr{E}_\alpha := \mathbf{e}_\alpha \rfloor V + (\mathbf{e}_\alpha \rfloor \Theta^\beta) \wedge \mathscr{H}_\beta + (\mathbf{e}_\alpha \rfloor \Omega^{\beta\gamma}) \wedge \mathscr{H}_{\beta\gamma} \tag{6.8}$$

and the *gravitational spin* 3-form

$$\mathscr{E}_{\alpha\beta} := \vartheta_{[\beta} \wedge \mathscr{H}_{\alpha]} \tag{6.9}$$

together with their corresponding material partners $\Sigma_\alpha$ and $\tau_{\alpha\beta}$ are the sources of the Poincaré gauge fields $\Theta^\alpha$ and $\Omega^{\alpha\beta}$, provided the $\mathscr{H}$'s do not vanish.

Whereas the field equations (6.6), (6.7) correspond to (6.1), the Bianchi identities

$$D\Theta^\alpha \equiv \Omega_\beta{}^\alpha \wedge \vartheta^\beta \tag{6.10}$$

$$D\Omega^{\alpha\beta} \equiv 0 \tag{6.11}$$

[7] For a fairly nontechnical presentation see Ref. 21 and its continuation Ref. 2. The theory was put into exterior calculus notation and further developed by McCrea,(43) Mielke,(44,47) Müller-Hoissen,(49) Schweizer *et al.*,(55,56) and Wallner.(68,69,70) A related but different approach was recently advanced by Edelen(11); compare also Baekler *et al.*,(4) Goenner,(15) Hayashi and Bregman,(17) Hayashi and Shirafuji,(18) Hehl,(20) Hennig and Nitsch,(24) Kopczyński,(31) Lenzen,(37,38) McCrea,(40,41) Mielke,(45,46) Ne'eman(50) and Szczyrba.(60)

parallel the homogeneous Maxwell equation (6.2). The energy-momentum and the angular momentum laws

$$D\Sigma_\alpha = (\mathbf{e}_\alpha \rfloor \Theta^\beta) \wedge \Sigma_\beta + (\mathbf{e}_\alpha \rfloor \Omega^{\beta\gamma}) \wedge \tau_{\beta\gamma} \tag{6.12}$$

$$D\tau_{\alpha\beta} = \vartheta_{[\beta} \wedge \Sigma_{\alpha]} \tag{6.13}$$

and their character as Noether identities[8] have already been discussed in Sect. 4.

It is not surprising to find the gauge currents (6.8), (6.9) of the non-Abelian Poincaré group as additional sources of the gauge fields in (6.6), (6.7)—in contrast to the Abelian $U(1)$ which does not provide such a term in (6.1). However, what is surprising is that these currents are well-behaved vector- and bivector-valued 3-forms which can be defined without taking recourse to the explicit form of the Lagrangian $V$.

The EC theory, treated in Sect. 5, is a degenerate case of the Poincaré gauge theory with

$${}^{EC}V = \frac{1}{l^2}\left(\Lambda\eta - \frac{1}{2}\Omega_{\alpha\beta} \wedge \eta^{\alpha\beta}\right) \tag{6.14}$$

and

$${}^{EC}\mathcal{H}_\alpha = 0, \qquad {}^{EC}\mathcal{H}_{\alpha\beta} = \frac{1}{2l^2}\eta^{\alpha\beta} \tag{6.15}$$

In this case automatic conservation applies, because the field equations (6.6), (6.7) turn out to be algebraic in $\Omega^{\alpha\beta}$ and $\Theta^\alpha$. Taking both gauge potentials as propagating, in the spirit of a conventional first-order Lagrangian field theory,[9] yields the quadratic Lagrangian,

$$V = \frac{\Lambda}{l^2}\eta + \frac{A_0}{4l^2}\Omega^{\alpha\beta} \wedge \eta_{\alpha\beta} - \frac{1}{2}\Theta^\alpha \wedge \mathcal{H}_\alpha - \frac{1}{2}\Omega^{\alpha\beta} \wedge \mathcal{H}_{\alpha\beta} \tag{6.16}$$

[8] If we were to derive (6.12), (6.13) from the left-hand sides of (6.6), (6.7), as we did in the Maxwell case, we would need the Bianchi identities. These identities are not necessary, however, if we derive (6.12), (6.13) by means of the Noether procedure.

[9] In this case the leading terms in (6.6), (6.7) are typically of the form $D{*}\Theta^\alpha$ and $D{*}\Omega^{\alpha\beta}$, respectively. Recently, in a very interesting paper, Aldrovandi and Stédile[(1)] tried to guess the explicit form of the left-hand sides of (6.6), (6.7) by applying a hypothetical *duality symmetry* (which they attribute to Popov and Daikhin[(53)]) to the Bianchi identities (6.10), (6.11). This hypothesis seems to run into difficulties, however, since in (6.6) quadratic curvature terms appear in $\mathcal{E}_\alpha$, which are absent in (6.10). An alternative way of "deriving" a quadratic Lagrangian by means of a *Gordon type decomposition* of $\Sigma_\alpha$ and $\tau_{\alpha\beta}$ for a Dirac field was given in Ref. 20; see also Seitz.[(58,59)]

the momenta being

$$\mathscr{H}_\alpha = -{}^*\left(\frac{1}{l^2}\sum_{a=1}^{3} A_a{}^{(a)}\Theta_\alpha\right) \tag{6.17}$$

and

$$\mathscr{H}_{\alpha\beta} = -\frac{A_0}{2l^2}\eta_{\alpha\beta} - {}^*\left(\frac{1}{\kappa}\sum_{b=1}^{6} B_b{}^{(b)}\Omega_{\alpha\beta}\right) \tag{6.18}$$

with the constants $(l, \Lambda, \kappa)$ and $(A_0,..., A_3, B_1,..., B_6)$, where ${}^{(a)}\Theta_\alpha$, ${}^{(b)}\Omega_{\alpha\beta}$ are the irreducible parts of the torsion and curvature, respectively.

Equation (6.16) represents the most general parity-conserving polynomial Lagrangian yielding field equations linear in the second derivatives of the potentials $\vartheta^\alpha$ and $\omega^{\alpha\beta}$. Consistent gravitational models have been constructed without apparent physical difficulties. Nevertheless, the quadratic Poincaré gauge theory, which gives the connection $\omega^{\alpha\beta}$ and the torsion $\Theta^\alpha$ a truly independent role in gravitational physics, has no *automatic* conservation; in other words, the contracted Bianchi identities play no direct part in this context and are *not* equivalent, via the field equations (6.6), (6.7), to the energy-momentum and angular momentum laws (6.12), (6.13). It would seem therefore, in accordance with the structure of a Yang–Mills theory, that automatic conservation, a distinctive feature of general relativity and EC theory, is not a vital requirement for a gravitational gauge theory.

## APPENDIX A1

### The Irreducible Decompositions of Torsion and Curvature

The irreducible decompositions (under Lorentz transformations) of the torsion and the curvature have been given, in exterior form notation, by Wallner.[69]

For the torsion we have

$$\Theta^\alpha = {}^{(1)}\Theta^\alpha + {}^{(2)}\Theta^\alpha + {}^{(3)}\Theta^\alpha \tag{A1.1}$$

where

$${}^{(2)}\Theta^\alpha = \frac{1}{3}\vartheta^\alpha \wedge (\mathbf{e}_\beta \rfloor \Theta^\beta) \quad (\text{TRATOR}) \tag{A1.2}$$

$${}^{(3)}\Theta^\alpha = \frac{1}{3}{}^*\{{}^*(\Theta^\beta \wedge \vartheta_\beta) \wedge \vartheta^\alpha\} \quad (\text{AXITOR}) \tag{A1.3}$$

$${}^{(1)}\Theta^\alpha = \Theta^\alpha - {}^{(2)}\Theta^\alpha - {}^{(3)}\Theta^\alpha \quad (\text{TENTOR}) \tag{A1.4}$$

The decomposition of the curvature is given by

$$\Omega_{\alpha\beta} = \sum_{n=1}^{6} {}^{(n)}\Omega_{\alpha\beta} \tag{A1.5}$$

where

$$ {}^{(2)}\Omega_{\alpha\beta} = -{}^*(\vartheta_{[\alpha} \wedge \Psi_{\beta]}) \quad (\text{PAIRCOM}) \tag{A1.6}$$

$$ {}^{(3)}\Omega_{\alpha\beta} = -\frac{1}{12} X^*(\vartheta_\alpha \wedge \vartheta_\beta) \quad (\text{PSCALAR}) \tag{A1.7}$$

$$ {}^{(4)}\Omega_{\alpha\beta} = -\vartheta_{[\alpha} \wedge \Phi_{\beta]} \quad (\text{RICSYMF}) \tag{A1.8}$$

$$ {}^{(5)}\Omega_{\alpha\beta} = \frac{1}{2} \vartheta_{[\alpha} \wedge \{\mathbf{e}_{\beta]} \rfloor (\Omega^\gamma \wedge \vartheta_\gamma)\} \quad (\text{RICANTI}) \tag{A1.9}$$

$$ {}^{(6)}\Omega_{\alpha\beta} = -\frac{1}{12} \Omega\, \vartheta_\alpha \wedge \vartheta_\beta \quad (\text{SCALAR}) \tag{A1.10}$$

$$ {}^{(1)}\Omega_{\alpha\beta} = \Omega_{\alpha\beta} - \sum_{n=2}^{6} {}^{(n)}\Omega_{\alpha\beta} \quad (\text{WEYL}) \tag{A1.11}$$

with

$$\Omega^\alpha = \mathbf{e}_\beta \rfloor \Omega^{\alpha\beta}, \qquad \Omega = \mathbf{e}_\beta \rfloor \Omega^\beta, \qquad X^\alpha = {}^*(\Omega^{\beta\alpha} \wedge \vartheta_\beta), \qquad X = \mathbf{e}_\alpha \rfloor X^\alpha$$

$$\Psi_\alpha = X_\alpha - \frac{1}{4} X\vartheta_\alpha + \frac{1}{2}\mathbf{e}_\alpha \rfloor (X^\beta \wedge \vartheta_\beta), \qquad \Phi_\alpha = \Omega_\alpha - \frac{1}{4}\Omega\vartheta_\alpha + \frac{1}{2}\mathbf{e}_\alpha \rfloor (\Omega^\beta \wedge \vartheta_\beta) \tag{A1.12}$$

We note the following symmetries for the various irreducible parts of the curvature. For $a = 1, 4, 6$, $b = 1, 2, 3$, $c = 1, 2, 4, 5, 6$, and $d = 1, 2, 3, 4, 5$,

$$\begin{aligned} {}^{(a)}\Omega^{\alpha\beta} \wedge \vartheta_\beta = 0, &\qquad {}^{(b)}\Omega^{\alpha\beta} \wedge \eta_{\alpha\beta\gamma} = 0, \\ {}^{(c)}\Omega^{\alpha\beta} \wedge \vartheta_\alpha \wedge \vartheta_\beta = 0, &\qquad {}^{(d)}\Omega^{\alpha\beta} \wedge \eta_{\alpha\beta} = 0 \end{aligned} \tag{A1.13}$$

## APPENDIX A2

### Irreducible Decomposition of a Vector-Valued 3-Form in 4 Dimensions

Let $V^\alpha$ be a vector-valued 3-form. From the point of view of the presentation in Sect. 2, it is more convenient to exhibit the irreducible components of the dual $*V^\alpha$. To do this, we define the (pseudoscalar) trace $T$ and the antisymmetric part $A$ by

$$T = \mathbf{e}_\alpha \rfloor *V^\alpha = *(V^\alpha \wedge \vartheta_\alpha) \tag{A2.1}$$

and

$$A = *V^\alpha \wedge \vartheta_\alpha = -*(\mathbf{e}_\alpha \rfloor V^\alpha) \tag{A2.2}$$

One may easily verify that $P^\alpha := *V^\alpha - (1/4)\,T\vartheta^\alpha$ satisfies $\mathbf{e}_\alpha \rfloor P^\alpha = 0$ and $Q^\alpha := *V^\alpha + (1/2)\mathbf{e}^\alpha \rfloor A$ satisfies $Q^\alpha \wedge \vartheta_\alpha = 0$. Finally, $R^\alpha := *V^\alpha - (1/4)\,T\vartheta^\alpha + (1/2)\mathbf{e}^\alpha \rfloor A$ is tracefree and symmetric, i.e., $\mathbf{e}_\alpha \rfloor R^\alpha = 0$ and $R^\alpha \wedge \vartheta_\alpha = 0$. The irreducible decomposition is therefore given by

$$*V^\alpha = R^\alpha + \frac{1}{4}\,T\vartheta^\alpha - \frac{1}{2}\mathbf{e}^\alpha \rfloor A \tag{A2.3}$$

$$16 = 9 \oplus 1 \oplus 6$$

In terms of tensor components, if we express $V^\alpha$ in the form $V^\alpha = V^\alpha{}_\beta \eta^\beta$, so that $*V^\alpha = V^\alpha{}_\beta \vartheta^\beta$, then

$$T = V^\alpha{}_\alpha \quad \text{and} \quad A = V_{[\alpha\beta]}\vartheta^\alpha \wedge \vartheta^\beta \tag{A.4}$$

## APPENDIX A3

### Irreducible Decomposition of a Bivector-Valued 3-Form in 4 Dimensions

Again, we choose to exhibit the irreducible components of $*U^{\alpha\beta}$ for a bivector-valued 3-form $U^{\alpha\beta}$. Let

$$Y^\alpha = \mathbf{e}_\beta \rfloor *U^{\alpha\beta} = *(U^{\alpha\beta} \wedge \vartheta_\beta) \tag{A3.1}$$

$$Z = *U^{\alpha\beta} \wedge \vartheta_\alpha \wedge \vartheta_\beta = *(\mathbf{e}_\beta \rfloor \mathbf{e}_\alpha \rfloor U^{\alpha\beta}) \tag{A3.2}$$

and

$$S^{\alpha\beta} = *U^{\alpha\beta} - \frac{2}{3}\,Y^{[\alpha}\vartheta^{\beta]} + \frac{1}{6}\,\mathbf{e}^\alpha \rfloor \mathbf{e}^\beta \rfloor Z \tag{A3.3}$$

Then $A^{\alpha\beta} := *U^{\alpha\beta} - (2/3)\,Y^{[\alpha}\vartheta^{\beta]}$ is tracefree $(\mathbf{e}_\beta \rfloor A^{\alpha\beta} = 0)$ and $B^{\alpha\beta} := *U^{\alpha\beta} + (1/6)\,\mathbf{e}^\alpha \rfloor \mathbf{e}^\beta \rfloor Z$ has no axial part $(B^{\alpha\beta} \wedge \vartheta_\alpha \wedge \vartheta_\beta = 0)$, while both $\mathbf{e}_\beta \rfloor S^{\alpha\beta}=0$ and $S^{\alpha\beta}\wedge\vartheta_\alpha\wedge\vartheta_\beta=0$. The irreducible decomposition is therefore

$$*U^{\alpha\beta}=S^{\alpha\beta}+\frac{2}{3}\,Y^{[\alpha}\vartheta^{\beta]}-\frac{1}{6}\,\mathbf{e}^\alpha \rfloor \mathbf{e}^\beta \rfloor Z \tag{A3.4}$$

$$24=16\ \oplus\ 4\ \oplus\ 4$$

In terms of tensor components, if $U^{\alpha\beta}=U^{\alpha\beta}{}_\gamma\eta^\gamma$, then

$$Y^\alpha=U^{\alpha\beta}{}_\beta \quad\text{and}\quad Z=U_{[\alpha\beta\gamma]}\vartheta^\alpha\wedge\vartheta^\beta\wedge\vartheta^\gamma \tag{A3.5}$$

## ACKNOWLEDGMENTS

One of the authors (FWH) is grateful for the invitation from the Department of Mathematical Physics of University College, Dublin, and from the Dublin Institute for Advanced Studies, where this article was written. His thanks go to Professors Michael Hayes (UCD) and John Lewis (DIAS). He also appreciated very much the splendid hospitality of the Franciscans at Broc House. We also thank Peter A. Hogan for interesting discussions.

## REFERENCES

1. R. Aldrovandi and E. Stédile, *Int. J. Theor. Phys.* **23**, 301 (1984).
2. P. Baekler and F. W. Hehl, in *From SU(3) to Gravity—Festschrift in Honor of Yuval Ne'eman*, E. Gotsman and G. Tauber, eds. (Cambridge University Press, Cambridge, 1985).
3. P. Baekler, F. W. Hehl, and H. J. Lenzen, "Vacuum Solutions with Double Duality Properties of the Poincaré Gauge Field Theory. II," in *Proceedings of the Third Marcel Grossmann Meeting on General Relativity*, Hu Ning, ed. (North-Holland, Amsterdam, 1983), p. 107.
4. P. Baekler, F. W. Hehl, and E. W. Mielke, "Vacuum Solutions with Double Duality Properties of a Quadratic Poincaré Gauge Field Theory," in *Second Marcel Grossmann Meeting on the Recent Progress of the Fundamentals of General Relativity*, R. Ruffini, ed. (North-Holland, Amsterdam, 1981), p. 413.
5. P. G. Bergmann and V. DeSabbata, eds., *Cosmology and Gravitation: Spin, Torsion, Rotation, and Supergravity* (Proceedings of the Erice School, May 1979) (Plenum Press, New York, 1980).
6. B. A. Bilby, *Prog. Solid Mech. (Amsterdam)* **1**, 331 (1980).
7. É. Cartan, *C. R. Acad. Sci. (Paris)* **174**, 583 (1922).
8. É. Cartan, *Ann. Ec. Norm. Sup.* **40**, 325 (1923); **41**, 1 (1924).
9. É. Cartan, *Ann. Ec. Norm. Sup.* **42**, 17 (1925).
10. É. Cartan, *Leçon sur la Géométrie des Espaces de Riemann* (Gauthier-Villars, Paris, 1963), 2nd edn.

11. D. G. B. Edelen, *Applied Exterior Calculus* (Wiley, New York, 1985).
12. A. Einstein, *The Meaning of Relativity* (Princeton University, Princeton, New Jersey, 1955), 5th edn.
13. A. Einstein, *Grundzüge der Relativitätstheorie* (Vieweg, Braunschweig, 1960), 2nd edn.
14. A. C. Eringen, *Nonlinear Theory of Continuous Media* (McGraw-Hill, New York, 1962).
15. H. Goenner, Invited plenary talk given at the Annual Conference of the German Physical Society (DPG) in Munich, 1985.
16. W. Günther, *Abhandl. Braunschweig. Wiss. Ges.* **6**, 207 (1954).
17. K. Hayashi and A. Bregman, *Ann. Phys. (N.Y.)* **75**, 562 (1973).
18. K. Hayashi and T. Shirafuji, *Prog. Theor. Phys.* **64**, 866, 883, 1435, 2222 (1980); **65**, 525, 2079(E) (1981); **66**, 318, 741(E), 258 (1981).
19. F. W. Hehl, "On the Energy Tensor of Spinning Massive Matter in Classical Field Theory and General Relativity," *Rep. Math. Phys. (Warsaw)* **9**, 55 (1976).
20. F. W. Hehl, "Four Lectures on Poincaré Gauge Field Theory," in *Proceedings of the 6th Course of the International School of Cosmology and Gravitation on Spin, Torsion and Supergravity*, P. G. Bergmann and V. de Sabbata, eds. (Plenum Press, New York, 1980).
21. F. W. Hehl, *Found. Phys.* **15**, 451 (1985).
22. F. W. Hehl, P. von der Heyde, G. D. Kerlick, and J. M. Nester, *Rev. Mod. Phys.* **48**, 393 (1976).
23. A. Held, ed. *General Relativity and Gravitation. One Hundred Years after the Birth of Albert Einstein*, Vol. 1 (Plenum, New York, 1980).
24. J. Hennig and J. Nitsch, *Gen. Relativ. Gravit.* **13**, 947 (1981).
25. R. Hojman, C. Mukku, and W. A. Sayed, *Phys. Rev. D* **22**, 1915 (1980).
26. W. Jaunzemis, *Continuum Mechanics* (Macmillan, New York, 1967).
27. G. D. Kerlick, "Spin and Torsion in General Relativity: Foundations and Implications for Astrophysics and Cosmology," Ph.D. thesis, Princeton, 1975.
28. T. W. B. Kibble, *J. Math. Phys.* **2**, 212 (1961).
29. H. Kleinert, *Gauge Theory of Stresses and Defects*, to be published.
30. K. Kondo, ed., *RAAG Memoirs of the Unifying Study of Basic Problems in Engineering and the Physical Sciences by Means of Geometry* (Gakujutsu Bunken Fukyu-Kai, Tokyo, 1955, 1958, 1962, 1968), Vols. 1–4.
31. W. Kopczyński, *J. Phys. A* **15**, 493 (1982).
32. E. Kröner, *Ann. Phys. (Leipzig)* **11**, 13 (1963).
33. E. Kröner, ed., *Mechanics of Generalized Continua.* IUTAM Symposium (Springer, Berlin, 1968).
34. E. Kröner, in *Physique de Defauts* (Les Houches School), R. Balian *et al.*, eds. (North-Holland, Amsterdam, 1981).
35. L. D. Landau and E. M. Lifshitz, *Theory of Elasticity.* Vol. 7 of *Course of Theoretical Physics* (Pergamon, Oxford, 1981), 2nd English edn.
36. H.-J. Lenzen, Diploma Thesis, University of Cologne (1982).
37. H.-J. Lenzen, *Nuovo Cimento B* **82**, 85 (1984).
38. H.-J. Lenzen, *Gen. Relativ Gravit.* **17**, 1137 (1985).
39. M. Mathisson, *Acta Phys. Pol.* **6**, 163 (1937).
40. J. D. McCrea, *J. Phys. A* **16**, 997 (1983).
41. J. D. McCrea, *Phys. Lett. A* **100**, 397 (1984).
42. J. D. McCrea, "The Use of REDUCE in Finding Exact Solutions of the Quadratic Poincaré Gauge Field Equations," in *Classical General Relativity*, W. B. Bonnor, J. N. Islam, and M. A. MacCallum, eds. (Cambridge University Press, Cambridge, 1984), p. 173.
43. J. D. McCrea, Invited lecture at the XIV International Conference on Differential Geometric Methods in Mathematical Physics, University of Salamanca, Spain, 1985).

44. E. W. Mielke, "Über die Hypothesen, welche der Geometrodynamik zugrunde liegen," Habilitation thesis, University of Kiel, 1982.
45. E. W. Mielke, *J. Math. Phys.* **25**, 663 (1984).
46. E. W. Mielke, *Fortschr. Phys.* **32**, 639 (1984).
47. E. W. Mielke, English version of Ref. 44, to be published (Akademie Verlag, Berlin, 1986).
48. C. W. Misner, K. S. Thorne, and J. A. Wheeler, *Gravitation* (Freeman, San Francisco, 1973).
49. F. Müller-Hoissen, *Ann. Inst. H. Poincaré A* **40**, 21 (1984).
50. Y. Ne'eman, "Gravity, Groups and Gauges," in *General Relativity and Gravitation. One Hundred Years after the Birth of Albert Einstein*, A. Held, ed. (Plenum Press, New York), Vol. 1, Chap. 10.
51. L. O'Raifeartaigh, *Rep. Progr. Phys.* **42**, 159 (1979).
52. A. Papapetrou, *Proc. R. Soc. London A* **209**, 248 (1951).
53. D. A. Popov and L. I. Daikhin, *Sov. Phys. Dokl.* **20**, 818 (1976).
54. J. A. Schouten, *Ricci Calculus* (Springer, Berlin, 1954), 2nd edn.
55. M. Schweizer and N. Straumann, *Phys. Lett. A* **71**, 493 (1979).
56. M. Schweizer, N. Straumann, and A. Wipf, *Gen. Relativ. Gravit.* **12**, 951 (1980).
57. D. W. Sciama, "On the Analogy between Charge and Spin in General Relativity," in *Recent Developments in General Relativity* (Pergamon and PWN, Oxford, 1962), p. 415.
58. M. Seitz, *Ann. Phys. (Leipzig)* **41**, 280 (1984).
59. M. Seitz, *Class. Quantum Grav.* **2**, 919 (1985).
60. W. Szczyrba, *Phys. Rev. D* **25**, 2548 (1981).
61. W. Thirring, *A Course in Mathematical Physics*, Vol. 2. *Classical Field Theory* (Springer, New York, 1979).
62. A. Trautman, *Symp. Math.* **12**, 139 (1973).
63. A. Trautman, "Comments on a Paper by Elie Cartan," in Ref. 5, p. 493.
64. A. Trautman, "Fiber Bundles, Gauge Fields and Gravitation," in *General Relativity and Gravitation. One Hundred Years after the Birth of Albert Einstein*, A. Held, ed. (Plenum, New York, 1980), Vol. 1, Chap. 9, pp. 287–308.
65. A. Trautman, *Differential Geometry for Physicists. Stony Brook Lectures* (Bibliopolis, Naples, 1984).
66. C. Truesdell and R. A. Toupin, "The Classical Field Theories," in *Encyclopedia of Physics III/1*, S. Flügge, ed. (Springer, Berlin, 1960), p. 226.
67. R. Utiyama, *Prog. Theor. Phys.* **64**, 2207 (1980).
68. R. P. Wallner, *Gen. Relativ. Gravit.* **12**, 719 (1980).
69. R. P. Wallner, *Acta Phys. Aust.* **55**, 67 (1983).
70. R. P. Wallner, *Gen. Relativ. Gravit.* **17**, 1081 (1985).
71. J. A. Wheeler, "Gravitation as Geometry. II," in *Gravitation and Relativity*. H.-Y. Chiu and W. F. Hoffmann, eds. (Benjamin, New York, 1964).
72. J. A. Wheeler, "Particles and Geometry," in *Lecture Notes in Physics* (Springer-Verlag, New York, 1982), Vol. 160, p. 189.

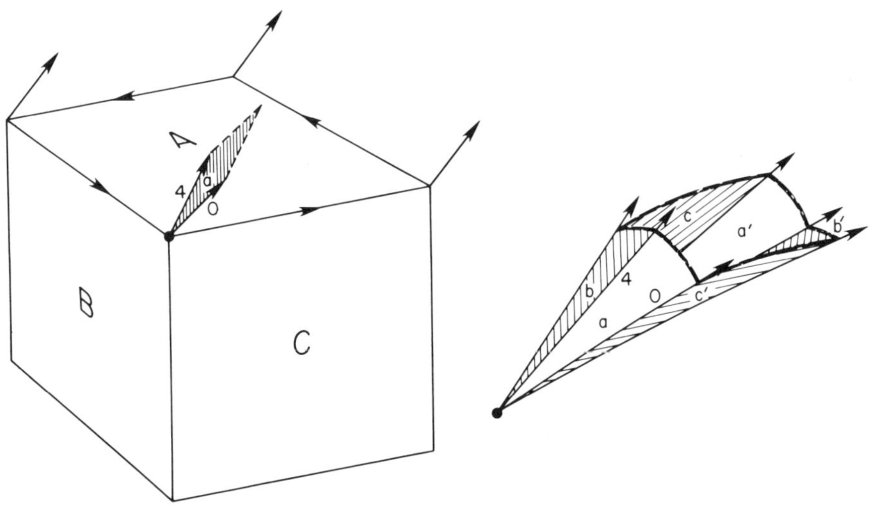


*Plate 20*

# The Boundary of a Boundary Principle: A Unified Approach

**Arkady Kheyfets**[1]

*Received August 23, 1985*

*The boundary of a boundary principle in field theories is described. The difference in treatment of the principle in electrodynamics and general relativity is pointed out and reformulated in terms of underlying mathematical structure of the theories. The problem of unifying the treatment is formulated and solved. The role of E. Cartan's concept of the moment of rotation associated with the curvature of a Levi–Civita connection on a frame bundle is shown to be crucial for the unification. The analysis of the boundary of a boundary principle in Kaluza–Klein theory is performed and the recipe for a unified treatment of the principle in electrodynamics and general relativity is shown to follow from the analysis. It is pointed out that the unification can be extended to Yang–Mills fields easily.*

## 1. MOTIVATIONS: ELECTRODYNAMICS VERSUS GENERAL RELATIVITY

The boundary of a boundary principle as a manifestation of the idea of austerity in physics—or law without law—was emphasized by J. A. Wheeler.(1) By that time applications of the principle in electrodynamics and general relativity had been worked out.(2,3) The principle itself can be considered as a topological one in its nature and written down in the form

$$\partial\partial = 0 \tag{1}$$

where $\partial$ is the boundary operator on a chain complex associated with a manifold. A detailed description of this construction can be found in any book on differential manifolds containing the description of integration

[1] Center for Theoretical Physics, The University of Texas at Austin, Austin, Texas 78712.

theory.[4] However, the details are not important for our purpose. The manifolds we are going to consider will be smooth, homotopically trivial and orientable, and we can think of the operator $\partial$ in a simple minded way as acting on a manifold and giving as a result the boundary of the manifold. Only one important idea needs to be recalled: For any $n$-dimensional manifold and any scalar valued $(n-2)$-form $\alpha$, we have

$$\int_{\partial\partial M} \alpha = 0 \tag{2}$$

Thus, we can use Stoke's theorem twice to get

$$\int_{M} dd\alpha = \int_{\partial M} d\alpha = \int_{\partial\partial M} \alpha = 0 \tag{3}$$

We can write (3) at each point $x \in M$ for any neighborhood of $x$ in $M$, which gives

$$dd\alpha = 0 \tag{4}$$

and, since it is true for any form $\alpha$, this makes relation

$$dd = 0 \tag{5}$$

just another expression of the boundary of a boundary principle (1). This fact can be expressed mathematically as follows: operation $d$ on cochains is dual to operation $\partial$ on chains and in our situation they express the same result. Relation (5) is an expression of the boundary of a boundary principle as good as (1), but it is more convenient to use. Therefore everywhere below in this paper whenever we mention the boundary of a boundary principle we should think of (5) rather than of (1).

Actually, when one uses (5) in electrodynamics and general relativity one does it twice, using the first time

$$dd\alpha = 0 \tag{6}$$

where $\alpha$ is a 1-form on spacetime, and another time using again (6) but assuming that $\alpha$ is a 2-form on spacetime. In the first case, (6) is usually called the 1-2-3 form of the boundary of a boundary principle, and in the second case it is called the 2-3-4 form of the principle.

The application of the boundary of a boundary principle in electrodymanics is straightforward. The electromagnetic field is assumed to be described by electromagnetic 2-form $F = \frac{1}{2} F_{\mu\nu}\, dx^{\mu} \wedge dx^{\nu}$ (a scalar valued 2-form on spacetime), the form being generated by 4-potential 1-form $A = A_{\mu}\, dx^{\mu}$ via

$$F = dA \tag{7}$$

The boundary of a boundary principle in the 1-2-3 form provides then equations for $F$:

$$dF = ddA = 0 \tag{8}$$

These equations are not enough to determine $F$, and additional equations are introduced:

$$d^*F = J \tag{9}$$

where * is the Hodge star operator. Equation (9) can be considered for the purpose of the paper as the definition of current density 3-form $J$. Since $^*F$ is a 2-form, application of the boundary of a boundary principle in 2-3-4 form gives

$$dJ = dd^*F = 0 \tag{10}$$

i.e., 2-3-4 form of the principle provides the conservation law for $J$ as an automatic consequence of field equations (9).

The application of the boundary of a boundary principle in general relativity is a little bit fancier and can be described as follows (for detailed mathematical description one is referred to Refs. 3 and 5, the last being considered by us as one of the best if not the best source for anybody who desires to reproduce calculations with complete understanding of what is really happening).

Gravitational effects are assumed to be described by Riemannian curvature form $e_\mu \wedge e_\lambda R^{|\mu\lambda|}{}_{|\nu\alpha|}\, dx^\nu \wedge dx^\alpha$ (a bivector valued 2-form on spacetime), the form being generated by Levi–Civita connection 1-form $\vartheta$ (also bivector-valued) via

$$R = \nabla\vartheta \tag{11}$$

where $\nabla$ means the exterior covariant derivative associated with $\vartheta$. This derivative has the properties that for metric tensor $g$ considered as a tensor-valued function on spacetime

$$\nabla g = 0 \tag{12}$$

and for the canonical soldering form $\varphi = e_\mu\, dx^\mu$ (Ref. 3 uses the notation $d\mathscr{P}$ instead of $\varphi$) considered as a vector-valued 1-form on spacetime

$$\nabla\varphi = 0 \tag{13}$$

Notice here that

$$S = \nabla\varphi \tag{14}$$

is nothing but the spacetime torsion, so that (13) means that there is no torsion. The 1-2-3 form of the boundary of a boundary principle together with some simple geometric properties of Levi–Civita connection provides then

$$\nabla R = \nabla\nabla\vartheta = 0 \tag{15}$$

which is nothing but Bianchi identities rewritten in symbolic form. The identities have a geometric interpretation. Roughly speaking they say that the sum over a closed two-dimensional surface of rotations induced by Riemannian curvature is equal to zero. One is directed to Ref. 3 for more precise formulation and details.

To consider the boundary of a boundary principle in 2-3-4 form, we introduce the density of the moment of rotation induced by Riemannian curvature, i.e., expression

$$^\star(\varphi \wedge R) = {}^\star(d\mathscr{P} \wedge R) \tag{16}$$

where $^\star$ means the Hodge star operator in the space of values for forms $\varphi$, $R$, or $\varphi \wedge R$, i.e., in the space of polyvectors. The expression (16) can be rewritten as follows:

$$\begin{aligned} ^\star(\varphi \wedge R) &= {}^\star(e_\nu \wedge e_\mu \wedge e_\lambda R^{|\mu\lambda|}{}_{|\alpha\beta|}\, dx^\nu \wedge dx^\alpha \wedge dx^\beta) \\ &= e_\kappa \varepsilon_{\nu\mu\lambda}{}^{\kappa} R^{|\mu\lambda|}{}_{|\alpha\beta|}\, \varepsilon^{\nu\alpha\beta\gamma}\, d^3\Sigma_\gamma = e_\kappa {}^*R^*{}_\nu{}^{\kappa\nu\gamma}\, d^3\Sigma_\gamma = e_\kappa G^{\kappa\gamma}\, d^3\Sigma_\gamma \end{aligned} \tag{17}$$

where $d^3\Sigma_\gamma = (1/3!)\, \varepsilon_{\gamma\nu\alpha\beta}\, dx^\nu \wedge dx^\alpha \wedge dx^\beta$, and $G^{\kappa\gamma}$ are the components of Einstein tensor.

Applying $\nabla$ to (16), we get

$$\nabla^\star(\varphi \wedge R) = {}^\star(\nabla(\varphi \wedge R)) = {}^\star(\nabla\varphi \wedge R - \varphi \wedge \nabla R) \tag{18}$$

and, since $\nabla\varphi = S = 0$ and $\nabla R = 0$ according to (15),

$$\nabla^\star(\varphi \wedge R) = 0 \tag{19}$$

Geometrically this means that the density of the moment of rotation induced by Riemannian curvature is equal to zero automatically. It has been shown in Ref. 3 that (19) is an expression of the boundary of a boundary principle in 2-3-4 form. Introducing the notation $G$ for the vector-valued 3-form associated with Einstein tensor,

$$G = e_\alpha G^{\alpha\nu}\, d^3\Sigma_\nu \tag{20}$$

and $T$ for the vector-valued 3-form associated with the momentum-energy tensor,

$$T = e_\alpha T^{\alpha\nu}\, d^3\Sigma_\nu \tag{21}$$

the Einstein equations can be rewritten as

$$G = T \tag{22}$$

where it is assumed that gravitational constant $\kappa = 1$.

Thus, from (17) and (19)–(22), it follows that

$$\nabla T = 0 \tag{23}$$

i.e., in general relativity, in the way analogous to that in electrodynamics, the boundary of a boundary principle in 2-3-4 form provides the conservation law for the source of gravitational field as an automatic consequence of the field equations (22).

Comparing the two situations—electrodynamics and general relativity—we find out that, whereas the boundary of a boundary principle in 1-2-3 form works in similar ways in both theories, the principle in 2-3-4 form is applied in rather different ways or, at least, it appears that it works differently.

Indeed, if we consider electrodynamics as the geometry of a $U(1)$ principal bundle with spacetime as its base and connection determined by electromagnetic field 4-potential form $A$, relation (8) has exactly the same interpretation in this geometry as (15) in general relativity, i.e., it represents Bianchi identities in described geometry. However, if we compare form $*F$ in (10) and $G = {}^{\star}(\varphi \wedge R)$ in (19), it appears, at least at first sight, that the boundary of a boundary principle in 2-3-4 form works absolutely differently in electrodynamics and in general relativity, and a nice geometric interpretation for $G$ as the moment of rotation density appears to be missing for $*F$ in electrodynamics.

J. A. Wheeler always used to consider this situation, when different theories would employ common simple principles but apply them in completely different ways, as something which was not exactly in correspondence with the general idea of austerity. The question was whether it was possible to correct the situation and, if so, how to go about it.

## 2. BASIC IDEAS AND THEIR MATHEMATICAL FORMULATION

Summarizing the last section, we can say that the task was to learn how to handle the boundary of a boundary principle in 2-3-4 form in a way unified for different field theories.

The basic ideas were developed in discussions between J. A. Wheeler and the author and were initially formulated by J. A. Wheeler in the form of questions. It is necessary to notice here that J. A. Wheeler always had an intuitive impression that the approach to the 2-3-4 form of the boundary of a boundary principle in general relativity was more fundamental and deep than the one in electrodynamics and, if we wanted to develop a unified approach, we should try to transfer the geometric construction from general relativity to electrodynamics, with correspondent modifications, of course. This boiled down to question 1:

*Question 1.* The moment of what is to play in electrodynamics the role analogous to that of the moment of rotation induced by curvature in general relativity?

This question, no matter how simple and remote from the solution it appears at first sight, turned out, when carefully mathematically formulated and analyzed, to give all the clues necessary to solve the problem and to develop an approach unified for different field theories. The straight answer to the question can be found in Sec. 4. The answer turns out to be simple and can be understood right away. In fact, it does not require anything but some mathematical refinement in the formulation of Maxwell equations. However, to be able to work out the guidelines of understanding the situation in different field theories and to back up the suggested recipe, it would be better to follow the line of reasonings described in Sec. 3.

It was not clear at that time if question 1 could be answered satisfactorily, and question 2 was asked by J. A. Wheeler.

*Question 2.* Will the consideration of a generalization of electrodynamics such as the Yang–Mills theory improve our understanding of the matter and provide some additional hints or ideas that can help in the solving of the problem?

It turned out to be much easier to answer the second question, and the answer was both yes and no. Yes, it did improve our understanding of the matter, but, no, it did not help to solve the problem. The generalization of the reasonings in electrodynamics to the case of Yang–Mills fields with non-Abelian gauge groups was too straightforward, and the way of applying the boundary of a boundary principle in the Yang–Mills theory was as far from that in general relativity as the one in electrodynamics. However, due to the same reason, it became clear that things in electrodynamics and in the Yang–Mills theory are so similar that, once one learns how to work in electrodynamics, one will be able to get it done for any Yang–Mills field, and this provided the basis for considering question 1 in its simplest form rather than in the most general one.

The two Wheeler questions can be reformulated in the statements about underlying mathematical structures in electrodynamics and general relativity. It turns out that they contain indications of very important differences in the structures of these two theories.

As it has already been mentioned, electrodynamics can be interpreted as geometry on a principal bundle with spacetime as its base, $U(1)$ as its structure group, and the electromagnetic 4-potential form as the connection form. General relativity also can be formulated as geometry on a principal bundle. There are different opinions on the question of what bundle should be considered as the principal bundle in general relativity. A very good review of the topic can be found in Ref. 6. But, no matter how the bundle is chosen, the structure group of the principal bundle is always non-Abelian, whereas the structure group $U(1)$ of electrodynamics is Abelian. If this difference was responsible for the difference in using the 2-3-4 boundary of a boundary principle in the two theories, then clarification of the matter in the Yang–Mills theory could give some additional information on the problem. This possibility has been considered,[7] and it turns out that all the procedures of electrodynamics can be duplicated in the Yang–Mills theory with a minimal amount of complications. The distinctions were connected with the fact that the field potential and strength became Lie-algebra valued instead of scalar-valued in electrodynamics (actually, also Lie-algebra valued but the Lie algebra of $U(1)$ was one-dimensional), gauge-covariant derivatives took the place of partial derivatives (the structure group became non-Abelian), and the strength of the field became gauge covariant rather than gauge invariant. The lesson learnt can be expressed as follows: there is almost no difference in the application of the boundary of a boundary principle between electrodynamics and Yang–Mills theory, leaving aside the fact that gauge covariance takes the place of gauge invariance, and the changes are almost trivial.

In short, even though question 2 indicates a very important difference between electrodynamics and general relativity, it does not help in solving the problem of the approach unification.

For the purpose of our paper, the principal bundle of general relativity can be considered as a frame bundle over a four-dimensional base manifold endowed with pseudo-Riemannian metric, so that its structure group is $GL(4)$. This enables one to construct the canonical soldering form on the bundle and thereby to endow the bundle with geometry richer than the one used in electrodynamics or Yang–Mills theory. We are talking here about the Levi–Civita connection. This connection is determined uniquely by the conditions that the covariant derivative of the metric is zero and the covariant derivative of the canonical soldering form (torsion) is equal to zero. If one takes another look at Sec. 1, one will notice that the concept of the soldering form is vital in 2-3-4 form of the boundary of a boundary principle and its geometric interpretation. Indeed, the moment of rotation induced by Riemannian curvature is expressed as the left dual of the exterior product of the canonical soldering form and Riemannian curvature form.

J. A. Wheeler's question 1 tells us that it would be desirable to have an analogous construction in electrodynamics. But the standard fiber-bundle description of electrodynamics does not provide it, because the corresponding bundle is not a frame bundle, a canonical soldering form on it does not exist, and, consequently, the moment cannot be constructed. This gives the idea that we should consider a theory that includes the description of electromagnetism and has a frame bundle as the underlying structure. An obvious choice is the Kaluza–Klein theory. The idea is to consider the same (with obvious changes) constructions as in general relativity and then to separate the conclusions about the electromagnetic field from those for gravitation. It can be done only to some extent (for example, the current conservation law will contain covariant instead of partial derivatives in curved spacetime), but it will be quite enough. As a result, we can perform the whole procedure in the same way as in general relativity to get the conservation laws for both momentum-energy of the electromagnetic field and the current of the source of the electromagnetic field. Since we start the analysis from the construction which is nothing but the generalization of the moment considered in general relativity, and the Riemannian curvature form on the unified space of the Kaluza–Klein theory incorporates also the electromagnetic field, we can hope to get an expression for the current density as the moment of something constructed of electromagnetic field, and this can provide the clues to the unified interpretation of the 2-3-4 form of the boundary of a boundary principle.

Section 3 contains a short account of what actually happens in Kaluza–Klein theory. Most of the mathematical and calculational details are omitted. The reason for this is not as much their length as the fact that they are rather refined and their explanation would not only be lengthy but also make the reasonings too bulky which could confuse the picture rather than clarify it. However, if one wants to reproduce the details, Ref. 5 provides all necessary background for this.

We are going to restrict ourselves to the discussion of the 2-3-4 form of the boundary of a boundary principle only because, as we have already seen, there are no problems with the 1-2-3 form of the principle. Also, as it has already been mentioned above, the consideration of general Yang–Mills fields does not contribute anything significant in the context of our problem, so that it is natural to consider here the "electromagnetism + gravity" Kaluza–Klein theory rather than the more general "Yang–Mills + gravity" case, although short comments on the general case will be provided.

## 3. THE MOMENT OF ROTATION INDUCED BY CURVATURE: KALUZA–KLEIN THEORY

It would be too lengthy to give a more or less rigorous description of the Kaluza–Klein theory. Fortunately, all our considerations are local, so that even a rather rough, oversimplified description with all kinds of language abusing will work fine.

We consider here Kaluza–Klein theory as geometry determined by Levi–Civita connection on a frame bundle with five-dimensional base. The base itself has the structure of the $U(1)$-bundle over a four-dimensional base (spacetime) with geometry determined by the electromagnetic potential $A$ as the connection. This means that at each point of the five-dimensional space mentioned above the space tangent to it can be split into a $4+1$ space, the 4-space being isomorphic to Minkowski space and form $A$ determining the projection onto the remaining direction. On the other hand, the five-dimensional space can be represented locally in the neighborhood of point $p$ as the product of a spacetime region and a space isomorphic to $U(1)$. This allows us to introduce metric on the five-dimensional space in such a way that for a pair of vectors $X$, $Y$ of the space tangent to the five-dimensional space

$$\bar{g}(X, Y) = g(hX, hY) + AX \cdot AY \tag{24}$$

where $h = 1 - A$ means the projection of the tangent five-dimensional space onto its four-dimensional subspace tangent to spacetime, $A$ is the projection onto its one-dimensional subspace isomorphic to the Lie algebra of $U(1)$, $g$ is the spacetime metric, and $\bar{g}$ is the metric on the five-dimensional space of Kaluza–Klein theory (notice that language already has been abused, but it is all right locally). Since expressions containing four-dimensional and five-dimensional entities will be used extensively in this section, we introduce the notation that the dash over the symbol means that it has to be understood as a five-dimensional one, otherwise it is four-dimensional. Metric $\bar{g}$ determines thereby the Hodge star operators $\bar{*}$ and $\bar{\star}$, Levi–Civita connection $\bar{\vartheta}$, and Riemannian curvature $\bar{R}$ on the five-dimensional space whereas $g$ determines $*$, $\star$, $\vartheta$, and $R$ on spacetime.

We can pick up a basis in the following way (convenient for calculations): take four orthonormal vector fields on spacetime, then take their horizontal lifts with respect to connection determined by $A$ (notations $e_\mu$, $\mu = 0, \ldots, 3$, will be used for the lifts) and pick up the fifth basic vector $e_c$ as the unit vector orthogonal to all $e_\mu$. We will also use notations $\xi^\mu$, $\xi^c$ for the 1-forms dual to $e_\mu$, $e_c$. In case we do not want to distinguish between spacetime components and the "charge" component, we will use Latin indices, say $\{e_i\}_i$, $i=0,\ldots,4$, means $\{e_\mu, e_c\}$ and $e_4=e_c$. Using (24) and structure equations we can calculate the matrix of connection forms and the components of the Riemannian curvature with respect to the basis described above $\{e_i\}=\{e_\mu, e_c\}$ and its dual $\{\xi^i\}=\{\xi^\mu, \xi^c\}$. Without any restrictions on generality the fields $\{e_\mu\}$ can be picked up in such a way that the four-dimensinal Levi–Civita connection components are zero over the original point $x$ (projection onto spacetime of point $p$).

We will only write down the results

$$\bar{\vartheta}^{\mu\nu} = \vartheta^{\mu\nu} - \tfrac{1}{2}F^{\mu\nu}\xi^c \tag{25}$$

$$\bar{\vartheta}^{c\mu} = -\bar{\vartheta}^{\mu c} = \tfrac{1}{2}F^\mu{}_\nu\xi^\nu \tag{26}$$

$$\bar{\vartheta}^{cc} = 0 \qquad \text{(obvious)} \tag{27}$$

The only nonzero components of the Riemannian curvature (up to the symmetries of the curvature form) are

$$\bar{R}^{\lambda\mu}{}_{\nu\kappa} = R^{\lambda\mu}{}_{\nu\kappa} + \tfrac{1}{2}F^{\lambda\mu}{}_{\kappa\nu} - \tfrac{1}{4}(F^\lambda{}_\nu F^\mu{}_\kappa - F^\lambda{}_\kappa F^\mu{}_\nu) \tag{28}$$

$$\bar{R}^{\lambda\mu}{}_{\nu c} = -\tfrac{1}{2}e_\nu[F^{\lambda\mu}] \tag{29}$$

$$\bar{R}^{c\lambda}{}_{\nu\mu} = \tfrac{1}{2}(e_\nu[F^\lambda{}_\mu] - e_\mu[F^\lambda{}_\nu]) \tag{30}$$

$$\bar{R}^{c\lambda}{}_{c\mu} = \tfrac{1}{2}e_c[F^\lambda{}_\mu] + \tfrac{1}{4}F_{\alpha\mu}F^{\alpha\lambda} \tag{31}$$

where the symbol $e_i[\cdots]$ means the action of the differential operator associated with vector $e_i$.

Equations (28)–(31) are the curvature components in decomposition

$$\bar{R} = e_i \wedge e_j \bar{R}^{|ij|}{}_{|km|}\xi^k \wedge \xi^m \tag{32}$$

The canonical soldering form can be written as

$$\bar{\varphi} = e_i\xi^i = e_\mu\xi^\mu + e_c\xi^c = \varphi + e_c\xi^c \tag{33}$$

The subsequent procedure will go in the way analogous to that for general relativity, i.e., the moment of rotation induced by curvature is taken

$$\bar{\star}(\bar{\varphi} \wedge \bar{R}) \tag{34}$$

Due to the fact that we consider here five-dimensional space, the moment will be a bivector-valued 3-form (in general relativity it was vector valued).

Then we take the exterior covariant derivative of (34) to get

$$\overline{\nabla\star}(\bar{\varphi} \wedge \bar{R}) = \bar{\star}(\overline{\nabla\varphi} \wedge \bar{R} - \bar{\varphi} \wedge \overline{\nabla R}) = 0 \tag{35}$$

It is obvious that the components of $\bar{R}$ given by (28)–(31) can be analyzed separately. We are not going to discuss here the terms arising from (30) and (31). We notice only that they give the relations that are automatically satisfied due to the 1-2-3 form of the boundary of a boundary principle and gauge invariance of $F$.

In analysis of the terms arising due to (28) we will write down only the terms providing nontrivial information, i.e., those which do not vanish after taking the derivative due to the 1-2-3 form of the boundary of a boundary principle and gauge invariance only. Thus, after some calculations, simplified by the circumstance that basis $\{e_\mu, e_c\}$ is orthonormal, one gets, for (34),

$$\bar{\star}(e_i \wedge e_\lambda \wedge e_\mu \bar{R}^{|\lambda\mu|}{}_{|\nu\kappa|}\xi^i \wedge \xi^\nu \wedge \xi^\kappa)$$
$$= e_\lambda e_c G^{\lambda\mu}\, d^3\Sigma_\mu + e_\lambda \wedge e_c(-F^{\lambda\alpha}F^\mu{}_\alpha + \tfrac{1}{4}g^{\lambda\mu}F_{\alpha\beta}F^{\alpha\beta})\, d^3\Sigma_\mu$$
$$+ \text{“extra terms”} \tag{36}$$

The second term in (36) can be easily recognized: it is the electromagnetic momentum-energy tensor

$$T^{\lambda\mu} = F^{\lambda\alpha}F_\alpha{}^\mu - \tfrac{1}{4}g^{\lambda\mu}F^{\alpha\beta}F_{\alpha\beta} \tag{37}$$

Introducing the notations

$$G = e_\lambda G^{\lambda\mu}\, d^3\Sigma_\mu \tag{38}$$

and

$$T = e_\lambda T^{\lambda\mu}\, d^3\Sigma_\mu \tag{39}$$

for Einstein and momentum-energy 3-forms, (36) can be rewritten in the form

$$G \wedge e_c + T \wedge e_c + \text{“extra terms”} \tag{40}$$

Applying $\bar{\nabla}$ to (40) and using (25)–(27), we get

$$\nabla G \wedge e_c - \nabla T \wedge e_c + \text{“extra terms”} = 0 \tag{41}$$

where “extra terms” vanish trivially (in the sense explained above), and we arrive at

$$\nabla G - \nabla T = 0 \tag{42}$$

and, since

$$\nabla G = 0 \tag{43}$$

due to obvious reasons, we automatically get the momentum-energy conservation law.

The result is in agreement with that of general relativity. It is natural and absolutely not surprising. Neither is it impressive. Everything was all right in general relativity anyway.

To analyze the curvature component produced by (29) the same procedure should be used, but the term to be considered is now

$$X = \tfrac{1}{2} e_\lambda \wedge e_\mu (e_\nu [F^{\lambda\mu}])\, \xi^\nu \wedge \xi^c \tag{44}$$

Having multiplied (44) by $\bar{\varphi} = e_\kappa \xi^\kappa + e_c \xi^c$, we get

$$\tfrac{1}{2} e_\kappa \wedge e_\lambda \wedge e_\mu (e_\nu [F^{\lambda\mu}])\, \xi^\kappa \wedge \xi^\nu \wedge \xi^c \tag{45}$$

so that, after taking the left dual, we arrive at

$$\tfrac{1}{2} e_\alpha \wedge e_c (e_\nu [{}^*F^\alpha{}_\kappa])\, \xi^\kappa \wedge \xi^\nu \wedge \xi^c \tag{46}$$

Expression (46) is not very convenient for analysis and comparison. We transform it first in the obvious way to

$$\tfrac{1}{2} (e_c \xi^c) \wedge (e_\alpha (e_\nu [{}^*F^\alpha{}_\kappa])\, \xi^\kappa \wedge \xi^\nu) \tag{47}$$

and consider the second factor in the exterior product (47), i.e.,

$$\tfrac{1}{2} e_\alpha (e_\nu [{}^*F^\alpha{}_\kappa])\, \xi^\kappa \wedge \xi^\nu \tag{48}$$

Notice that expression (48) represents a vector-valued 2-form. It is natural to call coefficients $e_\nu[{}^*F^\alpha{}_\kappa]$ of this form contravariant components. To be able to use antisymmetry of the electromagnetic tensor, and for the sake of convenience in comparison with standard results, it would be better to consider covariant components.

Indeed, any finite-dimensional vector space is isomorphic to its dual and, in the case where a metric on the space is given, the isomorphism can be chosen in a canonical way using the metric, i.e., it can be defined on basic vectors by

$$e_\alpha \to \xi^\beta g_{\beta\alpha} \tag{49}$$

Applying this isomorphism to (48), we come up with the 1-form valued form

$$Y = \xi^\beta (e_\nu [{}^*F_{\beta\kappa}])\, \xi^\kappa \wedge \xi^\nu \tag{50}$$

and, due to the antisymmetry of $*F_{\beta\kappa}$, it can be rewritten as

$$Y = \frac{1}{3!} e_\nu [*F_{\beta\kappa}] \, \xi^\beta \wedge \xi^\kappa \wedge \xi^\nu \tag{51}$$

Using (51) and the technique developed in Ref. 5, one can write instead of (47) the expression

$$e_c \xi^c \wedge \nabla *F + \text{"extra terms"} \tag{52}$$

and then take derivative $\bar{\nabla}$ of this expression. It can be shown that the result can be expressed in the form

$$e_c \xi^c \wedge \nabla\nabla *F + \text{"extra terms"} = 0 \tag{53}$$

where, just as before, "extra terms" give zero contribution due to the 1-2-3 form of the boundary of a boundary principle and gauge invariance of $F$ only.

Thus (53), which represents another part of the 2-3-4 form of the boundary of a boundary principle in Kaluza–Klein theory, gives

$$\nabla\nabla *F = 0 \tag{54}$$

and, since according to definition (9) $\nabla *F = J$, this gives

$$\nabla J = 0 \tag{55}$$

i.e., conservation for the source of electromagnetic field. This was expected.

But actually we got something more. If one takes another look at Eqs. (44)–(51), one will be able to give a short list of operations performed

$$Y = g \cdot \bar{*}(\bar{\varphi} \wedge X) \tag{56}$$

where symbol $g\cdot$ is used to write down the action of isomorphism (49), and $X$ can be considered very roughly as the gradient of the electromagnetic field tensor, the last being considered as a bivector-valued function. Expression (56) looks very much like the moment of $X$. The only new thing in (56) is operation $g\cdot$ but it could be used also in calculations for (28), in which case we would get covariant Einstein and momentum-energy form instead of contravariant with no major changes in anything essential.

The symbolic expression (56), together with comments following it, is fundamental in establishing the unified formulation of the boundary of a boundary principle in the 2-3-4 form for all basic field theories.

To conclude this section we want to make one remark. From comments in Sec. 1, it is clear that the description of the more general situation

"Yang–Mills field + gravitation" Kaluza–Klein theory would be essentially the same as that for "electromagnetism + gravitation," apart from complications in calculations and the need to use gauge covariance instead of gauge invariance. However, nobody cares about actually performing this procedure, because it will be seen in the next section that the unified procedure already has been worked out and its application in the Yang–Mills theory can be done as straightforwardly as in electrodynamics.

## 4. THE 2-3-4 FORM OF THE BOUNDARY OF A BOUNDARY PRINCIPLE: A UNIFIED APPROACH

Sec. 3 provides all the guidelines for formulating the 2-3-4 form of the boundary of a boundary principle in electrodynamics in the form very much analogous to that in general relativity [cf. Eqs. (29) and (56) and comments to (56)]. We give complete and detailed calculations in this section. There are two reasons for this. The first reason is that the calculations are very simple and elegant and they can be done in exactly the same way in the Yang–Mills theory. The second reason is that, due to lack of room, calculations in Sec. 3 are compressed and truncated to the edge of comprehensibility, and detailed calculations here are supposed to compensate for this.

We will follow (29) and (56) but assume now that everything in the equations has to be understood literally. We are going to consider electrodynamics in flat spacetime in orthonormal coordinate system, so that $g_{\mu\nu} = \eta_{\mu\nu}$. Equation (29) tells us that we have to start from the gradient of electromagnetic tensor, the last being considered as a bivector-valued function on spacetime, rather than a scalar-valued 2-form. We will use notation $\tilde{F}$ for it

$$\tilde{F} = e_\mu \wedge e_\nu F^{|\mu\nu|} \tag{57}$$

and the corresponding gradient will be

$$d\tilde{F} = e_\mu \wedge e_\nu F^{|\mu\nu|}{}_{,\lambda}\, dx^\lambda \tag{58}$$

Now we can calculate the moment of $d\tilde{F}$, using (56) with $\varphi = e_\mu\, dx^\mu$:

$$\begin{aligned} g\cdot\star(\varphi \wedge d\tilde{F}) &= \tfrac{1}{2} g\cdot\star(e_\kappa \wedge e_\mu \wedge e_\nu F^{\mu\nu}{}_{,\lambda}\, dx^\kappa \wedge dx^\lambda) \\ &= \tfrac{1}{2} g\cdot(e_\alpha {*F^\alpha}_{\kappa,\lambda}\, dx^\kappa \wedge dx^\lambda) \\ &= \frac{1}{3!}\, {*F}_{\alpha\kappa,\lambda}\, dx^\alpha \wedge dx^\kappa \wedge dx^\lambda = d{*F} \end{aligned} \tag{59}$$

or, briefly,

$$g \cdot {}^\star(\varphi \wedge d\tilde{F}) = d{}^*F = J \tag{60}$$

so that the 2-3-4 form of the boundary of a boundary principle automatically gives conservation for $J$:

$$dJ = d\{g \cdot {}^\star(\varphi \wedge d\tilde{F})\} = dd{}^*F = 0 \tag{61}$$

and $J$ is expressed by (60) as the moment of electromagnetic strength gradient, exactly as in (17), together with (22), momentum-energy $T$ is expressed in terms of gravitational strength gradient (indeed, Riemannian curvature describes the effects produced by gravitational strength gradient).

In short, Wheeler's idea of a unified geometric interpretation for the 2-3-4 form of the boundary of a boundary principle, when mathematically formulated and analyzed (this involves an analysis of the principle in Kaluza–Klein theory at an intermediate step), provides both the universal mathematical procedure of applying the principle in different field theories and an understanding of what is going on in physics. Above all, it helps us to understand what the idea of austerity in physics, of law without law, really means and seeks.

## 5. ACKNOWLEDGMENT

The author thanks M. Demianski, C. McCarter, W. Miller, and W. Schleich for helpful discussions, Z. Davis for her assistance in preparing this paper for publication, J. A. Wheeler, whose stimulating ideas and influence cannot be overestimated, and the Center for Theoretical Physics at the University of Texas at Austin. Publication of this paper was supported by NSF Grants PHY 8205717 and PHY 8503890.

## REFERENCES

1. J. A. Wheeler, *Physics and Austerity* (Anhui Science and Technology Publications, China, 1982).
2. É. Cartan, *Leçons sur la Géométrie des Espaces de Riemann*, (Gauthier-Villars, Paris, 1946).
3. C. W. Misner, K. S. Thorne, and J. A. Wheeler, *Gravitation* (W. H. Freeman, New York, 1973), Chap. 15.
4. S. Sternberg, *Lectures on Differential Geometry* (Chelsea, New York, 1983).
5. D. Bleecker, *Gauge Theory and Variational Principles* (Addison-Wesley, Reading, Massachusetts, 1981).
6. R. Percacci, "Role of Soldering in Gravity Theory," in *Proceedings of the XIII International Conference on Differential Geometric Methods in Theoretical Physics*, Shumen, Bulgaria, 1984.
7. A. Kheyfets and J. A. Wheeler, "The Boundary of a Boundary Principle in Gauge Field Theories," working paper, The University of Texas at Austin, 1985.

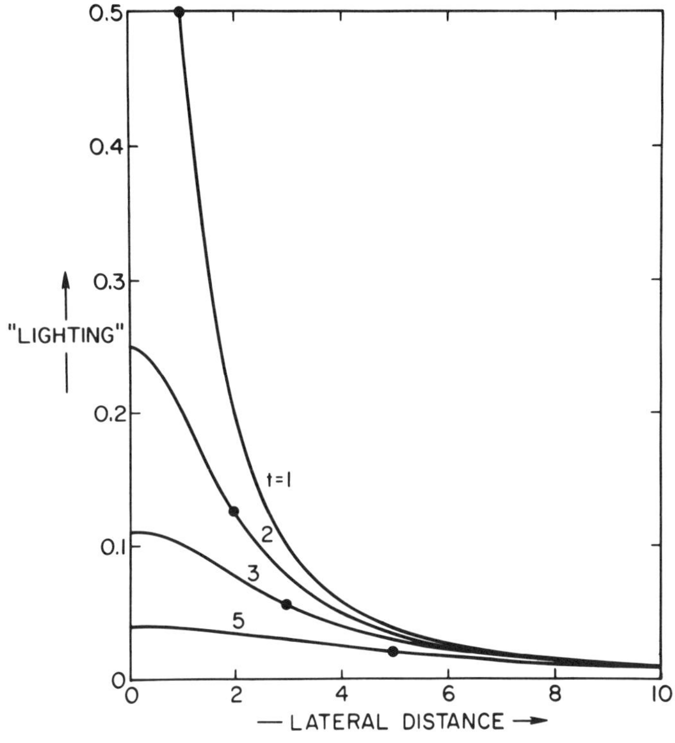


*Plate 21*

# The Gravitational Field at Spatial Infinity[1]

**Matthew Alexander**[2] **and Peter G. Bergmann**[3]

*Received June 6, 1985*

*This paper treats the formulation of the gravitational field variables and the equations obeyed by them at spatial infinity. The variables consist of a three-dimensional tensor and a scalar, which satisfy separate field equations, which in turn can be obtained from two distinct Lagrangians. Aside from Lorentz rotations, the symmetry operations include an Abelian gauge group and an Abelian Lie group, leading to a number of conservation laws and to differential identities between the field equations.*

## 1. INTRODUCTION

In previous papers[(1-3)] we have motivated our exploration of spatial infinity of asymptotically flat Einstein manifolds. Briefly, if on Cauchy surfaces the sources of the gratitational field have compact support, the gravitational field at spatial infinity obeys well-defined causal laws, which are much easier to deal with than the full gravitational field equations in space-time. They do retain some of the latters' peculiarities, so that they may be used as a training ground for the examination of some of the conceptual problems bedeviling the quantization of gravity.

In Ref. 3 we developed a projective construction of infinity for Minkowski space-time. We noted that this type of construction has certain conceptual advantages over the usual conformal one; a similar treatment of asymptotically flat Einstein manifolds thus recommends itself.

---

[1] This work was partially supported by the National Science Foundation through Grant PHY-8541793 to Syracuse University.

[2] Department of Mathematics, Fordham University.

[3] Departments of Physics, Syracuse University and New York University.

To begin with, we establish a topology on the set of all geodesics of the manifold. We define the *neighborhood* of a geodesic by way of the following construction. First, select a particular geodesic by specifying a point through which it passes and its direction at that point. Next, by means of a small displacement of the point, and by small Lorentz rotations, we construct a tube that encloses the given geodesic. A geodesic that has been "thickened" in this way (Geroch) thus corresponds to our intuitive notion of a geodesic and its associated neighborhood. Two geodesics will be considered equivalent (i.e., belonging to the same *equivalence class*) if each is contained in the neighborhood of the other for all but some finite length. A *projective completion* of the asymptotically flat manifold is then achieved by identifying each equivalence class of geodesics with a distinct ideal point at infinity. As a result of this construction, null infinity will have the same structure as in the conformal treatment, but timelike and spatial infinity will both be three-dimensional.

Spatial infinity consists of those ideal points that represent equivalence classes of spacelike geodesics. Intuitively, if a given spacelike geodesic enters and remains in the neighborhood of another spacelike geodesic, then these two curves are *asymptotically parallel.* A single equivalence class is thus comprised of a family of asymptotically parallel spacelike geodesics. In the case of Minkowski space-time, it can be easily demonstrated (with the help of hyperbolic coordinates) that the union of all such equivalence classes is a three-dimensional manifold which is homeomorphic and isometric to a timelike unit hyperboloid of constant curvature $-1$. The treatment of space-times that have nonzero mass is computationally more involved; but in this case as well an examination of the structure of the vacuum Einstein equations near spatial infinity leads to the same result, that spatial infinity is a three-dimensional hyperboloid, with all the properties mentioned above (see, e.g., Refs. 1 and 2).

Incidentally, in this discussion, the term "projective" is used somewhat loosely.[1-5] We do not mean to suggest any resemblance between our construction and the projective sphere $P^2$. If in our definition of equivalence, we were to identify opposite directions, "infinity" (and in particular, spatial infinity) would not be simply connected. Leaving aside the question of how one is to deal with timelike infinity (where the world-lines of material particles terminate), it would appear that a construction which resulted in spatial infinity being multiply connected would be best suited to the description of physical systems in which the sources extend to infinity (e.g., Dirac monopoles). For sources having compact support on each spacelike slice, such a construction is not particularly useful. For the purposes of this paper, we shall confine ourselves to systems of compact sources, and thus we shall employ our previous construction.

In order to illustrate the details of the construction, and to examine the limits of various fields at spatial infinity, it is convenient to adopt a particular type of coordinate system. A natural choice of coordinates will employ the spacelike geodesics themselves. To establish such a system, we first foliate the space-time with timelike hypersurfaces which we label $\Omega = \text{const}$. For our purposes, it is not necessary to have a global foliation (in fact, for space-times containing singularities it may not be possible within some finite region around a singularity to introduce a foliation); we will merely require that far enough away from the matter distribution we may construct a foliation that extends to infinity without mishap. Given the scalar field $\Omega$, we may locally define a congruence of curves that are everywhere normal to the hypersurfaces $\Omega = \text{const}$. This congruence of non-parallel curves will be comprised of geodesics if the normals to the $\Omega = \text{const}$ surfaces satisfy the "eikonal equation,"

$$g^{\mu\nu}\Omega_{,\mu}\Omega_{,\nu} = F(\Omega)$$

Here $F(\Omega)$ is some arbitrary function of $\Omega$, whose choice determines the method of indexing the hypersurfaces of the foliation, but which has no effect on the choice of foliation itself. The individual geodesics normal to the foliation will be identified by means of three parameters $u^k = \text{const}$ $(k = 0, 1, 2)$. The four variables $(u^k, \Omega)$ will serve as coordinates of space-time.

In the presence of massive matter far away from spatial infinity it turns out that our construction needs to be modified if the metric field is to be at least twice differentiable. We need to replace the geodesics by curves that are asymptotically geodesic, in that their curvatures go to zero at the rate $\sigma^{-1}$, where $\sigma$ is the proper distance from some fixed point on each curve. The asymptotic geodesics lie, however, within the "thickening" of a corresponding geodesic; the structure of spatial infinity is thus not affected by this loosening of the construction.

With these understandings the metric of space-time will take the form

$$\begin{gathered} ds^2 = -\Omega^{-4}(1 - \Omega\Psi)\, d\Omega^2 + \Omega^{-2}\varepsilon_{mn}\, du^m\, du^n \\ \Psi = \Psi(\Omega, u^k), \qquad \varepsilon_{mn} = \varepsilon_{mn}(\Omega, u^k) \end{gathered} \tag{1}$$

With this fixation the coordinate $\Omega$ is reciprocal to $\sigma$. Spatial infinity corresponds to $\Omega = 0$.

We require that at spatial infinity the function $\Psi$ exist, along with its derivatives with respect to $u^k$, and we subject the functions $\varepsilon_{mn}$ and their partial derivatives with respect to $\Omega$ to the same requirements. It turns out that these quantities suffice to formulate at spatial infinity Einstein's vacuum field equations. In what follows we shall use the notations

$$\begin{aligned} \Psi(0, u^k) &\equiv \psi(u^k) \equiv \psi \\ \varepsilon_{mn}(0, u^k) &\equiv e_{mn}(u^k) \equiv e_{mn} \\ \left.\frac{\partial \varepsilon_{mn}}{\partial \Omega}\right|_{\Omega=0} &\equiv q_{mn}(u^k) \equiv q_{mn} \end{aligned} \tag{2}$$

The tensor $e_{mn}$, the "background metric" at spatial infinity ("spi"), turns out to be the metric of a Lorentz hyperboloid by virtue of Einstein's field equations, whereas the scalar $\psi$ and the tensor $q_{mn}$ obey hyperbolic field equations on spi; they may be regarded as the dynamical field variables that describe the gravitational field at spi.

The field equations have been presented in Refs. 1 and 2. They are found to separate if the tensor $q_{kl}$ is replaced by a tensor $\alpha_{kl}$ defined as

$$\begin{aligned} \alpha_{kl} &= q_{kl} + e_{kl}(2\psi - q), \qquad q \equiv e^{mn} q_{mn} \\ q_{kl} &= \alpha_{kl} + e_{kl}(\psi - \tfrac{1}{2}\alpha), \qquad \alpha \equiv e^{mn}\alpha_{mn} \end{aligned} \tag{3}$$

With this replacement of variables the field equations assume the form

$$\nabla^2\psi - 3\psi = 0 \tag{4a}$$

$$\left.\begin{aligned} \nabla^2\alpha_{kl} - \tfrac{1}{2}e_{kl}\nabla^2\alpha - (\alpha^s{}_{k|sl} + \alpha^s{}_{l|sk}) + \tfrac{1}{2}a_{|kl} + \tfrac{1}{2}e_{kl}\alpha^{mn}{}_{|mn} \\ + 3\alpha_{kl} - e_{kl}\alpha = 0 \end{aligned}\right\} \tag{4b}$$

$$\alpha^s{}_{k|s} = 0 \tag{4c}$$

Here the symbol $\nabla^2$ and the vertical bar denote three-dimensional covariant differentiation in terms of the "background metric" $e_{mn}$. Likewise, $e^{mn}$ and $e_{mn}$ are used to raise and lower indices.

The field equations are linear with respect to the variables $\psi$ and $\alpha_{kl}$. The three equations (4c) are not independent of (4b); they are obtained from the latter as its (covariant) divergence. With its help several terms of (4b) may be eliminated, and this is sometimes advantageous.

The field equations can be obtained from two variational principles. With $e$ standing for the determinant of $e_{mn}$, the two Lagrangians are, respectively,

$$\begin{aligned} \Lambda_\psi &= -\tfrac{1}{2}\sqrt{e}\,(\psi^{|s}\psi_{|s} + 3\psi^2) \\ \Lambda_\alpha &= \sqrt{e}\,[\tfrac{1}{4}\alpha^{|s}\alpha_{|s} - \tfrac{1}{2}\alpha^{mn|s}\alpha_{mn|s} + \alpha^{km}{}_{|m}\alpha^n{}_{k|n} - \tfrac{1}{2}\alpha^{ms}{}_{|s}\alpha_{|m} \\ &\quad + \tfrac{1}{2}(\alpha^{mn}\alpha_{mn} - \alpha^2)] \end{aligned} \tag{5}$$

## 2. SYMMETRY TRANSFORMATIONS

As pointed out previously, the formalism presented in this paper is covariant with respect to various transformations. Considering that we are dealing with space-times that are asymptotically flat, it stands to reason that the formalism will not change under (asymptotic) Poincaré transformations. As a matter of fact, as one point of spi represents a whole congruence of asymptotically parallel asymptotic geodesics, the Poincaré group is reduced to the (homomorphic) Lorentz group, and the formalism is invariant with respect to this six-parametric Lie group. The quantity $e_{mn}$ is isometric with respect to Lorentz rotations.

There is a second group of transformations. If the foliation of space-time by hypersurfaces $\Omega = \text{const}$ is changed in that the origin of each curve $u^k = \text{const}$ is displaced by an amount $\frac{1}{2}f(u^k)$, such a distortion will have no effect on $e_{mn}$, nor on the scalar field $\psi$, but will change the tensor field $\alpha_{mn}$ according to the formula

$$\delta\alpha_{kl} = e_{kl}\nabla^2 f - f_{|kl} - 2e_{kl}f \tag{6}$$

As the function $f(u^k)$ is arbitrary, though assumed to be continuous and differentiable to several orders, the Abelian group (6) is a gauge group, and the transformation law of $\alpha_{kl}$ is homogeneous in the order of differentiation.

Finally there is a third group of transformations. Whereas the second group arises from mappings of space-time on itself that on each outgoing ray (asymptotic geodesic) tend to a constant displacement as spi is approached, this group arises from space-time mappings that diverge logarithmically on approach to spi, but this divergent factor, say $\log \Omega$, is multiplied by a vector that has a constant direction throughout spi. This notion of a constant vector is legitimate because on spi parallelity of world vectors (i.e., four-dimensional vectors) is integrable.

Suppose such a constant world vector is denoted by $\boldsymbol{\beta}$, and the unit vector normal to the hypersurface spi by $\hat{n}$, then a scalar $g(u^k)$ can be formed,

$$g = \boldsymbol{\beta} \cdot \hat{n} \tag{7}$$

which is, of course, not constant on spi, but is a linear combination of but four such scalars, each of which represents one possible choice of $\boldsymbol{\beta}$.

Under these mappings of space-time on itself the tensor field $\alpha_{mn}$ is invariant, but the scalar field $\psi$ transforms according to the formula

$$\delta\psi = 2g \tag{8}$$

The second and third groups of transformations all commute with each other; they have nonvanishing Lie brackets with the Lorentz rotations, and these Lie brackets belong, respectively, to the second and third groups. Accordingly, these two different kinds of "gauge transformations" form normal subgroups within the totality of symmetry transformations. The corresponding factor groups are obvious.

It should be noted that whereas the second group is a function group ($f$ is an arbitrary continuous differentiable function on spi), the third group is a four-parametric Lie group, in that the function $g$ is a linear combination of four fixed functions on spi. It satisfies the highly overdetermined differential equations

$$g_{|mn} - e_{mn}\, g = 0 \tag{9}$$

## 3. DIFFERENTIAL IDENTITIES

According to Noether's theorem the product of Euler–Lagrange equations by an infinitesimal symmetry transformation of the field variables equals identically a divergence. If the symmetry transformation contains arbitrary functions of the coordinates, this theorem implies the existence of differential identities between the left-hand sides of the field equations. For instance, in the case of Maxwell's equations these differential identities imply the vanishing divergence of the charge-current density, and in Einstein's equations of the gravitational field they lead to the covariant continuity equations for the energy-stress tensor.

Denote the left-hand side of Eqs. (4b) by the symbol $A_{k1}$. Then, By Eq. (6) the expression $A^{k1}\delta\alpha_{k1}$ must be a divergence. Integrating by parts we find that

$$A^{kl}(e_{kl}\nabla^2 f - f_{|kl} - 2e_{kl} f) \equiv Q^m{}_{|m} + (\nabla^2 A - A^{kl}{}_{|kl} - 2A)\, f \equiv 0 \tag{10}$$

where $Q^m{}_{|m}$ is a divergence. As the second term, with $f$ arbitrary, cannot be a divergence, it follows that the parenthesis vanishes identically,

$$\nabla^2 A - A^{kl}{}_{|kl} - 2A \equiv 0 \tag{11}$$

As

$$-2A \equiv \alpha^{mn}{}_{|mn} \tag{12}$$

the differential identity between the field equations, (11), may be written in the form

$$(A^{|l} - A^{ls}{}_{|s} + \alpha^{ls}{}_{|s})_{|l} \equiv 0 \tag{13}$$

This identity lies at the root of the relationship between Eqs. (4b) and (4c).

## 4. CONSERVATION LAWS

Noether's theorem associates every infinitesimal symmetry operation with an equation of continuity, which in turn implies a conservation law. In view of the topology of spi, $S^2 \times R$, these conservation laws lead to true constants of the motion, in that in each case the two-dimensional integral over a two-sphere intersecting the three-dimensional hyperboloid of spi has a value that does not depend on the details of the cut, and is in particular constant if the whole cut is displaced in the third, time-like direction.

Incidentally, though the transformations (7), (8) geometrically differ from the rigid displacements of the four-dimensional Poincaré group, together with the Lorentz rotations they form an abstract 10-parametric Lie group that is isomorphic to the (abstract) Poincaré group.

Let us begin with the transformations (7) and (8). If Noether's theorem is applied in the usual manner, the equation of continuity that is associated with that symmetry group turns out to be

$$[\sqrt{e}\, e^{kl}(\psi_{,l}\, g - \psi g_{,l})]_{,k} = 0 \tag{14}$$

The comma denotes partial differentiation, which in this equation happens to be equivalent to covariant differentiation. As Eq. (8) has four linearly independent solutions for the function $g(u^k)$, Eq. (14) corresponds to four linearly independent conserved quantities.

Next we turn to the symmetry transformations (6), which involve the arbitrary function $f(u^k)$. If the expressions $A_{k1}$, Eq. (4b), are multiplied by (6) [cf. Eq. (10)], we obtain an equation of continuity,

$$(\sqrt{e}\, A^{ks} f_{|s})_{,k} = 0 \tag{15}$$

which appears disappointing, as the conserved quantities vanish by virtue of the field equations themselves. It is well known, however, that similar results are obtained for most symmetry transformations involving arbitrary functions (usually called gauge transformations). Nontrivial conserved quantities can be obtained if an appropriately chosen curl (exterior derivative) is added to the conserved expression, whose divergence vanishes identically. This strategy is successful in the present case as well.

The exterior derivative to be added is

$$\begin{gathered}(U^{[ks]l} f_{|l})_{|s} \\ U^{[ks]l} = \sqrt{e}\, [\alpha^{lk|s} - \alpha^{ls|k} + \tfrac{1}{2}(e^{ls}\alpha^{|k} - e^{lk}\alpha^{|s})]\end{gathered} \tag{16}$$

By this addition the conservation law (15) is changed into

$$\begin{gathered}B^{k}{}_{,k} = 0 \\ B^k = \sqrt{e}\, [(\alpha^{mk|n} - \alpha^{mn|k}) f_{|mn} + \tfrac{1}{2}(\alpha^{|k}\nabla^2 f - \alpha^{|n} f^{|k}{}_{n})]\end{gathered} \tag{17}$$

As there are no restrictions on the choice of $f$, except for continuity and differentiability, the number of conservation laws implied by Eq. (17) is infinite.

The conservation laws (14) and (17) are both linear in the field variables involved. The third set of conservation laws, those associated with the Lorentz group, is quadratic. As the Lagrangians for $\psi$ and for $\alpha_{kl}$ are entirely separate, there are two sets of six conserved quantities each. For the scalar field the conservation law may be cast into the form

$$\begin{gathered} U^k{}_{,k} = 0 \\ U^k = \sqrt{e}\,[\psi^{|k}\psi_{|s}\xi^s - \tfrac{1}{2}(\psi^{|s}\psi_{|s} + 3\psi^2)\,\xi^k] \end{gathered} \tag{18}$$

where $\xi^k$ is a linear combination of the six vector fields that satisfy Killing's equation

$$\xi_{m|n} + \xi_{n|m} = 0 \tag{19}$$

A similar conservation law is associated with the tensor field $\alpha_{mn}$.

## 5. SPECIAL COORDINATES

Up to now the choice of coordinates on spi, $u^k$, has been left open. For applications it is of course frequently desirable to make some particular choice. One such choice is represented by pseudo-spherical, i.e., hyperbolic, coordinates. For the spacelike spherical cuts we introduce ordinary spherical coordinates $\phi$ and $\theta$, and for the timelike direction the hyperbolic coordinate $\chi$.

With this choice the "background metric" $e_{mn}$ is diagonal and takes the form

$$\begin{gathered} g_{00} = 1 \\ g_{11} = -\cosh^2\chi, \qquad g_{22} = -\cosh^2\chi\,\sin^2\theta \end{gathered} \tag{20}$$

The four independent solutions satisfying Eqs. (8) are

$$\begin{gathered} \hat{n}_{(0)} = \sinh\chi, \qquad \hat{n}_{(1)} = \cosh\chi\,\sin\theta\,\cos\phi \\ \hat{n}_{(2)} = \cosh\chi\,\sin\theta\,\sin\phi, \qquad \hat{n}_{(3)} = \cosh\chi\,\cos\theta \end{gathered} \tag{21}$$

All four expressions are scalars. The subscripts in parentheses indicate that they correspond to displacements in the four respective directions of a symbolic Minkowski–Lorentz tetrad.

Finally, we shall give the expressions for the six independent vector fields that occur in Eq. (18); they obey Eqs. (19). To begin with, there are three purely spatial rotations. They correspond to vector fields with vanishing components $\xi^0$. Their remaining components are

$$\begin{aligned} \xi^1_{(1)} &= \sin\phi, \qquad \xi^2_{(1)} = \cot\theta\cos\phi \\ \xi^1_{(2)} &= -\cos\phi, \qquad \xi^2_{(2)} = \cot\theta\sin\phi \\ \xi'_{(3)} &= 0, \qquad \xi^2_{(3)} = 1 \end{aligned} \tag{22}$$

For boosts we also have three vector fields, with the components

$$\begin{gathered} \xi^0_{(4)} = \sin\theta\cos\phi, \qquad \xi^1_{(4)} = \tanh\chi\cos\theta\cos\phi \\ \xi^2_{(4)} = -\tanh\chi\sin^{-1}\theta\sin\phi \\ \xi^0_{(5)} = \sin\theta\sin\phi, \qquad \xi^1_{(5)} = \tanh\chi\cos\theta\sin\phi \\ \xi^2_{(5)} = \tanh\chi\sin^{-1}\theta\cos\phi \\ \xi^0_{(6)} = \cos\theta, \qquad \xi^1_{(6)} = -\tanh\chi\sin\theta, \qquad \xi^2_{(6)} = 0 \end{gathered} \tag{23}$$

## 6. CONCLUDING REMARKS

The treatment of spatial infinity provides a "muscle-flexing" exercise (J. A. Wheeler) for the treatment of the quantization of the gravitational field in space-time. In common with the four-dimensional field, it possesses a function group of symmetry operations, which render the propagation of the field variables off a Cauchy surface subject to the arbitrariness of choice of a gauge frame. Like the field equations in space-time, there is an action principle (actually two action principles!), and Noether's theorem implies not only the existence of conserved quantities, but leads to Bianchi-like identities between the field equations. Remarkably, the field variables, and their equations, at spatial infinity are closed in themselves: they propagate from their own Cauchy data, without reference to the other degrees of freedom of the field in space-time.

Of course, the variables at infinity lack the immediate physical relevance of space-time fields. But they obey linear equations; thus they are much easier to discuss than their four-dimensional analogs. We hope that eventually their treatment, including quantization, will teach us something about the four-dimensional field.

We dedicate this paper to John A. Wheeler on the occasion of his seventy-fifth birthday in friendship and with all good wishes for the years to come.

## REFERENCES

1. M. Alexander, "Hamiltonian Dynamics at Spatial Infinity," Ph. D. thesis, 1982, available from University Microfilms, Ann Arbor, Michigan.
2. M. Alexander and P. G. Bergmann, in *Proceedings of the Third Marcel Grossman Meeting on General Relativity*, Hu Ning, ed. (Science Press and North-Holland Publishing Co., 1983), p. 575.
3. M. Alexander and P. G. Bergmann, *Found. Phys.* **14**, 925 1984.
4. P. Sommers, *J. Math. Phys.* **19**, 549 (1978).
5. R. Beig and B. G. Schmidt, "Einstein's Equations Near Spatial Infinity," Preprint MPA2, Max-Planck-Institut für Physik und Astrophysik, Garching bei München, 1982.

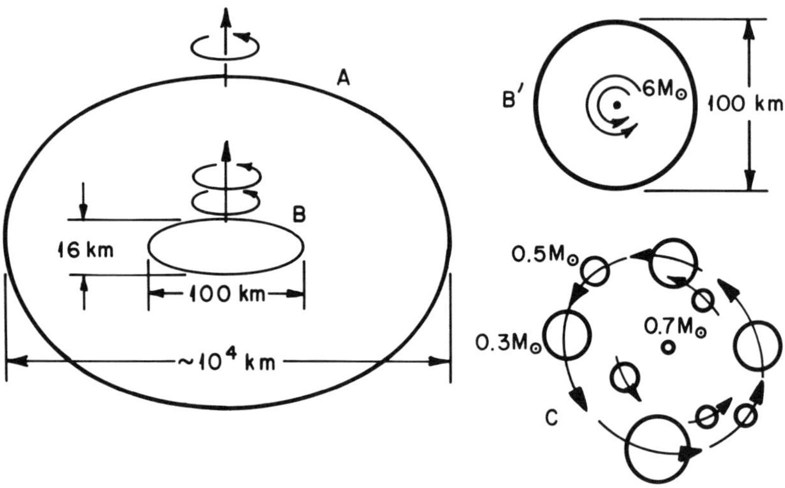


*Plate 22*

# Gravitational Radiation Reaction on the Motion of Particles in General Relativity[1]

**P. A. Hogan**[2] **and I. Robinson**[3]

*Received March 21, 1985*

*We examine the problem of deducing the geodesic motion of test particles from Einstein's vacuum field equations and its extension to include gravitational radiation reaction. In the latter case we obtain an equation of motion for a particle which incorporates radiation reaction of the electrodynamical type, but due to shearing radiation, together with a mass-loss formula of the Bondi–Sachs type.*

## 1. INTRODUCTION

For many purposes it is convenient to treat particles as singularities in the field. This approach has been used extensively in the derivation of equations of motion from the Einstein and Einstein–Maxwell equations.(1) It is not always easy, however, to understand the results of such calculations in geometrical terms. The difficulty arises partly from the fact that the metric is undefined precisely where it would be most useful: along the world-lines of the particles. Infeld and Schild(2) deal elegantly with this problem by decomposing the metric into a background metric, which is nonsingular along the world-line of the particle, and a singular contribution from the particle itself. The nonsingular background is used to provide a geometrical characterization of the path of the singularity. There is an indeterminacy in the process of decomposing the metric, which manifests itself as a gauge transformation. Infeld and Schild use—and

[1] Work supported in part by the National Science Foundation under Grant PHY-8306104.

[2] Mathematical Physics Department, University College, Dublin and School of Theoretical Physics, Dublin Institute for Advanced Studies.

[3] Mathematical Sciences Program, The University of Texas at Dallas, Richardson, Texas 75080.

restrict—gauge transformations by imposing De Donder conditions on the field of the particle. This is a very natural restriction to impose in a wide variety of problems.

In the present case, however, one is particularly interested in the wave fronts of radiation emanating from the particle. With respect to the full metric, these form a congruence of future-directed null cones; and it seems natural to require that they should satisfy the same conditions with respect to the background metric. It is convenient, moreover, to build the coordinate system around the congruence of null cones. We have used this combination of gauge and coordinate conditions in a previous paper[(3)] to investigate the motion of an electrically charged particle, in an approximation which takes no account of the particle's interaction with its own gravitational field.

The present investigation is concerned specifically with that interaction. We confine our attention to an uncharged particle; and we make a number of additional assumptions in order to simplify the calculation. There is no attempt at generality: the object is to exhibit a special case of a gravitational monopole deviating from geodesic motion under the influence of its own gravitational radiation.

## 2. GEODESIC MOTION

To make it possible to read this paper independently of our previous paper[(3)] we begin this section by repeating, in a slightly condensed form, the basic formulas of our method.

We visualize a "background" space-time containing a nonsingular, timelike world-line. The particle is introduced in the form of a perturbation of this space-time which is singular only on the nonsingular world-line (which refer to as the history of the particle). The line-element of the background space-time is given by[(7)]

$$ds^2 = -(\theta^1)^2 - (\theta^2)^2 + 2\theta^3\theta^4 \tag{1}$$

where

$$\theta^1 = rp^{-1}(e^\alpha \cosh\beta\, dx + e^{-\alpha}\sinh\beta\, dy + a\, du) \tag{1a}$$

$$\theta^2 = rp^{-1}(e^\alpha \sinh\beta\, dx + e^{-\alpha}\cosh\beta\, dy + b\, du) \tag{1b}$$

$$\theta^3 = dr + \tfrac{1}{2}c\, du \tag{1c}$$

$$\theta^4 = du \tag{1d}$$

This line-element is completely general in the sense that it involves six functions $\alpha$, $\beta$, $a$, $b$, $p$, $c$ of the four coordinates $(x, y, r, u)$. We assume that these functions can be expanded in powers of $r$, near $r=0$, as follows:

$$\alpha = \alpha_2 r^2 + \alpha_3 r^3 + \cdots \tag{2a}$$

$$\beta = \beta_2 r^2 + \beta_3 r^3 + \cdots \tag{2b}$$

$$a = a_1 r + a_2 r^2 + \cdots \tag{2c}$$

$$b = b_1 r + b_2 r^2 + \cdots \tag{2d}$$

$$c = 1 - 2\underset{0}{h} r + A_2 r^2 + \cdots \tag{2e}$$

$$p = \underset{0}{P}(1 + q_2 r^2 + q_3 r^3 + \cdots) \tag{2f}$$

The coefficients of the various powers of $r$ above are functions only of $x$, $y$, $u$. It follows from (2) that the shear and expansion of the future-pointing null vector field $\partial/\partial r$ are given by

$$\sigma = 2(\alpha_2 + i\beta_2)r + 3(\alpha_3 + i\beta_3)r^2 + \cdots \tag{3a}$$

$$\rho = \frac{1}{r} - 2q_2 r - 3q_3 r^2 + \cdots \tag{3b}$$

and thus the null hypersurfaces $u = \text{const}$ resemble future null-cones with vertices on $r=0$ as $r \to 0$. Assuming the metric to be regular on $r=0$ means that we can take it to be Minkowskian up to terms of order $r^2$. Thus in (2) we may take (cf. Ref. 8)

$$\underset{0}{P} = xv^1 + yv^2 + (1 - \tfrac{1}{4}(x^2 + y^2))\, v^3 + (1 + \tfrac{1}{4}(x^2 + y^2))\, v^4 \tag{4a}$$

$$\underset{0}{h} = \eta_{ij} a^i k^j = a^i k_i = \frac{\partial}{\partial u} (\log \underset{0}{P}) \tag{4b}$$

$$k^i = \underset{0}{P}^{-1} \{ -x\delta^i_1 - y\delta^i_2 - (1 - \tfrac{1}{4}(x^2 + y^2))\, \delta^i_3 + (1 + \tfrac{1}{4}(x^2 + y^2))\, \delta^i_4 \} \tag{4c}$$

In these formulas $v^i(u)$ is the 4-velocity of $r=0$ in rectangular Cartesian coordinates and time, $\eta_{ij}$ is the Minkowskian metric tensor in these coordinates, $v^i v_i = 1$, $a^i = \dot{v}^i$ (where the dot indicates differentiation with respect to $u$) and thus $a^i v_i = 0$ and $\dot{a}^i v_i = -a^2$ where $a^2 = a^i a_i$. Finally $k^i$ is tangent to $u = \text{const}$ near $r=0$.

We introduce our particle of mass $m = \text{const}$ by replacing (2e) by

$$c = -\frac{2m}{r} + 1 - 2\underset{0}{h} r + A_2 r^2 + \cdots \tag{5}$$

We leave the remaining expansions (2) unchanged except for allowing the functions of $x$, $y$, $u$ in them to acquire, in principle, terms involving $m$. Our task now is to impose Einstein's vacuum field equations on this perturbed space-time. The components of the Ricci tensor have expansions in powers of $r$ which, when equated to zero, result in many of the functions appearing in (2) vanishing, and we finally obtain from the leading $r^{-2}$ term in $R_{44}=0$ the important equation

$$m\underset{0}{h}=0 \tag{6}$$

with $\underset{0}{h}$ given by (4b). It thus follows that since $m \neq 0$ we have

$$a^i=0 \tag{7}$$

and the history of our test particle is a timelike geodesic of the background space-time. This is the conclusion reached by Infeld and Schild.[2]

## 3. VIOLATION OF GEODESIC MOTION

Guided by the well-established fact that shearing gravitational radiation reduces the mass of the source,[4,5] we look for a violation of geodesic motion via the presence of shear in the radiation given off by the mass particle. We shall also make a number of assumptions motivated by a desire to achieve our aim with the greatest possible simplicity. The first of these is to choose Minkowskian space-time as our background space-time. (Technical difficulties arise in our calculations if we persist with the more general background used in the previous section.) This entails taking

$$\alpha=\beta=a=b=0, \qquad p=\underset{0}{P}, \qquad c=1-2\underset{0}{h}r \tag{8}$$

in (1) with $\underset{0}{P}$ and $\underset{0}{h}$ given by (4). We perturb these functions by replacing them with

$$\alpha=\frac{\alpha_1}{r}+\frac{\alpha_2}{r^2}+\cdots \tag{9a}$$

$$\beta=\frac{\beta_1}{r}+\frac{\beta_2}{r^2}+\cdots \tag{9b}$$

$$a=\frac{a_2}{r^2}+\frac{a_3}{r^3}+\cdots \tag{9c}$$

$$b=\frac{b_2}{r^2}+\frac{b_3}{r^3}+\cdots \tag{9d}$$

$$p = \underset{0}{P}\left(1 + \frac{q_2}{r^2} + \frac{q_3}{r^3} + \cdots\right) \tag{9e}$$

$$c = -\frac{2m}{r} + 1 - 2\underset{0}{h}r + \cdots \tag{9f}$$

Although we have used the same notation, the functions of $x$, $y$, $u$ appearing above are not to be considered the same as those in the expansions (2) apart from $\underset{0}{P}$ and $\underset{0}{h}$. We may also assume for the moment that $m = m(u, x, y)$. We shall introduce an approximation in the following sense: In the units we are using, in which the gravitational constant and the velocity of light are both unity, the quantity $ma$, where $a = (a^i a_i)^{1/2}$, is dimensionless. We shall assume that this quantity is small of first order and write

$$ma = O_1 \tag{10}$$

Moreover we shall assume, allowing for refinements to appear later, that the $\alpha$, $\beta$, $a$, $b$, and $q$ functions in (9) are all small of first order.

We note that the shear of the null vector field $\partial/\partial r$ in the perturbed space-time is

$$\sigma = -\frac{\gamma}{r^2} + O\left(\frac{1}{r^3}\right) \tag{11}$$

where

$$\gamma = \alpha_1 + i\beta_1 \tag{12}$$

We shall assume for simplicity that

$$\alpha_1 = \alpha_1(u), \qquad \beta_1 = \beta_1(u) \tag{13}$$

Einstein's vacuum field equations for the line-element given by (1) and (9) can be obtained from Ref. 7. Those of particular interest to us at this stage yield (see Appendix)

$$q_2 = \tfrac{1}{2}(\alpha_1^2 + \beta_1^2) = \tfrac{1}{2}\,|\gamma|^2 = O_2 \tag{14a}$$

$$a_2 = -2\alpha_1 \underset{0}{P}\frac{\partial \underset{0}{P}}{\partial x} - 2\beta_1 \underset{0}{P}\frac{\partial \underset{0}{P}}{\partial y} \tag{14b}$$

$$b_2 = 2\alpha_1 \underset{0}{P}\frac{\partial \underset{0}{P}}{\partial y} - 2\beta_1 \underset{0}{P}\frac{\partial \underset{0}{P}}{\partial x} \tag{14c}$$

$$\dot{M} - 3\underset{0}{h}M + |\dot{\gamma} - \underset{0}{h}\gamma|^2 = 0 \tag{14d}$$

where

$$M = m - \dot{q}_2 + 2\underset{0}{h}q_2 - \frac{1}{2}\left\{\frac{\partial}{\partial x}(\underset{0}{P}^{-2}a_2) + \frac{\partial}{\partial y}(\underset{0}{P}^{-2}b_2)\right\} \tag{15}$$

If $\langle f \rangle$ denotes the average of some $f(u, x, y)$ over the spheres with line-element

$$\underset{0}{P}^{-2}(dx^2 + dy^2) \tag{16}$$

then, using (4), we have

$$\frac{d}{du}\langle fk^i \rangle = \langle (\dot{f} - 3\underset{0}{h}f)\, k^i \rangle \tag{17}$$

Putting $f = M$ and using (14d) and (13) gives

$$\begin{aligned} \frac{d}{du}\langle Mk^i \rangle &= -\tfrac{2}{3}(\alpha_1\dot{\alpha}_1 + \beta_1\dot{\beta}_1)\, a^i + \tfrac{1}{3}(\alpha_1^2 + \beta_1^2)\, a^2 v^i - |\dot{\gamma}|^2\, v^i \\ &= -\tfrac{2}{3}\dot{q}_2 a^i + \tfrac{1}{3}|\gamma|^2\, a^2 v^i - |\dot{\gamma}|^2\, v^i \end{aligned} \tag{18}$$

where the final equality results from (14a). By (14b), (14c) the fourth term in (15) gives no contribution to $\langle Mk^i \rangle$. If we now make our final simplifying assumption,

$$m = m(u) \tag{19}$$

we find that

$$\langle Mk^i \rangle = \langle M \rangle\, v^i - \tfrac{2}{3}q_2 a^i \tag{20}$$

Substituting this into the left-hand side of (18) and using (14a) again yields

$$\frac{d}{du}(\langle M \rangle\, v^i) = \tfrac{1}{3}|\gamma|^2(\dot{a}^i + a^2 v^i) - |\dot{\gamma}|^2\, v^i \tag{21}$$

Thus we have an equation of motion with gravitational radiation reaction due to the presence of shear in the radiation, namely,

$$\langle M \rangle\, a^i = \tfrac{1}{3}|\gamma|^2(\dot{a}^i + a^2 v^i) \tag{22}$$

together with a mass-loss formula

$$\frac{d}{du}\langle M \rangle = -|\dot{\gamma}|^2 \tag{23}$$

which, bearing in mind (13), is of the Bondi–Sachs[(4,5)] type.

## 4. DISCUSSION

If we neglect $O_2$ terms in the previous section we notice that (22) becomes the geodesic equation and, since it follows from (15) and (19) that

$$\langle M \rangle = m + O_2 \tag{24}$$

we have from (23) that $m = \text{const}$ in this approximation. The vacuum field equations not used in Sec. 3 involve, for example, $a_3$, $b_3$, $\alpha_3$ and $\beta_3$ which are $O_1$ quantities which have played no role in our derivation of (21) (see Appendix for the differential equations to be satisfied by these functions).

In addition we note that, after the simplifications of Sec. 3 have been made, the radiative part of the Riemann tensor ($\psi_4$ in the notation of Newman and Penrose[10]) is given by

$$\psi_4 = -\frac{1}{r} \underset{0}{P}^2 \frac{\partial}{\partial u}\left( \underset{0}{P}^{-1} \frac{\partial}{\partial u} (\underset{0}{P}^{-1} \bar{\gamma}) \right) + O\left(\frac{1}{r^2}\right) \tag{25}$$

This term appears in the lowest approximation. Its effect on equations of motion is visible only in the next. Thus our result is consistent with the geodesic equations of motion found, for that approximation, by Infeld and Schild.[2] Qualitatively it appears consistent with the radiation reaction derived by Lind *et al.*,[6] whose approach, howcvcr, is different; but we have not attempted a detailed comparison with their results.

## APPENDIX

We indicate here the origin of Eqs. (14) given in the text. To this end we first generalize (9e), (9f) to read

$$p = p_0 \left( 1 + \frac{q_2}{r^2} + \frac{q_3}{r^3} + \cdots \right) \tag{A.1}$$

$$c = r c_1 + c_0 - \frac{2m}{r} + \cdots \tag{A.2}$$

and we will specialize to (9e), (9f) later. Here the coefficients of the various powers of $r$ are functions of $u$, $x$, $y$. The vacuum field equations for the line-element (1) with (9) and (A.1), (A.2) are given in Ref. 7. In the terminology of Sachs[9] the "main equations" yield, for example,

$$q_2=\tfrac{1}{2}(\alpha_1^2+\beta_1^2), \qquad q_3=0$$

$$c=\Delta\log p_0, \qquad c_1=-2p_0^{-1}\dot{p}_0=-2H$$

$$a_2=p_0^4\left\{\frac{\partial}{\partial x}(p_0^{-2}\alpha_1)+\frac{\partial}{\partial y}(p_0^{-2}\beta_1)\right\}$$

$$b_3=p_0^4\left\{\frac{\partial}{\partial x}(p_0^{-2}\beta_1)-\frac{\partial}{\partial y}(p_0^{-2}\alpha_1)\right\}$$

$$\begin{aligned}\frac{\partial a_3}{\partial y}+\frac{\partial b_3}{\partial x}=\;&8(\dot{\beta}_3-3H\beta_3)-4m\beta_1-20\alpha_1\dot{\alpha}_1\beta_1-4\alpha_1^2\dot{\beta}_1\\&-8\beta_1^2\dot{\beta}_1+8\beta_1(\beta_1^2+3\alpha_1^2)H\\&-\beta_1\left(\frac{\partial a_2}{\partial x}+\frac{\partial b_2}{\partial y}\right)+3\alpha_1\left(\frac{\partial b_2}{\partial x}-\frac{\partial a_2}{\partial y}\right)\\&+a_2\left(\frac{\partial\alpha_1}{\partial y}-3\frac{\partial\beta_1}{\partial x}\right)-b_2\left(\frac{\partial\alpha_1}{\partial x}+3\frac{\partial\beta_1}{\partial y}\right)\\&+4p_0^{-1}\beta_1\left(a_2\frac{\partial p_0}{\partial x}+b_2\frac{\partial p_0}{\partial y}\right)+4p_0^{-2}a_2b_2\end{aligned} \tag{A.3}$$

$$\begin{aligned}\frac{\partial a_3}{\partial x}-\frac{\partial b_3}{\partial y}=\;&8(\dot{\alpha}_3-3H\alpha_3)-4m\alpha_1+12\alpha_1\beta_1\dot{\beta}_1+12\beta_1^2\dot{\alpha}_1\\&-8\alpha_1^2\dot{\alpha}_1+8\alpha_1(\alpha_1^2-3\beta_1^2)H\\&-\alpha_1\left(\frac{\partial a_2}{\partial x}+\frac{\partial b_2}{\partial y}\right)+3\beta_1\left(\frac{\partial a_2}{\partial y}-\frac{\partial b_2}{\partial x}\right)\\&-a_2\left(\frac{\partial\beta_1}{\partial y}+3\frac{\partial\alpha_1}{\partial x}\right)+b_2\left(\frac{\partial\beta_1}{\partial x}-3\frac{\partial\alpha_1}{\partial y}\right)\\&+4p_0^{-1}\alpha_1\left(a_2\frac{\partial p_0}{\partial x}+b_2\frac{\partial p_0}{\partial y}\right)+2p_0^{-2}(a_2^2-b_2^2)\end{aligned}$$

where the dot indicates differentiation with respect to $u$ and

$$\Delta=p_0^2\left(\frac{\partial^2}{\partial x^2}+\frac{\partial^2}{\partial y^2}\right) \tag{A.4}$$

The "supplementary conditions" yield

$$\dot{M}-3HM+|\dot{\gamma}-H\gamma|^2=\frac{1}{4}\Delta c_0+4\,\mathrm{Re}\left\{\gamma\frac{\partial}{\partial z}\left(p_0^2\frac{\partial H}{\partial z}\right)\right\} \tag{A.5}$$

where

$$M = m - \dot{q}_2 + 2Hq_2 - \frac{1}{2} p_0^2 \left\{ \frac{\partial}{\partial x} (p_0^{-2} a_2) + \frac{\partial}{\partial y} (p_0^{-2} b_2) \right\} \tag{A.6}$$

and

$$\gamma = \alpha_1 + i\beta_1, \qquad z = x + iy \tag{A.7}$$

and also

$$\begin{aligned}
&\frac{3}{2} p_0^{-1} (\dot{A}_3 - 4HA_3) - p_0 \frac{\partial m}{\partial x} \\
&\quad = p_0^3 \left\{ \frac{\partial}{\partial x} (p_0^{-2} \alpha_1 c_0) + \frac{\partial}{\partial y} (p_0^{-2} \beta_1 c_0) \right\} \\
&\qquad - b_2 \frac{\partial^2 p_0}{\partial x \, \partial y} - a_2 \frac{\partial^2 p_0}{\partial x^2} - (\dot{\beta}_1 - H\beta_1) p_0^2 \left\{ \frac{\partial}{\partial x} (p_0^{-1} \beta_1) - \frac{\partial}{\partial y} (p_0^{-1} \alpha_1) \right\} \\
&\qquad + \frac{\partial p_0}{\partial y} \frac{\partial a_2}{\partial y} + 2\beta_1 (\dot{\alpha}_1 - H\alpha_1) \frac{\partial p_0}{\partial y} \\
&\qquad - (\dot{\alpha}_1 - H\alpha_1) p_0^2 \left\{ \frac{\partial}{\partial x} (p_0^{-1} \alpha_1) + \frac{\partial}{\partial y} (p_0^{-1} \beta_1) \right\} \\
&\qquad - \frac{\partial p_0}{\partial x} \left( \dot{q}_1 - 2Hq_2 + \frac{\partial b_2}{\partial y} - b_2 \frac{\partial}{\partial y} (\log p_0) - a_2 \frac{\partial}{\partial x} (\log p_0) \right) \\
&\qquad - p_0 \frac{\partial}{\partial y} \left( \frac{1}{2} \frac{\partial a_2}{\partial y} - \frac{1}{2} \frac{\partial b_2}{\partial x} + 2\beta_1 (\dot{\alpha}_1 - H\alpha_1) \right) - \alpha_1 \beta_1 p_0 \frac{\partial H}{\partial y} \\
&\qquad - (\dot{\alpha}_1 - H\alpha_1) p_0^{-1} a_2 - (\dot{\beta}_1 - H\beta_1) p_0^{-1} b_2
\end{aligned} \tag{A.8}$$

where

$$A_3 = a_3 - \frac{1}{3} (b_2 \beta_1 + a_2 \alpha_1) - \frac{4}{3} q_2 p_0 \frac{\partial p_0}{\partial x} - \frac{2}{3} p_0^3 \frac{\partial}{\partial y} (p_0^{-1} \alpha_1 \beta_1) \tag{A.9}$$

We obtain one further supplementary condition by interchanging $x$ with $y$, $a_3$ with $b_3$, $a_2$ with $b_2$, and $\alpha_1$ with $-\alpha_1$ in (A.8), (A.9).

The simplifications introduced in Sect. 3 amount to choosing

$$p_0 = \underset{0}{P}, \qquad \alpha_1 = \alpha_1(u), \qquad \beta_1 = \beta_1(u), \qquad m = m(u) \tag{A.10}$$

and also assuming that $\alpha_1$, $\beta_1$, $\alpha_3$, $\beta_3$, $a_3$, $b_3$, $m$ are small of first order. Then from (A.3) we have

$$c_0=1, \qquad c_{-1}=-2\underset{0}{h}$$

$$\dot{\beta}_3-3\underset{0}{h}\beta_3=\frac{1}{8}\left(\frac{\partial a_3}{\partial y}+\frac{\partial b_3}{\partial x}\right)+O_2 \tag{A.11}$$

$$\dot{\alpha}_3-3\underset{0}{h}\alpha_3=\frac{1}{8}\left(\frac{\partial a_3}{\partial x}-\frac{\partial b_3}{\partial y}\right)+O_2$$

where $\underset{0}{P}$, $\underset{0}{h}$ are given by (4). The remaining equations in (A.3) yield (14a), (14b), (14c) while (A.5), and (A.6) reduce to (14d) and (15) respectively. In the case of (A.5) we note that the right-hand side gives no contribution now since $c_0=1$ and

$$\frac{\partial}{\partial z}\left(\underset{0}{P}^2\frac{\partial \underset{0}{h}}{\partial z}\right)=0 \tag{A.12}$$

The two remaining supplementary conditions are enormously simplified to read

$$\begin{aligned}\dot{a}_3-4\underset{0}{h}a_3&=-\frac{2}{3}\underset{0}{P}^2\frac{\partial F}{\partial y}+O_2\\ \dot{b}_3-4\underset{0}{h}b_3&=\frac{2}{3}\underset{0}{P}^2\frac{\partial F}{\partial x}+O_2\end{aligned} \tag{A.13}$$

where

$$F=2\alpha_1\frac{\partial \underset{0}{P}}{\partial x}\frac{\partial \underset{0}{P}}{\partial y}-\beta_1\left\{\left(\frac{\partial \underset{0}{P}}{\partial x}\right)^2-\left(\frac{\partial \underset{0}{P}}{\partial y}\right)^2\right\} \tag{A.14}$$

## REFERENCES


1. L. Infeld and J. Plebanski, *Motion and Relativity* (Pergamon, New York, 1960).
2. L. Infeld and A. Schild, *Rev. Mod. Phys.* **21**, 408 (1949).
3. P. A. Hogan and I. Robinson, "The Motion of Charged Test Particles in General Relativity," *Found. Phys.* **15**, 617 (1985).
4. H. Bondi, M. G. J. van der Burg, and A. W. K. Metzner, *Proc. R. Soc. London A* **269**, 21 (1962).
5. R. K. Sachs, *Proc. R. Soc. London A* **270, 103 (1962).**
6. R. W. Lind, J. Messmer, and E. T. Newman, *J. Math. Phys.* **13**, 1884 (1972).
7. P. A. Hogan and A. Trautman, "On Gravitational Radiation from Bounded Sources," in *Gravitation and Geometry*, W. Rindler and A. Trautman, eds. (Bibliopolis, Naples, 1985).
8. E. T. Newman and T. Unti, *J. Math. Phys.* **3**, 891 (1962).
9. R. K. Sachs, *Proc. R. Soc. London A* **264**, 309 (1961).
10. E. T. Newman and R. Penrose, *J. Math. Phys.* **3**, 566 (1962).

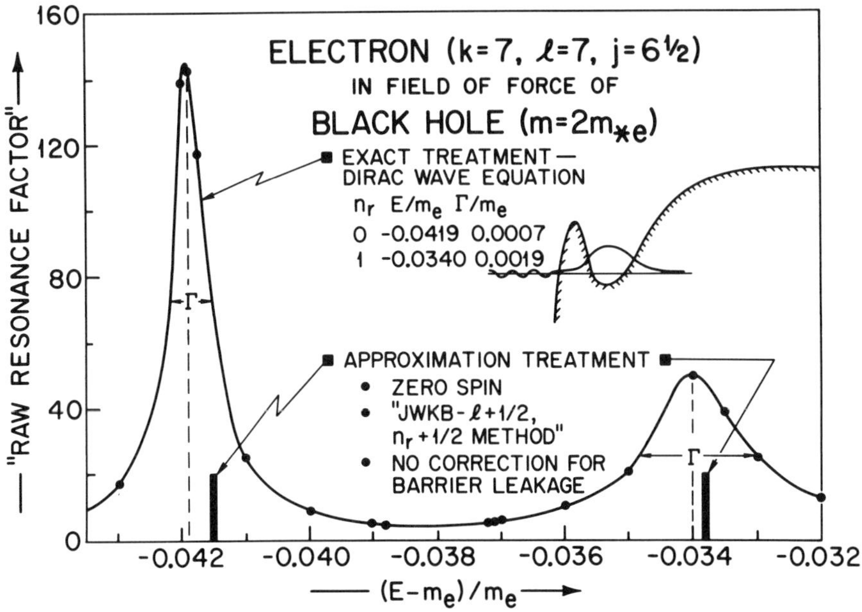
ELECTRON (k=7, ℓ=7, j=6½)
IN FIELD OF FORCE OF
BLACK HOLE ($m=2m_{*e}$)
EXACT TREATMENT—
DIRAC WAVE EQUATION
$n_r$ $E/m_e$ $\Gamma/m_e$
0 -0.0419 0.0007
1 -0.0340 0.0019
APPROXIMATION TREATMENT
ZERO SPIN
"JWKB-ℓ+1/2,
$n_r$+1/2 METHOD"
NO CORRECTION FOR
BARRIER LEAKAGE
Γ
Γ
"RAW RESONANCE FACTOR"
0
40
80
120
160
-0.042
-0.040
-0.038
-0.036
-0.034
-0.032
$(E-m_e)/m_e$

*Plate 23*

# Equivalent Lagrangians in Classical Field Theory

**Sergio Hojman**[1] **and L. C. Shepley**[2]

*Received June 24, 1985*

*Two Lagrangians L and L′ are equivalent if the equations of motion derived from them have the same set of solutions. In that case, a matrix* $\Lambda$ *may be defined which has the property that the trace of any analytic function of* $\Lambda$ *is a constant of the motion. We extend this trace theorem to the case of classical field theory and discuss some of the implications for quantum theory and for procedures for finding equivalent Lagrangians.*

## 1. INTRODUCTION

Consider two Lagrangians $L(q^i, \dot{q}^i, t)$ and $L'(q^i, \dot{q}^i, t)$ for a classical mechanical system with a finite number $n$ of degrees of freedom. The equations of motion are

$$E_i L = 0 \qquad \text{and} \qquad E_i L' = 0, \qquad \text{where} \quad i = 1, ..., n \tag{1}$$

and where $E_i$ is the Euler–Lagrange derivative:

$$E_i \equiv \frac{d}{dt}\frac{\partial}{\partial \dot{q}^i} - \frac{\partial}{\partial q^i} \tag{2}$$

If $L = \rho L' + dF/dt$ ($\rho$ is constant and $dF/dt$ is a total time derivative) or if $E_i L$ and $E_i L'$ merely have the same solution set, then $L$ and $L'$ are called *equivalent* (or *s*-equivalent).

From two equivalent Lagrangians $L$ and $L'$ may be formed two Hamiltonians $H$ and $H'$. A quantized theory may be constructed by the

[1] Centro de Estudios Nucleares, Universidad Nacional Autónoma de México, Circuito Exterior, C. U., México 04510, D.F., México.

[2] Center for Relativity, Physics Department, The University of Texas, Austin, Texas 78712.

usual canonical quantization scheme. These two quantized theories will not in general be equivalent.[1-3] This fact is a problem in the relationship between classical and quantum theories. Our interest in it is in great part inspired by the problems John Wheeler has so successfully dealt with in his work on the foundations of quantum mechanics.

The purpose of this paper is to extend some of the results obtained for equivalent Lagrangians for particle systems to the case of field theory. We will be treating classical fields in one spatial dimension only, although the results may be extended to other dimensions. In particular, for particle systems, if $L$ and $L'$ are equivalent, a matrix $\Lambda$ can be formed from $L$ and $L'$ which has the following property: The trace of any analytic function of $\Lambda$ is a constant of the motion.[1,4-10]

In Sect. 2, we extend the *trace theorem* to the case of classical field theory (some work on this subject appears in Ref. 11). We will deal with Lagrangians (compare Ref. 12), not Lagrangian densities, using a special notation. We will separately treat Lagrangians whose Euler–Lagrange equations are of second order and of first order in their time derivatives.[7,8] We do restrict ourselves to unconstrained systems and to Lagrangians which depend only on the first time derivatives of the fields. Constrained systems and singular field theories are currently under study by us.

We present two concrete examples in Sect. 3: an analog of the Dirac spinor field and massless scalar waves (both in two-dimensional space-time). Part of the discussion of the latter system is presented using Fourier decomposition. In Sect. 4, we discuss consequences of the trace theorem, implications of equivalent Lagrangians for quantum theory and for Noether's theorem,[9,13] and possible linkages between $\Lambda$ and the matrices used in the Lax method[14] for integration of nonlinear systems.

## 2. THE TRACE THEOREM

We consider a field having $N$ real-valued components $\phi^\alpha$. Each component $\phi^\alpha$ is taken to be a function of the Cartesian spatial coordinate $x$ (for a one-dimensional model space) and of time $t$. Our notation will be to write

$$\phi^\alpha{}_x \equiv \phi^\alpha(x, t) \tag{3}$$

We will also be considering multi-index functions of several spatial points, time $t$ being considered as a universal time in the manner of Newton. Consider such a function, say $B^{\alpha\beta}(x, y, z, t)$: Here $B^{\alpha\beta}$ is a function of three separate points in space and of course a function of time $t$. For this function we write

$$B^{\alpha\beta}{}_{xyz} \equiv B^{\alpha\beta}(x, y, z, t) \tag{4}$$

Of special importance is the delta function

$$\delta^{\alpha\beta}{}_{x-y} \equiv \delta^{\alpha\beta}\delta(x-y) \tag{5}$$

where $\delta(x-y)$ is the Dirac delta function and $\delta^{\alpha\beta}$ is the Kronecker delta. Note that $\delta^{\alpha\beta}{}_{x-y}$ has two indices $\alpha$, $\beta$ but is a function of the single space point $x-y$.

We denote integration by use of a summation convention. Summation over repeated Greek indices is implied as usual. A repeated subscript, denoting two identical points, implies integration over space. An example is

$$B^{\alpha\beta}{}_{xyz}C^{\gamma\beta\delta}{}_{yz} \equiv \sum_{\beta=1}^{N} \iint B^{\alpha\beta}{}_{xyz}C^{\gamma\beta\delta}{}_{yz}\, dy\, dz \tag{6}$$

which involves a double integration. We will in general assume that the integrals do exist. They might not, of course; for example, the function of two points $\Delta_{xy}$, defined by

$$\Delta_{xy} = \delta_{x-y} \tag{7}$$

has an undefined trace $\Delta_{xx}$.

The time derivative of a variable will be denoted by

$$\dot{\phi}^{\alpha}{}_{x} \equiv \frac{\partial}{\partial t}\phi^{\alpha}{}_{x} \tag{8}$$

that is, the super-dot denotes partial time derivative. We will find it useful in the examples to allow the subscript, which denotes a point in space, to be itself a function of $t$: $x(t)$. In that case, the super-dot means

$$\dot{\phi}^{\alpha}{}_{x} \equiv \left.\frac{\partial\phi^{\alpha}(y, t)}{\partial t}\right|_{y=x(t)} \tag{9}$$

Spatial derivatives are handled by use of the derivatives of the delta function:

$$\delta'^{\alpha\beta}{}_{x} = \delta^{\alpha\beta}\frac{d\delta(x)}{dx},\ \delta''^{\alpha\beta}{}_{x} = \delta^{\alpha\beta}\frac{d^2\delta(x)}{dx^2},\ldots \tag{10}$$

The spatial derivatives of a field or other function will then be written in the form

$$\delta'^{\alpha\beta}{}_{x-y}\phi^{\beta}{}_{y}=\frac{\partial}{\partial x}\phi^{\alpha}{}_{x},\ \delta''^{\alpha\beta}{}_{x-y}\phi^{\beta}{}_{y}=\frac{\partial^{2}}{\partial x^{2}}\phi^{\alpha}{}_{x},\ldots \tag{11}$$

Needless to say, we presume that the integrations by parts which are used in these formulas are always possible.

A Lagrangian functional or other functional of the field $\phi^{\alpha}{}_{x}$ and its time derivatives will have to be derived with respect to its arguments. These derivatives are related to variational derivatives. In order to make clear what a derivative with respect to a field is, we consider functionals $L(\phi, \dot{\phi}, t)$ of the form

$$\begin{aligned}L(\phi,\dot{\phi},t)={}&{}_{0}A+{}_{1}A^{\alpha}{}_{x}\phi^{\alpha}{}_{x}+{}_{2}A^{\alpha\beta}{}_{xy}\phi^{\alpha}{}_{x}\phi^{\beta}{}_{y}+\cdots\\&+{}_{0}B^{\alpha}{}_{x}\dot{\phi}^{\alpha}{}_{x}+{}_{1}B^{\alpha\beta}{}_{xy}\dot{\phi}^{\alpha}{}_{x}\phi^{\beta}{}_{y}+{}_{2}B^{\alpha\beta\gamma}{}_{xyz}\dot{\phi}^{\alpha}{}_{x}\phi^{\beta}{}_{y}\phi^{\gamma}{}_{z}+\cdots\end{aligned} \tag{12}$$

including terms of all orders in $\phi^{\alpha}{}_{x}$ and $\dot{\phi}^{\alpha}{}_{x}$. The numerical presubscript indicates the number of factors of $\phi^{\alpha}{}_{x}$. The capital letter indicates the number of factors of $\dot{\phi}^{\alpha}{}_{x}$. Note that in (12) the functional $L$ does not depend on space points, as indicated by the absence of a subscript.

The derivative of $L$ with respect to $\phi^{\alpha}{}_{x}$ is then

$$\begin{aligned}\partial L/\partial\phi^{\alpha}{}_{x}={}&{}_{1}A^{\alpha}{}_{x}+2{}_{2}A^{\beta\alpha}{}_{yx}\phi^{\beta}{}_{y}+\cdots\\&+{}_{1}B^{\beta\alpha}{}_{yx}\dot{\phi}^{\beta}{}_{y}+2{}_{2}B^{\beta\gamma\alpha}{}_{yzx}\dot{\phi}^{\beta}{}_{y}\phi^{\gamma}{}_{z}+\cdots\end{aligned} \tag{13}$$

Note that we will be presuming that such coefficients as ${}_{2}B$ are symmetric in the appropriate indices. The partial derivative of $L$ with respect to $\dot{\phi}^{\alpha}{}_{x}$ is similarly defined.

Now consider a system, described by the field $\phi^{\alpha}{}_{x}$, whose equations of motion are derivable by use of a Lagrangian $L(\phi, \dot{\phi}, t)$ depending at most on the first time derivative $\dot{\phi}^{\alpha}{}_{x}$. This variational principle reads

$$\delta\int L(\phi,\dot{\phi},t)\,dt=0 \tag{14}$$

The Euler–Lagrange equations are

$$E^{\alpha}{}_{x}L=0 \tag{15}$$

where the Euler derivative operator is

$$E^{\alpha}{}_{x}=\frac{d}{dt}\frac{\partial}{\partial\dot{\phi}^{\alpha}{}_{x}}-\frac{\partial}{\partial\phi^{\alpha}{}_{x}} \tag{16}$$

The time derivative $d/dt$ in this equation is

$$\frac{d}{dt}=\frac{\partial}{\partial t}+\dot{\phi}^{\alpha}{}_{x}\frac{\partial}{\partial\phi^{\alpha}{}_{x}}+\ddot{\phi}^{\alpha}{}_{x}\frac{\partial}{\partial\dot{\phi}^{\alpha}{}_{x}} \tag{17}$$

and $\partial/\partial t$ must be defined as we have above in (8) and (9).

The equations of motion are regular if $E^{\alpha}{}_{x}L=0$ may be put in the form

$$W^{\alpha\beta}{}_{xy}(\ddot{\phi}^{\beta}{}_{y}-F^{\beta}{}_{y})=0 \tag{18}$$

where $W^{\alpha\beta}{}_{xy}$ is a nonsingular matrix. By nonsingular is simply meant that (18) implies

$$\ddot{\phi}^{\beta}{}_{y}-F^{\beta}{}_{y}=0 \tag{19}$$

In the cases that interest us, we assume that $W^{\alpha\beta}{}_{xy}$ has a well-defined inverse $U^{\alpha\beta}{}_{xy}$ such that

$$U^{\alpha\gamma}{}_{xz}W^{\gamma\beta}{}_{zy}=\delta^{\alpha\beta}{}_{x-y} \tag{20}$$

We note that $W^{\alpha\beta}{}_{xy}$ and $U^{\alpha\beta}{}_{xy}$ are symmetric in the appropriate indices and that $U^{\alpha\beta}{}_{xy}$ is both a left and a right inverse in general. The formula for $W^{\alpha\beta}{}_{xy}$ is

$$W^{\alpha\beta}{}_{xy}=\frac{\partial^2 L}{\partial\dot{\phi}^{\alpha}{}_{x}\partial\dot{\phi}^{\beta}{}_{y}} \tag{21}$$

but an explicit form for $U^{\alpha\beta}{}_{xy}$ is often hard to obtain. The force vector $F^{\alpha}{}_{x}$ is obtained from the relation

$$W^{\alpha\beta}{}_{xy}F^{\beta}{}_{y}=-\frac{\partial^2 L}{\partial\dot{\phi}^{\alpha}{}_{x}\partial\phi^{\beta}{}_{y}}\dot{\phi}^{\beta}{}_{y}-\frac{\partial^2 L}{\partial\dot{\phi}^{\alpha}{}_{x}\partial t}+\frac{\partial L}{\partial\phi^{\alpha}{}_{x}} \tag{22}$$

A second Lagrangian $L'$ for the field $\phi^{\alpha}{}_{x}$ will in general yield different equations

$$W'^{\alpha\beta}{}_{xy}(\ddot{\phi}^{\beta}{}_{y}-F'^{\beta}{}_{y})=0 \tag{23}$$

$L'$ is subordinate to $L$ if $F'^{\alpha}{}_{x}\equiv F^{\alpha}{}_{x}$ (as a functional of $\phi^{\alpha}{}_{x}$, $\dot{\phi}^{\alpha}{}_{x}$, $t$). In this case any solution of (18) is a solution of (23). $L'$ is equivalent to $L$ if, also, $W'^{\alpha\beta}{}_{xy}$ is nonsingular.

We define the operator $D/dt$ by

$$\frac{D}{dt}\equiv\frac{\partial}{\partial t}+\dot{\phi}^{\alpha}{}_{x}\frac{\partial}{\partial\phi^{\alpha}{}_{x}}+F^{\alpha}{}_{x}\frac{\partial}{\partial\dot{\phi}^{\alpha}{}_{x}} \tag{24}$$

The commutator of $D/dt$ and $\partial/\partial\dot{\phi}^\alpha{}_x$ is

$$\left[\frac{D}{dt},\frac{\partial}{\partial\dot{\phi}^\alpha{}_x}\right]=\frac{\partial F^\beta{}_y}{\partial\dot{\phi}^\alpha{}_x}\frac{\partial}{\partial\dot{\phi}^\beta{}_y}+\frac{\partial}{\partial\phi^\alpha{}_x} \tag{25}$$

The utility of $D/dt$ is that both $L$ and $L'$ obey

$$\frac{D}{dt}\frac{\partial L}{\partial\dot{\phi}^\alpha{}_x}-\frac{\partial L}{\partial\phi^\alpha{}_x}\equiv 0 \tag{26}$$

$$\frac{D}{dt}\frac{\partial L'}{\partial\dot{\phi}^\alpha{}_x}-\frac{\partial L'}{\partial\phi^\alpha{}_x}\equiv 0 \tag{27}$$

identically. The commutator (25) allows us easily to calculate the time derivative of $W^{\alpha\beta}{}_{xy}$.

The result is most easily written in a matrix notation, and it is in this notation that we will complete the proof of the *trace theorem*. Let $\mathbf{W}$ stand for the matrix whose components are $W^{\alpha\beta}{}_{xy}$. Define the matrix $\mathbf{F}$ by

$$\mathbf{F}\equiv\left(\frac{\partial F^\alpha{}_x}{\partial\dot{\phi}^\beta{}_y}\right)=(F^{\alpha\beta}{}_{xy}) \tag{28}$$

Matrix multiplication $\mathbf{WF}$ is

$$\mathbf{WF}=(W^{\alpha\gamma}{}_{xz}F^{\gamma\beta}{}_{zy}) \tag{29}$$

Superscript T denotes matrix transpose; thus $\mathbf{W}^\mathsf{T}=\mathbf{W}$, but $\mathbf{F}$ is not in general symmetric.

The results for the time derivatives of $\mathbf{W}$ and $\mathbf{W}'$ are

$$D\mathbf{W}/dt=-\tfrac{1}{2}(\mathbf{WF}+\mathbf{F}^\mathsf{T}\mathbf{W}) \tag{30}$$

$$D\mathbf{W}'/dt=-\tfrac{1}{2}(\mathbf{W}'\mathbf{F}+\mathbf{F}^\mathsf{T}\mathbf{W}') \tag{31}$$

where we have used the fact that $\mathbf{F}'=\mathbf{F}$ (because $L'$ is subordinate to $L$), but have not so far required that $\mathbf{W}$ or $\mathbf{W}'$ be invertible.

The *trace theorem* deals with the matrix $\mathbf{\Lambda}$ defined by

$$\mathbf{\Lambda}=\mathbf{UW}' \tag{32}$$

where $\mathbf{U}$ is the inverse of $\mathbf{W}$, as defined above. Because of Eq. (30) for the time derivative of $\mathbf{W}$, we have

$$D\mathbf{U}/dt=\tfrac{1}{2}(\mathbf{FU}+\mathbf{UF}^\mathsf{T}) \tag{33}$$

Therefore

$$D\mathbf{\Lambda}/dt = \tfrac{1}{2}[\mathbf{\Lambda}, \mathbf{F}^{\mathrm{T}}] \tag{34}$$

The time derivatives of powers of $\mathbf{\Lambda}$ are thus

$$D\mathbf{\Lambda}^k/dt = \tfrac{1}{2}[\mathbf{\Lambda}^k, \mathbf{F}^{\mathrm{T}}] \tag{35}$$

where $k$ is some integer. Now if $f(z)$ is an analytic function of the complex variable $z$, we define $f(\mathbf{\Lambda})$ as a power series. Then

$$Df(\mathbf{\Lambda})/dt = \tfrac{1}{2}[f(\mathbf{\Lambda}), \mathbf{F}^{\mathrm{T}}] \tag{36}$$

The right side of (36) clearly has zero trace. If the trace of $f(\mathbf{\Lambda})$ does exist, then $D/dt$ and trace commute. Therefore

$$D(\operatorname{tr} f(\mathbf{\Lambda}))/dt = 0 \tag{37}$$

this being the conclusion of the trace theorem.

The crucial equation (34) holds even for the case where $\mathbf{W}'$ may be singular. The case where $\mathbf{W}'$ is nonsingular, however, is particularly important, for it is then that $L$ and $L'$ are equivalent. The trace of $f(\mathbf{\Lambda})$ may not exist, on the other hand, especially in this latter case. A good example is $\mathbf{W} = \mathbf{W}'$: When $\mathbf{\Lambda}$ is the unit matrix, with components $\delta^{\alpha\beta}{}_{xy}$, its trace does not exist.

As in the finite-dimensional case, the trace theorem is necessary but not sufficient in the following sense: If there exists an $L'$ subordinate to $L$, and if $\operatorname{tr} f(\mathbf{\Lambda})$ exists, then $\operatorname{tr} f(\mathbf{\Lambda})$ is a constant of the motion. However, the existence of a matrix $\mathbf{\Lambda}$ such that $\operatorname{tr} f(\mathbf{\Lambda})$ exists and is a constant of the motion for all $f$ does not imply that a subordinate $L'$ exists.

In the above discussion we assumed that $\mathbf{W}$ was nonsingular. The extreme opposite of this situation is when $\mathbf{W}$ vanishes. This case is when $L = L(\phi, \dot{\phi}, t)$, but the equations of motion are of first order in the time derivatives (spatial derivatives of any order are allowed). We consider an $L$ of the first-order form

$$L = l^{\alpha}{}_{x}\dot{\phi}^{\alpha}{}_{x} + m \tag{38}$$

(Under reasonable conditions, this form follows from $\mathbf{W} = 0$.)

The equations for $\phi^{\alpha}{}_{x}$ in this case are

$$\omega^{\alpha\beta}{}_{xy}(\dot{\phi}^{\beta}{}_{y} - g^{\beta}{}_{y}) = 0 \tag{39}$$

where

$$\omega^{\alpha\beta}{}_{xy} = \frac{\partial l^{\alpha}{}_{x}}{\partial \phi^{\beta}{}_{y}} - \frac{\partial l^{\beta}{}_{y}}{\partial \phi^{\alpha}{}_{x}} \tag{40}$$

We assume this matrix is nonsingular. In that case the force vector $g^{\alpha}{}_{x}$ is defined by

$$\omega^{\alpha\beta}{}_{xy}\, g^{\beta}{}_{y} = -\frac{\partial l^{\alpha}{}_{x}}{\partial t} + \frac{\partial m}{\partial \phi^{\alpha}{}_{x}} \tag{41}$$

The equations of motion become identities when written as

$$\frac{\Delta}{dt}\frac{\partial L}{\partial \dot{\phi}^{\alpha}{}_{x}} - \frac{\partial L}{\partial \phi^{\alpha}{}_{x}} = 0 \tag{42}$$

where $\Delta/dt$ has the form

$$\frac{\Delta}{dt} = \frac{\partial}{\partial t} + g^{\alpha}{}_{x}\frac{\partial}{\partial \phi^{\alpha}{}_{x}} \tag{43}$$

When the identity (42) is derived with respect to $\phi^{\gamma}{}_{z}$, the result is

$$\frac{\partial(\omega^{\alpha\beta}{}_{xy}\, g^{\beta}{}_{y})}{\partial \phi^{\gamma}{}_{z}} + \frac{\partial^2 l^{\alpha}{}_{x}}{\partial t\, \partial \phi^{\gamma}{}_{z}} - \frac{\partial^2 m}{\partial \phi^{\alpha}{}_{x}\, \partial \phi^{\gamma}{}_{z}} + \omega^{\alpha\beta}{}_{xy}\frac{\partial g^{\beta}{}_{y}}{\partial \phi^{\gamma}{}_{z}} = 0 \tag{44}$$

Upon reversing the values of $\alpha$ and $\gamma$ and of $x$ and $z$ and then subtracting the resulting equation from (44), we find

$$\Delta\mathbf{\Omega}/dt + \mathbf{\Omega}\mathbf{G} - \mathbf{G}^{\mathrm{T}}\mathbf{\Omega} = 0 \tag{45}$$

where $\mathbf{\Omega}$ is the matrix with components $\omega^{\alpha\beta}{}_{xy}$, and $\mathbf{G}$ is the matrix with components $\partial g^{\beta}{}_{y}/\partial\phi^{\gamma}{}_{z}$. Notice that we used the fact that $\mathbf{\Omega}$, being the curl of the vector $l^{\alpha}{}_{x}$, has zero curl itself.

For a subordinate Lagrangian $L'$, the equations of motion are

$$\omega'^{\alpha\beta}{}_{xy}(\dot{\phi}^{\beta}{}_{y} - g^{\beta}{}_{y}) = 0 \tag{46}$$

where $g^{\beta}{}_{y}$ is the same as above. The matrix $\mathbf{\Omega}'$ obeys

$$\Delta\mathbf{\Omega}'/dt + \mathbf{\Omega}'\mathbf{G} - \mathbf{G}^{\mathrm{T}}\mathbf{\Omega}' = 0 \tag{47}$$

The analog of the matrix $\mathbf{\Lambda}$ is the matrix $\mathbf{M}$ defined by

$$\mathbf{M} = \mathbf{\Omega}'\mathbf{\Omega}^{-1} \tag{48}$$

Then $\mathbf{M}$ obeys the equation

$$\Delta\mathbf{M}/dt = -[\mathbf{M}, \mathbf{G}^{\mathrm{T}}] \tag{49}$$

As before it follows that $\operatorname{tr} f(\mathbf{M})$ is a constant of the motion for any analytic $f$.

Before discussing these results, we illustrate them with two simple examples.

## 3. EXAMPLES

### 3.1. Dirac Equation Analog

Our first example is of a Lagrangian which yields first-order equations of motion. The Lagrangian is linear in the first time derivatives and involves a pair of fields $\phi_x$, $\psi_x$:

$$L = \phi_y \dot{\psi}_y \tag{50}$$

The equations of motion are

$$E^1{}_x L = \frac{d}{dt}(\partial L/\partial \dot{\phi}_x) - \partial L/\partial \phi_x = -\dot{\psi}_x = 0 \tag{51}$$

$$E^2{}_x L = \frac{d}{dt}(\partial L/\partial \dot{\psi}_x) - \partial L/\partial \psi_x = \dot{\phi}_x = 0 \tag{52}$$

A subordinate Lagrangian is $L'$ given by

$$L' = (\phi_y)^2 \dot{\psi}_y \tag{53}$$

The equations of motion for $L'$ are related to those for $L$ by the relations

$$E^1{}_x L' = 2\delta_{x-y} \phi_y E^1{}_y L \tag{54}$$

$$E^2{}_x L' = \delta_{x-y} \phi_y E^2{}_y L \tag{55}$$

These last equations define the matrix $M$, whose components are constants of the motion: The field $\phi_y$ is independent of time, according to (52), and so clearly any trace of any function of $\mathbf{M}$ will also be independent of time.

### 3.2. Massless Scalar Waves

Our second example is a scalar wave $\phi(x, t)$, which in our notation is $\phi_x$. We consider the set of solutions of the wave equation

$$\frac{\partial^2 \phi}{\partial t^2} - \frac{\partial^2 \phi}{\partial x^2} = 0 \tag{56}$$

We write $\delta'_x$ and $\delta''_x$ for the first two derivatives of the one-dimensional delta function; the wave equation, in our notation, reads

$$\ddot{\phi}_x - \delta''_{x-y}\phi_y = 0 \tag{57}$$

A second form of this equation stems from the fact that

$$\delta''_{x-y} = \delta'_{x-z}\delta'_{z-y} \tag{58}$$

The wave equation is then written in matrix form as

$$\ddot{\phi}_x - (\mathbf{A}^2)_{xy}\phi_y = 0 \tag{59}$$

and it has the immediate general solution

$$\phi_x = (\exp \mathbf{A}t)_{xy}\alpha_y + (\exp -\mathbf{A}t)_{xy}\beta_y \tag{60}$$

where $\alpha_y$ and $\beta_y$ are constants in time (that is, functions only of the one space variable), and the exponential function is defined as a power series. Here $(\mathbf{A})_{xy} = \delta'_{x-y}$, so that

$$\phi_x(t) = \alpha_{x+t} + \beta_{x-t} \tag{61}$$

The standard Lagrangian is

$$L = \tfrac{1}{2}\int [(\partial\phi/\partial t)^2 - (\partial\phi/\partial x)^2]\, dx = \tfrac{1}{2}\dot{\phi}_x\dot{\phi}_x + \tfrac{1}{2}\delta''_{x-y}\phi_x\phi_y \tag{62}$$

This Lagrangian happens to be the integral of a Lagrangian density, of course, but it will not necessarily be true that an equivalent Lagrangian will involve such a single integral over space.

Examples of constants of the motion are $\alpha_x$ and $\beta_x$. These are vectorial in the sense that a constant of the motion exists for each point $x$: $\alpha_x$ and $\beta_x$ are simply numbers, independent of $t$. However, these functions, along with their derivatives $\alpha'_x$ and $\beta'_x$, are used to form values of $\phi_x$. With this point in mind, we can find expressions involving $\phi_x$ and $\dot{\phi}_x$ which are constants of the motion. Thus

$$2\alpha'_x = \dot{\phi}_{x-t} + \delta'_{x-y-t}\phi_y \tag{63}$$

$$2\beta'_x = -\dot{\phi}_{x+t} + \delta'_{x-y+t}\phi_y \tag{64}$$

are two examples of vector constants. A scalar constant is obtained when a vector constant is integrated over space with a weighting function $w_x$ (also independent of $t$). If $w_x$ and $v_x$ are weighting functions of one variable only, then two scalar constants are

$$A = w_{y+t}(\dot{\phi}_y + \delta'_{y-z}\phi_z) \tag{65}$$

$$B = v_{y-t}(\dot{\phi}_y - \delta'_{y-z}\phi_z) \tag{66}$$

Notice that our definition of constant of the motion is a functional (of field variables and their derivatives) whose time derivative vanishes. This functional may involve multiple integrals over space and thus may in general be nonlocal. It is more usual in field theory to construct constants of motion $Q$ by integrating the time component $J^0$ of a conserved current $J^\mu$ (here $\mu = 0, 1$, with $x^0 = t$, $x^1 = x$, and the comma used below denotes partial derivative):

$$Q = \int J^0 \, dx, \qquad \text{where} \quad J^\mu{}_{,\mu} = 0 \tag{67}$$

Our definition includes constants $Q$ of this sort. For instance, the energy $E$ of the field is in our notation

$$E = \tfrac{1}{2}(\dot{\phi}_x\dot{\phi}_x - \delta''_{x-y}\phi_x\phi_y) \tag{68}$$

Clearly $E$ may be expressed as

$$E = \int J^0 \, dx \tag{69}$$

where $J^\mu$, given by

$$J^0 = \tfrac{1}{2}[(\phi_{,0})^2 + (\phi_{,1})^2], \qquad J^1 = -\phi_{,0}\phi_{,1} \tag{70}$$

is a conserved current.

The constants $A$ and $B$ of (65) and (66) may likewise be expressed as integrals of time components of conserved currents. However, we conjecture that for general nonlocal Lagrangians our definition of constant of the motion is broader than the usual one.

One method of trying to discover an equivalent Lagrangian $L'$ is to construct a matrix $\Lambda$ such that the trace of any function of $\Lambda$ is a constant of the motion (even though the existence of $\Lambda$ does not guarantee the existence of $L'$). One such $\Lambda$ has components which are constants similar to $A$ above:

$$\Lambda_{xy} = \delta_{x+y+t+z}(\dot{\phi}_z + \delta'_{z-w}\phi_w) \tag{71}$$

In this case an $L'$ does exist:

$$L' = \tfrac{1}{6}\delta_{u+v+z+t}\dot{\phi}_u\dot{\phi}_v\dot{\phi}_z - \tfrac{1}{2}\delta'_{u+v+z+t}\phi_u\dot{\phi}_v\dot{\phi}_z - \tfrac{1}{3}\delta'''_{u+v+z+t}\phi_u\phi_v\phi_z \quad (72)$$

The equation of motion derived from $L'$ is

$$\Lambda_{xy}(\ddot{\phi}_y - \delta''_{y-z}\phi_z) = 0 \quad (73)$$

and so $L'$ is at least subordinate to $L$. Actually $\Lambda$ is invertible unless

$$\dot{\phi}_x + \delta'_{x-y}\phi_y = 0 \quad (74)$$

but any such $\phi_x$ is a solution of the wave equation. Hence $L'$ is equivalent to $L$.

We note that the Fourier transform of $\phi_x$ may be considered a vector $\psi_k$ defined by

$$\phi_x = \int e^{ikx}\,\psi_k\,dk \quad (75)$$

Each $\psi_k$ satisfies

$$\frac{d^2\psi_k}{dt^2} + k^2\psi_k = \ddot{\psi}_k + k^2\psi_k = 0 \quad (76)$$

These equations may be considered as a set of uncoupled ordinary differential equations, each the equation of motion for a one-dimensional system. Any one-dimensional system has an infinite number of equivalent Lagrangians.[9] The Fourier synthesis of a collection of these Lagrangians may be used to form equivalent Lagrangians for $\phi_x$.

The usual Lagrangian $L_k$ which gives rise to Eq. (76) for the simple harmonic oscillator is

$$L_k = \tfrac{1}{2}(\dot{\psi}_k{}^2 - k^2\psi^2) \quad (77)$$

To investigate equivalent Lagrangians, we rewrite $L_k$ by integrating the first term in $\int L_k\,dt$ by parts to obtain $\tilde{L}_k$, in which the Lagrangian is expressed as a function $\mu$ times the equation of motion[15]:

$$\tilde{L}_k = \mu(\ddot{\psi}_k + k^2\psi_k), \qquad \mu = -\tfrac{1}{2}\psi_k \quad (78)$$

Now suppose a second function $\mu'$ is formed as the product of $\mu$ and a constant of the motion $C$:

$$\mu' = C\mu \quad (79)$$

with

$$\frac{DC}{dt} = -k^2\psi_k\left(\frac{\partial C}{\partial \dot{\psi}_k}\right) + \dot{\psi}_k\left(\frac{\partial C}{\partial \psi_k}\right) + \frac{\partial C}{\partial t} = 0 \tag{80}$$

Then $\mu'$ may be used to form a Lagrangian $\tilde{L}'_k$ equivalent to $\tilde{L}_k$ (which is equivalent to $L_k$):

$$\tilde{L}'_k = \mu'(\ddot{\psi}_k + k^2\psi_k) \tag{81}$$

This form of the equivalent Lagrangian may be changed into one which does not explicitly depend on $\ddot{\psi}_k$. The procedure[7] is first to solve the equation

$$\partial v/\partial \dot{\psi}_k = \mu' \tag{82}$$

and then to write

$$\begin{aligned} L'_k &= -\frac{Dv}{dt} = -\left[-k^2\psi_k\left(\frac{\partial v}{\partial \dot{\psi}_k}\right) + \dot{\psi}_k\left(\frac{\partial v}{\partial \psi_k}\right) + \frac{\partial v}{\partial t}\right] \\ &= \mu' k^2\psi_k - \dot{\psi}_k\left(\frac{\partial v}{\partial \psi_k}\right) + \frac{\partial v}{\partial t} \end{aligned} \tag{83}$$

It is interesting at this point to compute the momentum $p'_k$ conjugate to $\psi_k$:

$$\begin{aligned} p'_k &= \frac{\partial L'_k}{\partial \dot{\psi}_k} = k^2\psi_k\left(\frac{\partial^2 v}{\partial \dot{\psi}_k{}^2}\right) - \frac{\partial v}{\partial \psi_k} - \dot{\psi}_k\left(\frac{\partial^2 v}{\partial \dot{\psi}_k\,\partial \psi_k}\right) - \frac{\partial^2 v}{\partial \dot{\psi}_k\,\partial t} \\ &= k^2\psi_k\left(\frac{\partial \mu'}{\partial \dot{\psi}_k}\right) - \frac{\partial v}{\partial \psi_k} - \dot{\psi}_k\left(\frac{\partial \mu'}{\partial \psi_k}\right) - \frac{\partial \mu'}{\partial t} \end{aligned} \tag{84}$$

Since $\mu' = C\mu$, the momentum may be written as

$$\mathrm{P}'_k = C\left[k^2\psi_k\left(\frac{\partial \mu}{\partial \dot{\psi}_k}\right) - \dot{\psi}_k\left(\frac{\partial \mu}{\partial \psi_k}\right) - \frac{\partial \mu}{\partial t}\right] - \frac{\partial v}{\partial t} = -\tfrac{1}{2}C\dot{\psi}_k - \frac{\partial v}{\partial t} \tag{85}$$

The Hamiltonian $H'_k$ obtained from $L'_k$ then has the simple value

$$H'_k = p'_k\dot{\psi}_k - L'_k = -CE_k - \frac{\partial v}{\partial t} \tag{86}$$

where the energy $E_k$ is

$$E_k = \tfrac{1}{2}(\dot{\psi}_k)^2 + \tfrac{1}{2}k^2(\psi_k)^2 \tag{87}$$

If $C$ does not explicitly depend on $t$, then the solution $v$ may be taken to be independent of $t$; then clearly $\partial L'/\partial t = 0$ (since $\partial v/\partial t = 0$). In our case, $C$ must be a function of the energy $E_k$ if $\partial C/\partial t$ is to vanish:

$$C = C(E_k) \tag{88}$$

We will concentrate on this case, as an example. Note that the Hamiltonian then has the value

$$H'_k = -C(E_k)\,E_k \tag{89}$$

The Lagrangian $L'_k$ of (83) is to be taken as the Lagrangian for the $k$th mode of the scalar field; the full Lagrangian and Hamiltonian are then

$$L' = \sum_k L_k, \qquad H' = \sum_k H_k \tag{90}$$

When the system is quantized, the expectation value of the Hamiltonian will then be

$$\langle H \rangle = \sum_k \langle H_k \rangle = -\sum_k \langle C(E_k)\,E_k \rangle \tag{91}$$

The ground state of the field is customarily taken to be that with the lowest value of $\langle H \rangle$. The ground state found in this way may therefore differ markedly from that derived from minimizing the expectation value of the total energy $E = \sum_k E_k$.

We conclude by sketching the derivation of (72). Start with (77), the usual Lagrangian for the $k$th mode. One constant of the motion, corresponding to $A$ in (65), is

$$K = e^{ikt}(\dot{\psi}_k - ik\psi_k) \tag{92}$$

Notice that $K$ appears to be complex as does the equivalent Lagrangian $L'_k$ derived below. However, the result for $L'$ when the Fourier synthesis is complete will be real. The new $L'_k$ is obtained by requiring

$$\begin{aligned} \frac{d}{dt}\left(\frac{\partial L'_k}{\partial \dot{\psi}_k}\right) - \frac{\partial L'_k}{\partial \psi_k} &= K\left[\frac{d}{dt}\left(\frac{\partial L_k}{\partial \dot{\psi}_k}\right) - \frac{\partial L_k}{\partial \psi_k}\right] \\ &= e^{ikt}(\dot{\psi}_k - \psi_k)[\ddot{\psi}_k + k^2\psi_k] \end{aligned} \tag{93}$$

The result (up to addition of a total time derivative) is

$$L'_k = e^{ikt}[\tfrac{1}{6}(\dot{\psi}_k)^3 - \tfrac{1}{2}ik\psi_k(\dot{\psi}_k)^2 + \tfrac{1}{3}ik^3(\psi_k)^3] \tag{94}$$

When $L'_k$ is integrated with respect to $k$, the $L'$ of (72) is recovered directly.

## 4. CONCLUSIONS

When two Lagrangians $L$ and $L'$ exist for a classical system, they define a matrix $\mathbf{\Lambda}$ which relates the equations of motion derived from $L$ to those derived from $L'$. The trace theorem states that if $f$ is an analytic function, then $\operatorname{tr} f(\mathbf{\Lambda})$ is a constant of the motion (as defined by $L$) when $L'$ is subordinate or equivalent to $L$. We have extended this theorem, which had been known for systems with a finite number of degrees of freedom, to field theories where $L$ and $L'$ depend at most on the first time derivatives of the fields. We proved the theorem for nonsingular theories where the equations of motion (field equations) are of second order and where they are of first order in the time derivatives.

The theorem has meaning only when $\operatorname{tr} f(\mathbf{\Lambda})$ exists, but there is a great deal of freedom in the choice of $f$. In treating this theorem, we have been careful not only to include theories where $L$ is an integral of a Lagrangian density, but also to include those where $L$ is defined by multiple integrals. Constants of the motion, too, will in general be defined by multiple integrals, and it is important that this sort of structure be included in the theory.

The trace theorem only sets necessary conditions, not sufficient conditions, for the existence of Lagrangians $L'$ equivalent or subordinate to a given one. In two examples, we showed that nontrivial equivalent Lagrangians do exist. The cases where $L$ is essentially unique and those where $L$ is not deserve further study.

The quantum theory based on an equivalent Lagrangian $L'$ will typically be quite different from that based on $L$.[1,4–6] This difference is well known for particle systems and has been recognized in field theory. In particular, studies of the $\sigma$-model have led to nonlocal transformations from one Lagrangian system to an equivalent system, the two systems having quite different behavior as far as asymptotic freedom is concerned.[16] Thus some of the equivalent Lagrangians may have desirable quantum properties and others not.

It is worth emphasizing here that our concept of constant of the motion differs somewhat from the usual idea of constructing such a constant from a conserved current. Our concept includes the latter, as we illustrated in the case of massless scalar waves. Since our definition includes the possibility of multiple integrals over space, we conjecture that some of our nonlocal constants may not be included in the usual formalism. This possibility indicates the desirability of extending Noether's theorem to include nonlocal currents.

The matrices involved in our work suggest still another line of study. The so-called Lax method has been of great help in solving nonlinear systems of ordinary as well as partial differential equations.[14] The key ingredient of that method is very similar to the equation of motion our $\Lambda$ matrix satisfies. Further studies are needed to decide whether any relationship between the Lax method and the trace theorem exists (see Ref. 17). If a relation can be found, the trace theorem will, in principle, provide a systematic way of integrating nonlinear systems when one Lagrangian and symmetries of the field equations are known.

By symmetries in the above comment, we mean something somewhat more general than the usual Noetherian notion. A general symmetry is a mapping which preserves the space of solutions to the equations of motion. Noether transformations are a subset of these transformations; a general symmetry may alter the equations of motion and a variational principle which gives them. The altered structure is equivalent to the original, but it is seen that the study of equivalent Lagrangians is important for understanding symmetries in this general sense.[13]

## ACKNOWLEDGMENT

This work was partially supported by Consejo Nacional de Ciencia y Tecnologia Grant PCCBCEU-000392 and National Science Foundation Grant INT 8201927.

## REFERENCES

1. S. Hojman and R. Montemayor, "S-Equivalent Lagrangians for Free Particles and Canonical Quantization," *Hadronic J.* **3**, 1644–1657 (1980).
2. M. Henneaux, "Equations of Motion, Commutation Relations and Ambiguities in the Lagrangian Formalism," *Ann. Phys. (N.Y.)* **140**, 45–64 (1982).
3. M. Henneaux and L. C. Shepley, "Lagrangians for Spherically Symmetric Potentials," *J. Math. Phys.* **23**, 2101–2107 (1982).
4. F. J. Kennedy, Jr., and E. H. Kerner, "Note on the Inequivalence of Classical and Quantum Hamiltonians," *Am. J. Phys.* **33**, 463–466 (1965).
5. G. Marmo and E. J. Saletan, "*q*-Equivalent Particle Hamiltonians, III. The Two-Dimensional Quantum Oscillator," *Hadronic J.* **1**, 955–965 (1978).
6. A. P. Balachandran, T. R. Govindarajan, and B. Vijayalakshmi, "Particles of Half-Integral or Integral Helicity by Quantization of a Nonrelativistic Free Particle, and Related Topics," *Phys. Rev. D* **18**, 1950–1959 (1978).
7. S. Hojman and H. Harleston, "Equivalent Lagrangians: Multidimensional case," *J. Math. Phys.* **22**, 1414–1419 (1981).
8. S. Hojman and L. F. Urrutia, "On the Inverse Problem of the Calculus of Variations," *J. Math. Phys.* **22**, 1896–1903 (1981).
9. S. Hojman and L. C. Shepley, "Lagrangianos Equivalentes," *Rev. Mex. Fis.* **28**, 149–205 (1982).

10. W. Sarlet, "The Helmholtz Conditions Revisited. A New Approach to the Inverse Problem of Lagrangian Dynamics," *J. Phys. A: Math. Gen.* **15**, 1503–1517 (1982).
11. J. R. Farias and N. L. Teixeira, "Equivalent Lagrangians in Field Theory," *J. Phys. A: Math. Gen.* **16**, 1517–1521 (1983).
12. P. A. M. Dirac, *Lectures on Quantum Field Theory* (Belfer Graduate School of Science, Yeshiva University, New York, 1966).
13. S. Hojman, "Symmetries of Lagrangians and of Their Equations of Motion," *J. Phys. A: Math. Gen.* **17**, 2399–2412 (1984).
14. P. Lax, "Integrals of Nonlinear Equations of Evolution and Solitary Waves," *Commun. Pure Appl. Math.* **21**, 467–490 (1968).
15. R. Hojman, S. Hojman, and J. Sheinbaum, "Shortcut for Constructing any Lagrangian from Its Equations of Motion," *Phys. Rev. D* **28**, 1333–1336 (1983).
16. C. R. Nappi, "Some Properties of an Analog of the Chiral Model," *Phys. Rev. D* **21**, 418–420 (1980).
17. J. F. Cariñena and L. A. Ibort, "Bilagrangian Connections and the Reduction of Lax Equations," *Lett. Math. Phys.* **8**, 359–365 (1984).

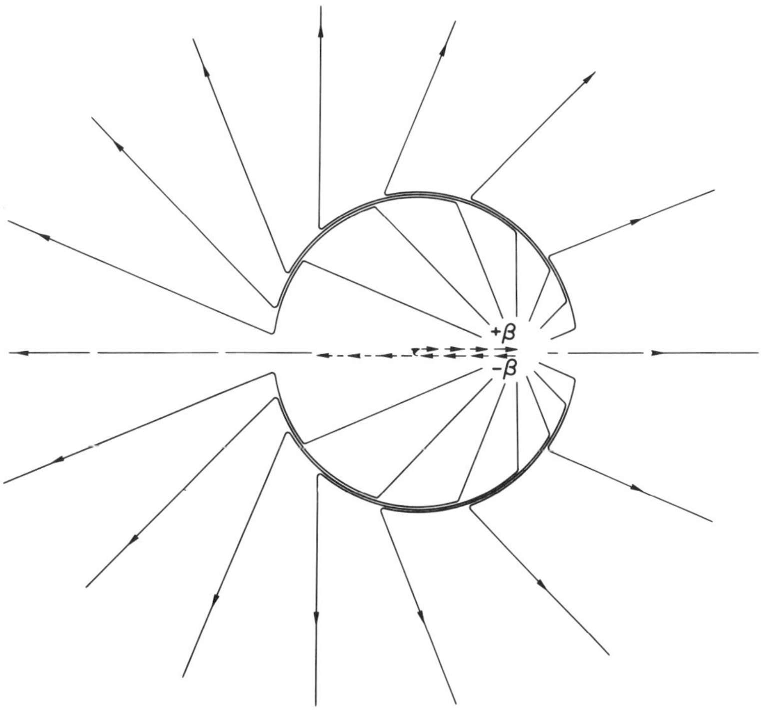


*Plate 24*

# $p$-Form Electrodynamics

**Marc Henneaux**[1,2] **and Claudio Teitelboim**[2,3]

*Received November 22, 1985*

*A generalization of gauge theory in which the gauge potential 1-form is replaced by a p-form is studied. Charged particles are then replaced by elementary extended objects of dimension $p-1$. It is shown that this extension is compatible with space-time locality only if the gauge group is $U(1)$. A source which is a closed $p-1$ surface has zero total charge and corresponds to a particle–antiparticle pair. Its quantum rate of production in an external uniform field is evaluated semiclassically. The analog of the Dirac magnetic pole is constructed. It is another extended object, of dimension $n-p-3$, where n is the dimension of space-time. The "electric" and "magnetic" charges obey the Dirac quantization condition. This condition is derived in two different ways. One method makes use of local gauge patches and the other brings in singular gauge transformations. A topological mass term is introduced and it is shown that it can coexist with a magnetic pole when $n=2p+1$, provided the topological mass is quantized.*

## 1. INTRODUCTION

This article is dedicated to John Archibald Wheeler, decisive teacher and friend, on his 75th birthday.

The paper is devoted to analyzing an extension of electrodynamics in which the electromagnetic field potential 1-form $A_\mu$ is replaced by a $p$-form. The sources of the field are then not any longer charged particles but become elementary extended objects of dimension $p-1$.

The plan of the report is the following. Sec. 2 extends the ordinary notion of gauge invariance under an internal symmetry to extended objects.

[1] Faculté des Sciences, Université Libre de Bruxelles, Campus Plaine C.P. 231, B-1050 Bruxelles, Belgium.

[2] Centro de Estudios Cientificos de Santiago, Casilla 16443, Santiago 9, Chile.

[3] Center for Theoretical Physics, The University of Texas at Austin, Austin, Texas 78712.

It is proved that the generalization is compatible with space-time locality only if the gauge group is $U(1)$. The general formalism is developed for this case and it is shown, in particular, that the total charge of a closed (compact without boundary) object is zero. Hence a closed object appears as the analog of a particle–antiparticle pair. The corresponding quantum production rate in a uniform external field is evaluated semiclassically. The results discussed in this section have previously been reported in Ref. 1.

Section 3 introduces the analog of the Dirac magnetic pole for $p$-form electrodynamics. It is shown that the product of the "electric" and "magnetic" charges obey the Dirac quantization condition. The discussion follows a generalization of the method of local gauge patches and phase factors of Wu and Yang.

Next, in Sec. 4, another derivation of the charge quantization rule by the method of the singular gauge transformation of Goddard and Olive is given.

Finally, Sec. 5 deals with the introduction of a Chern–Simons topological mass term. It is observed that in the special case in which the dimension of space-time equals $2p+1$ the equations are still linear. In that case, and seemingly only then, one can make a topological mass coexist with a magnetic pole, provided the topological mass is quantized.

## 2. GAUGE INVARIANCE FOR EXTENDED OBJECTS

### 2.1. Points Replaced by Extended Objects

Objects of dimensionality higher than that of points, among them strings, appear to play an increasingly fundamental role in physics. It is then natural to attempt transferring to these extended systems ideas and theoretical constructions which have been fruitful when applied to points.

One such idea is gauge invariance under an internal symmetry. The idea has been highly successful when applied to points. One can construct in that case a Yang–Mills gauge theory for any semisimple Lie group.

In order to extend the idea of gauge invariance to objects of dimensionality $p-1$ $(p \geqslant 1)$, one first recalls that in a "first quantized" theory of such systems, one considers "wave functionals" $\phi[x(\sigma)]$ defined over a configuration space in which each point is a coordinatized $(p-1)$-manifold $x^\mu(\sigma)$. Here, the index $\mu$ ranges over the space-time coordinates $(\mu = 0, 1, \ldots, n-1)$ whereas $x_\mu = x^\mu(\sigma)$, with $\sigma = (\sigma^1, \ldots, \sigma^{p-1})$, are the parametric equations of a manifold of dimension $p-1$ embedded in a Minkowski space-time of dimension $n$.

Next, one notices that without further structure, the quantum theory possesses a global $U(1)$-invariance, namely

$$\phi[x(\sigma)] \to e^{i\alpha}\phi[x(\sigma)] \tag{1}$$

where $\alpha$ is a real number, constant over configuration space, i.e., independent of $x^\mu(\sigma)$.

What one wants to do is to extend (1) into a local gauge invariance, based on a more general internal group $G$. In other words, the goal is to build a theory invariant under "gauge transformations of the second kind" given by

$$\phi^A[x(\sigma)] \to T^A{}_B[x(\sigma)]\,\phi^B[x(\sigma)] \tag{2}$$

where $T^A{}_B[x(\sigma)]$ is a $x^\mu(\sigma)$-dependent matrix belonging to the $\phi^A$-representation of the internal symmetry group $G$ under consideration.

In order to implement gauge invariance under (2), one defines a connection $A$ over the infinite-dimensional configuration space of $x^\mu(\sigma)$. This connection permits one to transport the field $\phi^A[x(\sigma)]$ along any path $x^\mu(\sigma;\tau)$, and enters the theory through the parallel transport equation,

$$\frac{d\phi}{d\tau}(\tau) = -qA(\tau)\,\phi(\tau) \tag{3}$$

where $q$ is a coupling constant and

$$A(\tau) \equiv A[x(\sigma;\tau)] \tag{4a}$$

$$\phi(\tau=0) = \phi[x(\sigma;\tau=0)] \tag{4b}$$

Here, $A[x(\sigma)] = A^a[x(\sigma)]\,t_a$ is a one-form in $x(\sigma)$-space which is $G$-Lie-algebra-valued (the matrices $(t_a)^A{}_B$ are the infinitesimal generators of the $\phi^A$-representation).[4]

## 2.2. Locality Brings in $p$-form

So far, nothing prevents, of course, the existence of the connection form $A$. What puts strong restrictions on the construction is the demand that $A$ be described by a local space-time field. This requirement is justified

[4] This means that $A$ associates with each vector in $x(\sigma)$-space, i.e., with each infinitesimal displacement, an infinitesimal $G$-transformation. The mathematical structure considered here is a "$G$-principal bundle" over the space of $x^\mu(\sigma)$ (not over space-time, unless $p=1$, in which case these spaces coincide).

by the desire to treat ultimately the connection as a dynamical entity, within the known context of field theory.

Now, if the $(p-1)$-manifold is displaced from an initial configuration $x_1(\sigma)$ to a final one $x_2(\sigma)$, it sweeps a $p$-dimensional path $\Gamma_p$ in space-time. The locality requirement on $A$ states that $A^a\, d\tau$ should be given by

$$A^a[x(\sigma)]\, d\tau = \frac{1}{p!} \int_{x(\sigma)} A^a_{\mu_1 \cdots \mu_p}\, dx^{\mu_1} \wedge \cdots \wedge dx^{\mu_p} \tag{5}$$

where the components $A^a_{\mu_1 \cdots \mu_p}$ are totally antisymmetric functions of the space-time coordinates $x^\mu$, and where the integral is extended over the $(p-1)$-manifold $x^\mu(\sigma)$. ($p$-Forms have appeared in a variety of contexts in recent years. Some references are listed in Ref. 2.)

In (5), the exterior product, $dx^{\mu_1} \wedge \cdots \wedge dx^{\mu_p}$, is the tangent to the path $\Gamma_p$. In the system of coordinates $(\xi^1, ..., \xi^p) \equiv (\sigma^1, ..., \sigma^{p-1}, \tau)$ on $\Gamma_p$, one has in component form

$$dx^{\mu_1} \wedge \cdots \wedge dx^{\mu_p} = \frac{\partial x^{[\mu_1}}{\partial \xi^1} \cdots \frac{\partial x^{\mu_p]}}{\partial \xi^p}\, d\xi^1 \wedge \cdots \wedge d\xi^p \tag{6}$$

where the square bracket denotes antisymmetrization (without dividing by $p!$).

It should be pointed out here, before going further, that one should not expect the local form (5) of the connection to be preserved under arbitrary gauge transformations (2). Rather, the gauge transformations will also have to obey appropriate locality requirements (see below). But this is all right, since we want to preserve manifest space-time locality of the theory.

With (5), the solution of the parallel transport equations reads

$$\phi_\Gamma[x_2(\sigma)] = P \exp\left[ -q \int_{\Gamma_p} A \right] \phi[x_1(\sigma)] \tag{7}$$

where the symbol $P$ denotes path ordering in the $\tau$-parameter, which is necessary because the operators $A(\tau)$ at different "times" do not commute.

### 2.3. Gauge Group Must Be Abelian

It is here that a major obstacle appears. Indeed, an extra requirement is that the operator in (7) must be invariant under reparametrizations of $\Gamma_p$, that is, it must remain unchanged if $(\xi^1, ..., \xi^p)$ undergo a general coordinate transformation which may be continuously connected with the identity.

This is impossible[1] if $p>1$.

To carry on the argument, it is useful to recast the demand of general covariance of (7) in more geometrical terms. What one has in mind is considering different paths in configuration space which go from $x_1(\sigma)$ to $x_2(\sigma)$, but which correspond—when viewed in space-time— to cutting the same manifold $\Gamma_p$ by different one-parameter sequences of $(p-1)$-dimensional sections $x^\mu(\sigma, \tau)$. For those paths the transport (7) should be path independent ("zero curvature" along those directions).

The condition for path independence is familiar from many-time field theory. It simply says that the Hamiltonian densities $\mathscr{H}_\perp$ at different space points must commute. Here, this means

$$[\mathscr{H}_\perp(\sigma, \tau), \mathscr{H}_\perp(\sigma', \tau)] = 0 \tag{8}$$

with

$$\mathscr{H}_\perp = \frac{q}{p!} A^a_{\mu_1 \cdots \mu_p} \frac{\partial x^{[\mu_1}}{\partial \sigma^1} \cdots \frac{\partial x^{\mu_{p-1}}}{\partial \sigma^{p-1}} n^{\mu_p]} t_a \tag{9}$$

where $n$ is the unit normal to $x(\sigma)$ which lies in $\Gamma_p$. [In general,[3] one must allow for the generator $\mathscr{H}_i$ $(i=1,\ldots, p-1)$ of spatial reparametrizations to appear on the right side of (8). However, in the present case, that generator is obtained by replacing $n^\mu$ in (9) by $\partial x^\mu/\partial \sigma^i$ and vanishes identically.]

Now, if one computes explicitly (8), one finds that, in order for (7) to hold for any choice of the $A^a_{\mu_1 \cdots \mu_p}(x)$, one must have $C_{ab}{}^c = 0$, where the $C_{ab}{}^c$ are the internal group structure constants. That is, the group must be Abelian. But this implies that (8) holds also for different times $\tau$, which in turn makes the path ordering in (7) superfluous.

Notice that the argument fails when $p=1$, in which case (8) is empty. When $p=1$ and for any gauge group the path ordering in (7) is automatically invariant under changes $\tau \to f(\tau)$ of the path parameter provided $f'(\tau)>0$, that is, provided the reparametrization does not reverse the orientation of the path.

Actually, the above analysis is unnecessary for a spacelike path $\Gamma_p$. In that case it is sufficient to recall that there is no ordering relation between points of a Riemannian space of dimension greater than 1 endowed with a positive definite metric. If $p$ has a timelike direction, the light cone introduces an ordering relation, but only when (8) holds.

One thus sees that it is not possible to construct a non-Abelian Yang–Mills gauge field theory for extended objects in space-time: the only allowed possibility, which we now pass to discuss, corresponds to the Abelian $U(1)$ group.[5]

In that case, the form $A$ may be taken to be real valued, provided an imaginary unit is inserted in the exponent in (7) which becomes

$$\phi_\Gamma[x_2(\sigma)] = \exp\left[-iq\int_{\Gamma_p} A\right]\phi[x_1(\sigma)] \tag{10}$$

with $\phi$ being a single complex-valued functional of $x(\sigma)$.

### 2.4. Gauge Transformation of *p*-Form

Similarly, the gauge transformation (2) reduces to

$$\phi[x(\sigma)] \to \exp\left[-iq\int_{x(\sigma)} \mu\right]\phi[x(\sigma)] \tag{11}$$

where $\mu$ is an arbitrary $p-1$ form. The dependence of the gauge transformation on the configuration point $x^\mu(\sigma)$ is thus quite simple, and given by the region over which one integrates the $p-1$ form $\mu$. This particular form of the $U(1)$ group element in (11) reflects the space-time locality requirement on the gauge transformations discussed above.

Under (11), the potential $A$ transforms as

$$A \to A + d\mu \tag{12}$$

This transformation law is a rewriting of the known transformation law for a connection over a principal bundle ($\delta A \sim g\, dg^{-1}$ in the Abelian case, where $d$ is now the exterior derivative in the "base" manifold, i.e., the configuration space of the $x^\mu(\sigma)$), provided appropriate conditions are imposed at the boundary of $x(\sigma)$.

[5] It appears to be common knowledge among researchers that attempts to build a non-Abelian theory for $p > 1$ have not worked out. The above proof provides a geometrical understanding of this failure.

The necessity of the Abelian structure reminds one of what happens with homotopy groups of a manifold. Only the first homotopy group—associated with loops—can be non-Abelian, again because there is no way to order the product of two $p$-spheres when $p \geqslant 2$.

Indeed, one finds

$$g\,\delta g^{-1}=\int_{x(\sigma)+\delta x(\sigma)}\mu-\int_{x(\sigma)}\mu \tag{13a}$$

$$=\oint_{x\cup(x+\delta x)}\mu \tag{13b}$$

$$=\int_{\Sigma_p}d\mu \tag{13c}$$

where $\Sigma_p$ is the $p$-surface spanned by the variation of $x(\sigma)$. In order to go from (13b) to (13c), we have used Stokes theorem, while the transition from (13a) to (13b) assumes vanishing integrals over the sides (if any).

If the surface $x(\sigma)$ has no boundary, there is no "side," and (13b) always holds. If $x(\sigma)$ extends to infinity, one must impose appropriate fall-off of $\mu$ at infinity. Finally, if $x(\sigma)$ is compact with boundary, one must take the projection of $\delta x$ perpendicular to the boundary to vanish (this situation will be met in the last section).

Once (13c) is established, one easily gets (12) by recalling the relation (5) between the connection $A$ on the bundle and the $p$-form $A$ which appears in (7) (both denoted by the same symbol).

### 2.5. Action Principle

In a first quantized version (relativistic quantum mechanics) of the theory of the extended object $x(\sigma)$, the functional field $\phi$ is the wave functional of $x(\sigma)$. If one passes to a field theory of extended objects, $\phi$ becomes a field operator.

The corresponding classical theory of one extended object interacting with a dynamical field $A$ may be derived from an action which is a straightforward extension of the standard one for classical electrodynamics of one point particle. In that action we shall reserve the letter $x$ for a generic space-time point and we will denote by $z(\sigma^1,\ldots,\sigma^{p-1},\tau)=z(\xi^1,\ldots,\xi^p)$ the parametric equations of the $p$-dimensional history of the object $z(\sigma)$. Also the exterior derivative $F$ (field strength) of $A$ will come in:

$$F=dA=\frac{1}{(p+1)!}F_{\mu_1\cdots\mu_{p+1}}\,dx^{\mu_1}\wedge\cdots\wedge dx^{\mu_{p+1}} \tag{14}$$

with

$$F_{\mu_1\cdots\mu_{p+1}}=\partial_{\mu_1}A_{\mu_2\mu_3\cdots\mu_{p+1}}-\partial_{\mu_2}A_{\mu_1\mu_3\cdots\mu_{p+1}}-\cdots-\partial_{\mu_{p+1}}A_{\mu_2\mu_3\cdots\mu_1} \tag{15}$$

The action, which is a functional of $A(x^1,\ldots,x^n)$ and $z(\xi^1,\ldots,\xi^p)$, reads[6]

$$S[A,z] = -\frac{1}{2}\frac{1}{(p+1)!}\int F^2\,d^nx - m\int (-{}^{(p)}g)^{1/2}\,d^p\xi$$

$$-\frac{q}{p!}\int A_{\mu_1\cdots\mu_p}(z)\,dz^{\mu_1}\wedge\cdots\wedge dz^{\mu_p} \tag{16}$$

where $F^2$ stands for $F_{\mu_1\cdots\mu_{p+1}}F^{\mu_1\cdots\mu_{p+1}}$ and ${}^{(p)}g$ is the determinant of the metric $g_{ab}$ on the history of the extended object

$$g_{ab} = \frac{\partial z^\mu}{\partial\xi^a}\frac{\partial z_\mu}{\partial\xi^b} \tag{17}$$

If one has several objects interacting with the field $A$, one must include in (16) one $m$-term and one $q$-term for each object. One assumes classically that for each object the $p$th direction is timelike, so that ${}^{(p)}g$ is negative. This restriction is only temporary and is lifted in quantum mechanics where spacelike and null histories are also admitted in the path integral. As a consequence, the objects may coalesce, be created from the vacuum, or annihilated (see below).

## 2.6. Special Dimensions

The coupling constant $q$ in (16) is analogous to the elementary electric charge. It has dimensions of a length raised to the $(\frac{1}{2})[n-2(p+1)]$th power. On the other hand, $m$ is a measure of the inertia of the object—analogous to the rest mass of a particle—and has dimensions of a length to the $(-p)$. The field components $A_{\mu_1\cdots\mu_p}$ have dimensions of a length to the $(2-n)/2$. (We use units where $\hbar=1$ and assume that the Minkowskian coordinates $x^\mu$, $z^\mu$ have dimensions of a length.)

Of particular interest are those dimensions for which the coupling constant is a pure number. This happens when

$$n = 2(p+1) \tag{18}$$

Equation (18) also gives the dimensions for which $F$ and $*F$ have the same number of components.

In that case the action (16) is duality invariant away from the charges.

[6] Note that because of the presence of the potential $A$ in the last term of the action, quantum electric objects do interact directly with $A$ (through a "covariant functional derivative") and not only with $F$. This will be reemphasized in the next section.

Thus, one finds a relation between duality invariance and the pure number character of the coupling constant.

Another noteworthy related property is that in the dimensions (18) the $F^2$ term in (16) is invariant under pointwise conformal (Weyl) rescalings of the metric in the $n$-dimensional space-time.

If one deals with particles ($p=1$), Eq. (18) yields $n=4$, as one knows. On the other hand, it is interesting to notice what happens if one sets instead $p=4$. In that case, one may look at the history of the object as being physical space-time embedded in a higher-dimensional flat space. From (18), that space must have dimension $n=10$, which is precisely the minimum number of dimensions which makes the (local) embedding geometrically possible for a generic 4-manifold. [The Hilbert action for the $p$-manifold may be added to (16). This addition modifies (20) below and makes it consistent with Einstein's equations. Reference 4 deals with this problem.]

Finally one should recall that $n=10$ is the critical dimension in superstring theory. Moreover a 4-form appears explicitly in a limiting case of string theory, chiral $N=2$ supergravity in ten space-time dimensions (see the last of Refs. 2).

## 2.7. Equations of Motion

The equations of motion derived from (16) are

$$\partial_\mu F^{\mu\mu_1\cdots\mu_p} = j^{\mu_1\cdots\mu_p} \tag{19}$$

$$m^{(p)}[(-g)^{1/2}\square]\, z^\mu = -qF^\mu{}_{\mu_1\cdots\mu_p}\frac{\partial z^{\mu_1}}{\partial\xi^1}\cdots\frac{\partial z^{\mu_p}}{\partial\xi^p} \tag{20}$$

where

$$j^{\mu_1\cdots\mu_p} = q\int\delta^{(n)}(x-z)\, dz^{\mu_1}\wedge\cdots\wedge dz^{\mu_p} \tag{21}$$

is the current associated with the history $z(\xi)$ and ${}^{(p)}\square$ denotes the D'alambertian in the $p$-dimensional curved space with metric (17),

$${}^{(p)}[(-g)^{1/2}\square]\, z^\mu = \frac{\partial}{\partial\xi^a}\left[(-g)^{1/2}\, g^{ab}\frac{\partial\xi^\mu}{\partial\xi^b}\right] \tag{22}$$

Note that since $g_{ab}$ given by (17) involves the $z^\mu$, the left side of (22) is a nonlinear function of $z^\mu$ (except for the case $p=2$, when it can be made linear by an appropriate choice of the coordinates $\xi$).

## 2.8. Conservation of Charge

It follows from (19) that the current (21) must be conserved,

$$\partial_{\mu_1} j^{\mu_1 \cdots \mu_p} = 0 \tag{23}$$

an equation which may also be obtained by demanding invariance of the action (17) under the gauge transformation (12),

$$\delta A = d\mu, \qquad \delta z^\mu = 0 \tag{24}$$

for an arbitrary $(p-1)$-form $\mu$.

Equation (2.23) may be converted in a conservation law by means of Stoke's theorem.

$$\int_M d\omega = \int_{\partial M} \omega \tag{25}$$

where $\omega$ is a $k$-form and $M$ is a manifold of dimension $k+1$ with $k$-dimensional boundary $\partial M$.

The following analysis will rely repeatedly on (25), on the fact that one integrates $k$-forms only over $k$-dimensional manifolds and on the identity relating the divergence with the exterior derivative.

$$\partial_{\mu_1} \omega^{\mu_1}{}_{\mu_2 \cdots \mu_k} = (-1)^{nk} [{}^*d^*\omega]_{\mu_2 \cdots \mu_k} \tag{26}$$

where the star denotes the dual

$${}^*\omega_{\mu_1 \cdots \mu_{n-k}} = \frac{1}{k!} \varepsilon_{\mu_1 \cdots \mu_n} \omega^{\mu_{n-k+1} \cdots \mu_n} \tag{27}$$

Using (26), one may rewrite (23) as

$$d * j = 0 \tag{28}$$

an equation which, upon integration over $n-p+1$ dimensional manifold $M_{n-p+1}$ with boundary $(\partial M)_{n-p}$, yields, from (25)

$$\int_{(\partial M)_{n-p}} * j = 0 \tag{29}$$

Now, one may consider, in particular, $M$ to be formed by two spacelike $(n-p)$ surfaces $\Sigma_{n-p}$ and a remote timelike tube. If the current $j$

is localized in space, then the integral over the remote tube vanishes and one finds from (29) that the charge

$$Q = \int_{\Sigma_{n-p}} * j \tag{30}$$

is independent of the spacelike section $\Sigma_{n-p}$. This implies that $Q$ is Lorentz invariant and conserved.

In the particular case $p = 1$, (30) gives the usual integration over an $(n-1)$ surface of the component of $j$ normal to that surface. [The normal comes in from the dual in (30).] One sees, however, that the higher the dimension of the extented object the lesser the dimension of the cut $\Sigma_{n-p}$ that comes into the definition of the total charge.

If the current is of the form (21) (possibly summed over several objects), then contributions to (30) come only from the intersection of $\Sigma_{n-p}$ with the $p$-dimensional history of the object. That intersection has dimension zero and consists, therefore, of a number of points with each point giving a contribution $\pm q$.

Finally, one can find an expression for $Q$ in terms of the field strength $F$ which is the generalization of Gauss's law in electromagnetism. This is done by observing that (19) and (26) imply

$$d * F = (-1)^{(n-p-1)} * j \tag{31}$$

Therefore, if one denotes by $(\partial\Sigma)_{n-p-1}$ the $n-p-1$ dimensional boundary of $\Sigma_{n-p}$, one has the expression

$$Q = (-1)^{(n-p-1)} \int_{(\partial\Sigma)_{n-p-1}} * F \tag{32}$$

for the total charge within $(\partial\Sigma)_{n-p-1}$.

## 2.9. Total Charge of Closed Object Vanishes

The preceding discussion of the total charge parallels closely that for electrodynamics. However, an important additional property appears for extended objects. It is this: the total charge of an object which is not infinitely extended in all its directions is equal to zero. Thus, the analog of a single point charge in electrodynamics is an infinitely extended object. On the other hand, a closed object is the analog of a particle–antiparticle pair.

This property follows also from the divergence-free character of the current. In fact, if one evaluates the divergence of (21) by differentiation under the integral sign, one finds

$$\partial_{\mu_1} j^{\mu_1 \cdots \mu_p}(x) = -q \int d[\delta^{(n)}(x - z(\xi))\, dz^{\mu_2} \wedge \cdots \wedge dz^{\mu_p}] \tag{33}$$

where the exterior derivative acts on $z$. Next, one rewrites the right side of (2.33)—which is an integral over the history of the object—as an integral over the history's boundary. It follows that, since (33) must vanish for every $x$, the history must have no boundary. This is an alternative statement of the conservation of charge in this formalism.[7]

Now, if the history has no boundary, it must be either closed or infinitely extended, in each direction. If there is at least one spatial direction in which the history is closed, one may find a boundary $(\partial\Sigma)_{n-p-1}$ located at a finite distance which totally surrounds the object. One then finds that the intersection of $\Sigma_{n-p}$ with the history consists of an even number of points, all of them lying in the interior of $(\partial\Sigma)_{n-p-1}$ and with the $(+q)$ contributions exactly cancelling the $(-q)$ ones.

This situation is similar to what happens with pair annihilation in electrodynamics of point charges. There, one has an electron first moving forward in time and then coming back as a positron. The electron describes a continuous history which crosses each spacelike plane either twice or never, so that the total charge is conserved and equal to zero.

Here, given the spatially extended nature of the object, the same phenomenon happens in spacelike directions. If the object is localized, for each time that the history crosses $\Sigma_{n-p}$ with a $(+q)$ contribution, there is a "return" piece that gives $(-q)$. If the object is infinitely extended in all directions, some of the return portions may be thought of as having moved out to infinity, leaving an unbalanced net contribution to the total charge.

## 2.10. Quantum Mechanical Production of Closed Object in Uniform Extended Field

Since an infinitely extended object is somewhat unnatural from the physical point of view, it appears that one should focus one's attention on closed objects. It should be stressed that closed objects do interact with the $A$-field, even though they have a vanishing total charge. In particular, they may be produced in an external field just as electron–positron pairs are created.

[7] If one introduces simultaneously magnetic poles and a topological mass (Sec. 5), Eq. (23) no longer holds and this conclusion ceases to be valid. Electric objects with a boundary are then brought in.

The semiclassical calculation of the production probability in a uniform external field is quite simple. It has the additional virtue of illustrating that not only closed objects, but also closed histories, are of interest and use. For that reason it will be given here as a last point.

The probability in question may be written as

$$A \exp[-B] \tag{34}$$

where both $A$ and $B$ are slowly varying functions of the parameters of the problem (the charge $q$, the mass $m$, and the field strength). Only the coefficient $B$ will be dealt with here. It is given to a first approximation by the Euclidean version of the action (16) evaluated on the extremal history of highest symmetry.[(5)] [Since one deals with an external field, the $F^2$ term in (16) is omitted from the action in this calculation.]

If the field lies in a $(p+1)$ dimensional hyperplane, one writes

$$F^{\mu_1 \cdots \mu_{p+1}} = E\varepsilon^{\mu_1 \cdots \mu_{p+1}} \tag{35}$$

where $E$ is a constant, and takes for the history a $p$-sphere of radius $R$ contained within that plane. After applying Stoke's theorem to the interaction term, the Euclidean action $S_E$ for the sphere reads

$$S_E = mA_p - qEV_{p+1} \tag{36}$$

where $V_{p+1} = \{\pi^{(p+1)/2}/\Gamma\,[(p+3)/2]\}\,R^{p+1}$ is the volume contained within the $p$-sphere and $A_p = (\partial V_{p+1}/\partial R)$ is the surface area.

The extremal radius $R_0$ is obtained by demanding that the derivative of (36) with respect to $R$ vanishes. This yields

$$R_0 = \frac{pm}{qE} \tag{37}$$

a result which, inserted back into (36), leads to

$$B = S_E(R_0) = \left[\pi^{(p+1)/2}/\Gamma\left(\frac{p+3}{2}\right)\right] m(R_0)^p \tag{38}$$

In Minkowski space, the object materializes as a $(p-1)$ sphere of radius $R_0$ [the equatorial intersection of the $p$-sphere with a spacelike $(n-1)$ plane] and then proceeds expanding at a speed approaching that of light with the $(p-1)$ sphere sweeping a $p$-hyperboloid.

## 3. PHASE FACTORS AND MONOPOLES OF HIGHER RANK

### 3.1. Monopoles of Higher Rank Analogous to the Dirac Magnetic Pole

As shown in Ref. 6, one can also consider magnetic objects as sources for the $p$-form field.

In a space-time of dimension $n$, the magnetic pole is another extended object whose history has dimension $n-p-2$, dual to that of the electric pole. It should be emphasized that this monopole, although extended, corresponds to the Dirac point-monopole of ordinary electrodynamics and not to the singularity-free 't Hooft–Polyakov object found in non-Abelian Yang–Mills theories.[7] Indeed, it has been shown above that non-Abelian $p$-form Yang–Mills gauge theories cannot be constructed.

In Ref. 6, the theory of higher-rank monopoles was developed by introducing the higher-dimensional analog of the Dirac string,[8] along which the $p$-form potential is singular. This singularity of the potential is classically fictitious, however, for the total $(p+1)$-electromagnetic field—to which classical objects couple—is regular on the string. Its only singularities are located at the electric or magnetic sources.

To ensure that the string singularity is still of no physical relevance quantum mechanically is less immediate, for quantum objects do couple directly to the potential, and not only to the field strength. As shown in Ref. 6, the demand that the generalized Dirac string—treated as a dynamical object along the lines of Ref. 8—be a pure gauge quantity leads to the quantization condition

$$q\bar{q} = 2\pi\hbar \text{ (integer)} \tag{39}$$

on the electric and magnetic strengths $q$, $\bar{q}$.

### 3.2. Phase Factors and Local Gauge Patches

We rederive here the quantization condition (39) by following a somewhat different route, inspired by the string-free description of Wu and Yang.[9] As it is known, their method is based on the concept of "phase factor."

The phase factor for $p$-form electrodynamics is defined as in the Maxwell case. To each $p$-surface $\Sigma_p$, one associates a $U(1)$ group element defined as

$$\exp\left[-i\frac{q}{\hbar}\int_{\Sigma_p} A\right] \tag{40}$$

If $\Sigma_p$ has a boundary consisting of two $(p-1)$-dimensional electric configurations $x_1(\sigma)$ and $x_2(\sigma)$, (40) yields the phase factor (7) of Sec. 2, by which the parallel transported functional $\phi_\Gamma[x_2(\sigma)]$ at $x_2(\sigma)$ differs from $\phi_\Gamma[x_1(\sigma)]$ at $x_1(\sigma)$. But (40) can be considered for more general $p$-surfaces.

Of particular interest are closed $p$-surfaces. The phase factor (40) is then a gauge-invariant object which can be interpreted as giving the $U(1)$ element generated by parallel transport along a closed loop in $x(\sigma)$-space.

If the closed surface is the boundary of a $(p+1)$-surface, the integral of $A$ can be converted by means of Stokes' theorem into a $(p+1)$-integral involving only the field strength ["magnetic flux through the $(p+1)$-surface"]. However, if the closed $p$-surface under consideration is not the boundary of a $(p+1)$-surface, i.e., if it is not contractible to a point, one cannot express the phase factor in terms of the electromagnetic field strength alone. This indicates that the phase factor contains more information than the field strength. As tested by the Bohm–Aharonov experiment,[(10)], this additional information is relevant in quantum mechanics.

The above definition of the phase factor assumes that the $A$-field is well defined everywhere [except, of course at the sources, where it generically blows up, but one does not consider surfaces $\Sigma_p$ in (40) which cross them]. It is well known that this regularity assumption cannot be true if one allows magnetic sources. Indeed, one finds in that case that

$$dF = (-1)^{n(p+1)+1} * \bar{j} \tag{41}$$

where $*\bar{j}$ is the dual to the magnetic current density $\bar{j}$, while $F = dA$ leads to

$$dF = 0 \tag{42}$$

One can tell the difference between (41) and (42) even outside the sources, by considering integrals of $F$ on closed $(p+1)$-surfaces $\Sigma_{p+1}$ which "link" the magnetic pole.[8] For such surfaces, (42) implies $\int F = 0$, whereas (41) implies a nonvanishing result.[9]

This means that if one tries to solve $F = dA$ for $A$ in the field of a monopole, one finds singularities of the string-type in $A$, even outside the

[8] By "link," it is meant: (i) that $\Sigma_{p+1}$ is the boundary of at least one $(p+2)$-surface and (ii) that any $(p+2)$-surface with $\Sigma_{p+1}$ as a boundary intersects the monopole an odd number of times, i.e., the intersection consists of $(2k+1)$ points. (This is the nonvanishing "linking number" in the mathematical literature; see, for example, Ref. 11.)

[9] The magnetic current is constructed in the same way as the electric one, with the magnetic strength $\bar{q}$ and the magnetic pole history in place of the electric ones. The electric coupling $q$ has units of length to the power $\frac{1}{2}[n-2(p+1)]$, whereas $\bar{q}$ has units of length to the power $-\frac{1}{2}[n-2(p+1)]$. The product $q\bar{q}$ is always a pure number.

sources. The potential $A$ cannot be defined regularly everywhere. Rather, one must divide space in various patches if one insists on having a well-defined potential. On each of these patches, one has a different, regular $A$. In the overlapping regions, the different potentials are related by a gauge transformation ("transition functional"). The transition functional must be a well-defined $U(1)$ element. This leads to the quantization condition.

### 3.3. Sources with Topology of a Sphere

For concreteness, let us illustrate these concepts by considering compact electric and magnetic poles possessing the topology of spheres (electric spheres of dimension $p-1$, magnetic spheres of dimension $n-p-3$). The history of a magnetic pole is then an $(n-p-2)$-dimensional cylinder, which we link with a $(p+1)$-sphere $S_{p+1}$.

The linking sphere $S_{p+1}$ is divided into an "upper region" and a "lower region." The upper region contains the "North side" of the "equatorial $p$-sphere" $S_p$ and extends below it, but without including the South Pole. The "lower region" is obtained by a reflection about the equator.

One then defines a potential $A^{(\mathrm{I})}$ regular everywhere in the upper region (this potential would have a string singularity if extended all the way down to the South Pole), and a potential $A^{(\mathrm{II})}$ regular everywhere in the lower region.

Since $A^{(\mathrm{I})}$ and $A^{(\mathrm{II})}$ describe the same physical situation, they must obey

$$F^{(\mathrm{I})} = dA^{(\mathrm{I})} = F^{(\mathrm{II})} = dA^{(\mathrm{II})}$$

i.e.,

$$d(A^{(\mathrm{I})} - A^{(\mathrm{II})}) = 0 \tag{43}$$

in the overlapping region.

Classically, (43) is all that is required on $A^{(\mathrm{I})}$ and $A^{(\mathrm{II})}$. Identical field strengths imply indeed the same classical physics. There is no quantization condition on the monopole strength. Quantum mechanically, however, the condition (43) is no longer sufficient, since quantum electric objects do couple to the potential.

In order for the potentials $A^{(\mathrm{I})}$ and $A^{(\mathrm{II})}$ to describe the same physical situation, they must be related by a gauge transformation, i.e., one must have

$$A^{(\mathrm{II})} - A^{(\mathrm{I})} = d\mu \tag{44}$$

where the $(p-1)$-form $\mu$ is not necessarily "single valued" everywhere on the overlap ($A^{(\mathrm{II})}-A^{(\mathrm{I})}$ is not expected to be exact), but is such that the phase transformation

$$\exp\left[-i\frac{q}{\hbar}\int_{x(\sigma)}\mu\right] \tag{45}$$

is well defined for any configuration $x(\sigma)$ of the $(p-1)$-electric spheres. This is necessary and sufficient to guarantee that the wave functionals $\phi_{\mathrm{I}}[x(\sigma)]$ in the gauge $A_{\mathrm{I}}$ and $\phi_{\mathrm{II}}[x(\sigma)]$ in the gauge $A_{\mathrm{II}}$ are indeed related by a well-defined $U(1)$ gauge transformation,

$$\phi_{\mathrm{II}}[x(\sigma)]=\exp\left[-i\frac{q}{\hbar}\int_{x(\sigma)}\mu\right]\phi_{\mathrm{I}}[x(\sigma)] \tag{46}$$

Now, the region common to both the upper and lower parts of the $(p+1)$-sphere which links the magnetic pole is topologically the product of a $p$-sphere (the equator $S_p$) and a one-dimensional interval $[-\varepsilon, +\varepsilon]$. Its only nonexact, closed $p$-form is given by the $p$-sphere volume form $V^{(p)}$. So one has

$$A^{(\mathrm{II})}-A^{(\mathrm{I})}=kV^{(p)}+d\lambda \tag{47}$$

where $\lambda$ is a well-defined $(p-1)$-form and $k$ is so far an arbitrary real number.

### 3.4. Quantization and Homotopy

Taking for $x(\sigma)$ an electric configuration which lies entirely on the equator, one sees that the ambiguity in the integral of $\mu$ over $x(\sigma)$ is given by the integral of $A^{(\mathrm{II})}-A^{(\mathrm{I})}$ over the equator, i.e., by $k$ (the integral of $V^{(p)}$ is normalized to 1). Hence, (47) defines an acceptable gauge transformation if and only if $k$ is quantized according to

$$qk=2\pi\hbar\,(\text{integer}) \tag{48}$$

The number $k$ con only take discrete values.

It results from this analysis that, even though the $p$-form $A$ cannot be viewed as a connection on a principal $U(1)$-bundle over space-time where $p>1$ (parallel transport is not defined in space-time, which does not coincide with the space of the extended objects in that case), the type of mathematical structure which appears is very similar to what arises when $p=1$. As it is known, magnetic poles in ordinary electromagnetism, i.e.,

$U(1)$-bundles over space-time minus the monopole location, are also classified by the group of the integers.

The analogy goes actually far, for one can interpret the number $k$ (which is equal to the magnetic source strength, see below) as defining an element of the homotopy group $\pi_1(U(1))$ of the gauge group $U(1)$. Hence $k$ is a topological charge and, as such, is conserved under any continuous deformation of the fields.

The association is made as follows. First, one defines the phase factor for any $p$-surface $\Sigma_p$ lying on the $(p+1)$-sphere $S_{p+1}$ which links the pole under consideration as

$$\exp\left[-\frac{iq}{\hbar}\int_{\Sigma_p^{\mathrm{II}}} A^{\mathrm{II}}\right]\exp\left[-\frac{iq}{\hbar}\int_{\Sigma_{p-1}} \mu\right]\exp\left[-\frac{iq}{\hbar}\int_{\Sigma_p^{\mathrm{I}}} A^{\mathrm{I}}\right] \tag{49}$$

Here $\Sigma_p^{\mathrm{I}}$ ($\Sigma_p^{\mathrm{II}}$) is the part of $\Sigma_p$ which lies in the Northern (Southern) hemisphere of the $(p+1)$-sphere $S_{p+1}$ and $\Sigma_{p-1}$ is its intersection with the equator. The phase factor (49) satisfies the group property.

Next, one notices that if $\Sigma_p$ is a closed surface, the phase factor (49) reduces again to the gauge-invariant quantity

$$\exp\left[-\frac{iq}{\hbar}\int_{\Sigma_{p+1}} F\right] \tag{50}$$

where $\Sigma_{p+1}$ is a $(p+1)$-surface on the $(p+1)$-sphere whose boundary is $\Sigma_p$. Two different choices of $\Sigma_{p+1}$ might differ by

$$\exp\left[-\frac{iq}{\hbar}\int_{S_{p+1}} F\right] \tag{51}$$

where the integral is extended to the whole $(p+1)$-sphere. By means of Stokes' theorem and (41), this last expression can be rewritten as

$$\exp\left[-\frac{iq}{\hbar}\int_{V_{p+2}} dF\right]=\exp\left[-\frac{iq\bar{q}}{\hbar}\right] \tag{52}$$

But the magnetic pole strength $\bar{q}$ is nothing but the constant $k$ introduced in (47), so that (50) is well defined as a result of the quantization condition (48).

The identification of $k$ with $\bar{q}$ is simply made by observing that the following chain of identities hold:

$$\bar{q} = \int_{V_{p+2}} dF = \int_{S_{p+1}} F$$
$$= \int_{\mathrm{I}} F + \int_{\mathrm{II}} F$$
$$= \int_{S_p} (A^{\mathrm{II}} - A^{\mathrm{I}})$$
$$= k \int_{S_p} v^{(p)} = k \tag{53}$$

In (53), $S_p$ is the equator, and the relation (47) has been used.

Once (53) is established, one considers a one-parameter family of $p$-spheres $S_p(\tau)$ such that $S_p(0) = S_p(1)$ is the North Pole, which completely wraps around the $(p+1)$-sphere $S_{p+1}$ (see Ref. 12 for a picture in the $p = 1$ case). As $\tau$ varies from 0 to 1, the associated group element (49) describes a closed curve in $U(1)$, which starts and ends at the identity. Hence, it defines an element of $\pi_1(U(1)) = Z$. It is now easy to see, using (50) and (51), that the integer in

$$q\bar{q} = 2\pi\hbar \text{ (integer)} \tag{54}$$

precisely characterizes the homotopy class of that loop.

The interpretation of the magnetic strength of the pole as a topological charge explicitly displays its independence from all continuous changes of the parameters in the problem. In particular, it shows that the field is necessarily singular on the monopole for otherwise the loop constructed above should be trivial (as seen by shrinking the $(p+1)$-sphere to a point). This means that there is no hope to construct everywhere regular magnetic pole fields in the context of the Abelian theory alone (a well-known result for $p = 1$).

### 3.5. State as Collection of Patch Wave Functionals

As a final point, we note that the concept of "wave functional" $\phi[x(\sigma)]$ has to be generalized in the field of a monopole since the potential $A$ cannot be globally defined in a single gauge.[13] This is done as for $p = 1$. A wave functional is now the collection $\phi^{\mathrm{I}}[x(\sigma)]$, $\phi^{\mathrm{II}}[x(\sigma)]$, $\phi^{\mathrm{III}}[x(\sigma)] \cdots$ of ordinary wave functionals defined on space-time patches which admit a singularity-free $A$. These wave functionals obey the obvious compatibility conditions in the overlap of two patches.

## 4. SINGULAR GAUGE TRANSFORMATIONS

The Wu–Yang approach for deriving the quantization condition is equivalent to the so-called "singular gauge transformation" method.[14] That method is not as mathematically rigorous as the above approach, but possesses the advantage that one can work with a single (singular) potential defined all over space-time, instead of various (regular) potentials which depend on the space-time patch on which they are defined. For this reason, we briefly describe it here.

### 4.1. Introduction of Arbitrary String

In order to be able to write the $(p+1)$-field strength $F$ as the exterior derivative of a potential $A$, one introduces a string of dimension $n-p-2$, whose space-time history is an $(n-p-1)$-surface.[6] The purpose of the string is to take out the unwanted magnetic flux. Hence, one requires that the string be a surface whose only boundary is the magnetic source, so that any $(p+1)$-surface which links the pole crosses it. This leaves a lot of freedom in the choice of the string location, which might even be compact (with boundary) when $n-p\geqslant 1$ and when the pole is compact (without boundary).[10]

One then defines an $(n-p-1)$-form field $G$, entirely localized on the string history (and hence singular) obeying

$$d^*G=(-1)^p * \bar{j} \tag{55}$$

The explicit expression for $^*G$ can be found in Ref. 6 and will not be needed here.

All that is required is the fact that the integral of the $(p+1)$-form $^*G$ over any $(p+1)$-surface is equal to the magnetic charge $\bar{q}$ times the number of times the surface intersects the string, each intersection being counted with the appropriate sign.

[10] The string cannot intersect an electric object ("Dirac veto"). This excludes an interesting possibility, namely the existence of a topologically stable "composite dyon": an object formed by an electric pole and a magnetic one (not necessarily of the same dimension) which do not mutually intersect but are intertwined with a nonzero linking number. However, in such a configuration the electric pole would necessarily cross the Dirac string whose only boundary is the magnetic pole. This argument does not exclude an "elementary dyon," formed by a magnetic (electric) pole whose history is contained in the history of an electric (magnetic) one. Note that in this elementary dyon the electric and magnetic objects do not have to be of the same dimension. In the traditional case $n=4$, $p=1$ this reduces to the usual dyon which is a particle carrying both electric and magnetic charge.

For a given location of the string, one defines (up to a regular gauge transformation) the potential $A$ through

$$F = dA + (-1)^{(p+1)(n+1)} {*G} \tag{56}$$

Thanks to the addition of the $*G$-term in the right-hand side of (56), there is now no contradiction between (41) and $d^2A = 0$.

### 4.2. Displacement of String as Singular Gauge Transformation

Two different string locations I, II lead to two different fields $*G^{\mathrm{I}}$ and $*G^{\mathrm{II}}$ and to two different potentials $A^{\mathrm{I}}$ and $A^{\mathrm{II}}$ related by

$$d(A^{\mathrm{II}} - A^{\mathrm{I}}) = {*G^{\mathrm{II}}} - {*G^{\mathrm{I}}} \tag{57}$$

and

$$d({*G^{\mathrm{II}}} - {*G^{\mathrm{I}}}) = 0 \tag{58}$$

Because $*G^{\mathrm{II}} - *G^{\mathrm{I}}$ has support on the boundary-free surface I + II composed of the combined strings, $A^{\mathrm{II}} - A^{\mathrm{I}}$ is a closed form in the region which excludes the strings. Hence one defines

$$A^{\mathrm{II}} - A^{\mathrm{I}} = d\mu \text{ (away from the strings)} \tag{59}$$

and one thinks of the transition from the first string location to the second one as a gauge transformation. However, because of (57), the $(p-1)$-form $\mu$ cannot be globally defined (it must be "multiple-valued") and one says that the displacement of the strings is generated by a "singular gauge transformation."

The multiple-valuedness of $\mu$ is measured by its integral over $(p-1)$-closed surfaces (which are boundaries of $p$-surfaces). One finds

$$\begin{aligned}\int_{\partial\Sigma_p} \mu &= \int_{\Sigma_p} d\mu \\ &= \int_{\Sigma_p} (A^{\mathrm{II}} - A^{\mathrm{I}}) \end{aligned}\tag{60}$$

Given $\partial\Sigma_p$, there are many $p$-surfaces which have $\partial\Sigma_p$ as boundary. Two different choices $\Sigma_p$ and $\Sigma'_p$ define a closed $p$-surface. If that closed surface links the strings, one gets a nonvanishing result for the integral

$\int_{\Sigma_p - \Sigma'_p} (A^{\mathrm{II}} - A^{\mathrm{I}})$, since it follows from (57) and the definition of $*G^{\mathrm{II}}$ and $*G^{\mathrm{I}}$ that

$$\begin{aligned}\int_{\Sigma_p - \Sigma'_p} (A^{\mathrm{II}} - A^{\mathrm{I}}) &= \int_{\Sigma_{p+1}} d(A^{\mathrm{II}} - A^{\mathrm{I}}) \\ &= \int_{\Sigma_{p+1}} (*G^{\mathrm{II}} - *G^{\mathrm{I}}) \\ &= \bar{q}\,(\text{linking number of strings and } \Sigma_{p+1}) \end{aligned} \tag{61}$$

(with $\partial \Sigma_{p+1} = \Sigma_p - \Sigma'_p$). Accordingly, the multiple-valuedness of the integral (60) is just given by the magnetic charge $\bar{q}$.

### 4.3. Group Element Must Be Single Valued: Quantization of Charge

It is essential in order to interpret a $(p-1)$-form as generating a permissible gauge transformation that the associated $U(1)$ group element

$$\mathrm{rxp}\left[-\frac{iq}{\hbar}\int_{x(\sigma)} \mu\right] \tag{62}$$

be single-valued for any configuration $x^\mu(\sigma)$ of the electric objects. By considering loops $x^\mu(\sigma, \tau)$ in $x^\mu(\sigma)$-space which start and end at $x^\mu(\sigma)$, and which are such that the spanned surface $x^\mu(\sigma, \tau)$ in spacetime is a boundary,[11] one sees that the possible ambiguity in (62) is given by

$$\exp\left[-\frac{iq\bar{q}}{\hbar}\right] \tag{63}$$

The absence of ambiguity in the $U(1)$ gauge transformation (63) leads thus again to the quantization condition (39). It appears accordingly that whichever way one looks at higher-rank monopoles, the quantization condition (39) must be imposed on the magnetic and electric couplings.

[Yet another method was developed in Ref. 6 along Dirac's lines.[(8)] It treats the string as a dynamical object and not as an external parameter in the quantum theory of electric objects in a monopole background. The string location is then a gauge freedom. The corresponding quantum first-class constraints possess nontrivial solutions—i.e., the gauge freedom

[11] Such surfaces can be generated by electric objects which are closed, infinitely extended or, even, possess a boundary (see footnote 7). In the latter cases, one has to keep the "ends" of the electric objects fixed. Hence, the quantization condition manifestly applies to all electric "charges." This remark will find an application in the next section.

inherent in the location of the string is realized quantum mechanically—only if the quantization condition holds.]

## 5. MAGNETIC POLES IN TOPOLOGICALLY MASSIVE ELECTRODYNAMICS

It is well known that if the dimension of spacetime is odd, one can add a topological mass term to the Maxwell equations.[15]

This mass term has a very simple structure when the space-time dimension $n$ is related to the rank $p$ of the potential form $A$ by

$$n = 2p + 1 \tag{64}$$

In this case, to which we shall restrict the discussion, the field equations are linear even after the mass term is included. They read

$$d{}^*F + 2\alpha F = (-1)^p \, {}^*j \tag{65}$$

and

$$dF = (-1)^p \, {}^*\bar{j} \tag{66}$$

where $j$ and $\bar{j}$ are the electric and magnetic "currents," respectively. [When (64) holds, ${}^*F$ is a $p$-form, as is $j$, whereas $\bar{j}$ is a $(p-1)$-form.]

As shown in Ref. 16, Eqs. (65) and (66) derive from a variational principle which is invariant under gauge transformations. Furthermore, the "topological mass" $2\alpha$ must be quantized in order to have a consistent theory.

### 5.1. Magnetic Pole as Source of Electric Charge

The quantization of $\alpha$ can be understood in the light of the previous discussion as follows.

First, one notes that in the field of a monopole of strength $\bar{q}$, all electric charges remain quantized according to

$$q\bar{q} = 2\pi\hbar \text{ (integer)} \tag{67}$$

Indeed, the topological considerations of Secs. 3 and 4, based on the unchanged equation (66), are still valid.

Next, one realizes that the formalism is inconsistent, unless there is an electric source of strength $2\alpha\bar{q}$. This is because (65) and (66) imply

$$d^*j = 2\alpha^*\bar{j} \tag{68}$$

The electric current is not conserved.[12]

This means that there is an electric surface, of coupling strength $2\alpha\bar{q}$, which possesses the magnetic pole as boundary, so as to fulfill (68). [Magnetic objects have one dimension less than electric ones when (64) holds.[13]] That surface is analogous to the Dirac string in its geometric properties, and can be taken to be compact (with boundary!) if the pole is a closed surface. Furthermore, even though it possesses a boundary, it can consistently be coupled to the $A$-field in a way that maintains invariance under (11), (12), since its boundary lies on the monopole.

### 5.2. Topological Mass Is Quantized

Now, as all electric couplings, the electric strength $2\alpha\bar{q}$ must be quantized according to (67). As pointed out in the previous section, this result holds even for surfaces with a boundary. Here, one finds

$$2\alpha\bar{q}^2 = 2\pi\hbar \text{ (integer)} \tag{69}$$

which is the result announced above. [Notice that $q$ has units of length to the power $-\frac{1}{2}$, while $\bar{q}$ has units of length to the power $\frac{1}{2}$; $\alpha$ is a mass and has units of length to the power $-1$.]

### 5.3. Dirac String Is Charged

As a final point, we note that when $\alpha \neq 0$, one can think of the Dirac string—which has the same dimensionality as the electric surfaces—as carrying electric charge. When that additional charge is taken into account, one finds that the total electric current is conserved (see Ref. 16 for the details).

[12] In contrast, $d^*\bar{j} = 0$. There is a strong difference between electric and magnetic sources, because the mass terms only occur in the "electric equations" (65).

[13] When $n \neq 2p+1$, the electric current is also nonconserved, but the equation corresponding to (68) is more complicated. It appears that there is no easy way to solve (68) analogous to the $n = 2p+1$ case, because of a mismatch of dimensions between boundaries of electric objects and magnetic objects.

## ACKNOWLEDGMENT

This work was supported in part by N.S.F. Grant No. PHY-8216715 to The University of Texas at Austin, by research funds of the Center for Theoretical Physics of The University of Texas at Austin, by a grant of the Tinker Foundation to the Centro de Estudios Cientificos de Santiago, and by the Fonds National Belge de la Recherche Scientifique.

## REFERENCES

1. C. Teitelboim, *Phys. Lett. B* **167**, 63 (1986).
2. V. I. Ogievetsky and I. V. Polubarinov, *Sov. J. Nucl. Phys. (Yadernaya Fizika)* **4**, 210 (1968) (in Russian); P. Townsend, K. Pilch and P. van Nieuwenhuizen, *Phys. Lett. B* **139**, 38 (1984); E. Cremmer, B. Julia, and J. Scherk, *Phys. Lett. B* **76**, 409 (1978); D. Z. Freedman and P. K. Townsend, *Nucl. Phys. B* **177**, 282 (1981); J. Thierry-Mieg and Y. Ne'eman, *Proc. Natl. Acad. Sci. USA* **79**, 7068 (1982); E. Cremmer and J. Scherk, *Nucl. Phys. B* **72**, 117 (1974); M. Kalb and P. Ramond, *Phys. Rev. D* **9**, 2273 (1974); Y. Nambu, *Phys. Rep.* **23**, 250 (1976); T. Curtright and P. G. O. Freund, *Nucl. Phys. B* **172**, 431 (1980); J. Schwarz and P. C. West, *Phys. Lett. B* **126**, 301 (1983); P. A. Marchetti and R. Percacci, *Lett. Math. Phys.* **6**, 405 (1982).
3. C. Teitelboim, *Ann. Phys. (N.Y.)* **79**, 542 (1973).
4. T. Regge and C. Teitelboim, in *Proceedings of the First Marcel Grossmann on General Relativity*, R. Ruffini, ed. (North-Holland, Amsterdam), p. 77.
5. S. Coleman, *Phys. Rev. D* **15**, 2929 (1977); A. Vilenkin, *Phys. Rev. D* **27**, 2848 (1983).
6. C. Teitelboim, *Phys. Lett. B* **167**, 69 (1986).
7. G. 't Hooft, *Nucl. Phys. B* **79**, 276 (1974); A. M. Polyakov, *JETP Lett.* **20**, 194 (1974).
8. P. A. M. Dirac, *Phys. Rev.* **74**, 817 (1948).
9. T. T. Wu and C. N. Yang, *Phys. Rev. D* **12**, 3845 (1975); O. Alvarez, *Comm. Math. Phys.* **100**, 279 (1985).
10. Y. Aharonov and D. Bohm, *Phys. Rev.* **115**, 485 (1959).
11. H. Flanders, *Differential Forms* (Academic Press, New York), Chap. 6.
12. S. Coleman, in *Le monopôle magnétique cinquante ans après* R. Stora, ed., *Les Houches 1981* (North-Holland, Amsterdam).
13. T. T. Wu and C. N. Yang, *Nucl. Phys. B* **107**, 365 (1976).
14. P. Goddard and D. Olive, *Rep. Prog. Phys.* **41**, 1357 (1978).
15. W. Siegel, *Nucl. Phys. B* **156**, 135 (1979); J. Schonfeld, *Nucl. Phys. B* **185**, 157 (1981); R. Jackiw and S. Templeton, *Phys. Rev. D* **23**, 2291 (1981); S. Deser, R. Jackiw, and S. Templeton, *Phys. Rev. Lett.* **48**, 975 (1982); *Ann. Phys. (N.Y.)* **140**, 372 (1982); R. Jackiw, in *Les Houches 1983* (North-Holland, Amsterdam); C. Aragone, *Phys. Lett. B* **131**, 69 (1983).
16. M. Henneaux and C. Teitelboim, *Phys. Rev. Lett.* **56**, 689 (1986).

$$-\frac{\nabla^2 \psi}{\left(\delta\, {}^{(3)}\mathcal{G}\right)^2} + {}^{(3)}R\, \psi = 0$$

*Plate 25*

# General Covariance and Quantum Theory

**Bahram Mashhoon**[1]

*Received November 15, 1985*

*The extension of the principle of relativity to general coordinate systems is based on the hypothesis that an accelerated observer is locally equivalent to a hypothetical inertial observer with the same velocity as the noninertial observer. This hypothesis of locality is expected to be valid for classical particle phenomena as well as for classical wave phenomena but only in the short-wavelength approximation. The generally covariant theory is therefore expected to be in conflict with the quantum theory which is based on wave–particle duality. This is explicitly demonstrated for the frequency of electromagnetic radiation measured by a uniformly rotating observer. The standard Doppler formula is shown to be valid only in the geometric optics approximation. A new definition for the frequency is proposed, and the resulting formula for the frequency measured by the rotating observer is shown to be consistent with expectations based on the classical theory of electrons. A tentative quantum theory is developed on the basis of the generalization of the Bohr frequency condition to include accelerated observers. The description of the causal sequence of events is assumed to be independent of the motion of the observer. Furthermore, the quantum hypothesis is supposed to be valid for all observers. The implications of this theory are critically examined. The new formula for frequency, which is still based on the hypothesis of locality, leads to the observation of negative energy quanta by the rotating observer and is therefore in conflict with the quantum theory.*

This paper is based on my attempts to understand wave phenomena in a gravitational field. It would appear at first sight that the general theory of relativity, though purely local, could naturally accommodate wave phenomena. It is the thesis of this paper that this is not so. General relativity provides an elegant and—to the extent that observational data

[1] Institut für Theoretische Physik, Universität zu Köln, D-5000 Cologne 41, Federal Republic of Germany, and Department of Physics and Astronomy, University of Missouri-Columbia, Columbia, Missouri 65211.

are available—correct description of classical particle phenomena in a gravitational field. This is true for classical wave phenomena as well, but only in the short-wavelength ("ray") approximation. Waves are characteristically nonlocal, as opposed to classical test particles and rays which are basically local. These are complementary aspects of a particle, and the dichotomy has been resolved in the quantum theory through the Heisenberg uncertainty relations. The quantization of general relativity would then have to accommodate the principle of complementarity. It is proposed here that since general relativity in its present form cannot accommodate classical wave phenomena in a natural way, it cannot be brought into harmony with the quantum theory. It appears that classical general relativity would have to be modified in order to bring about a reconciliation of the theory of gravitation with the fundamental principles of the quantum theory.

It is a great pleasure for me to dedicate this paper to John A. Wheeler on the occasion of his seventy-fifth birthday.

## 1. INTRODUCTION

Since the development of the quantum theory, many attempts have been made at a quantum theory of the gravitational field. The goal has been to bring about the reconciliation of the general theory of relativity and the quantum theory. It would appear that no formal quantization of the relativistic theory of gravitational field would be successful unless it simultaneously brings about a synthesis of the fundamental principles of quantum mechanics with those of general relativity. In this regard, it has occurred to many physicists that the principle of general covariance poses grave difficulties (see, for instance, the paper of Wigner in Ref. 1 and the remarks of Born in the discussion following it). It is the purpose of this paper to analyze these difficulties in the context of a specific example.

The principle of general covariance is based on the observation that the results of measurements should not depend on the choice of coordinates assigned to events, i.e., observables are scalar invariants. Any local theory can be written in a generally covariant form[(2)]; however, the manner in which this is done has important consequences for the measurement of physical quantities by accelerated observers. This is, in fact, the *physical basis* of the principle of general covariance. That is, general relativity as a consistent theory determines how physical quantities should be measured. In fact, in extending the Poincaré-invariant theory to a generally covariant theory, an underlying hypothesis is employed which determines the results of measurements performed by accelerated observers. Without this prescription, Einstein's principle of equivalence, which is the cornerstone of the theory, would lose its operational significance. That is, local equivalence of gravitational and inertial forces would be useless as a postulate if one does not know what accelerated observers would measure. This basic assumption may be stated as follows:

*Hypothesis of Locality—An accelerated observer is locally equivalent to a hypothetical inertial observer which has the same velocity at the given point in space-time.*

*All observers are accelerated*; therefore, the hypothesis of locality forms the basis for the interpretation of many experiments that constitute the observational foundation of the special theory of relativity (e.g., the Michelson–Morley experiment). In fact, the hypothesis of locality is necessary for the success of the Poincaré-invariant theory. The principle of general covariance follows directly from the hypothesis of locality together with Lorentz invariance: Under arbitrary choices of coordinates, an observer undergoes arbitrary acceleration and is therefore equivalent, in each case, to a member of a set of hypothetical inertial observers with different velocities at the same point in space and time. But these hypothetical inertial observers are physically equivalent by Lorentz invariance. Hence the results of measurements by the observer should be independent of the choice of coordinates. The fundamental significance of the hypothesis of locality was first explicitly recognized by Einstein.

The thesis of this paper is based on the observation that the hypothesis of *locality* involves a length $\mathscr{A}$ which is the smallest length that can be formed from the acceleration of the observer and the speed of light in vacuum ($c$). For example, $\mathscr{A}=c^2/g$ for a linearly accelerated observer with acceleration $g$, while $\mathscr{A}=c/\Omega$ for a rotating observer with angular frequency $\Omega$. It appears plausible that the hypothesis of locality should hold for all phenomena with characteristic lengths that are negligible compared to $\mathscr{A}$. However, there is no reason why this hypothesis should hold for phenomena with characteristic wavelengths $\lambdabar$ comparable or longer than $\mathscr{A}$. One is therefore led to the assumption that the principle of general covariance as well as the general theory of relativity should hold for all phenomena involving "classical particles," which are, by definition, point-like. And the question naturally arises whether the principle of general covariance and the general theory of relativity are valid for "classical wave phenomena" beyond the geometric optics approximation. This question is of central importance for the problem of quantization since quantum theory is based on the idea of wave–particle duality. It is shown in Appendix A that all the experimental tests of the theory involving wave phenomena have verified the theory only in the limit $\lambdabar \ll \mathscr{A}$. It is therefore necessary to examine critically *wave* effects in the generally covariant theory.

In this paper attention is focused on the question: *What is the frequency of electromagnetic waves received on the Earth*? This problem is of fundamental physical interest for the investigation of astronomical phenomena. Theoretically, it involves both the rotation of the Earth and its gravitational field. It will prove interesting to consider a simpler problem first: *What is the frequency of electromagnetic waves measured by a uniformly rotating observer*? The theory of relativity provides unique answers to the questions formulated here. The next section is devoted to the exposition of the relativistic result.

## 2. DOPPLER EFFECT

Consider first two inertial observers $\mathcal{O}$ and $\mathcal{O}'_0$ such that $\mathcal{O}'_0$ moves with velocity $\mathbf{v}$ with respect to $\mathcal{O}$ (considered at rest). A packet of electromagnetic waves can be Fourier analyzed into its elementary components. Observers $\mathcal{O}$ and $\mathcal{O}'_0$ measure the frequency of a monochromatic component and assign to it the values $\omega$ and $\omega'_0$, respectively. The relationship between $\omega$ and $\omega'_0$, i.e., the relativistic Doppler formula, was first developed by Einstein[3] on the basis of the *invariance of the phase* of the wave. The treatment of this problem in terms of 4-vectors is contained in many standard sources. The elementary measurements of the theory are in terms of clocks and light signals and what is measured is a scalar invariant. Thus the measured value of the frequency can be expressed as

$$\text{frequency} = -k \cdot u \tag{1}$$

where $k$ and $u$ are the wave-vector and the observer 4-velocity, respectively. A Cartesian system of coordinates is introduced in the inertial frame and the signature of the Minkowski metric is assumed to be $+2$ throughout this paper. The result is then

$$\omega'_0 = \frac{\omega - \mathbf{v} \cdot \mathbf{k}}{(1-\beta^2)^{1/2}} \tag{2}$$

where $\beta = v/c$ and $\mathbf{k}$ is the wave vector according to the observer $\mathcal{O}$.

Consider now the same situation as before except that the inertial observer $\mathcal{O}'_0$ is replaced by an *accelerated* observer $\mathcal{O}'$. What is the relationship between $\omega$ and the frequency measured by the noninertial observer ($\omega'$)? The answer can be found from the hypothesis of locality which postulates a certain local equivalence between accelerated and inertial observers (Fig. 1). It follows from this hypothesis that

$$\omega' = \frac{\omega - \mathbf{v}(t) \cdot \mathbf{k}}{(1-\beta^2)^{1/2}} \tag{3}$$

Let the accelerated observer rotate uniformly with angular velocity $\Omega$ about the $x^3$-axis of the inertial system of coordinates $(t, x^i; i = 1, 2, 3)$. The magnitude of $\omega'$ can be calculated explicitly as follows: In the inertial frame

$$c\mathbf{k} = \omega(\sin\theta_k \cos\phi_k, \sin\theta_k \sin\phi_k, \cos\theta_k) \tag{4}$$

with respect to the Cartesian frame, and the position of the observer $\mathcal{O}'$ at any given time $t$ is given by

$$x^1 = R\cos(\Omega t + \phi_0), \qquad x^2 = R\sin(\Omega t + \phi_0), \qquad x^3 = z_0 \tag{5}$$

where $R$ is the radius of the circular orbit and $\phi_0$ and $z_0$ are constants. It follows from Eqs. (3)–(5) that

$$\omega' = \gamma\omega[1 + \beta\sin\theta_k \sin(\Omega t + \phi_0 - \phi_k)] \tag{6}$$

where $\gamma = (1-\beta^2)^{-1/2}$. This is the standard result of the generally covariant theory and some of its consequences should be noted: (i) For each value of $\omega$, there is only one value of $\omega'$; (ii) $\omega'$ depends explicitly upon time; (iii) $\omega'$ is independent of the state of polarization of the wave; (iv) $\omega'$ is

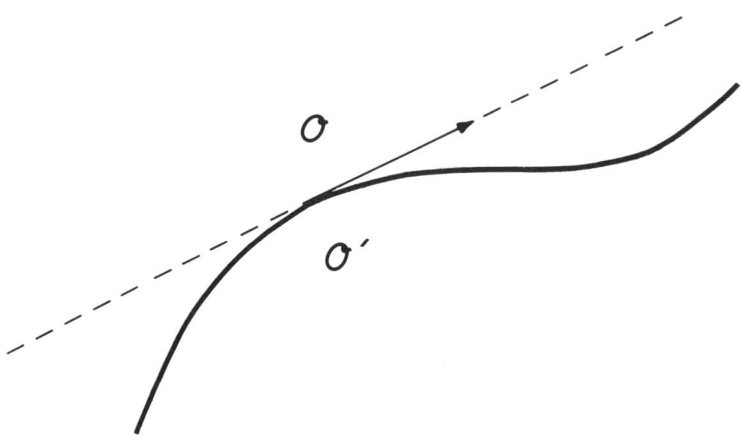


Fig. 1. The hypothesis of locality states that at a given point in space-time an accelerated observer $\mathcal{O}'$ and a hypothetical inertial observer $\mathcal{O}$ with the same velocity as $\mathcal{O}'$ obtain the same results for physical measurements performed at that point.

always positive; (v) for $\theta_k = 0$ (normal incidence), $\omega' = \gamma\omega$ is independent of time and expresses the transverse Doppler effect.

These results, which were obtained from a simple application of the hypothesis of locality, are in conflict with elementary expectations from the classical theory of electrons developed by Lorentz. Before turning to this subject, it is important to note that the aspects of this theory that are needed here are in fact consistent with the quantum theory.

## 3. INTERPRETATION OF THE ZEEMAN EFFECT

The splitting of spectral lines by an external magnetic field was first observed by Zeeman in 1896. According to the *normal* Zeeman effect, for instance, an emission line would split into two lines by an external field if the radiation were observed along a direction parallel to the magnetic field. Thus in the presence of a uniform magnetic field a line of frequency $\omega$ would be split into a right circularly polarized (positive helicity) component of frequency $\omega + \Omega_L$ and a left circularly polarized (negative helicity) component of frequency $\omega - \Omega_L$, where $\Omega_L > 0$ is proportional to the strength of the magnetic field.

The normal Zeeman effect was explained by Lorentz in 1897 on the basis of the classical theory of electrons.[(4)] The motion of the electrons in the magnetic field is described by the Lorentz force law. The radiation emitted by the perturbed electrons along the direction of the external magnetic field ($B$) consists of two circularly polarized components with frequencies $\omega \pm eB/(2m_e c)$, just as observed by Zeeman in his early experiments. In fact, Zeeman's discovery led to the determination of the charge ($-e$) to mass ($m_e$) ratio for the electron.

The results of the Lorentz theory may be interpreted in a simple way (Larmor, 1899) by using Larmor's theorem (1897) which, for the present purposes, may be stated as follows: The classical motion of a particle of mass $m$ and charge $q$ in a weak magnetic field $\mathbf{B}$ is equivalent to the motion of the particle with respect to a coordinate system rotating uniformly with angular frequency $\mathbf{\Omega}_L = q\mathbf{B}/(2mc)$. The significance of Larmor's theorem lies in the fact that it relates the measurements of an *inertial observer* in the presence of a weak magnetic field to those of a *uniformly rotating observer*. Let the magnetic field be along the $x^3$-axis, so that the Larmor frequency of the noninertial observer is in the negative $x^3$-direction with magnitude $\Omega_L = eB/(2m_e c)$. Electromagnetic radiation of frequency $\omega$ traveling along the $x^3$-axis can be decomposed into two components each of frequency $\omega$ and with opposite states of circular polarization. The rotating observer will therefore measure two components with circular polarization: a positive helicity component of frequency $\omega + \Omega_L$ and a negative helicity component of frequency $\omega - \Omega_L$. The inverse process is also expected to hold, and this would be consistent with the reciprocity between the emission and absorption of radiation. An interesting historical account of these researches is contained in Ref. 5.

The key result of this interpretation of the Zeeman effect can now be stated as follows: An observer rotating with frequency $\Omega$ about the direction of propagation of a circularly polarized electromagnetic wave of frequency $\omega$ will measure a frequency $\omega - \Omega$ if the wave is right circularly polarized and a frequency $\omega + \Omega$ if the wave is left circularly polarized. These results follow essentially from the elementary addition of angular frequencies since for a circularly polarized wave the electric and magnetic fields are rotating in the same (opposite) sense as the propagation of the wave if the wave is right (left) circularly polarized. It is therefore clear that the frequency "measured" by the rotating observer is given here in units of the proper time of the inertial observer. If the proper time of the accelerated observer is used instead, the observed frequency becomes

$$\omega' = \frac{\omega \mp \Omega}{(1-\beta^2)^{1/2}} \tag{7}$$

where $c\beta$ is the speed of the rotating observer. For $\beta \ll 1$, or along the axis of rotation, Eq. (7) reduces to the Newtonian result. The new result must be compared and contrasted with the transverse Doppler effect: Since

$$\omega' = \gamma\omega(1 \mp \lambda\!\!\!^{-}\Omega/c) \tag{8}$$

it is clear that the new formula reduces to the transverse Doppler formula for $\Omega/\omega = \lambda\!\!\!^{-}/\mathscr{A} \to 0$, where $\mathscr{A} = c/\Omega$ in this case. Hence the new result reduces to the transverse Doppler formula in the geometric optics limit. A hypothetical inertial observer at the same space-time point and with the same velocity as the rotating observer does *not* measure the frequency given by Eq. (7), and the hypothesis of locality is violated. This conflict can be resolved by a modification of the theory which would go beyond the hypothesis of locality, but would still be consistent with the invariance of an observable under arbitrary coordinate transformations.

## 4. NEW DEFINITION OF FREQUENCY

*The frequency of electromagnetic radiation is basically inferred from the manifestations of periodic behavior it induces in the mechanical motion of charged particles.* In general relativity, the acceleration of a charged particle in an external electromagnetic field is given by the covariant form of the Lorentz force law. Hence this notion of frequency may be naturally incorporated into the framework of the generally covariant theory by assuming that the observed frequencies correspond to the Fourier components of the *measured* values of the electromagnetic field tensor. More specifically, let $\lambda^{\mu}_{(\alpha)}$ be a tetrad frame associated with the motion of an observer and let $F_{(\alpha)(\beta)} = F_{\mu\nu}\lambda^{\mu}_{(\alpha)}\lambda^{\nu}_{(\beta)}$ be the measured components of the field. Greek indices run from 0 to 3 throughout this paper. Let $F_{(\alpha)(\beta)}$ be expressed as a function of $\tau$, the proper time along the path of the observer. The frequencies associated with the Fourier resolution of the resulting function will be defined to be the observed frequencies. This definition is manifestly covariant and it appears to be consistent with the theoretical framework of Einstein's theory.

The new definition of frequency incorporates a feature that is absent from the standard general relativistic theory of measurement (based on the hypothesis of locality): The number of values assigned to a physical quantity depends on the observer. Thus according to the Doppler formula only a single value is assigned to the frequency of a ray for each observer, while the more general formula (7) for the transverse Doppler effect does not have this property: For each frequency measured by the inertial observer, there are in general two frequencies measured by the rotating observer.

It is interesting to consider again the problem of the relationship between $\omega$ and $\omega'$ as measured by observers $\mathcal{O}$ and $\mathcal{O}'$ (cf. Sec. 2) on the basis of the definition of frequency formulated above. The mathematical analysis is given in Appendix B. It is shown there that the frequency observed in the rotating frame is

$$\omega' = \frac{\omega - M\Omega}{(1-\beta^2)^{1/2}} \tag{9}$$

where $M$ is the angular momentum parameter along the axis of rotation and it can take on the values $M = 0, \pm 1, \pm 2, \ldots$. In Eq. (9), $\beta = v/c$, where $\mathbf{v} = \mathbf{\Omega} \times \mathbf{r}$ and $\mathbf{r}$ is the position vector of the rotating observer in the inertial frame. Equation (7) is recovered for an electromagnetic wave incident along the axis of rotation. It is important to compare these results with those of the standard theory: (i) For each value of $\omega$, there is in general a countably infinite number of values for $\omega'$; (ii) this set of values for $\omega'$ is independent of time; (iii) $\omega'$ depends explicitly on the angular momentum of the radiation field; (iv) $\omega'$ can be zero or negative; (v) for the case of normal incidence, $\omega' = \gamma(\omega \mp \Omega)$ is a generalization of the transverse Doppler formula consistent with expectations from the classical theory of electrons.

The angular momentum of the electromagnetic radiation is a global property of the field and this is consistent with the nonlocal character of the frequency defined above. For a scalar field of radiation, the application of the above method implies that

$$\omega' = \frac{\omega - \mu\Omega}{(1-\beta^2)^{1/2}} \tag{10}$$

where $\mu = 0, \pm 1, \pm 2, \ldots$, has the same interpretation as before. However, for radiation incident along the direction of rotation we have $\omega' = \gamma\omega$ just as in the transverse Doppler effect. In the geometric optics limit $\mu\Omega = \boldsymbol{l} \cdot \boldsymbol{\Omega}$, where $\boldsymbol{l}$ is proportional to the orbital angular momentum of the ray with wave vector $\mathbf{k}$, $\boldsymbol{l} \approx \mathbf{r} \times \mathbf{k}$, and $\mathbf{r}$ indicates the position vector of the ray. Thus $\boldsymbol{l} \cdot \boldsymbol{\Omega} \approx \mathbf{v} \cdot \mathbf{k}$ and Eq. (10) reduces to $\omega' = \gamma(\omega - \mathbf{v} \cdot \mathbf{k})$, which is identical with the Doppler result. Hence the new definition of frequency reduces to the standard Doppler principle in the geometric optics approximation. This fact explains the time dependence of the *Doppler* frequency: The waves interfere constructively along a ray reaching the observer. Thus the measured frequency of the ray would depend on the velocity of the observer and the direction of the ray.

## 5. DISCUSSION OF THE CLASSICAL THEORY

It has been shown thus far—by means of a specific difficulty regarding the measured value of frequency according to accelerated observers—that the standard generally covariant theory based on the hypothesis of locality is applicable to wave phenomena only in the short-wavelength (i.e., ray) approximation. In the framework of general relativity, where curvature invariants already provide a covariant length scale, the geometric optics approximation must be interpreted in a rather extreme form: That electromagnetic rays follow *null geodesics* is a consequence of Maxwell's equations only in the $\lambda \to 0$ limit.

In the context of the classical theory a new definition of frequency has been introduced which goes beyond the hypothesis of locality and which appears to be satisfactory from the point of view of actual measurements in the laboratory. It is also consistent with intuitive expectations from the classical theory of electrons. The new definition of frequency, although nonlocal, can still be combined with invariance under general coordinate transformations. It must be emphasized that the hypothesis of locality has not been banished from the theory in this way since it still forms the physical basis of writing Maxwell's equations in generally covariant form:

$F_{\mu\nu}\, dx^{\mu} \wedge dx^{\nu}$ is an invariant so that the hypothesis of locality is still at work in the interpretation of the measurable components of the electromagnetic field. The nonlocality of the new definition of frequency arises from the Fourier expansion of these components over *proper time*, so that the nonlocality is temporal (rather than spatial). Thus although the new definition of frequency is expected to be generally valid, the actual formula for frequency is still implicitly based on the hypothesis of locality (as applied to the electromagnetic field).

A significant feature of the new formula for frequency is that it can be zero when $\omega/\Omega$ is an integer, and that the measured frequency can be negative if $\omega < M\Omega$. Notice that negative frequencies can occur for high enough multipoles even when $\omega \gg \Omega$. In the classical theory negative wave frequency may be interpreted as propagation in the opposite direction. For example, consider right circularly polarized electromagnetic waves propagating along the axis of rotation with frequency $\omega \leqslant \Omega$. For $\omega = \Omega$, the observer measures a constant field and hence $\omega' = 0$. For $\omega < \Omega$, the rotating observer can interpret the result of measurement of frequency as follows: The frequency is positive and equal to $\gamma(-\omega + \Omega)$, but the wave is propagating in the opposite direction. Thus the problem here is precisely as in the prerelativistic Doppler formula which is based on the Newtonian addition of linear velocities just as Eq. (7) is due to the addition of angular frequencies. The relativistic result that an observer traveling with a beam of electromagnetic radiation could never be at rest with the wave is contradicted here for an accelerated observer. This is the case, for instance, if the observer rotates in the same sense and at the same rate as the electric and magnetic fields in a right circularly polarized beam. Hence the new results must be regarded as approximate if the postulate that the speed of light is independent of the motion of the observer is found to be generally valid.

With these reservations in mind, one may conclude that the fact that frequency can become negative in Eqs. (9) and (10) does not by itself contradict the fundamental laws of classical physics. That is, at the purely theoretical level, these results do not point to a basic conflict between the requirement of general covariance and the laws of electrodynamics. This is not, however, the case with respect to quantum physics where the negative frequency of radiation would imply that the photon energy (as measured by an accelerated observer) would be negative. This subject is treated in detail in the following section.

## 6. QUANTUM THEORY OF THE SPIN-ROTATION COUPLING

Consider a system $S$ in an inertial reference frame and an observer rotating with respect to the inertial frame with angular velocity $\Omega$. Let the system undergo a transition from an energy state $E_2$ to a lower energy state $E_1$ by the emission of a photon of frequency $\omega$,

$$\hbar\omega = E_2 - E_1 \tag{11}$$

It will be assumed that the states of the system are eigenstates of the component of the total angular momentum operator along the direction of rotation of the accelerated observer. Let the corresponding eigenvalues be $J_2$ and $J_1$. Thus the photon carries an angular momentum component equal to $\hbar M$, where

$$\hbar M = J_2 - J_1 \tag{12}$$

Furthermore, it will be assumed that the rotating observer performs measurements on the system before and after the transition. Let the rotating observer assign the values $E_2'$ and $E_1'$ to the energies of the system before and after the emission, respectively. If the rotating observer detects the photon, what is the value of the frequency that this observer assigns to the photon? To answer this question, the following postulate will be introduced:

*Generalized Bohr Frequency Condition—The frequency as measured by the noninertial observer is given by $\hbar\omega' = E_2' - E_1'$, where $E_2'$ and $E_1'$ are the measured values of the energies of the two levels according to the accelerated observer.*

This postulate reduces to the Bohr frequency condition in the limit of vanishing acceleration. Its validity is based on energy conservation and on the requirement that the concept of light *quantum* be independent of the motion of the observer. Furthermore, there is the implicit assumption that causality is preserved in the transformation from the inertial to the accelerated observer.

It now remains to determine $E'$. This could be done in two different, but essentially equivalent, ways:

(a) Using the correspondence principle, the relationship between $E'$ and $E$ must be the same as in classical mechanics in the limit of large quantum numbers. Let $H$ and $H'$ denote the Hamiltonians for the inertial and rotating observers, respectively. It is a well-known result of classical mechanics that

$$H' = H - \mathbf{\Omega} \cdot \mathbf{J} \tag{13}$$

where **J** is the (invariant) angular momentum of the system.[6] In a stationary state the time dependence is of the form $\exp(-iEt/\hbar)$; therefore, the energy $E'$ in terms of *proper time* is given by

$$E' = \gamma(E - \mathbf{\Omega} \cdot \mathbf{J}) \tag{14}$$

in the quasi-classical limit. It follows from the generalized Bohr frequency condition that

$$\omega' = \gamma(\omega - M\Omega) \tag{15}$$

as before. It is clear from this derivation that Eq. (15) holds quite generally for any radiation field independently of the helicity of the particle since the nature of the photon does not enter this analysis.

(b) Alternatively, Eq. (14) can be derived from the quantum mechanical version of the hypothesis of locality, namely, the equality of probabilities for the accelerated and inertial observers at the same space-time event. Specifically, let $\psi'$ and $\psi$ denote the amplitudes of a state of the system $S$ according to the rotating and inertial observers, respectively. According to the hypothesis of locality, the relationship between $\psi'$ and $\psi$ is the same as that between the amplitudes determined by two *inertial* observers in two systems of coordinates one rotated with respect to the other (in the passive sense) by the same amount as with the frames attached to $\mathcal{O}'$ and $\mathcal{O}$. If this angle is $\Omega t$, then according to this hypothesis

$$\psi' = \exp(it\mathbf{\Omega} \cdot \mathbf{J}/\hbar)\, \psi \tag{16}$$

up to a phase factor, where **J** is the total angular momentum operator of the system. This hypothesis, which uses the isotropy of space for inertial observers, is the starting point of the standard treatment of uniformly rotating observers in nonrelativistic quantum mechanics.[7] In this way one recovers the same result as before, namely, that the energy as determined by the rotating observer in terms of proper time is given by Eq. (14). This result implies that the energy of the photon, as measured by a rotating observer, is influenced by its angular momentum. More specifically, for the case of normal incidence, the intrinsic spin of the photon is involved. These results apply generally to any field of radiation.

It follows from these considerations that the hypothesis of locality leads directly to the conclusion that the frequency (and energy) of a photon as measured by a rotating observer could be negative. This appears to contradict one's intuitive notions concerning a quantum of energy. To be sure, the hypothesis that the exchange of energy takes place in quanta of *positive energy* $\hbar\omega$ was formulated implicitly in terms of inertial observers. However, all actual observers are accelerated. In particular, observers on the Earth are rotating, yet no one has ever detected a negative-energy photon. In view of this, the following hypothesis may be formulated:

*The quantum hypothesis is valid for all observers regardless of their states of motion.*

Thus when a rotating observer states that a quantum of electromagnetic energy has been detected, then according to the above hypothesis the quantum must have positive energy (as measured by the rotating observer). It is in this sense that while Eq. (15) has a wider range of validity than the usual Doppler formula, it is still considered defective since it leads to negative energy quanta. In this way the hypothesis of locality—without which the principle of general covariance cannot be formulated—is in conflict with the quantum theory.

It must be emphasized that this conclusion is based on the assumption that *the description of the causal sequence of events is independent of the motion of the observer.* Suppose that an inertial observer reports the emission of a photon by an atom and the subsequent absorption of the photon by an accelerated detector. The accelerated observer (comoving with the detector) is expected to describe the same sequence of events. Alternatively, for $E'_2 < E'_1$ the accelerated observer may report the emission of a photon by the detector and the absorption of the photon by the atom. In the classical theory, the negative frequency of radiation may be interpreted as indicating propagation of positive-frequency radiation in the opposite sense; similarly, in the quantum theory one may interpret absorption of a negative-energy photon as being tantamount to emission of a positive-energy photon. This would imply a breakdown of causality for $E'_2 < E'_1$ in the transformation from the inertial to the accelerated observer. However, one would expect that such a violation of causality, if it were to occur, would be independent of the details of the processes under observation.

It would be of great interest to subject Eq. (15) to experimental test. Consider this equation in the form

$$\omega' = \gamma\omega(1 - M\Omega/\omega) \tag{17}$$

one might speculate that a more general formula might include terms quadratic, etc., in $\Omega/\omega$ and thereby avoid the difficulty regarding negative frequencies associated with Eq. (17). For a disk of radius $R = 10$ cm rotating at a rate of a thousand rounds per second, $\beta \simeq 2 \times 10^{-6}$ at the rim and $(\Omega/\omega)^2 \simeq 10^{-15}$ for radiation of wavelength $\lambda \simeq 1$ cm. It therefore appears that deviations from Eq. (17) would be at the limit of observability with the present techniques using highly stable frequency standards.

If it is determined that Eq. (17) is indeed defective, the hypothesis of locality would have to be completely abandoned except in the short-wavelength approximation. This would have far-reaching consequences for the theory of gravitation since the interpretation of measurements performed by observers in a gravitational field is inextricably connected—via the principle of equivalence—with the description of phenomena according to accelerated observers. The validity of the principle of equivalence for general wave phenomena would be severely restricted and classical general relativity would have to be extensively modified in order to bring it into harmony with the quantum theory.

## APPENDIX A

The experimental tests of the general theory of relativity involving wave effects are all based on phenomena with wavelengths $\lambdabar \ll \mathscr{A}$, where $\mathscr{A}$ is a characteristic length scale related to the acceleration of gravity. For the various solar system tests of the theory, let $g = G\mathscr{M}/\mathscr{L}^2$ be the acceleration of gravity at the surface of a body with mass $\mathscr{M}$ and radius $\mathscr{L}$. It follows that $\mathscr{A} = \mathscr{L}/(G\mathscr{M}/c^2\mathscr{L}) \gg \mathscr{L}$. For instance, in the case of the acceleration of gravity at the surface of the Earth $\mathscr{A} \simeq 1$ light-year. It is therefore evident that the relation $\lambdabar \ll \mathscr{A}$ is satisfied for various experiments, such as the measurement of the bending of electromagnetic waves in the field of the Sun (as well as the gravitational lens effect in cosmology), radar time-delay measurements, and gravitational redshift observations.

In the precise tests of the equality of gravitational and inertial masses, the relevant intrinsic length scales are de Broglie wavelengths of the particles involved. That is, a momentum $p_g = \hbar g/c^2$ may be ascribed to a field of gravitational acceleration $g$. All particles with momenta $p \gg p_g$ fall with the same acceleration. Any deviation from this behavior would be expected to be of order $p_g/p$ which is too small to be of interest at present in connection with the tests of the principle of equivalence.

Observations of the binary pulsar PSR 1913+16 have provided indirect evidence for the existence of gravitational waves. The observed rate of period decay for this system compared to the prediction of general relativity on the basis of the standard quadrupole formula is[8] $1.00 \pm 0.04$.

If $P$ and $\varepsilon$ are the binary period and the orbital eccentricity, respectively, the dominant wavelength of the emitted gravitational waves is expected to be

$$\lambda_g \simeq \tfrac{1}{2}(1-\varepsilon)^{3/2}\, cP \tag{A1}$$

It follows from this relation that $\lambdabar_g/\mathscr{A} \simeq 10^{-4}$ for the binary pulsar and any deviation of this form from the standard general relativistic result would probably be too small to be detectable at present.

Thus all experiments involving wave phenomena that provide observational support for the general theory of relativity have tested the theory in the short-wavelength approximation. The possibility of testing general relativity beyond the geometric optics limit has been suggested previously[9,10] in connection with the phenomenon of differential gravitational deflection of electromagnetic radiation by a rotating mass.

## APPENDIX B

Let $F_{\mu\nu}(x)$ be the components of the electromagnetic field tensor as measured by static observers in the Cartesian space. An observer moving with velocity $\lambda^{\mu}_{(0)}$ measures the components of this tensor with respect to a frame $\lambda^{\mu}_{(i)}$, $i=1, 2, 3$, where $\lambda^{\mu}_{(\alpha)}$, $\alpha=0, 1, 2, 3$, constitute an orthonormal tetrad frame that is the reference frame of the moving observer. The measured components of the field are given by $F_{(\alpha)(\beta)} = F_{\mu\nu}\lambda^{\mu}_{(\alpha)}\lambda^{\nu}_{(\beta)}$. Alternatively, let the coordinate transformation from the Cartesian system to the rest system of the moving observer be given by $x \to x'(x)$; then the components of the tensor in the moving system are

$$F'^{\mu\nu}(x') = \frac{\partial x'^{\mu}}{\partial x^{\alpha}}\frac{\partial x'^{\nu}}{\partial x^{\beta}} F^{\alpha\beta}(x) \tag{B1}$$

The observed frequencies of the electromagnetic radiation in a frame that is stationary with respect to the moving system correspond to the frequencies of the Fourier components of the field tensor when expressed in terms of the proper time of the moving observer.

Consider a packet of electromagnetic radiation in the inertial frame characterized by the vector potential $A_{\mu}(x)$. A gauge may be chosen such that $A_0(x)=0$. Let

$$\mathbf{A}(x) = \mathrm{Re} \sum_{\mathbf{k}} \mathbf{a}(\mathbf{k})\, e^{i(\mathbf{k}\cdot\mathbf{r} - \omega t)} \tag{B2}$$

where $\omega = |\mathbf{k}| c$ is the frequency of each Fourier component as measured in the inertial frame, and $\mathbf{k} \cdot \mathbf{a}(\mathbf{k}) = 0$. It follows from the expansion of a plane wave in terms of spherical waves that $(c = 1)$

$$\mathbf{A}(x) = 4\pi \operatorname{Re} \sum_{\mathbf{k}JlM} i^l [\mathbf{Y}^*_{JlM}(\hat{\mathbf{k}}) \cdot \mathbf{a}(\mathbf{k})] \, j_l(\omega r) \, \mathbf{Y}_{JlM}(\hat{\mathbf{r}}) \, e^{-i\omega t} \tag{B3}$$

where $j_l$, $l = 0, 1, 2, \ldots$, are spherical Bessel functions, $\hat{\mathbf{r}}$ is the unit vector indicating position $\mathbf{r} = r\hat{\mathbf{r}}$, $J = 1, 2, 3, \ldots$, and $M = 0, \pm 1, \pm 2, \ldots$.

The transformation of coordinates to the rest system of the rotating observer is given by

$$t' = t \tag{B4}$$

$$x'^1 = x^1 \cos \Omega t + x^2 \sin \Omega t \tag{B5}$$

$$x'^2 = -x^1 \sin \Omega t + x^2 \cos \Omega t \tag{B6}$$

$$x'^3 = x^3 \tag{B7}$$

For polar coordinates $(r, \theta, \phi)$, the spatial transformations amount to $r \to r'$, $\theta \to \theta'$, and $\phi \to \phi' + \Omega t$. It follows that

$$\mathbf{Y}_{JlM}(\theta, \phi) \to e^{iM\Omega t} \mathbf{Y}_{JlM}(\theta', \phi') \tag{B8}$$

hence the vector potential in the rotating system is given by $A'_0(x') = 0$ and

$$\mathbf{A}'(x') = 4\pi \operatorname{Re} \sum_{\mathbf{k}JlM} i^l [\mathbf{Y}^*_{JlM}(\hat{\mathbf{k}}) \cdot \mathbf{a}(\mathbf{k})] \, j_l(\omega r) \, \mathbf{Y}_{JlM}(\hat{\mathbf{r}}') \, e^{-i(\omega - M\Omega)t} \tag{B9}$$

In terms of proper time $\tau$, $dt = \gamma \, d\tau$, the frequencies observed by the rotating observer are $\omega' = \gamma(\omega - M\Omega)$ for $M = 0, \pm 1, \pm 2, \ldots$.

This general result for the observed frequencies reduces to $\omega' = \gamma(\omega \mp \Omega)$ for radiation propagating along the $x^3$-axis. It is interesting to derive this formula in detail in order to see explicitly the contribution of the vectorial character of the electromagnetic radiation. This should be contrasted with the Doppler formula which is a consequence of the invariance of the *phase*, and hence independent of the nature of the radiation field. Consider circularly polarized radiation of frequency $\omega$ propagating along the $x^3$-axis with

$$\mathbf{A} = \operatorname{Re}[a_0(\hat{\mathbf{x}}^1 \pm i\hat{\mathbf{x}}^2) \, e^{i\omega(x^3 - t)}] \tag{B10}$$

where $a_0$ is a constant. The transformation law for the vector potential implies that

$$A'^1 = A^1 \cos \Omega t + A^2 \sin \Omega t \tag{B11}$$

$$A'^2 = -A^1 \sin \Omega t + A^2 \cos \Omega t \tag{B12}$$

$$A'^3 = A^3 \tag{B13}$$

so that according to the rotating observer the vector potential is

$$\mathbf{A}' = \mathrm{Re}[a_0(\hat{\mathbf{x}}'^1 \pm i\hat{\mathbf{x}}'^2)\, e^{i\omega x'^3 - i(\omega \mp \Omega)t'}] \tag{B14}$$

Thus the frequency of the radiation measured by the rotating observer is $\omega' = \gamma(\omega \mp \Omega)$, where the upper (lower) sign refers to right (left) circularly polarized radiation.

## REFERENCES


1. E. Wigner, in *Jubilee of Relativity Theory* (Bern, 1955), *Helv. Phys. Acta*, Suppl. IV (Birkhäuser Verlag, Basel, 1956), pp. 210–226.
2. E. Kretschmann, *Ann. Phys. (Leipzig)* **53**, 575 (1917).
3. A. Einstein, *Ann. Phys. (Leipzig)* **17**, 891 (1905).
4. H. A. Lorentz, *The Theory of Electrons* (2nd edn., 1915) (Dover, New York, 1952).
5. E. T. Whittaker, *A History of the Theories of Aether and Electricity* (Nelson, London, 1951), Vol. 1, Chap. XIII.
6. L. D. Landau and E. M. Lifshitz, *Mechanics* (Pergamon Press, Oxford, 1960), Chap. VI.
7. I. I. Rabi, N. F. Ramsey, and J. Schwinger, *Rev. Mod. Phys.* **26**, 167 (1954).
8. J. M. Weisberg and J. H. Taylor, *Phys. Rev. Lett.* **52**, 1348 (1984).
9. B. Mashhoon, *Nature (London)* **250**, 316 (1974).
10. B. Mashhoon, *Phys. Rev. D* **11**, 2679 (1975).

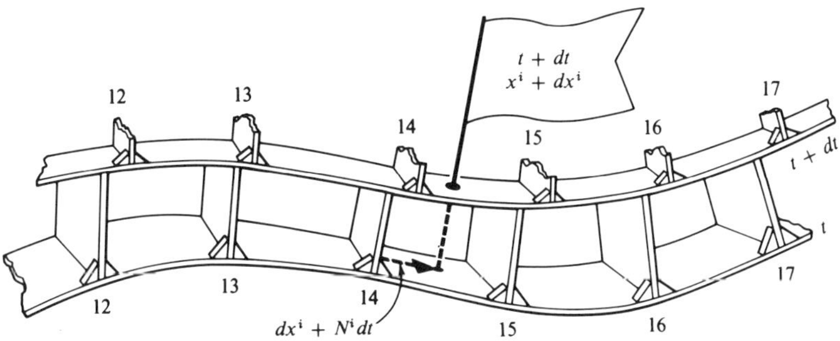


*Plate 26*

# Steering the Universe

**James Isenberg**[1]

*Received January 17, 1986*

*We discuss the present status of the Cauchy problem for Einstein's classical field theory of gravity.*

"These are the handles that steer the universe!" This quote appears on a scrap of paper, dated October, 1972, which I have saved. Written in John Wheeler's usual bold black fountain pen script, partly covering the equations which it is ennobling, this quote reminds me of the tremendously stimulating weekly meetings I had with John back then during my last years as an undergraduate at Princeton University. I always came out of those sessions stuffed with questions to think about (many times more questions than I came in with) yet believing that I was almost ready to put my hands on those handles. This simultaneous feeling of strong understanding yet deep puzzlement which John so often instilled in me and his other students is the product of his unique style of teaching. I am most pleased to see that, approaching age 75, John Wheeler is stimulating a new crop of students with this same wonderful gift. The physics community, as well as the inquiring public as a whole, is very much indebted to John Wheeler.

What exactly are those "handles" that John was telling me about back then, and how do they "steer" the universe? Roughly speaking, the handles are the freely specifiable conformal initial data out of which one can construct space-time model solutions of Einstein's equations. They "steer" the universe in the sense that, by choosing them, one determines mathematically what the model universe will be like, physically. The study of exactly how this works is called the Cauchy (or initial value) problem of

[1] Department of Mathematics, University of Oregon, Eugene, Oregon 97403.

general relativity. John Wheeler has always been a strong proponent of this approach to studying gravity. Thus, in this short article, I wish to briefly review the status of the Cauchy problem of general relativity, pointing out its successes (with emphasis on the recent developments) and discussing some questions which remain open.

## 1. A WELL-POSED CAUCHY PROBLEM

Not all systems of partial different equations (covariantly written or not) can be formulated as Cauchy problems which have solutions. For example, while the standard Klein–Gordon system[2]

$$\Box_\mu \Box^\mu \varphi = 0 \tag{1}$$

admits a Cauchy formulation, the similar system

$$(1-\varphi^2)\,\Box_\mu \Box^\mu \varphi = 0 \tag{2}$$

does not (since clearly the future and past evolution of any data with $\varphi$ set equal to 1 is not determined.) Systems which do admit a Cauchy formulation are said to be "well-posed."

For a complicated system like the Einstein field equations, determining well-posedness is not simple. Fortunately, at least for the vacuum Einstein equations, the issue was settled a number of years ago.[(2)] The vacuum Einstein system *is* well-posed; and in the course of the proof one shows not only that proper[3] initial data can be evolved into the future and the past, but one also shows that the evolved space-time solutions (known as the "space-time developments" of the data) have a well-defined local causal cone structure.

What about the Einstein field equations with other fields coupled in? In most physically interesting cases, the proof of well-posedness for the vacuum case extends to the coupled (nonvacuum) system. This was demonstrated years ago for the Einstein–Maxwell, Einstein–Yang–Mills, Einstein–Klein–Gordon, and Einstein–fluid field equations. Much of this work is summarized in Ref. 4. Recent results show that in spite of certain difficulties encountered in dealing with "classical fermion" fields, the stan-

[2] Our notation conventions generally follow those of Ref. 1. Note in particular that Greek indices $\mu$ run from 0 to 3 and indicate space-time tensors, while Latin indices $a$ run from 1 to 3 and indicate spatial tensors. In addition, we use $\Box_\mu$ to denote the space-time covariant derivative while $\nabla_a$ denotes the spatial covariant derivative.

[3] By "proper" initial data, we mean data lying in appropriate function spaces. See Ref. 2 and also the work of Hughes, Kato, and Marsden.[(3)]

dard models for supergravity are well-posed as well.[5] Note that in the course of examining supergravity, we have verified that Einstein's theory is well-posed in the tetrad-connection formulation as well as in the usual metric language.

While well-posedness guarantees that there is a space-time development of appropriate initial data, it does not guarantee that the development will last very long. This is not surprising, especially in view of the well-known proclivity of space-times to develop singularities. For some systems, like the Yang–Mills equations in Minkowski background, one can prove long-time ("global") as well as short-time ("local") existence of solutions.[6] But this important issue is unresolved—indeed not yet very well formulated—for the Einstein system. It ties in with the problem of "cosmic censorship" and the question of the nature of the boundary of space-time developments. There has been some progress recently in these areas, and we shall return to consider them in Sec. 6.

## 2. DYNAMICAL FORMULATION

To exploit the Cauchy problem of a well-posed space-time covariant system of field equations, it is useful to cast the system into dynamical (or "$3+1$") form. Specifically, one wishes to re-express the space-time covariant fields in terms of spatially covariant fields which are time-dependent, and in addition one wishes to rewrite the space-time covariant field equations as evolution equations governing the time-development of the fields together perhaps with constraint equations restricting the choice of the fields as initial data.

For general relativistic field theories, the dynamical formulation is somewhat complicated by the lack of a fixed background space-time with preferred families of equal time slices. However, following the work of Wheeler,[7] Arnowitt *et al.*,[8] and others, one easily deals with these complications and obtains the following set up:

Base manifold:

$\Sigma^3$ three-dimensional Hausdorff orientable parallelizable $C^3$ manifold

Dynamical fields:

$\gamma_{ab}$ Riemannian metric on $\Sigma^3$

$K^c_d$ tensor field on $\Sigma^3$ (with $\gamma_{ac}k^c_d = \gamma_{dc}k^c_a$)

Atlas fields:

$N$ scalar field on $\Sigma^3$

$M^a$ vector field on $\Sigma^3$

Constraint equations:

$$\nabla_a K^a{}_b - \nabla_b(\mathrm{tr}\, K) = 0 \tag{3}$$

$$R + (\mathrm{tr}\, K)^2 - K^m{}_n K^n{}_m = 0 \tag{4}$$

Evolution equations:

$$(d/dt)\,\gamma_{ab} = -2NK_{ab} + \mathfrak{L}_M \gamma_{ab} \tag{5}$$

$$(d/dt)\,K^c{}_d = N(R^c{}_d + (\mathrm{tr}\, K)\, K^c{}_d) - \nabla^c \nabla_d N + \mathfrak{L}_M k^c{}_d \tag{6}$$

(Here $\mathfrak{L}_M$ is the Lie derivative operator along the vector field $M$.)

To build a space-time using this setup, one proceeds as follows:

(a) One chooses $\Sigma^3$, noting that the manifold of the generated space-time will necessarily be $M^4 = \Sigma^3 \times \mathbb{R}^1$.

(b) One picks initial data $(\gamma_{ab}, K^c{}_d)$—for time $t=0$—on $\Sigma^3$, requiring that this data satisfy the constraint equations (3) and (4). In the constructed space-time, $\gamma_{ab}$ will be the intrinsic geometry of the $t=0$ hypersurface, and $K^c{}_d$ will be its extrinsic geometry (second fundamental form).

(c) One freely chooses the atlas fields $N$ (the "lapse") and $M^c$ (the "shift") for time $t=0$ as well as for all time $t$ future and past. Note that the evolution of $N$ and $M^c$ is *not* determined by the system; nor do these fields appear in the constraints. Indeed, the free choice of the lapse $N(t)$ determines where (in proper time) the constant time slices will be in the constructed space-time, and the free choice of the shift $M^c(t)$ determines what the coordinate charts will be (hence the name "atlas fields").

(d) One evolves $\gamma^{ab}(t)$ and $K^c{}_d(t)$ into the future and the past using the evolution equations (5) and (6) (in which the chosen functions $N(t)$ and $M^c(t)$ appear).

(e) One sets

$$g = \gamma_{ab}\,dx^a\,dx^b - N^2(dt + M_a dx^a)(dt + M_b dx^b) \tag{7}$$

This is a space-time metric on the manifold $M^4 = \Sigma^3 \times \mathbb{R}^1$. It satisfies the covariant vacuum Einstein field equations

$$G_{\mu\nu} = 0 \tag{8}$$

This discussion has of course focused on the vacuum Einstein equations. The generalization to include other fields (e.g., Maxwell, Yang–Mills, Dirac, supergravity) is straightforward and so we shall not include it here.[(9)]

What are the virtues of the dynamical formulation of the Einstein theory? Firstly, and most importantly, it is a practical way of building space-times of physical interest.[4] Of course one must figure out how to solve the nonlinear system of constraints (3) and (4), and certain problems must be faced in carrying out the evolution; however, much is known about both of these aspects, as we shall discuss in Sec. 3 and Sec. 5. With a large computer at one's disposal, one can readily build interesting space-time solutions of Einstein's equations, and this is being done by many groups throughout the world.[(10)]

Secondly, the dynamical formulation allows one to reformulate many questions of interest in general relativity in terms of Riemannian geometry on $\Sigma^3$ rather than pseudo-Riemannian geometry on $M^4$. In many cases, this reformulation is crucial to answering the question. We cite as examples (a) the question of linearization stability of solutions of Einstein's equations[(11)] (Do all solutions of the linearized Einstein equations generate solutions of the full system?), (b) the problem of positive mass in general relativistic space-times,[(12,13)] and (c) recent results which show that conformal symmetries in spatially closed or asymptotically flat solutions of Einstein's equations are almost always full isometries.[(14)]

Before proceeding to discuss further the various steps of the space-time construction scheme outlined above, we wish to point out that the dynamical formulation is in fact (with slight modifications) a canonical one.[(8,9)] That is, if one replaces $K^c{}_d$ (which we use because it leads to the simplest form for the field equations) by

$$\pi^{\mathrm{ef}} := \sqrt{\gamma}\,[\gamma^{\mathrm{ef}}(\operatorname{tr} K) - K^{\mathrm{ef}}] \tag{9}$$

then one can treat the dynamical fields $(\gamma_{ab}, \pi^{cd})$ as points in a phase space $\mathbb{P} = T^*\mathfrak{M}(\Sigma^3)$ with a well-defined symplectic structure

$$\Omega = \int_{\Sigma^3} d\pi^{cd} \wedge d\gamma_{cd} \tag{10}$$

(Here $\mathfrak{M}(\Sigma^3)$ is the space of Riemannian metrics on $\Sigma^3$.) The evolution in $\mathbb{P}$ is generated (via the symplectic and its associated Poisson bracket) by a Hamiltonian

$$H = \int_{\Sigma^3} [\nu\mathfrak{H} + \xi^a\mathfrak{H}_a] \tag{11}$$

[4] Note that only *globally hyperbolic* space-times can be constructed in this way. A globally hyperbolic space-time is one which contains a spacelike hypersurface which is intersected once and only once by every future and past endless curve. (Such a hypersurface is called a Cauchy surface.) One might argue that nonglobally hyperbolic space-times are unphysical since physical predictability breaks down in such a space-time.

where $v$ is an arbitrary scalar field, $\xi^a$ an arbitrary vector field, and $\mathfrak{H}$ and $\mathfrak{H}^a$ are the constraints (3) and (4) written in terms of $\pi$ instead of $K$. Note that this Hamiltonian has the form—a linear combination of constraints—characteristic of all field theories with gauge freedom.[15]

## SOLVING THE CONSTRAINTS

If the dynamical formulation is to be useful in generating solutions, one must have a practical scheme for obtaining data $(\gamma_{ab}, K^c{}_d)$ which satisfy the constraint equations. This is provided by the Lichnerowicz–Choquet-Bruhat–York (LCY) method[16] which works as follows:[5]

(1) Choose an arbitrary Riemannian metric $\lambda_{ab}$ and an arbitrary symmetric tensor $\mu^{cd}$ such that $\bar{\nabla}_c \mu^{cd} = 0$ (transverse) and $\lambda_{cd}\mu^{cd} = 0$ (traceless). [We denote by $T^*{}_{TT}\mathfrak{M}(\Sigma^3)$ the set of all such $(\lambda_{ab}, \mu^{cd})$; note that $\bar{\nabla}$ is the spatial covariant derivative compatible with $\lambda_{ab}$.]

(2) Pick an arbitrary spatial constant $\tau$. [This will be the mean curvature $\operatorname{tr} K = \tau$ of the data $(\gamma, K)$; note that the LCY procedure only produces constant mean curvature data.]

(3) Solve the quasilinear elliptic equation

$$\bar{\nabla}^2 \varphi = (1/8)\, \bar{R}\varphi - (1/8)\mu^c{}_d \mu^d{}_c \varphi^{-7} + (1/12)\tau^2 \varphi^5 \tag{12}$$

for $\varphi$.

(4) Construct

$$\gamma_{ab} = \varphi^4 \lambda_{ub} \tag{13}$$

and

$$K^{ab} = \varphi^{-10}\mu^{ab} + (1/3)\, \varphi^{-4}\lambda^{ab}\tau \tag{14}$$

In describing the LCY method this way, we have been slightly misleading. It turns out that not all choices of $(\lambda_{ab}, \mu^{cd}) \in T^*{}_{TT}\mathfrak{M}(\Sigma^3)$ permit solutions of Eq. (11) to be found. However, solutions do exist (and therefore the LCY method does work) for an open dense subset of $T^*\mathfrak{M}_{TT}(\Sigma^3)$ whose characterization has been complicated this past year. Most of the work was done by Y. Choquet-Bruhat, N. O'Murchadha, and J. York.[16] The last step was completed by J. Isenberg (unpublished work). The characterization relies upon the recently proven Yamabe Theorem,[18]

[5] Again, we focus on the vacuum Einstein equations for convenience. The generalization to handle the Einstein equations with other fields coupled in is straightforward.[17]

which states that every Riemannian metric on a 3-manifold may be conformally mapped to one which has constant scalar curvature of value either $+1, 0$, or $-1$. Therefore the set $\mathfrak{M}(\Sigma^3)$ partitions into $\mathfrak{M}_{(+)}(\Sigma^3)$, $\mathfrak{M}_{(0)}(\Sigma^3)$, and $\mathfrak{M}_{(-)}(\Sigma^3)$. We then find that for $\tau \neq 0$,

$$\mathfrak{A}_{\tau \neq 0}(\Sigma^3) = T^*{}_{TT}\mathfrak{M}(\Sigma^3) - \{(\lambda_{ab}, \mu^{cd}) | \mu^{cd} = 0 \text{ everywhere and } \lambda_{ab} \in \mathfrak{M}_{(+)}(\Sigma^3) \cup \mathfrak{M}_{(0)}(\Sigma^3)\} \tag{15}$$

is the set for which Eq. (12) has solutions, while for $\tau = 0$,

$$\begin{aligned}\mathfrak{A}_{\tau = 0}(\Sigma^3) = T^*{}_{TT}\mathfrak{M}(\Sigma^3) &- \{(\lambda_{ab}, \mu^{cd}) | \mu^{cd} = 0 \text{ everywhere and } \\ &\qquad \lambda_{ab} \in \mathfrak{M}_{(+)}(\Sigma^3) \cup \mathfrak{M}_{(-)}(\Sigma^3)\} \\ &- \{(\lambda_{ab}, \mu^{cd}) | \mu^{cd} \text{ is nonzero everywhere and } \\ &\qquad \lambda_{ab} \in \mathfrak{M}_{(0)}(\Sigma^3) \cup \mathfrak{M}_{(-)}(\Sigma^3)\}\end{aligned} \tag{16}$$

is the set for which this is true.

Thus we may view the LCY method in terms of a map

$$\mathscr{Y}_\tau : \mathfrak{A}(\Sigma^3) \to \mathscr{C}_\tau(\Sigma^3) \tag{17}$$

where $\mathfrak{A}(\Sigma^3)$ refers to $\mathfrak{A}_{\tau \neq 0}(\Sigma^3)$ or $\mathfrak{A}_{\tau = 0}(\Sigma^3)$ depending upon $\tau$, and $\mathscr{C}_\tau(\Sigma^3)$ refers to the sets of $(\gamma_{ab}, K^c{}_d)$ which satisfy the constraints (3) and (4) and have $\operatorname{tr} K = \tau$.

One is tempted at this point to identify $\mathfrak{A}(\Sigma^3)$ as the sought-for handles of the universe, and with certain modifications which we discuss in the next section, this identification is correct. To be fully justified, however, it must be true not only that solutions to Eq. (12) *exist*, but moreover that they *can in practice be found.* While Eq. (12) is nonlinear enough to generally preclude finding a large family of solutions in closed form, it is well-behaved[6] enough to allow numerical solutions to be readily obtained. This remains true in the nonvacuum case.

[6] The existence of a functional

$$S_{[\varphi]} = \int [(\nabla\varphi)^2 + (1/16)\, R\varphi^2 + (1/48)\, \mu^c{}_d \mu^d{}_c \varphi^{-6} + (1/72)\, \tau^2 \varphi^6]$$

whose critical points satisfy Eq. (12) helps.

## 4. THE DEGREES OF FREEDOM OF THE GRAVITATIONAL FIELD

Roughly speaking, the "degrees of freedom" of a classical field theory of physics refers to that minimal information necessary to specify physically distinct solutions of the field equations. In an abstract sense, for a given classical field theory it is easy to identify the space of degrees of freedom. It is simply $\mathfrak{F} = \mathfrak{S}/\mathscr{G}$ where $\mathfrak{S}$ refers to the space of all solutions and $\mathscr{G}$ refers to the group whose orbits on $\mathfrak{S}$ link physically equivalent solutions (e.g., in Maxwell's theory $\mathscr{G}$ is the group of gauge transformations on space-time while in Einstein's theory $\mathscr{G}$ is the group of space-time diffeomorphisms $\mathscr{D}(\mathfrak{M}^4)$). However, if one is to do anything with $\mathfrak{F}$, one must rewrite $\mathfrak{S}$ in terms of information one can choose freely, and one must be able to carry out the quotient space operation in a concrete way.

For those field theories which have a well-posed Cauchy problem, the dynamical formulation permits one to identify $\mathfrak{S}$ with the initial data which solves the constraints, since each such set of data generates a space-time solution of the covariant field equations.[7] However, one is still faced with the task of finding a free parametrization of the solutions of the constraints, and of quotienting out by $\mathscr{G}$ in a manageable way.

The quotienting is particularly devious in Einstein's theory since it involves diffeomorphisms which move forward and backward in evolution time. The action of such diffeomorphisms is not readily represented on the space of initial data. However, for space-times with closed spacelike hypersurfaces and with constant mean curvature (CMC) foliations[8] there is a way to get around this. We recall the result of Brill and Flaherty[(19)] which shows that for such space-times (which in addition satisfy Einstein's equations), there is at most one hypersurface with constant mean curvature equal to a specified value of $\tau$. Thus by fixing $\tau$, one effectively factors out the diffeomorphisms which move in time. The result is that the space of degrees of freedom of the Einstein theory (for spatially closed space-times with CMC foliations) is given by

$$\tilde{\mathscr{C}}_\tau(\Sigma^3) := \mathscr{C}_\tau(\Sigma^3)/\mathscr{D}(\Sigma^3) \tag{18}$$

where $\mathscr{D}(\Sigma^3)$ is the spatial diffeomorphism group whose elements $\eta$ act via pullback

$$\eta\colon (\gamma_{ab}, K^c{}_d) \mapsto (\eta^*\gamma_{ab}, \eta^*K^c{}_d) \tag{19}$$

[7] We recall that this identification works only for globally hyperbolic space-times.

[8] $(\mathfrak{M}^4, g)$ admits a CMC foliation iff there exists a parametrized embedding $i_t\colon \Sigma^3 \to \mathfrak{M}^4$ such that each point in $\mathfrak{M}^4$ lies in the image of $i_t$ for one and only one value of $t$, and $i_t(\Sigma)$ has $\operatorname{tr} K = \text{const}$. One can show that if a space-time solution of Einstein's equations admits such a foliation, then $\operatorname{tr} K$ is a strictly monotonic function of $t$.

To obtain the free parametrization of $\tilde{\mathscr{C}}_\tau(\Sigma^3)$ one now relies upon the LCY map $\mathscr{Y}_\tau\colon \mathfrak{A}(\Sigma^3) \to \mathscr{C}_\tau(\Sigma^3)$ discussed in Sec. 3. Three properties of this map are crucial:[20] (i) It is surjective (so every set of data $(\gamma, K) \in \mathscr{C}_\tau(\Sigma^3)$ can be obtained as $\mathscr{Y}_\tau(\lambda, \mu)$ for some $(\lambda, \mu) \in \mathfrak{A}(\Sigma^3)$. (ii) It is equivariant under the action of $\mathscr{D}(\Sigma^3)$ (so $\mathscr{Y}_\tau[\eta^*(\lambda, \mu)] = \eta^*[\mathscr{Y}_\tau(\lambda, \mu)]$ for any $\eta \in \mathscr{D}(\Sigma^3)$. (iii) Its only invariance is the action of the conformal group $\Theta(\Sigma^3)$, whose elements $\theta$ act via

$$\theta\colon (\lambda, \mu) \mapsto (\theta^4 \lambda, \theta^{-4} \mu) \tag{20}$$

Using these properties, and defining

$$\tilde{\mathfrak{A}}(\Sigma^3) := \mathfrak{A}(\Sigma^3)/\mathscr{D}(\Sigma^3) \ltimes \Theta(\Sigma^3) \tag{21}$$

(where $\mathscr{D} \ltimes \Theta$) is the semidirect product of the two groups), one easily obtains a well-defined reduced map

$$\tilde{\mathscr{Y}}_\tau\colon \tilde{\mathfrak{A}}(\Sigma^3) \to \tilde{\mathscr{C}}_\tau(\Sigma^3) \tag{22}$$

This map is bijective $(1-1$ and onto) so it effectively represents $\tilde{\mathscr{C}}_\tau(\Sigma^3)$ by $\tilde{\mathfrak{A}}(\Sigma^3)$. Interestingly, one can also show[20] that $\tilde{\mathscr{Y}}_\tau$ is a canonical map, pulling back the symplectic and Poisson bracket structure on $\tilde{\mathscr{C}}(\Sigma^3)$ to that on $\tilde{\mathfrak{A}}(\Sigma^3)$. (Both $\tilde{\mathfrak{A}}(\Sigma^3)$ and $\tilde{\mathscr{C}}(\Sigma^3)$ are symplectic manifolds, as can be proven using reduction theory.)

One might object at this point that $\tilde{\mathfrak{A}}(\Sigma^3)$ is not as simple as one might like, since its elements are equivalence classes of data under the combined action of the conformal group $\Theta(\Sigma^3)$ and the spatial diffeomorphism group $\mathscr{D}(\Sigma^3)$. Note however, that one easily fixes the conformal action by requiring $\det \lambda = 1$; and the action of the spatial diffeomorphism group is a relatively familiar, relatively innocuous one. So $\tilde{\mathfrak{A}}(\Sigma^3)$ does provide a fairly practical parametrization of the degrees of freedom of the Einstein gravitational field for spatially closed space-times with CMC foliations.

We wish to note that there are other approaches to identify the degrees of freedom which look promising. That of Ashtekar[21] is especially interesting; and we note that it is complementary to the one discussed here, as it focuses on asymptotically flat space-times.

## 5. EVOLVING SPACE-TIMES: THE CHOICE OF TIME

Well-posedness guarantees that for any specified set of initial data $(\gamma_{ab}, K^c{}_d)$ there exists space-time developments of that data. That is, there exist space-times $(\mathfrak{M}^4, g)$ which (i) satisfy the Einstein equations (8), and

(ii) contain hypersurfaces with intrinsic metric $\gamma_{ab}$ and second fundamental form $K^c{}_d$.

One would like to use the evolution equations (5) and (6) to obtain these space-time developments, and in principle this can always be done, at least numerically, provided one has a way to control the behavior of the time slices. Recall that the evolution equations allow for complete freedom in the choice of the time slices, with the lapse function $N$ acting as the free parameter governing this choice. This freedom is nice, except in general there is nothing to prevent certain unpleasant occurrences. Firstly, the time slices might develop geometric kinks, by which we mean that the intrinsic and/or the extrinsic geometry might become singular even though the ambient space-time is perfectly well behaved. Such kinks, artificial though they might be, prevent further evolution. Secondly, the time slices might become null, in the sense that the tangent bundle to the evolving slices acquire null vectors. This, too, prevents further evolution of the space-time via Eqs. (5) and (6).[9] Finally, in a space-time with true singularities, a poor choice of time slicing might nonuniformly intersect the singularity long before it has to. Again, evolution stops, and much of the space-time remains undeveloped and unexplored.

It has long been believed that by choosing to evolve using exclusively maximal ($\operatorname{tr} K=0$) time slices in building asymptotically flat space-times and CMC time slices in space-times which are spatially compact, one could avoid all three problems. Numerical experience generally supports this contention, and now there is significant theoretical support as well. The key results are due to Gerhardt[(23)] in the spatially compact case, and Bartnik[(24)] in the asymptotically flat case. These results show that as long as one works in a region of space-time which is bounded away from singularities by certain "barrier hypersurfaces," then a foliation of the space-time by CMC space-like[10] slices exists. These slices have no kinks and are bounded away from the local null cones.

Thus CMC foliations conquer the first two problems. The third concern—whether or not CMC foliations have the desired behavior near singularities—is an open question, but there is evidence for this as well. For example, in space-times such that all singularities are of the "crushing type" defined by Eardley and Smarr,[(25)] the barriers required by Gerhardt and Bartnik are automatically in place, and so CMC foliations uniformly approach singularities as desired. The crushing singularity assumption is a

[9] There is a way to construct space-times from data specified on *null* hypersurfaces. It is usually referred to as the "characteristic initial value problem. We will not discuss it here.[(22)]

[10] For convenience, henceforth we shall use "CMC" to refer both to $\operatorname{tr} K\neq 0$ hypersurfaces and $\operatorname{tr} K=0$ (maximal) hypersurfaces.

strong one (hardly ever ascertainable from a given set of initial data), but this result is encouraging. So too is the theorem[(26)] which shows that in every space-time of the "Gowdy type"—space-times with two commuting spacelike Killing vector fields—CMC foliations exist and uniformly approach the singularity. Much work remains to be done on this problem, however.

Even before the Gerhardt–Bartnik results were obtained, many of those in the business of constructing space-times numerically used the CMC and maximal time slices for carrying out their evolution. There are time slicing choices which are more easily implemented[11]—one must solve a linear elliptic equation for $N$ on each time slice in order to generate CMC evolution—but these usually break down except for certain simple classes of space-times. The CMC and maximal schemes seem to be the most promising in general.

Of course it is possible that certain space-times do not contain even one such slice. The Gerhardt—Bartnik results bear on this question: If a space-time admits the appropriate barriers, then it contains a constant mean curvature hypersurface (along with a local foliation). But it is generally not easy to verify the existence of the barriers. If space-times containing no CMC spacelike hypersurfaces do exist, then many interesting theorems are lost (e.g., the linearization stability results[(12)]), and the space $\mathfrak{A}(\Sigma^3)$ is no longer the complete space of degrees of freedom. So the question of whether there exist space-time solutions of Einstein's equations with no CMC hypersurfaces is a very important one.

## 6. MAXIMAL DEVELOPMENTS AND COSMIC CENSORSHIP

In some cases, it is sufficient to know what a space-time is like in a small region (e.g., for studying stellar collapse). In other situations, however, one wishes to know about the "full space-time." An important result of Choquet-Bruhat and Geroch[(27)] tells us that in a certain restricted sense, for any given set of initial data, there exists a unique such "full spacetime." More specifically, their theorem says that for any chosen data $(\gamma, K) \in \mathscr{C}(\Sigma^3)$ there is a unique[12] globally hyperbolic space-time development $(\mathfrak{M}^4, g)_{\max}$ which is maximal in the sense that it properly contains as a submanifold every other development of the $(\gamma_{ab}, K^c{}_d)$.

An interesting and important question now arises: Is $(\mathfrak{M}^4, g)_{\max}$

[11] The simplest choice uses $N = 1$ and is called geodesic normal slicing. This scheme quickly develops kinks in all but the simplest cases.

[12] Strictly speaking, it is only unique up to diffeomorphism.

indeed the largest space-time containing an imbedded hyperspace with the given data $(\gamma, K)$? In asking this question, we are not expressing any doubts concerning the Choquet-Bruhat–Geroch theorem. Note, however, that $(\mathfrak{M}^4, g)_{max}$ is defined to be the largest *globally hyperbolic* space-time containing $(\gamma, K)$. There may be a further, *nonglobally hyperbolic* extension of $(\mathfrak{M}, g)_{max}$. This is the issue being raised.

It is interesting for a number of reasons. We cite two: Firstly, we recall that by definition of global hyperbolicity, $(\mathfrak{M}^4, g)_{max}$ lies in the domain of dependence of the surface and hence is completely determined by $(\gamma, K)$. Physics in $(\mathfrak{M}^4, g)_{max}$ is deterministic. In any extended portion of the space-time, however, determinism fails and most likely causality fails as well. Secondly, we recall that if a globally hyperbolic space-time (like $(\mathfrak{M}^4, g)_{max}$) admits an extension, that portion of its boundary across which the extension is done is a region of Cauchy horizon. Roughly speaking, the boundary (if it exists) of a globally hyperbolic space-time consists of regions of Cauchy horison and regions of curvature singularity—essentially nothing else. So determining extendibility tells one quite a bit about the nature of the singularities in a given globally hyperbolic space-time.

The conjecture of strong cosmic censorship[28,29] says that for generic sets of initial data, no extensions of $(\mathfrak{M}^4, g)_{max}$ can be made. This conjecture has strong appeal since if it were not true, then one would be faced with lots of solutions of Einstein's equations in which the predictability of physics fails.

Is it true? While the answer is not known, there is recent evidence supporting the conjecture. One can show that if a space-time solution of Einstein's equations is bounded by a compact Cauchy horizon, then it necessarily contains an isometry.[30] Generic space-times contain no isometries, so if a generic space-time has a Cauchy horizon, that horizon cannot be compact. This is far from a proof of strong cosmic censorship, but it indicates a promising direction of research.

## 7. PHYSICS AND THE CAUCHY FORMULATION OF EINSTEIN'S THEORY

The Cauchy analysis of Einstein's theory sets up a formalism for generating space-time solutions from specified initial data, and moreover provides a practical way to construct such solutions (at least numerically). How does this help one to understand gravitational physics and cosmology?

Before commenting on the utility of individually constructed model space-time solutions, we note that, in some cases, one can mathematically deduce important conclusions for large classes of space-times using the initial value formalism. Some of the most interesting of these are of the nature of predictions of global properties of space-times based just on knowledge of initial data. We cite the following examples: (a) The singularity theorems[(4)] of Hawking and Penrose which guarantee geodesic incompleteness (i.e., singularities) in space-times with certain types of initial data; (b) the Schoen–Yau result which shows that in any expanding globally hyperbolic space-time with spatial topology $T^3$ (or certain others) the expansion must be continue forever[(31)]; (c) Moncrief's work in Gowdy models which allows one to predict from the initial data whether a Cauchy horizon or a curvature singularity will form.[(32)]; (d) Christodoulou's recently initiated programs of studying the "global initial value problem"—i.e., the complete evolution—for spherically collapsing systems.[(33)]

Unfortunately, results of the sort just described are generally very hard to obtain. Thus much of what one learns about gravitational systems and cosmology must be gained from the study of specific models. These models, while involving much ingenuity and lots of hard work, are still quite limited. One is forced to impose large amounts of symmetry and one is forced to use simplified matter fields. What can one learn this way? Perhaps quite a bit. In this regard, it would be very interesting to study the extent to which the Einstein equations (with matter fields coupled in) can be "modulated" or "averaged" into simpler systems which can be handled. The idea is to prove that the behavior of the simpler (modulated) system approximates fairly accurately that of the more complicated original system, much as statistical mechanics approximates the behavior of many-particle systems. Such schemes have been examined in studies of plasmas[(34)] and solitons.[(35)] They have proven to be very useful. We are currently examining their possible application to Einstein's theory.

Whether this works or not, the enterprise of constructing space-time solutions of Einstein's equations from initial data seems to have a promising future. As computers become bigger and faster, as numerical techniques become more sophisticated and more reliable, and as relativists gain access to both, one will be able to set up models which are much better and closer approximations to real physical systems of interest.

In a sense, we have identified the gavitational "handles" of the universe. Learning how to manipulate them should teach us much about how the universe works.

## ACKNOWLEDGMENTS

This work was supported in part by the NSF under grant No. DMS 83-03998 at the University of Oregon.

## REFERENCES


1. C. Misner, K. Thorne, and J. A. Wheeler, *Gravitation* (Freeman, San Francisco, 1973).
2. Y. Choquet-Bruhat, *Acta. Math.* **88**, 141 (1952).
3. T. Hughes, T. Kato, and J. Marsden, *Arch. Ration. Mech. Anal.* **63**, 273 (1976).
4. S. Hawking and G. Ellis, *The Large-Scale Structure of Space-time* (Cambridge, University Press, Cambridge, 1973), Chap. 7.
5. D. Bao, Y. Choquet-Bruhat, J. Isenberg and P. Yasskin, *J. Math. Phys.* **26**, 329 (1985).
6. D. Eardley and V. Moncrief, *Commun. Math. Phys.* **83**, 192 (1982).
7. J. A. Wheeler, in *Relativity, Groups and Topology*, B. DeWitt and C. DeWitt, eds. (Gordon and Breach, New York, 1964).
8. R. Arnowitt, S. Deser, and C. Misner, in *Gravitation: An Introduction to Current Research*, L. Witten, ed. (Wiley, New York, 1962).
9. J. Isenberg and J. Nester, in *General Relativity and Gravitation* Vol. 1, A. Held, ed. (Plenum, New York, 1980).
10. L. Smarr, in *General Relativity and Gravitation*, B. Bertotti, F. deFelice, and A. Pascolini, eds. (Reidel, Dordrecht, 1984).
11. A. Fisher, J. Marsden, and V. Moncrief, *Ann. Inst. H. Poincaré* **33**, 147 (1980); J. Arms, J. Marsden, and V. Moncrief, *Ann. Phys.* **144**, 81 (1982).
12. R. Schoen and S.-T. Yau, *Phys. Rev. Lett.* **43**, 1457 (1979); E. Witten, *Commun. Math. Phys.* **80**, 381 (1981).
13. F. Flaherty, *Asymptotic Behavior of Mass and Space-time Geometry* (Springer, New York, 1984).
14. D. Eardley, J. Isenberg, J. Marsden, and V. Moncrief, "Homothetic and Conformal Properties of Solutions to Einstein's Equations," preprint, University of California, Santa Barbara, 1986.
15. P. A. M. Dirac, *Lectures on Quantum Mechanics* (Belfer Lectures)(Academic Press, New York, 1964).
16. Y. Choquet-Bruhat and J. York, in *General Relativity and Gravitation*, Vol. 1, A. Held, ed. (Plenum, New York, 1980).
17. J. Isenberg and J. Nester, *Ann. Phys.* **108**, 368 (1979).
18. R. Schoen, "Conformal Deformation of Riemannian Metric to a Constant Scalar Curvature," preprint, University of California Berkeley, 1984.
19. D. Brill and F. Flaherty, *Ann. Inst. H. Poincaré* **28**, 335 (1978); J Marsden and F. Tipler, *Phys. Rep.* **66**, 109 (1980).
20. J. Isenberg and J. Marsden, *J. Geom. Phys.* **1**, 85 (1984).
21. A. Ashtekar, "On the Hamiltonian of General Relativity," preprint, Syracuse University, 1984.
22. H. Friedrich, *Proc. R. Soc. London A* **375**, 169 (1981).
23. C. Gerhardt, *Commun. Math. Phys.* **89**, 523 (1983).
24. R. Bartnik, *Commun. Math. Phys.* **94**, 155 (1984).
25. D. Eardley and L. Smarr, *Phys. Rev. D.* **19**, 2239 (1979).
26. J. Isenberg and V. Moncrief, *Commun. Math. Phys.* **86**, 485 (1982).
27. Y. Choquet-Bruhat and R. Geroch, *Commun. Math. Phys.* **3**, 334 (1969).
28. R. Penrose, in *General Relativity: An Einstein Centenary*, S. Hawking and W. Israel, eds. (Cambridge University Press, Cambridge, 1979).
29. R. Geroch and G. Horowitz, in *General Relativity: An Einstein Centenary*, S. Hawking and W. Israel, eds. (Cambridge University Press, Cambridge, 1979).
30. V. Moncrief and J. Isenberg, *Commun. Math. Phys.* **89**, 387 (1983).

31. R. Schoen and S.-T. Yau, *Proc. Natl. Acad. Sci. USA* **75**, 2567 (1978); *Ann. Math.* **110**, 127 (1979).
32. V. Moncrief, *Phys. Rev. D* **23**, 312 (1981).
33. D. Christodoulou, "A Mathematical Theory of Gravitational Collapse," preprint, Syracuse University, 1984.
34. P. Similon, A. Kaufman, and D. Holm, "Ponderomotive Hamiltonian and Lyapunov Stability for Magnetically Confined Plasma in the Presence of r. f. Field," preprint, Lawrence Berkeley Laboratory, 1984.
35. N. Ercolani, M. Forest, and D. McLaughlin, "Oscillations and Instabilities in Near Integrable PDEs," preprint, Los Alamos, 1984.

*Plate 27*

# On Observing the Absence of an Atom

**R. H. Dicke**[1]

*Received June 10, 1985*

*An atom is confined to a box in its ground state. An attempt is made to observe it in the left half of the box by scattering photons out of a photon wave packet passing through this half of the box. If no photons are scattered, the atom is missing. It is located on the right side of the box and its wave function is changed. The expectation value of the combined atom and photon energy is increased. For the other alternative, that the atom is found on the left side, the expectation value is decreased. By including both alternatives, it is shown that the mean energy is conserved.*

It is a pleasure to contribute this brief note to honor a friend, my former colleague and teacher, John Archibald Wheeler. The subject of this note is particularly appropiate, for some 47 years ago Professor Wheeler, who supervised my senior thesis at Princeton, called my attention to the Einstein–Podolsky–Rosen paper. This was the start of a long and continued interest on my part in the role of measurements in quantum mechanics. Professor Wheeler's own interest in and important contributions to this subject are well known.

As a senior I was perplexed by an apparent paradox that has been resolved only recently.(1) In an honest bridge game the probability of one's partner having the ace of spades in his or her hand is 1/4. But this probability suddenly increases to 1/3 if the card is *observed* to be missing from one's own hand. Quantum mechanical analogues of this experiment seem paradoxical.

In quantum mechanics an *observation* of a quantum system plays two roles: (a) It introduces a physical interaction between the observing

[1] Joseph Henry Laboratories, Physics Department, Princeton University, Princeton, New Jersey 08544.

apparatus and the observed system, an interaction which may change the state of the system, forcing it into a state compatible with the observation. (b) It provides information about this new state of the system. Could there be a situation where the observation shows that an *interaction* did *not* take place and where this new knowledge implies that a *change* in the state of the system has taken place?[1,2]

A famous gedanken experiment of Heisenberg provides one illustration of the paradox. A particle is confined to a large box in the lowest energy state of the box. A microscope is used to observe if the particle is in the field of view of the microscope, a small region inside the box. As usually described, the particle scatters a photon and the scattered photon is detected to show where the particle is located in the field of view. But as a result, the interaction with the photon makes the particle's momentum uncertain.

But what if the scattering does not take place, i.e., the particle is not observed? This fact to some extent reduces the probability of the particle being found in the field of view of the microscope by an immediate, subsequent photon. Thus the wave function of the particle is changed without a photon being scattered.

The paradox can be cast in sharp relief in the following way. Assume that a sodium atom in its ground state is confined to a large, thin rectangular box ($\sim 1$ cm on a side in its large dimensions) in the lowest vibrational state, i.e., the lowest energy state of the box. Now assume that the left half of the box is illuminated by a short intense burst of sodium resonance radiation from a laser. The light pulse is so intense that many photons are scattered by the atom if it is found on the left side. But if the atom is on the right side, none are scattered. In the latter case the atom is thrown into a state of substantially zero probability for being found on the left side by a second radiation pulse immediately following the first.

This implies that the wave function for the center-of-mass coordinates of the atom has been changed. This is a change that increases the expectation value of the atom's energy, but without a photon being scattered. Something seems to have gone wrong.

How can the paradox be resolved? The following is a list of conceivable explanations:

(a) *It never happens. An interaction is required to change the state of the atom and the atom is always found on the left side if the light pulse is intense enough.* If the first photon does not find the atom (a probability, $p$) we might assume that the atom remains in its original state and the second photon also has a probability of $p$ of not finding the atom. For $n$ photons the probability of not find the atom is $p^n$ which for sufficiently large $n$ can be made arbitrarily small, if $p < 1$. Unfortunately this is not an explanation within the framework of quantum mechanics. The experiment provides a means of localizing the atom, and the probability of the atom being found on the left side is computed in quantum mechanics to be 0.5, not 1.

(b) *We cannot make the observation.* The light wave must be confined to the left side of the box. This implies that it contains photons with a distribution of momenta and we cannot determine if a photon has been scattered through a small angle. This is only partly true, which is almost worse than being wrong. If the radiation pulse is coherent, a multiple photon excitation of a single radiation mode, a mode in the form of a focused wave packet, scattering into some other (orthogonal) radiation mode should be observable. Photons need not necessarily be associated with plane wave modes and they need not have definite momenta and energy.[(3)] Furthermore, photon scattering is not the only possible way of locating the atom. For example, the box potential might be very shallow and the recoil momentum given to the atom might eject the atom from the box if the atom is found on the left side. Or an ultraviolet radiation pulse might be substituted for the resonance radiation. It would ionize the atom on the left side of box and the electron or ion might be detected.

(c) *An "interaction" does occur even when the atom is found on the right side of the box.* If this explanation seems intuitively wrong, it may be because of the pervasiveness of physical concepts having their origin in classical physics, or in certain vagueness in the meaning of the word "interaction." Intuitively we may be tempted to think of the atom as being either on the left or on the right side of the box before the observation is performed. If on the right side we note that an interaction cannot occur. But this initial separation into two mutually exclusive classes "atom on left" and "atom on right" is not meaningful under quantum mechanics, for the atom is in a pure vibrational state which permits the atom to be found on either side. The "interaction" is with the atom in this *pure state*, and only after the measurement do we conclude that the atom is on the right and the measurement irreversibly projects out the state "atom on right."

This example has been previously examined in considerable detail, treating both the atom and the radiation field quantum mechanically, using standard quantum measurement theory.[(1)] It is shown that the absence of a scattered photon does indeed reduce the probability of the atom being found on the left side of the box, implying that the expectation value of the energy of the atom is increased.

It is shown in lowest-order perturbation theory that this reduction of the probability that the atom is found on the left side of the box is associated with a second-order photon scattering process. The resonance radiation is assumed to be initially described by a multiple photon excitation of a focused wave packet of short duration. The second-order scattering process is one for which a photon is first absorbed by the atom and is then emitted back into the wave packet, leaving the photon field in its original state. This absorbed photon has neither its energy nor its momentum well defined. Thus the atom, by "handling" the photon, can aquire both momentum and energy, without violating conservation laws. The experiment is impossible unless a focused wave packet is used with the corresponding uncertainty in the momentum and energy of the photons.

The absence of a scattered photon does not imply that an "interaction" between the photon wave packet and the atom has not taken place, for the second-order scattering process can be considered an "interaction."

There is one slightly perplexing matter. In quantum mechanics the expectation value of the energy of a closed system should be conserved, even if the system is not in a state of definite energy. Before the interaction between the photon wave packet and the atom occurs, the expectation value of the total energy is that of the wave packet plus the ground-state energy of the atom. After the interaction, if the state of the photon packet is unchanged, its energy expectation value is also unchanged. But the expectation value of the atom's energy is increased as a result of the change in its center-of-mass wave function.

This difficulty is only an illusion. The conclusion that a photon has not been scattered cannot be drawn without introducing an irreversible decision process. This is represented in the theory by a projection operator which selects the part of the system/apparatus wave function compatible with the observation. This is sometimes called "the contraction of the wave packet." This introduction of a projection operator after the observation induces a change in the wave function which is incompatible with the quantum mechanical equations of motion.

To discuss the conservation of the expectation value of energy, we assume that the incident light pulse is of short duration compared with the radiation life of the atom. We simplify the problem by examining the atom after the light pulse has acted on it but before the atom, if it is excited, has a chance to decay. The measurment is used to ask if the atom is excited or not.

As mentioned above, we assume that the atom is confined to a box which is thin in the direction of propagation of the light pulse, but is otherwise large. The atom is assumed to have only two internal energy states, analogous to the two orientations of a spin 1/2 particle. The wave function of the combined atom/photon system can be written in the form $v(x)\,|m, n\rangle$. Here $v(x)$ is the wave function for the center-of-mass motion of the atom in the box. The first component of the ket vector, $m=0, 1$, represents the internal state of the atom and the second component is the photon number of the radiation pulse. The photon numbers for the other (orthogonal) radiation modes are not written for they are all 0.

For the initial state of the system the wave function is

$$\Psi(0) = v(x)\, |0, n\rangle \tag{1}$$

Here $v(x)$ is now the ground-state wave function of the center-of-mass coordinates, and $m=0$ is the internal ground state of the atom.

The Hamiltonian of the combined system is

$$H = p^2/2m + \hbar\omega\sigma^*\sigma + H_p + H_i \tag{2}$$

Here the first term represents the energy contribution from the center-of-mass motion of the atom, the second is the internal energy of the atom, the third is the energy of the photon field, and the last is the interaction energy of the field and atom. The operators $\sigma^*$ and $\sigma$ are the Pauli spin operators which excite and deexcite the atom. See Ref. 1 for details concerning the operators $H_p$ and $H_i$.

The effect of the interaction of the photons and atom has previously been computed.[1] After the interaction, and before photon emission by the atom occurs, the wave function becomes

$$\Psi = \Psi_0 + \Psi_1 \tag{3}$$

Here $\Psi_0$ and $\Psi_1$ represent the two possible outcomes of the measurement, after the interaction between the atom and the light pulse. The atom is observed to be either unexited or excited.

$$\Psi_0 = \cos(\theta/2)\, v(x)\, |0, n\rangle \tag{4}$$

$$\Psi_1 = \sin(\theta/2)\, v(x)\, |1, n-1\rangle \tag{5}$$

$$\theta = 2\sqrt{n}\, V\, |U(x, y)|\, t^* \tag{6}$$

$$t^* = \int f(t)\, dt \tag{7}$$

In Eqs. (4) and (5) $\theta$ is the analogue of the spin tilt angle in nuclear magnetic resonance. For a 180° flip the atom is excited. In Eq. (6) $U(x, y)$ is the light wave amplitude at $z=0$, where the interaction takes place, and $V$ is a constant. In (7) $f(t)$ is the (real) envelope function that describes the variation with time of the amplitude of the light pulse at the box.

After measuring the internal state of the atom the wave function becomes either (4) or (5) depending upon whether the atom is observed to be in its ground state or excited state. It should be noted that if the atom is in its ground state and $U(x, y)$ is uniform over the left side of the box and is zero on the right, the photon number $n$ can be chosen to yield $\cos(\theta/2) = 0$ on the left. The new center-of-mass wave function includes this cosine factor and is zero on the left side of the box. Then the atom must be absent from the left side immediately after the photon wave packet passes through the box and the atom is observed to be in its ground state.

The projection operators for the excited state and ground state of the atom are respectively

$$P = \sigma^* \sigma \tag{8}$$

$$1 - P = 1 - \sigma^* \sigma \tag{9}$$

These operators commute with all the terms in the Hamiltonian (2) except the interaction term.(1)

Before the interaction the expectation value of the energy of the combined system is

$$E = (\Psi(0), H\Psi(0)) \tag{10}$$

where the wave function is assumed to be normalized to unity.

After the photon wave packet has passed through the box and the internal state of the atom has been measured, the wave function is either

$$\Psi_0 = (1 - P)\,\Psi \tag{11}$$

or

$$\Psi_1 = P\Psi \tag{12}$$

depending upon the outcome of the observation. The probabilities of the atom being found in its ground state or excited state are respectively

$$p_0 = (\Psi_0, \Psi_0) \tag{13}$$

and

$$p_1 = (\Psi_1, \Psi_1) \tag{14}$$

The interaction term in the Hamiltonian contributes to the energy, but only while the photon wave packet is passing through the box. Consequently

$$H_i \Psi = H_i \Psi_0 = H_i \Psi_1 = 0 \tag{15}$$

The expectation values of the energy of the system for each of the two possible outcomes of the experiment can be expressed as

$$E_0 = (\Psi_0, (H - H_i)\,\Psi_0)/(\Psi_0, \Psi_0) \tag{16}$$

and

$$E_1 = (\Psi_1, (H - H_i)\,\Psi_1)/(\Psi_1, \Psi_1) \tag{17}$$

As noted above, $E_0 > E$ because of the increase in the energy expectation value of the center-of-mass motion of the atom.

Combining the contributions to the mean energy from both outcomes of the observation gives the combined mean energy

$$\underline{E} = p_0 E_0 + p_1 E_1 = (\Psi_0, (H - H_i)\,\Psi_0) + (\Psi_1, (H - H_i)\,\Psi_1) \tag{18}$$

Substituting Eq. (11) and (12) in (18) gives

$$\underline{E} = (\Psi, (H - H_i)(1 - P)\,\Psi) + (\Psi, (H - H_i)\,P\Psi) \tag{19}$$

In obtaining (19) use is made of the fact that $P$ commutes with $(H - H_i)$ and that all projection operators satisfy the characteristic equation $P^2 = P$. This equation combined with (15) yields

$$\underline{E} = (\Psi, H\Psi) = (\Psi(0), H\Psi(0)) \tag{20}$$

Here the second equality follows from the fact that the wave function satisfies the Schrödinger equation and that consequently $H$ is a constant of the motion. Thus $\underline{E} = E$ and the combined mean energy after the observation is the same as that before.

In conclusion, energy conservation is no problem. To use photon scattering to detect the presence or absence of an atom is to make a "measurement." If the mean energy is increased when the atom is observed to be missing, it is decreased when the atom is present. All debts are paid and the books are balanced.

It is a pleasure to acknowledge helpful conversations with John Wheeler and with P. J. E. Peebles.

## REFERENCES

1. R. H. Dicke, *Am. J. Phys.* **49**, 925–930 (1981).
2. D. M. Greenberger, *Rev. Mod. Phys.* **55**, 875–905 (1983).
3. R. J. Glauber, *Quantum Electronics III*, Vol. 1, Edited by P. Grivet and N. Bloembergen, eds. (Columbia University Press, New York, 1964), p. 111.

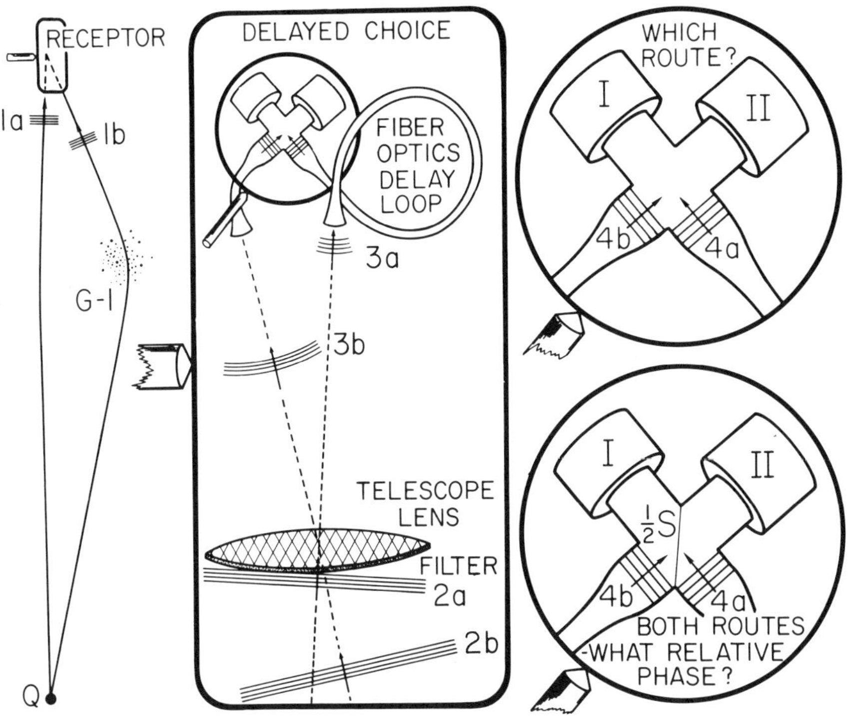
RECEPTOR
1a
1b
G-1
Q
DELAYED CHOICE
FIBER
OPTICS
DELAY
LOOP
3a
3b
TELESCOPE
LENS
FILTER
2a
2b
WHICH
ROUTE?
I
II
4b
4a
I
II
$\frac{1}{2}$S
4b
4a
BOTH ROUTES
-WHAT RELATIVE
PHASE?

*Plate 28*

# How to Evade the Confrontation with the Uncertainty Relations

**V. B. Braginsky**[1] **and F. Ya. Khalili**[1]

*Received June 13, 1985*

*It is demonstrated that one can in principle register an arbitrarily small force acting on a free particle by employing only measurements of its coordinates.*

It is usually accepted that the uncertainty relations limit the sensitivity in quantum measurements. During the last few years several authors have demonstrated that in many cases it is possible to evade the confrontation with the uncertainty relations and obtain an arbitrary high sensitivity (see review articles, Refs. 1 and 2).

During the development of very sensitive gravitational antennae, it became necessary to analyze the problem of the limited sensitivity in the test-body experiments. The main results of this analysis may be formulated as follows:

1. The macroscopic mechanical oscillators and free masses must be regarded as quantum objects even at high temperatures, provided that their coupling with the heat bath is weak enough.[3,4]

2. One can detect arbitrary small forces acting on the oscillator if quantum nondemolition measurement (QND) is realized. One can realize the QND measurements by coordinate measurements only.[1,2]

In this article we shall consider the possibility for detecting a small force acting on a free mass, provided that a coordinate of a free mass may be measured with an error not less than $(\hbar\tau_0/2m)^{1/2}$ and the momentum with the error not less than $(\hbar m/2\tau_0)^{1/2}$ if the coordinate was measured continuously (here $m$ stands for the mass and $\tau_0$ for the averaging time). These

[1] Department of Physics, Moscow State University, Moscow, USSR.

two limits are direct consequences of the Heisenberg uncertainty relation and of the continuity of the coordinate measurement. Therefore, the smallest detectable force acting on the mass has to be equal to

$$F \simeq \frac{\lambda}{\tau_F} \sqrt{\frac{\hbar m}{\tau}} \tag{1}$$

if there are zero frequency spectral components in the force spectrum, and

$$F \simeq \lambda \sqrt{\frac{\hbar m}{\tau_F^3}} \tag{2}$$

if there are no such components; $\tau_F$ is the duration of the force $F$, $\tau$ is the time of measurement, and $\lambda$ is a dimensionless factor of the order of unity which depends on the form of the force $F$ in time.

We shall show below that the limits (1) and (2) can be exceeded if two conditions are fulfilled. Qualitatively these conditions may be formulated as follows:

(1) It is necessary to measure the coordinate of the mass in a certain short interval of time $\tau_0 \ll \tau_F$ only.

(2) It is necessary to apply a focusing potential to the mass during a short period of time. This potential will create a quantum state with anticorrelated coordinate and momentum:

$$\langle \Delta p \, \Delta x \rangle < 0$$

Let us suppose that the coordinate of the mass is measured during a short interval $0 < t < \tau_0$ with an error $\sigma_1$. The error of the momentum is correspondingly equal to $\hbar/2\sigma_1$. After the interval $\tau_1 \gg \tau_0$, due to the wave-packet expansion, the uncertainty of the coordinate will be equal to

$$\Delta^2 = \sigma_1^2 + \left(\frac{\hbar\tau_1}{2m\sigma_1}\right)^2$$

Now let us suppose that during $\tau_1 < t < \tau_1 + \tau_2$ a parabolic potential $U = \frac{1}{2}m\omega^2 x^2$ is applied to the free mass. If $\sin \omega\tau_2 > [1 + (\omega\tau_1)^a]^{-1/2}$, then the mass will be in the state of anticorrelatd coordinate and momentum:

$$\langle \Delta p \, \Delta x \rangle < 0$$

during $t > \tau_1 + \tau_2$ and the wave packet will begin to squeeze. After the interval of time $\tau_1$, which is defined by the condition

$$\omega(\tau_1 + \tau_3) \cos \omega\tau_2 + (1 - \omega^2 \tau_1 \tau_3) \sin \omega\tau_2 = 0 \tag{3}$$

the dispersion of the coordinate will reach its minimal value

$$\sigma_2^2 = \sigma_1^2 \frac{1+(\omega\tau_3)^2}{1+(\omega\tau_1)^2} \tag{4}$$

If the coordinate of the mass is measured once more at the time $t=\tau_1+\tau_2+\tau_3$ during $\tau_0 \ll \tau_1$, then one can measure an arbitrary small force $F$ which was applied during $\tau_1+\tau_2+\tau_3$ if the value $\tau_0$ is small enough.

To simplify the final expressions, let us suppose that $\tau_1=\tau_3$ and that $\tau_2 \ll \tau_1$ (the asymmetry $\tau_1 \neq \tau_3$ does not lead to any principal new results). Then, if in the first measurement the coodinate of the mass was equal to $x_1 \pm \sigma_1$ and $F=0$ during $\tau_1+\tau_3$, the second measurement must give $-x_1 \pm \sigma_1$. But if the experimenter finds that

$$|x_1+x_2| > \sigma_1$$

he has to conclude that during $\tau_1+\tau_3$ a nonzero force $F$ was acting on the mass. The minimal detectable force is determined by the value $\sigma_1$. If $\tau_F \simeq \tau_1+\tau_3$, then

$$F_{\min} \simeq \frac{\lambda}{\tau_F^2}\sqrt{\hbar m\tau_1} = \lambda\sqrt{\frac{\hbar m}{\tau_F^3}}\sqrt{\frac{\tau_0}{\tau_F}} \tag{5}$$

It is evident that, the more the gain in sensitivity is compared with the limit (2), the less is the factor $\sqrt{\tau_0/\tau_F}$. It is not difficult to make calculations for a general case when $\tau_F \neq \tau_1+\tau_3$ and $\tau_1 \neq \tau_3$. But in this case the gain in sensitivity will also be approximately equal to $\sqrt{\tau_0/\tau_F}$ in comparison with the limit (2).

The main features of the described procedure of measurements are similar to those of the stroboscopic measurement with an oscillator.[6] As in the case of an oscillator, the "price" for the higher sensitivity is the higher uncertainty of the coordinate between two stroboscopic measurements and the loss of information on the form and duration of the force $F$.

The proposed procedure may be useful for gravitational wave antennae if it is possible to realize the superposition of a pure parabolic potential at large distances between the masses. Probably this procedure will also be useful when a weak AC electromagnetic field has to be measured by means of elementary particles.

In conclusion, we emphasize one important circumstance concerning the limit sensitivity of quantum measurements, particularly of the QND measurements. It is evident that the sensitivity in such measurements is important not only for the solution of the problem of the discovery of gravitational radiation bursts from astrophysical catastrophes in our Galaxy or in other galaxies. It is likely that new procedures of measurements will be proposed and developed to explore biological objects. It is also probable that a nonstandard procedure of measurements will be realized to make the so-called "quantum mechanical computer."[7] At present there is no coherent and consistent theory of quantum measurements with nonlinear and parametric interaction between the monitored object and the measuring device. In particular, for QND measurements a universally necessary and sufficient condition is the commutability of the evolution operator of the pair "measuring device + monitored object" with the operator of the measured variable. There are no simpler and sufficient conditions for QND measurements. Only the criteria for QND measurements in a few particular cases are known.

It is reasonable to expect that increasing the sensitivity and creating new method of measurements will permit us to observe such phenomena as background gravitational radiation[8] and the neutrino background in our Galaxy,[9] and to detect a small magnetic moment of the nonzero rest mass neutrino. Undoubtedly these phenomena are important for the understanding of the world in which we live.

## REFERENCES


1. C. M. Caves, K. S. Thorne, R. Drver, and V. D. Sandberg, *Rev. Mod. Phys.* **52**, 341 (1980).
2. V. B. Braginsky, Yu. I. Vorontsov, and K. S. Thorne, *Science* **209**, 547 (1980).
3. V. B. Braginsky, *Sov. Phys. JETP* **53**, 1434 (1967).
4. V. B. Braginsky and V. S. Nazarenko, *Sov. Phys. JETP* **57**, 1421 (1969).
5. V. B. Braginsky and Yu. I. Vorontsov, *Sov. Phys. Usp.* **114**, 41 (1974).
6. V. B. Braginsky, Yu. I. Vorontsov, and F. Ya. Kkalili, *Sov. Phys. JETP Lett.* **47**, 296 (1978).
7. R. P. Feynman, Quantum mechanical Computers, *Found. Phys.* **16**, (1986), to appear.
8. L. P. Grischuk, *Sov. Phys. JETP Lett.* **23**, 326 (1970).
9. V. F. Schvarzman, V. B. Braginsky, S. S. Gerstein, Ya. B. Zeldovich, and M. Iu. Khlopov, *Sov. Phys. JETP Lett.* **15**, 224 (1982).

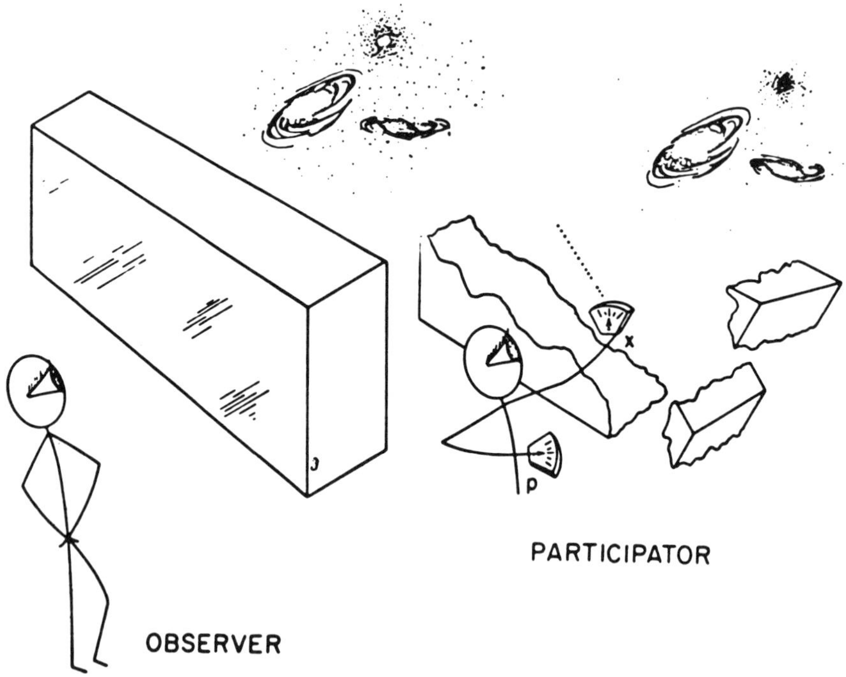


*Plate 29*

# Are the Quantum Rules Exact? The Case of the Imperfect Measurements[1]

**Bernard d'Espagnat**[2]

*Received April 3, 1985*

*Should we doubt the exactness of the predictive quantum rules of calculation? Although this question is sometimes raised in connection with the one on how to physically understand quantum mechanics, these two questions should not be mixed up. It is recalled here that even the first one is stil an object of controversy, and it is shown (a) that in one specific case the arguments put forward in support of such doubts are hardly cogent but (b) that, nevertheless, at least in one specific other context, the question is worth attention. This is the problem of repeated imperfect measurements. Relative to it, a theoretical possibility is shown of discriminating between the thesis that the quantum rules are exact and a powerful theory of which it is proved that it cannot be reconciled with the assumption.*

## 1. INTRODUCTION

Any discussion bearing on the basic ideas of quantum physics should first of all carefully distinguish between the axioms of this theory and its conceivable interpretations. It should admit that the former are but predictive "rules of computation" and that this is the very reason why they constitute the hard core of the theory in question: for the observation that these rules are well verified in a great number of experiments is just that of a plain fact, which no dispute over "the significance of it all" can falsify.

Once the necessity of making this distinction is acknowledged, the problems bearing on the foundations nicely separate—in a first approximation at least—in two sets. One of these is composed of the problems in which the rules in question are assumed to be both exact and

[1] Paper dedicated to Professor J. A. Wheeler.

[2] Laboratroire de Physique Théorique et Particules Élémentaires (associé au C.N.R.S.), Université de Paris-Sud, 91405 Orsay, France.

universal. Such problems are of the type "How can we interpret these rules in terms of an independently existing reality (of a 'hardware located out there' as Professor Wheeler puts it)?" or even, more radically, "Do we *have* to interpret them in such a way?." As is well known, Wheeler himself has taught us in a masterly way to look boldly at several perplexing possibilities, including even that of a negative answer to this last question. Since the vistas he opened thereby are such as to influence, in the long run, the whole of our thinking on the measurement problem, it is a great pleasure to dedicate to him an investigation such as this one.

The problems in the other set are those in which it is the universality of the quantum rules themselves that is questioned. Of course the possibilities along these lines are limited. Nevertheless they do exist due to the obvious fact that the actual experiments confirming the rules in question have but a finite accuracy and do not cover the infinite domain of what could conceivably be tested (and, of course, solving the problems in the first set would be less difficult if the idea of the universality of the quantum rules was given up). As implied by the title, the problems investigated in this article belong to the second set.

At this stage it is, however, necessary to be somewhat more explicit on one point. It concerns the special status of the "last" quantum mechanical axiom, namely the one concerning such things as the collapse of the state vector upon measurement ("projection postulate"). In fact, although this axiom is often useful in computing experimental predictions, for such a purpose it can always, at least in principle, be dispensed with if the Schrödinger equation is taken as universal. For this and for other reasons the problems concerning it can to a large extent be postponed until the question of the interpretation of the theory is considered, that is, these problems actually belong to the first set. Let us therefore take the last axiom out of the set of what we call "the rules of quantum mechanics." Let us also, by the way, restrict our considerations to the case of nonrelativistic quantum physics. Then, the hypothesis that the quantum rules are exact reduces to the assumption that (a) the Schrödinger time-dependent equation is an exact one, and (b) the probabilities of the various possible outcomes of a measurement are given by the usual probability rule.

For this second axiom to make sense, we have of course to admit that a classically describable "ultimate ideal measuring setup" can be thought of in connection with any of the physical phenomena we are interested in. It is with respect to the measurement results given by this setup, conceived of as placed at the end of the whole process and assumed to operate without failures of any kind, that the probability axiom is supposed to be valid.

Out of the many types of proposals implying or suggesting that the above-defined quantum rules are *not* exact, we here select two for discussion. One of them is the class of the proposals put forward by physicists hoping to thereby save Einsteinian local realism—or more generally "separability"—in spite of the contrary evidence provided by the Bell theorem and by the corresponding distant correlation experiments. We comment on it qualitatively in Sect. 2. The other one is an interesting idea essentially aimed at describing repeated imperfect experiments. In Sect. 3 (where our acceptation of the word "imperfect" is also given) some consequences of this idea are compared with those of a method strictly based on the quantum rules. The results are discussed in Sect. 4.

## 2. LOCAL REALISM VERSUS QUANTUM RULES

As noted above, many attempts of varied kinds were made at saving Einsteinian local realism in spite of recent evidence. Most of such attempts are based on the fact that the experimental tests of the Bell inequalities have up to now been made with detectors that are poor approximations to ideal ones (in particular, the number of undetected pairs far exceeds the number of detected ones). A consequence of this is that the experimental results actually disprove local realism only if an additional hypothesis is made, relative to the way the instruments work. Although some such hypotheses are very natural, still the upholders of local realism stress the fact that they all are but assumptions. And they go on by pointing out that—as one of them puts it—"most scientists, given a choice between abandoning [local] physical reality and abandoning such an *ad hoc* hypothesis ··· will unhesitatingly choose the latter course."

*A priori* such statements look sound. And—especially when they appear in print as natural conclusions to debates on these matters, as was recently the case[(1)]—they may well give an unprejudiced reader the impression that in truth the choice is between relinquishing some technical hypotheses and abandoning a very general principle which up to now has worked well. There would of course then be no doubt as to where the preference of a physicist would go. Actually, however, the dilemna is quite of a different kind: for keeping Einsteinian local realism does not only imply giving up a technical hypothesis about imperfect instruments. It also implies giving up the hypothesis that the rules of quantum mechanics are exact in the sense explicated in the introduction. For if they *are* exact in this sense then no experiment is needed. Bell's theorem alone disproves Einsteinian local realism. The choice therefore is *not* between a technical assumption on the one hand and a general principle on the other hand. Although the experiments made in connection with the Bell inequalities were most useful, their very existence has had the inconvenience of obscuring to some extent the point just made. The real choice is a sharp one and it is between two general basic ideas: the one of local realism (or separability, which amounts to the same within some here irrelevant small differences[2]) on the one hand and the one that the quantum rules are exact on the other hand. But then it must be admitted that the experimental evidence in favor of the latter view has been steadily growing during the seventies and is now really impressive.[3] In fact, if local realism were to be maintained, the presently available data could apparently be accounted for only by means of ingenious theoretical models (see, e.g., Ref. 4) which

(a) are purely *ad hoc* in several respects (they rely on special assumptions about the ways some particles escape detection and, above all, they are by no means derivable from a general physical theory) and

(b) are built up so as to lead to predictions coinciding more or less with the quantum ones as regards photon-pair correlations. This coincidence, however, could only be made approximate, the difference being fortunately not appreciably larger than the present experimental accuracy limit.

Few are anyway the scientists who are so fond of local realism that they would heartily give up, for that reason, the idea that the quantum rules are exact; but presumably fewer still will be those who will take this step when it is fully realized that we are merely offered *such* models as alternatives to the rules in question.

## 3. IMPERFECT MEASUREMENTS, COMPARING THE MILANO AND THE CONVENTIONAL APPROACHES

This section and the next one are concerned with the problem of how to predict the results of such "imperfect" measurements (definition below) are as likely to occur when an observable is repeatedly measured at small time intervals on a system. This problem happens to be related to the question whether or not the quantum rules are exact. This is due to the fact that one of the most powerful theoretical methods for dealing with such measurements, namely the Milano theory,[5] is based on a formalism in which the state reduction procedure occurring in a measurement is treated in an unconventional way. Now, as mentioned above, this state reduction should not *per se* be considered as being part of the "basic quantum rules" the exactness of which is here under examination. Nevertheless a change in the conventional description of it may conceivably imply also a change in what we agreed to call "the quantum rules" themselves. Specifically, it is the question whether or not the Milano theory compels us to introduce a change of such a basic type that is here to be examined.

From the theorist's point of view, the simplest of all conceivable quantum measurement processes is one with the following property. If before the measurement takes place the quantum system (let us call it $S$) already is in an eigenstate $|\varphi_i\rangle$ of the quantity $B$ which is to be measured, then, after the process is over, $S$ still is in the eigenstate $|\varphi_i\rangle$ *and* the instrument $A$ contains an unambiguous record of what this state is. If $A$ is considered as a quantum system, the second condition implies that $A$ is itself in a state orthogonal to any other state it would be in, had $S$ initially been in any state $|\varphi_j\rangle$ other than $|\varphi_i\rangle$. By convention let us say that such measurements are "perfect" and that other ones are imperfect. As a rule, measuring instruments are contrived so that we can think of them as being at least "approximately perfect" in this sense. However, in processes in which a measurement is repeated a great many times on a system during a short time interval it is readily seen that the constituting elementary measurements are likely to be imperfect. By considering, for example, a sequence of closely packed Stern–Gerlach devices the reader will easily verify this, but it is the merit of the physicists of the Milano group to have been among the first persons to call attention to the point.

Consequently, these physicists had to change—as we said—the projection postulate somewhat. To investigate whether or not this implies a basic change in what we called "the quantum rules"—and, if yes, to check whether this change is observable—we compare, in what follows, the predictions of the Milano theory, for a specific process, with those that follow from just the conventional "quantum rules." In this specific process the system $S$ is a spin-1/2 precessing around the $z$ axis under the influence of a constant mgnetic field and the $x$ component of which is measured, at one or several time(s), with imperfect instrument(s).

In the conventional approach, if $\xi$ is the "pointer coordinate" of an instrument $A$, we may assume the time evolution of the $S+A$ system to take place in accordance with the von Neumann scheme:

$$g(\xi)|\varphi_i\rangle \to g_i(\xi)|\varphi_i\rangle \tag{1}$$

where $g$ and $g_i$ are the wave functions of $A$ before and after the interaction respectively (for convenience we assume no degeneracy) and where

$$S_x|\varphi_i\rangle = (-1)^{i+1}|\varphi_i\rangle, \qquad i=1,2 \tag{2}$$

A known example of an $S+A$ system obeying this condition is the one described by the Hamiltonian

$$H = H_0 + H' \tag{3}$$

where $H_0$ incorporates the interaction of $S$ with the constant magnetic field and where

$$H' = \beta(t)\, S_x(-i\hbar)\, \partial/\partial\xi \tag{4}$$

$\beta(t)$ being equal to zero except during a short "measurement time interval" $(t', t'')$ during which it is so large that $H_0$ can be neglected. Setting

$$\alpha(t) = \int_{t'}^{t} \beta(\tau)\, d\tau \tag{5}$$

and

$$a_i = (-1)^{i+1} \int_{t'}^{t''} \beta(t)\, dt \tag{6}$$

it is then a trivial matter to check that during the measurement time interval the ket

$$|\psi_i\rangle = g[\xi - (-1)^{i+1}\, \alpha(t)]\, |\varphi_i\rangle \tag{7}$$

is, for any function $g(\xi)$, a solution of the Schrödinger equation. Hence the "von Neumann scheme" (1) holds, with

$$g_i(\xi) \equiv g(\xi - a_i) \tag{8}$$

The "imperfect measurement case" is the one in which the mutual orthogonality of the two $g_i$'s cannot be neglected. Equations (1), (7), and (8) show, for example, that if, immediately before they interact, $A$ is in state $g(\xi)$ and $S$ in state

$$\sum_i c_i\, |\varphi_i\rangle, \qquad i = 1, 2 \tag{9}$$

then immediately afterwards the $S + A$ system is in state

$$\sum_i c_i\, g(\xi - a_i)\, |\varphi_i\rangle \tag{9$'$}$$

From this result, the probability $p_+ (p_-)$ that, after the interaction, an "ultimate observer" would find the pointer coordinate $\xi$ lying in the positive (negative) half axis is easily found to be

$$p_\pm = \langle\psi_0|\, U^+ \left(\sum_i |\varphi_i\rangle\, f_i^{(\pm)} \langle\varphi_i|\right) U\, |\psi_0\rangle \tag{10}$$

with

$$f_i^{(+)} = \int_0^{\infty} |g(\xi - a_i)|^2 \, d\xi, \qquad f_i^{(-)} = \int_{-\infty}^{0} |g(\xi - a_i)|^2 \, d\xi \tag{11}$$

here $|\psi_0\rangle$ is the initial spin state and $U$ is the appropriate spin evolution operator.

Now, this formula coincides exactly with the one resulting from the Milano theory [use Eq. (1.28) of Ref. 5, replacing $|q\rangle \, dq \, \langle q|$ by $|\varphi_i\rangle\langle\varphi_i|$ and $\int dq$ by $\sum_i$) applied to the same process. And this might suggest that the latter theory is basically a "theory of imperfect measurements," in which the conventional quantum rules (in our sense, see introduction) are still exact. However, such an assumption must be submitted to further tests. For this purpose let us now consider the case in which $S$ is made to interact successively, at times $t_1$ and $t_2$ respectively, with two instruments $A_1$ and $A_2$ of type $A$. Let us ask for the probabilities $p_{\alpha\beta}$ ($\alpha, \beta = \pm$) that after both interaction processes are completed the pointer coordinates $\xi_1$ and $\xi_2$ are found with signs $\alpha$ and $\beta$ respectively (i.e., lying anywhere on the thus defined half-axes). If the quantum rules are exact, the calculation can be made, as above, without the help of a "projection postulate" of any kind, since the state vector $|\psi(t)\rangle$ of the composite $S + A_1 + A_2$ system can be determined for any value of $t$ by applying equations such as (9) and (9′), at each interaction time, to each of the two components of $|\psi\rangle$ at that time, and by using the usual time evolution operator in-between. The calculation is straightforward. Applying then the usual probability rule at a time $t > t_2$ (which of course implies summing over the spin directions since the "ultimate observer" is not supposed to directly observe the spin), one finds

$$p_{\alpha\beta} = \langle\psi_0| \, U^+ \left( \sum_{ii'} |\varphi_i\rangle \, K_{ii'}^{(\alpha,\beta)} \langle\varphi_{i'}| \right) U |\psi_0\rangle \tag{12}$$

with

$$K_{ii'}^{(\alpha,\beta)} = \sum_j \langle\varphi_i| \, \hat{U}^+ \, |\varphi_j\rangle \, f_j^{(\beta)} \langle\varphi_j| \, \hat{U} \, |\varphi_{i'}\rangle \, M_{ii'}^{(\alpha)} \tag{13}$$

and

$$M_{ii'}^{(+)} = \int_0^{\infty} g^*(\xi - a_i) \, g(\xi - a_{i'}) \, d\xi, \qquad M_{ii'}^{(-)} = \int_{-\infty}^{0} g^*(\xi - a_i) \, g(\xi - a_{i'}) \, d\xi \tag{14}$$

In these equations $U$ and $\hat{U}$ are the unitary operators describing the precession of $S$ during the time intervals $(0, t_1)$ and $(t_1, t_2)$ respectively and the other symbols have their above defined meanings.

Now, if the same problem is investigated within the realm of the Milano method the result is formally given by Eqs. (12), (13), and (11) again, but now with Eqs. (14′),

$$M_{ii'}^{(+)} = (f_i^{(+)} \cdot f_{i'}^{(+)})^{1/2}, \qquad M_{ii'}^{(-)} = (f_i^{(-)} \cdot f_{i'}^{(-)})^{1/2} \tag{14′}$$

replacing Eqs. (14). This follows from Eq. (2.7) of Ref. 5, that is, from applying the Milano "generalized projection postulate" to the interaction of $S$ with $A_1$ in the way explicated in the quoted reference [at this point also, formula (1.28) of this reference must, of course, be adapted to the case of an observable with a discrete spectrum by making the changes indicated here below Eq. (11)].

## 4. DISCUSSION

The $p_{\alpha,\beta}$ given by Eq. (12) are different according to whether Eqs. (14) (conventional approach, quantum rules "exact") or Eqs. (14′) (Milano method) are used. Since these $p_{\alpha\beta}$ are observable quantities, the question of principle is settled. The theory based on Eq. (2.7) of Ref. 5 with $\mathbf{I}_{1,2} = \mathbf{R}_{\pm}$ cannot be reconciled with the assumption that the quantum rules are exact, not even in the restrictive sense mentioned in the introduction. This, after all, is not very surprising since this theory was initially based on Ludwig's proposal of a modification of the quantum formalism aimed at removing the difficulties of the measurement theory. But *a priori* the above-considered conjecture that the only real novelty of the method lay in its way of generalizing the projection postulate was not absurd. We now have a proof that its departure with conventional quantum mechanics is in fact much more radical.

Of course this result raises the question "which one of the two methods is correct?" And this brings forward the further question "can the matter be settled experimentally?" On these two questions we can unfortunately make, up to now, but rather sketchy remarks. One of them is that, for such purposes, instruments isolated enough from their environment so as to be describable quantum mechanically would have to be found. This aim is perhaps not unreachable considering the fact that instruments parts of which at least must be so described seem to be necessary for such purposes as that of detecting gravitational waves, and that theoretical investigations incorporating such notions are presently well under way (see, e.g., Ref. 6). Our other remark is that anyhow the problem may be experimentally difficult for, in most circumstances, the observable differences between the predictions derived from Eqs. (14) and (14′) are very small. One way of maximizing the contrast between the predictions of the two theories might be to consider the difference

$$d = p_{+,+} - p_{+,-}$$

From formulas (10)–(14), (14′) $d$ is found to be (initial state $S_x = +1$)

$$d = \cos 2\theta (f_1^{(+)} - f_2^{(+)})(\cos^2 \alpha M_{1,1}^{(+)} - \sin^2 \alpha M_{2,2}^{(+)}) - \sin 2\theta \sin 2\alpha\, M_{1,2}^{(+)} \quad (15)$$

where $2\alpha$ and $2\theta$ are the precession angles corresponding to intervals $(0, t_1)$ and $(t_1, t_2)$ respectively. Since $M_{1,1}^{(+)}$ and $M_{2,2}^{(+)}$ are identical in the two theories, it is appropriate to choose $\theta = \pi/4$. Then, with

$$g(x) = \pi^{-1/4} \exp(-x^2/2) \quad (16)$$

one gets (with $a = a_1$, $a_2 = -a_1$ and assuming $a \gg 1$)

$$d \propto \exp(-a^2) \text{ in the conventional theory and}$$

$$d \propto a^{-1/2} \exp(-a^2/2) \text{ when Eq. (14′) is used}$$

With the same notations and still assuming $S_x = +1$ the probability $p$ that the first instrument pointer coordinate is found negative when both the magnetic field and the second intrument are removed is

$$p \propto a^{-1} \exp(-a^2)$$

hence, as a function of $a$, the ratio $d/p^{1/2}$ tends to a constant limit for $a \to \infty$ in the second approach while it decreases rapidly in the conventional approach [qualitatively this remains true also for other forms of $g(x)$]. Differences such as this one might conceivably serve to experimentally discriminate between the two theories. This, however, is, as we said, just a sketchy indication and it is clear that much more extensive studies would be necessary before it could be concluded that tests along such lines are feasible.

As a final remark we would like to point out that the theory of imperfect measurements is presumably still in its first stages, that it may become quite useful in connection with that of repeated measurements (with one or with several instruments), and that it may help to open new vistas concerning the question as to what "measurement" actually is.

Parts of the investigations reported here were supported by and carried out at the Institute of Theoretical Physics of the University of

California in Santa Barbara within the realm of the 1984 workshop on Quantum Noise in Macroscopic Systems. The author acknowledges with pleasure the hospitality extended to him by the Institute on that occasion.

## NOTE ADDED IN PROOF

The difference between the predictions of the two methods is seen to vanish when, in Eqs. (11) and (14), the summation ranges are shrinked to infinitesimal intervals, the $p_{\alpha\beta}$ being then obtained by adding all the contributions from the relevant intervals. The replacement of Eq. (2.7) by Eq. (2.21) in Ref. 5 amounts to the same. The thus restored agreement indicates that in a sense this replacement is equivalent to a come-back to the conventional quantum formalism.

## REFERENCES

1. T. W. Marshall, *Nature* **313**, 438 (1985).
2. B. d'Espagnat, *Phys. Rep.* **110**, 203 (1984).
3. A. Aspect, P. Grangier, and G. Roger, *Phys. Rev. Lett.* **49**, 91 (1982).
4. T. W. Marshall, E. Santos, and F. Selleri, *Phys. Lett. A* **98**, 5 (1983).
5. A. Barchielli, L. Lanz, and G. M. Prosperi, *Nuovo Cimento B* **72**, 79 (1982).
6. C. M. Caves, K. S. Thorne, R. W. P. Drever, V. O. Sandberg, and M. Zimmermann, *Rev. Mod. Phys.* **52**, 341 (1980).

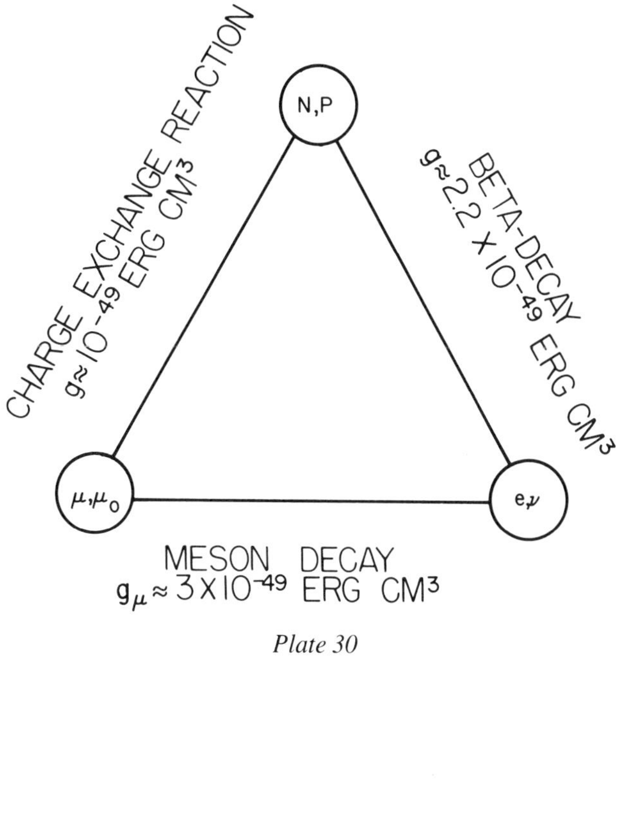
CHARGE EXCHANGE REACTION
g≈10⁻⁴⁹ ERG CM³
N,P
BETA-DECAY
g≈2.2 X 10⁻⁴⁹ ERG CM³
μ,μ₀
e,ν
MESON DECAY
g_μ ≈ 3 X 10⁻⁴⁹ ERG CM³

*Plate 30*

# The Problems in Quantum Foundations in the Light of Gauge Theories[1]

**Yuval Ne'eman**[2,3]

*Received May 1, 1985*

*We review the issues of nonseparability and seemingly acausal propagation of information in EPR, as displayed by experiments and the failure of Bell's inequalities. We show that global effects are in the very nature of the geometric structure of modern physical theories, occurring even at the classical level. The Aharonov–Bohm effect, magnetic monopoles, instantons, etc. result from the topology and homotopy features of the fiber bundle manifolds of gauge theories. The conservation of probabilities, a supposedly highly quantum effect, is also achieved through global geometry equations. The EPR observables all fit in such geometries, and space-time is a truncated representation and is not the correct arena for their understanding. Relativistic quantum field theory represents the global action of the measurement operators as the zero-momentum (and therefore spatially infinitely spread) limit of their wave functions (form factors). We also analyze the collapse of the state vector as a case of spontaneous symmetry breakdown in the apparatus-observed state interaction.*

## 1. INTRODUCTION

John Archibald Wheeler's contributions have spanned almost every area of twentieth century physics, and yet it is in the two fields of general relativity and the foundations of quantum mechanics that his influence has reached beyond the formalism and phenomena, down into the philosophical layer—the metaphysics, in the original sense. In his work on the general theory of relativity, it was Wheeler who emphasized the importance of the

[1] Work supported in part by U.S. DOE Grant DE-FG05-85ER40200.

[2] Center for Particle Theory, The University of Texas at Austin, Austin, Texas 78712, and Sackler Faculty of Exact Sciences, Tel Aviv University, Tel Aviv, Israel.

[3] Wolfson Chair Extraordinary in Theoretical Physics, TAUP N-161-85.

geometrical interpretation. His interest in "wormholes" and "black holes" first pointed to physical effects whose origin is purely topological—an idea that has matured into our present "instantons," etc. Coining the term "geometrodynamics," he kept physicists aware of the challenge of doing to electromagnetism and the other interactions that which Einstein had achieved for gravitation. The ideal of a fully geometrical theory of physics was always there is his teaching. Nowadays, with the verified unified weak-electromagnetic theory (quantum asthenodynamics or QAD) and with the somewhat less manifestly validated quantum chromodynamics (QCD) for the strong nuclear interactions, that ideal is partially realized by the geometrical character of these theories. As gauge theories, they are already geometrically realized as fiber bundle manifolds (FBM). And yet, beyond the present stage, we have glimpses of further extensive geometrization—supermanifolds in the unification with gravity through supergravity, spontaneous Kaluza–Klein compactification for the internal symmetries, and geometrical lattice-spanned manifolds for superstrings, the present candidate universal theories.

In the foundations of quantum mechanics (QM), Wheeler has again played a leading role. In the fifties he guided and sponsored the introduction of Everett's relative-state ("many-worlds") interpretation.[1] Since then he has mostly pressed for a more positivistic approach, insisting on the essential role of the (physical) observer in mapping out the limits of quantum physics' residual reality and its contents.[2] Here again, as in general relativity, he has put his conviction, imagination, and eloquence at the service of the philosophical foundations.

In the present work, we would like to connect the two areas. We shall utilize the great advances in the geometrization program to throw new light on the problems and prescriptions of quantum-measurement theory, paradoxical as they might appear. We show that the discrete (2-point) nonlocal features of the Einstein–Podolsky–Rosen (EPR) quantum situations are entirely similar to the infinite and continuous nonlocal correlations of gauge theories. Moreover, it seems very natural that such features should appear in measurement theory, based as that theory should be on parallel-transportable observables.

We also present a symmetry interpretation of the nature of the measurement-interaction inducing the collapse of the state vector. Our picture is inspired by the notion of spontaneous symmetry breakdown (SSB) with its geometrical-topological features.

## 2. EPR-TYPE NONLOCAL CORRELATIONS

As background for our geometrical discussion, we summarize current views of the EPR[3] "paradox." The EPR setup has been generalized in various suggested experiments,[4] especially processes involving the detection of spin correlations (these are more conclusive, due to their discrete properties). The key tool has been J. S. Bell's theorem,[5] providing testable inequalities[6] derived from the local-type deterministic alternatives to QM. The results[7,8] have provided an overwhelming refutation of the local-deterministic alternatives and also, at least to a very large extent, a confirmation of QM. The latter somewhat cautious phrase we use is only due to the remaining deterministic alternative of a nonlocal "hidden variables" theory, following de Broglie and Bohm.[9] To this author, however, it appears somewhat futile to replace QM by another (and perhaps more complicated) nonlocal nonlinear theory, seemingly with predictions everywhere identical to QM. Such a theory should thus anyhow be regarded at this stage as an *interpretation*, like the "relative-state interpretation,"[1,10] unless a Popperian falsifying experiment is suggested, capable of distinguishing between these ideas and "orthodox" QM.

The refuted local-deterministic theories make two postulates[11] which do not appear in the axiomatics of QM but were originally assumed by EPR.

(a) The existence of a *committed* "underlying reality," committed in the sense that it is immutably "there," whether or not it is observed; any probabilistic description provided by QM thus results only from our lack of information—rather than from a true undetermined state of affairs, lasting up to the instant of measurement, as claimed by orthodox QM.

(b) The separability of mechanically and spatially (or causally) isolated systems. This feature was especially tested in a recent experiment[8] in which the inexistence of a causal link between the measurements at different points was carefully ensured.

We eschew the discussion of those deeper issues relating to the uncommitted nature of the underlying reality, if any. J. A. Wheeler[2] has given us a beautiful description of the manner in which the elements of that reality are "created" through their interaction with instruments and recording observers, the "Abrahamic" view (see footnote 36 in Ref. 2). The recording which shapes and commits reality is Bohr's "closure by irreversible amplification." (Note that this does not entail "conscious" registration, as is sometimes suggested. Any such suggestion is yet another *interpretation*.) All that is implied by QM is a material interaction serving as a "measurement," something irreversibly recorded by the mere presence of a background of macroscopic matter in the universe.

Returning to the EPR setup, we quote B. d'Espagnat's resumé[12] of the proben status of nonseparability, the denial of EPR's second assumption: "When two quantum systems that originally could be considered as each having a complete set of definite properties of its own...." Again, "... in spite of the possible separation of the corresponding wave-packets [moving shutters may be introduced (see the optical switch in the experiment of Aspect *et al.*[8])], they are—at least as regards spin—to be considered as forming only one system...."

For the sake of clarity, we assume the EPR experiment to involve the emission at the origin of two particles with spins $J=1$ or $J=\frac{1}{2}$ from an original $J=0$ state, and the simultaneous measurement of spin components or polarizations of the emitted states $A$ and $B$ at $\pm x$ respectively (or at least measurements representing space-like separated events). The simplest such experiments exploit the decay of $\pi^0 \to 2\gamma$, or an atomic transition in $Ca^{40}$ with a similar emission of two gamma rays. The two states are found to have correlated helicities $J_3[A(+x)] = \pm\frac{1}{2}$, $J_3[B(-x)] = \mp\frac{1}{2}$. However, the violation of Bell's inequalities contradicts the possibility that this be the result of the definite helicities being there already prior to the measurement (as a *committed* "underlying reality"); and neither does the experimental setup allow the information about the measurement at $A$ to arrive in $B$ before the simultaneous measurement there!

## 3. GAUGE THEORIES AS FIBER BUNDLE MANIFOLDS

QED, the more recent QAD unifying it with the weak interactions, and the (as yet separate) QCD strong interaction theory—all of these are gauge Theories. H. Weyl[13] suggested this feature for QED, and C. N. Yang and R. Mills[14] generalized the concept to non-Abelian gauge groups. It was then shown[15] that such gauge theories should be regarded as specific geometric constructs, namely fiber bundle manifolds (FBM).

A principal fiber bundle[16] $\mathscr{P}(\mathscr{M}, G, \pi, \cdot)$ is a manifold defined by a base space $\mathscr{M}$ (space-time or some subspace of it, in gauge theories), a structure group $G$ [$U(1)$ in QED, $SU(2) \times U(1)$ in QAD, $SU(3)$ in QCD], a "vertical" projection $\pi$,

$$\pi\colon\ p(x, \Gamma) \to x, \qquad p \in \mathscr{P}, \qquad x \in \mathscr{M}, \qquad \Gamma \in G$$

*Locally*, this is just a direct product of the "fiber" $G$ with $x$

$$p(x, \Gamma) = x \otimes \Gamma$$

but if extended, that would trivialize $\mathscr{P}$. The fourth defining element is the right action $(\cdot)$ of $G$ on $\mathscr{P}$,

$$\begin{gathered}\forall \Gamma, \Gamma_1, \Gamma_2 \in G, \; p \in \mathscr{P} \\ \pi(p \cdot \Gamma) = \pi(p), \qquad p \cdot \Gamma = p' \in \mathscr{P} \\ p \cdot (\Gamma_1 \Gamma_2) = (p \cdot \Gamma_1) \cdot \Gamma_2\end{gathered} \tag{1a}$$

The right action is thus a mapping $t$ from $g$, the abstract Lie algebra of $G$, with the commutator as Lie bracket ("LB")

$$\Gamma = \exp i(\gamma, \mathbf{g}) = \exp ig \tag{1b}$$

into $\mathscr{P}_*$, the tangent-manifold of $\mathscr{P}$, i.e. the basis of "vector-field" differential operators realizing $\mathbf{g}$ through Poisson brackers (PB). Fiber bundles such as $\mathscr{P}$ are then endowed with a specific tool, the *connection one form* $\omega$, essential to our discussion. This is an (almost) inverse map from $\mathscr{P}_*$ onto $g$ (it cannot be a true inverse, as the dimensionality of $\mathscr{P}_*$ is larger than that of $G$)

$$\begin{gathered}t: g \to \mathscr{P}_* \\ \forall L, L' \in g, \qquad t(L) = L' \in \mathscr{P}_* \\ [L, L_1]^t_{\mathrm{LB}} = [L^t, L^t_1]_{\mathrm{PB}}\end{gathered} \tag{2}$$

The connection $\omega$ is Lie-algebra valued,

$$\omega = \omega^a L_a, \qquad \omega^a = \omega^a_M \, dz^M, \qquad L^t_a = l^M_a \frac{\partial}{\partial z^M} \tag{3a}$$

$$\omega^a_M \, dz^M \,\lrcorner\, l^N_b \frac{\partial}{\partial z^N} = \delta^N_M \, \delta^a_b \tag{3b}$$

so that we indeed get a mapping

$$\begin{gathered}\omega: \mathscr{P}_* \to g \\ \omega(L^t_a) = L^t_a \,\lrcorner\, \omega = L^t_a \,\lrcorner\, \omega^b L_b = L_a\end{gathered} \tag{3c}$$

As a result of the two mappings—the "dot" right action and the connection, the principal fiber bundle $\mathscr{P}$ is a strongly correlated manifold. The group $G$ and its Lie algebra acquire global characteristics: although there is a fiber $G(x)$ at every point $x \in \mathscr{M}$, all local algebraic generators $L^t_a(x)$ are assured to be correlated at all different points $x, x' \in \mathscr{M}$. This is guaranteed by the connection $\omega$, which reproduces the same abstract (and thus global) $L_a$ at all points.

The original definition of a "vertical" direction is inherent in FBM, as it is defined by the projection $\pi$. Now it is the connection that defines the subset $H$ of horizontal vector fields: those that yield nothing upon contraction with $\omega$, the kernel $H$ at a point $p \in \mathscr{P}_*$.

$$\omega_p(h') = h' \lrcorner\, \omega_p = 0, \qquad h' \in H_p \subset \mathscr{P}_*, \qquad H_{p \cdot g} = H_p \cdot g \tag{4}$$
$$\mathscr{P}_{*p} = H_p + V_p, \qquad H_p = \ker(\omega_p), \qquad V_p = \mathrm{Im}_t(g)$$

$\omega$ defines horizontal displacements via "covariant derivatives"—parallel displacement vector fields ($\lambda$ is a 1-form)

$$D_\mu = \partial_\mu - \omega_\mu^a(x) L_a, \tag{5}$$
$$D = dx^\mu D_\mu, \qquad D\lambda = d\lambda - [\omega, \lambda]$$

The curvature 2-form is given by

$$R = d\omega - \tfrac{1}{2}[\omega, \omega] \tag{6}$$

This is the field strength in a physical theory, while $\omega$ is the potential. The curvature results from the noncommutativity of the covariant derivatives,

$$[D, D] = R \tag{7}$$

It thus represents the changes when "moving" from $A$ to $B$ in a parallelogram $ACBD$ either via $C$ or via $D$. This curvature is then guaranteed to be horizontal and fulfills the Bianchi identity

$$L' \lrcorner\, R = 0 \tag{8}$$
$$DR = 0 \tag{9}$$

The FBM thus supports an observable (the Lie algebra $g$) which is guaranteed to be well defined everywhere over $\mathscr{M}$. It is thus omnipresent everywhere in $\mathscr{M}$, capable of "materializing" locally at any $x$. All of this does not involve special relativity or causality. The omnipresence and correlations are generated by the specific geometry of the FBM.

To rephrase these facts in a language reminiscent of EPR: *two measurements at $x$ and $x'$, at spacelike distances, and even over a curved space, have the same correlated definitions of observables such as spin and can define parallel directions* ("vertical" in their character here) *at both places, without any signals being transmitted.*

## 4. GAUGE THEORIES AND QUANTUM PROBABILITIES

In these definitions, the FBM is a "classical" theory, and yet it has these EPR-type correlations.[17] That "classical" theories should possess the nonseparability features generally associated with QM appears to us to clear the latter theory from some of the accusations often levelled at it.

Alternatively, *we can regard gauge theories—and for all we know these may be the only physical theories relevant to the physical world—as already containing some of the features characterizing the quantum picture.*

When electromagnetism or a Yang–Mills interaction were first described geometrically, *it seemed that only the nature of the fiber variable (the phase, in QED) was a reflection of QM*, the picture being otherwise classical. It then turned out that the *guaranteeing of the conservation of probability (unitarity)*, a highly quantum feature, is also already present in the geometry.[18] The RQFT systematics in $D$-dimensional space generate two unphysical components in the massless $A_\mu$ photon ($D$-vector potential) or for each of the $a = 1,\ldots, N$ internal dimensions of the Yang–Mills potential $A^a_\mu$ ($\mu = 0,\ldots, 3$ in space-time). This is because the physical Hilbert space carries the photon or YM quanta as massless irreducible representations of the Poincaré group. These correspond to the representations of the subgroup $SO(\mathrm{D}-2)$ of the $D-2$ transverse directions for a $D$-dimensional base manifold. In physical Minkowski space, $D = 4$ and the group is $SO(2)$. (We have used the more general definition, current in 1985 when the conjecture of a larger primordial dimensionality, à la Kaluza–Klein, is favored among particle physicists.) The Yang–Mills interaction thus suffers from the unphysical propagation of two inexistent modes, and "ghost fields" are introduced artificially so as to correct these unphysical effects. However, it turns out[18] that there is no need for any artificial intrusion—the "*ghost*" *fields are already present there automatically in the geometry!* Moreover, *the Curci–Ferrari equations*[19] *constraining the ghost fields and guaranteeing unitarity are just the Cartan–Maurer structural equations of the manifold.* The shock of this revelation may explain the attempts on the part of some theoreticians to search (hopelessly it seems) for alternative interpretations of these facts. The geometrical identification has, however, been extremely useful, for instance in the study of "anomalies," yet another RQFT feature of these geometrical gauge interactions.

The ghost fields $\chi^a$ constitute in essence the "vertical" component of the connection, which can be expanded locally once we select a section,

$$\omega = A^a_\mu\, dx^\mu + \chi^a_i\, dy^i = A^a_\mu\, dx^\mu + \chi^a$$

or, on a FBM where $G$ is doubled (for right and left action)

$$\omega = A^a_\mu\, dx^\mu + \chi^a + \bar{\chi}^a \tag{10}$$

and the $\chi^a$ and $\bar{\chi}^a$ fulfill the relevant structural equations, guaranteeing the horizontality of $R$ of (9). As a two-form over the entire FBM dimensionality, $R$ involves differentials in $dx^\mu \wedge dx^\nu$ (the physical field strength or Maxwell tensor $F^a_{\mu\nu}$), in $dx^\mu \wedge dy^i$, $dx^\mu \wedge d\bar{y}^i$, $dy^i \wedge dy^j$, $dy^i \wedge d\bar{y}^j$, $d\bar{y}^i \wedge d\bar{y}^j$. The structural equations are just the statement guaranteeing the vanishing of all but the Maxwell tensor $F^a_{\mu\nu}$.

$$\begin{aligned} &(dx \wedge dy)\colon\ sA = D\chi; && (dx \wedge d\bar{y})\colon\ \bar{s}A = D\bar{\chi} \\ &(dy \wedge dy)\colon\ s\chi = -\tfrac{1}{2}[\chi, \chi]; && (d\bar{y} \wedge d\bar{y})\colon\ \bar{s}\bar{\chi} = -\tfrac{1}{2}[\bar{\chi}, \bar{\chi}] \\ &(dy \wedge d\bar{y})\colon\ s\bar{\chi} = b; && (dy \wedge d\bar{y})\colon\ \bar{s}\chi = -b - [\chi, \bar{\chi}] \\ &\qquad sb = 0 && \qquad \bar{s}b = -[\bar{\chi}, b] \end{aligned} \tag{11}$$

$s$, $\bar{s}$ are vertical exterior derivatives,

$$\begin{aligned} &d_{\mathscr{P}} = d + s + \bar{s}, \qquad s^2 = \bar{s}^2 = 0 \\ &s\bar{s} + \bar{s}s = 0, \qquad sd + ds = 0, \qquad \bar{s}d + d\bar{s} = 0 \end{aligned} \tag{12}$$

$b$ is an auxiliary field.

These are precisely the Curci–Ferrari unitarity equations.[19]

The *FBM geometry of the gauge theories* (and these may be the only theories used by nature) thus "*knows*" *about the probabilistic aspect of QM and guarantees the conservation of unitarity.*

Note that even though we used a localization in (10), the *ghost fields are also really global objects, definable over the entire FBM.*

It thus appears to us that the key to the understanding of EPR is in the recognition of the global geometrical nature of the act of observation and measurement. If gauge theories and their FBM geometry require non-localized operators to define the conserved observable, it is natural that a measurement at $A$ should in effect have measured $B$ as well. The "commitment" at $B$ happens when we measure. These are nonlocal features that correspond to the same approximation as the entire definition of the system to be measured. In his "delayed choice experiment,"[2] Wheeler has even constructed an example on a cosmic scale. It all reflects a "reality" that is highly global in its geometry. What makes EPR appear unreal is the fact that (with Einstein, Podolsky, and Rosen) we tend to visualize the experiment in the sectional setting of Galilean 3-space or even Riemannian 4-space, whereas the true arena is one in which the latter are just the base manifolds. It is the additional dimensionality of the fibers—without which our measured observables would be meaningless—it is in this geometry that we should consider the measurement. *This is a geometry with essentially global features, and they are reflected in QM.*

## 5. THE AHARONOV–BOHM EFFECT

There are other nonlocal QM effects which are again the result of global geometric features in FBM. The simplest is the Aharonov–Bohm effect.[20] Fringes indeed appear on a screen[21] in the interference between two electron beams passing in an $F_{\mu\nu}=0$ region on either side of a solenoid containing a magnetic flux (note that this effect too was greeted with disbelief when it was first predictd). It is highly instructive to follow the derivation of the effect by considering the FBM geometry.[22] The cause is that the base manifold (coordinates $r$, $\theta$)

$$\mathscr{M}_{AB}=\mathbb{R}^2-\Delta^2 \tag{13}$$

is the plane with a finite disk extracted (even if the radius $r_\Delta \to 0$, i.e., only a point missing), a multiply connected region. Using the path integral formalism,[23] one sees that if a nonzero magnetic flux $\Phi$ is confined inside the disc, it induces a change in the one-dimensional unitary representation of the fundamental group of the space, the discrete first homotopy class $\pi_1=\mathbb{Z}$ of the windings $S^1$ around $\Delta^2$. The FBM has "pre-images" with coordinates $(r_S, \pi^{-1}(\theta_S)$, $\pi$ the projection)

$$\theta=\pi^{-1}(\theta_S)=\theta_S,\, \theta_S\pm 2\pi, ..., \theta_S\pm 2n\pi, \qquad n\in\mathbb{Z} \tag{14}$$

The (nonrelativistic) path integral[23] from the electron gun at $\mathbf{r}_G$ to the screen at $\mathbf{r}_S$. ($\mathscr{L}$ is the classical Lagrangian function, $H$ the Hamiltonian)

$$\begin{aligned} K(\mathbf{r}_S, \mathbf{r}_G; t) &:= \langle \mathbf{r}_S |\exp(-itH)| \mathbf{r}_G\rangle \\ &= \int \delta\mathbf{r}(t) \exp\left\{ i\int_0^t d\tau\, \mathscr{L}(\mathbf{r}(\tau), \dot{\mathbf{r}}(\tau); \tau)\right\} \end{aligned} \tag{15}$$

has a kernel resulting from a summation over all paths $n\in\mathbb{Z}$; $n<0$ refers to $|n|$ clockwise windings; $n=0$ just stays to the right of $\Delta$; $n=1$ to its left; $n>1$ implies $n-1$ counterclockwise windings.

$$K_{AB}(\mathbf{r}_S, \mathbf{r}_G; t)=\sum_{n=-\infty} \exp[i\varepsilon(\theta_S-\theta_G+2n\pi)]\, K_n(\mathbf{r}_S, \mathbf{r}_G; t) \tag{16a}$$

with a phase shift

$$\varepsilon=e\frac{\Phi}{hc} \tag{16b}$$

for the free-particle kernels

$$K_n(r_S, \theta_S, r_G, \theta_G; t)$$
$$= (r_S r_G)^{-1/2} \int_{-\infty} \frac{dk}{2\pi} \exp\{ik(\theta_S - \theta_G + 2n\pi)\} \chi_{|k|}(r_S, r_G; t) \tag{17a}$$

where the reduced kernel

$$\chi_{|k|}(r_S, r_G; t) = \int \delta r(\tau) \exp\left\{i \int_0^t d\tau \left[\frac{m}{2} \dot{r}^2(\tau) - \frac{k^2 - \frac{1}{4}}{2mr^2(\tau)}\right]\right\}$$
$$= \frac{m}{it} (r_S r_G)^{1/2} \exp\left[\frac{im}{2t}(r_S^2 + r_G^2)\right] (-i)^{|k|} J_{|k|}(mr_S r_G/t) \tag{17b}$$

$J_{|k|}(pr)$ the appropriate Bessel function. The calculation is that of a path integral over the universal covering space, the FBM of (11), using appropriate techniques.[24]

Summing up, the nonlocal QM fringes can be calculated directly from the FBM geometry, a sum over all paths on $\mathscr{P}_{AB}$ and not on $\mathscr{M}_{AB}$. *The effect results from a global feature of* $\hat{\mathscr{P}}_{AB}$.

Wu and Yang[15] have in fact *used these global features* of the Aharonov–Bohm effect to define a gauge theory. Working relativistically, it is the phase factor (integrating over a closed loop) or path integral

$$\exp\left\{\frac{ie}{\hbar c} \int_S^G A_\mu(x)\, dx^\mu\right\} = \exp\left\{-\frac{ie}{\hbar c} \Phi\right\} \tag{18a}$$

that should, on the one hand, cause fringes to appear at $Q$, but should also display the same fringes for two magnetic fluxes (18a), (18b) related by

$$\Delta\Phi = \Phi_a - \Phi_b = \frac{\hbar c}{e} n, \qquad n \in \mathbb{Z} \tag{18b}$$

This necessitates a gauge invariance of the left-hand side of (18a) for a $\Delta\Phi$ as in (18b). The electron wave function $\psi_a$ and potential $(A_\mu)_a$ have to be gauge transformable into $\psi_b$ and $(A_\mu)_b$ without affecting (18a). This is solved by

$$\psi_b = S^{-1} \psi_a, \qquad S = e^{-i\alpha(x)}$$
$$(A_\mu)_b = (A_\mu)_a - i\frac{\hbar c}{e} S \partial_\mu S^{-1} = (A_\mu)_a + \frac{\hbar c}{e} \partial_\mu \alpha(x) \tag{19a}$$

$S$ must be single-valued, but $\alpha$ need not be, provided

$$\Delta\alpha = \frac{e}{\hbar c} \oint [(A_\mu)_b - (A_\mu)_a]\, dx^\mu = \frac{e}{\hbar c} (\Phi_b - \Phi_a) \tag{19b}$$

is the gain for every winding.

It is thus through the global feature that the electromagnetic gauge invariance and the entire electromagnetic interaction are defined. Yang[(25)] has generalized this idea as a definition for any Yang–Mills gauge, thus emphasizing the significance of the *global* features.

In contradistinction to FBM-structured gauge theories, gravity was generally studied locally rather than with a global emphasis. However, Aharonov–Bohm effects have been pointed out in gravity recently,[(26)] and the global structure is sure to be even richer than in an FBM. The local structure is now understood to be that of a "soft" group manifold (SGM)[(27)] with spontaneous fibration[(28)] and factorization,[(27)] and these provide for large global variety.

## 6. MONOPOLES AND INSTANTONS

Another example of a global effect is the Dirac monopole. As reintroduced by Wu and Yang,[(15)] this corresponds to a base manifold

$$\mathscr{M}_D = \mathbb{R}^3 - \Delta^3 \tag{20}$$

3-space with a sphere ($r_\Delta \to 0$) at the origin removed. The Dirac monopole with strength $g$ is placed there, at the origin. Taking an uncontractable sphere $S^2$ in $\mathscr{M}_D$, they show first that it is impossible to have a singularity-free $A$ over the entire sphere. A loop integral

$$\Omega = \oint A_\mu\, dx^\mu \tag{21}$$

taken on a horizontal $(r, \theta)$ circle will, by Gauss' theorem, correspond to the flux through that circle

$$\Omega(r, \theta) = 2\pi g(1 - \cos\theta) \tag{22}$$

We start with a small $\theta$ at the North Pole (NP)

$$\Omega(r, 0) = 0$$

and keep sliding to the equator

$$\Omega(r, \pi/2) = 2\pi g$$

At $\theta = \pi - \varepsilon$ we have

$$\Omega(r, \pi - \varepsilon) \to 4\pi g \qquad \text{as} \quad \varepsilon \to 0$$

but with zero cross section

$$\Omega(r, \pi) = 0$$

i.e., the local description has to fail, at least at the South Pole (SP). This is typical of an FBM, in which indeed one only invokes regional coordinate charts $U_\alpha$, $U_\beta$. Taking $U_\alpha$ from the NP "down" all the way beyond the equator, and $U_\beta$ from the SP "up" to the equator and beyond, a band with $\theta = \pi/2 + \delta$ just $S$ of the equator can be described in both overlapping coordinate charts, in the region

$$(\pi/2) - \delta \leqslant \theta \leqslant (\pi/2) + \delta, \qquad \delta < \pi/2$$

Taking for $A_\mu$ at $t = 0$ the spherical components (axially symmetric)

$$\begin{aligned} U_\alpha&: A_t = A_r = A_\theta = 0, \qquad A_\phi = \frac{g}{r \sin \theta}(1 - \cos \theta) \\ U_\beta&: A_t = A_r = A_\theta = 0, \qquad A_\phi = \frac{-g}{r \sin \theta}(1 + \cos \theta) \end{aligned} \tag{23a}$$

we find that the gauge transformation, defined for an FBM as the transformation relating the $U_\alpha$ and $U_\beta$ descriptions in the overlap region, is given by

$$S_{\alpha\beta} = \exp(-i\alpha) = \exp\left(\frac{2ige\phi}{\hbar c}\right) \tag{23b}$$

which has to be single-valued, thus yielding Dirac's quantization condition,

$$\frac{2ge}{\hbar c} = n, \qquad n \in \mathbb{Z} \tag{24}$$

Therefore

$$S_{\alpha\beta} = \exp(in\phi) \tag{23c}$$

The phase factor (21) will have a unique value. All of that, of course, provided we stay out of $\Delta^3$ itself. In fact, homotopy tells us that when

$$\pi_1(U(1)) = \mathbb{Z} \tag{25}$$

we shall have quantized monopoles, corresponding to the *first Chern class* of $\mathscr{P}(\mathscr{M}_D, U(1), \pi, \cdot)$. For a Yang–Mills theory with $SU(2)$ as structure group, there are only two classes of monopoles, depending on whether one is on the sheet connected to $\pm 1$ in $SU(2)$.

'
t Hooft and Polyakov[29] have shown how a Yang–Mills theory generates magnetic monopoles under spontaneous symmetry breakdown (SSB). The SSB is achieved through the action of a Nambu–Goldstone SSB field, with a nonvanishing vacuum expectation value (VEV) according to Higgs and Kibble. Such a field $\chi$ behaves as a representation $\mathscr{D}(G)$ of the gauge group $G$. The nonzero VEV

$$\langle 0|\chi^a|0\rangle \neq 0 \tag{26}$$

breaks $G$-symmetry, leaving a stability subgroup $H$. For $\chi^3$, the third component of an $SO(3)$ vector, when $G = SO(3)$, $H = SO(2)$. The SSB is thus described by $S^2 = SO(3)/SO(2)$. The vacuum in Minkowski space (i.e., the base manifold $\mathscr{M}_{3.1}$) is given by a sphere $S^2$ at $r \to \infty$, so that SSB is given by the second homotopy group, which reduces according to a theorem,

$$\pi_2(G/H) = \pi_1(H) = \mathbb{Z} \tag{27}$$

Similarly, if we use the supergroup[30] $G = SU(2/1)$ as the simple unification of the Salam–Weinberg $SU(2) \times U(1)$ QAD, and take the SSB as given by the sixth component of an $SU(2/1)$ octet, $H = U(1)$ and (27) holds. (Note that it does not hold for $G = SU(2) \times U(1)$ and the usual choice of a Higgs–Kibble meson.)

Instantons[31] are yet another type of global objects, solutions to the Yang–Mills equations in a classical Euclidean space-time. It has been shown that tunnelling transitions between any two such solutions correspond to global quantum solutions.[32] In the Euclidean base space $\mathscr{M}_4$, the vacuum corresponds to the asymptotic equatorial 3-space $S^3$, and the instantons are given by the third homotopy group

$$\pi_3(SU(2)) = \mathbb{Z} \tag{28}$$

For $SU(3)$, the count is more complicated, with seven zero modes, but we shall not discuss this result here. We only mention that it is now believed that instantons (or inter-instanton transitions = QCD vacua) with their topological charges play an important role in QCD and quark confinement.

We have gone into some detail in relating the Aharonov–Bohm and monopole cases, less for the more complicated instantons. There are additional global effects relating to "Gribov ambiguities," etc. Summing up, we emphasize that *the geometrical-topological "classical" structure of our present field theories*—the gauge models—*is highly global and materializes in very nonlocal effects.*

## 7. APPARENT ACAUSALITY AND RQFT

Nonrelativistic QM cannot pretend to causal propagation of information, as it assumes Galilean covariance, with $c \to \infty$. To check on this aspect, we are required to use the relativistic theory in which QM is embedded, RQFT. This theory led to difficulties for a while in the 1960's, when the successes of QED seemed hard to generalize. After the successes of 1971 (renormalization of Yang–Mills theories, including SSB) we seem to have an excellent theory. Supersymmetry, the embedding of the Poincaré group in a supergroup, has even yielded finite four-dimensional field theories, such as "$N=4$ super Yang–Mills," with no ultraviolet divergences altogether.

In RQFT, the Yang–Mills theories have the gauge group algebra $L_a \in g$ of (1b), (2) realized through the Noether theorem,

$$L_a = \int d\sigma^r \frac{\partial \mathscr{L}}{\partial(\partial_\mu \phi^r)} [\mathscr{D}(L_a)\phi]^r \tag{29}$$

where $\sigma^\mu$ is a spacelike 3-hyperplane in $\mathscr{M}_{3,1}$, $\mathscr{L}$ the Lagrangian 4-density, and $\mathscr{D}(L_a)$ the unitary representation of $L_a \in g$ carried by the field components $\phi^r(x)$. The generator is thus a global feature, integrated over the entire space. To understand its global action in terms of QM, we note that (29) represents the *charge* in a Coulomb-like sense, for the Yang–Mills interaction. Dynamically, we know that the non-Abelian group results in the gauge-field potential itself carrying charge, so that the long-range behavior is $\sim r$ rather than $\sim r^{-1}$. However, the Coulomb charge can be regarded as the zero-momentum limit of the form factor,

$$j(\mathbf{k}) = \int d^3x \exp(i\mathbf{k} \cdot \mathbf{x})\, j^0(x) \tag{30}$$

where $j^0(x)$ is the time component of the current density, the integrand of (29). *With* $\mathbf{k} \to 0$, *the uncertainty principle spreads* (30) *over an ever larger region of space. The global action of the Lie algebra of the observables in the FBM is thus guaranteed by RQFT to be spread out over the entire space!*

Indeed, in the sixties, $S$-matrix theory, entirely represented in momentum space, was the favored theory. It had originally been advanced by Heisenberg to avoid future difficulties with space quantization. We observe that within RQFT, it is $\mathbf{k} \to 0$ momentum space that plays the most important conceptual role for QM.

Summing up, we have shown that although the measurement apparatus in EPR may be localized at $A$ as a single-body wave function, its action as a measurement device is just the localization $L'(x_A)$ [Eq. (2)] of a globally defined operator $L$ whose region of definition (resulting from Eqs. (29) and (30)) extends all the way to $x_B$; otherwise there would be no meaning in preserving the same definition of an observable at $A$ and $B$, a precondition in checking whether or not that measured variable (spin, etc.) maintains its correlation and how it is done. The only known alternative to RQFT as a relativistic generalization of QM is the method of path integrals, and we have seen in Sects. 5 and 6 that this also involves globally defined operations and is even more explicitly a topologically global theory.

## 8. THE COLLAPSE OF THE STATE VECTOR VIEWED AS SSB

The great remaining "mystery" of QM is the state-vector collapse, viewed as a physical process. This is the above-mentioned "closure by irreversible amplification"[(2)] of N. Bohr.

It would appear that this process is very much in the class of a spontaneous symmetry breakdown. Prior to the masurement, we have a quasi-symmetric state vector:

$$\alpha\psi^{\uparrow} + \beta\psi^{\downarrow} \tag{31}$$

for the two superposed spin-up and spin-down eigenstates, for instance. The measurement causes only one of the above to subsist (in our world, at least, if we take the relative-state interpretation). This is entirely similar to a Goldstone–Nambu or Higgs–Kibble situation, in which the stable solution is not the symmetric one. Similarly, if we should measure our electron's position, we would break a certain translation quasi-invariance in which all $x$-space is allowed, though constrained by the values of $\psi(x)$. In all cases, of course, the uncertainty principle compensates with a new quasi-invariance (=indeterminacy) in the conjugate variable.

Modern physics both in the particle domain and in condensed matter is very much dominated by SSB and corresponding phase transitions. In that sense, we should not be surprised at the nature of the measurement process. It just means that *the measuring apparatus is by definition a quantum-mechanical array with a Goldstone–Higgs-like interaction as its main required feature.*

This article has been in the nature of an effort at demystification. It seems to us that just as Newton's mechanics gained in insight from developments in the next century—from the work of d'Alembert, Laplace, Lagrange, Euler, Hamilton, and Jacobi—so can QM be better understood today. Gauge theories and spontaneous symmetry breakdown have acquired a key role in the actual physics—and have also clarified much of what is mysterious in QM. It could be that this does not bring the quantum picture of "reality" any closer to our everyday intuition, but it shows that it is in harmony with geometrical concepts that we seem to have accepted independently. It also exorcises any apparent contradiction between the quantum scene and the concepts which have evolved in classical (relativistic) field theory.

## REFERENCES


1. J. A. Wheeler and W. H. Zurek, *Quantum Theory of Measurement* (Princeton University Press, Princeton, New Jersey, 1983).
2. J. A. Wheeler, "*Delayed-Choice Experiments and the Bohr–Einstein Dialog*" (American Philosophical Society Publication, 1981).
3. A. Einstein, B. Podolsky, and N. Rosen, *Phys. Rev.* **47**, 777 (1935).
4. M. Schwarz, *Phys. Rev. Lett.* **6**, 556 (1961); B. d'Espagnat, *Nuovo Cimento* **20**, 1217 (1961); R. Armenteros *et al.*, *Proceedings, 1962 International Conference on High-Energy Physics (CERN)*, J. Prentki, ed. (CERN Publication, Geneva, 1962); D. Bohm and Y. Aharonov, *Phys. Rev.* **108**, 1070 (1957).
5. J. S. Bell, *Rev. Mod. Phys.* **38**, 447 (1966); *Physics* **1**, 195 (1964).
6. J. F. Clauser, M. A. Horne, A. Shimony, and R. A. Holt, *Phys. Rev. Lett.* **23**, 880 (1969).
7. J. F. Clauser and A. Shimony, *Rep. Prog. Phys.* **41**, 1881 (1978); A. Aspect, P. Grangier, and G. Roger, *Phys. Rev. Lett.* **49**, 91 (1982).
8. A. Aspect, J. Dalibard, and G. Roger, *Phys. Rev. Lett.* **49**, 1804 (1982).
9. H. Everett III, *Rev. Mod. Phys.* **29**, 454 (1957); J. A. Wheeler, *Rev. Mod. Phys.* **29**, 463 (1957); N. Graham, Ph.D. Thesis, University of North Carolina at Chapel Hill, 1970; B. S. DeWitt, *Phys. Today* **23**, 30 (1970).
10. D. Bohm, *Phys. Rev.* **85**, 166, 180 (1952); D. Bohm and B. Hiley, *Found. Phys.* **14**, 270 (1984); L. de Broglie, *Tentative d'Interpretation Causale et Nonlineaire de la Mechanique Ondulatoire* (Gauthier-Villars, Paris, 1956); J. S. Bell, *Found. Phys.* **12**, 989 (1982).
11. B. d'Espagnat, *Sci. Am.* **241/5**, 158 (1979).
12. B. d'Espagnat, *Conceptual Foundations of Quantum Mechanics* (W. A. Benjamin, Menlo Park, California, 1971), Chap. III, Sects. 7 and 9.
13. H. Weyl, *Z. Phys.* **56**, 330 (1929).
14. C. N. Yang and R. Mills, *Phys. Rev.* **95**, 631 (1954); **96**, 191 (1954).
15. E. Lubkin, *Ann. Phys. (N.Y.)* **23**, 233 (1963); *J. Math. Phys.* **5**, 1603 (1964); A. Trautman, *Rep. Math. Phys.* **1**, 29 (1970); H. G. Loos, *Phys. Rev. D* **10**, 4032 (1974); T. T. Wu and C. N. Yang, *Phys. Rev. D* **12**, 3845 (1975).
16. W. Drechsler and M. E. Mayer, *Differential Geometry and Gauge Theories*, Springer-Verlag Lecture Notes in Physics, *Vol.* 67 (Springer-Verlag, New York, 1977); S. Sternberg, *Lectures on Differential Geometry* (Prentice-Hall, Englewood Cliffs, New Jersey, 1964);

Y. Ne'eman, *Symétries Jauges et Variétés de Groupe* (University of Montréal Press, Montréal, 1979).

17. Y. Ne'eman, *Proc. Natl. Acad. Sci. U.S.A.* **80**, 7051 (1983).
18. J. Thierry-Mieg, *J. Math. Phys.* **21**, 2834 (1980); *Nuovo Cimento A* **56**, 396 (1980); Y. Ne'eman and J. Thierry-Mieg, *Proc. Natl. Acad. Sci. U.S.A.* **77**, 720 (1980); *Nuovo Cimento A* **71**, 104 (1982); L. Beaulieu and J. Thierry-Mieg, *Nucl. Phys. B* **197**, 477 (1982).
19. G. Curci and R. Ferrari, *Nuovo Cimento A*, **35**, 1, 273 (1978).
20. Y. Aharonov and D. Bohm, *Phys. Rev.* **115**, 485 (1959); **123**, 1511 (1961); **125**, 2192 (1962); **130**, 1625 (1963); W. Ehrenberg and R. E. Siday, *Proc. Phys. Soc. London B* **62**, 8 (1949).
21. R. G. Chambers, *Phys. Rev. Lett.* **5**, 3 (1960).
22. G. Morandi and E. Menossi, *Eur. J. Phys.* **5**, 49 (1984).
23. R. P. Feynman, *Rev. Mod. Phys.* **29**, 337 (1948); R. P. Feynman and A. R. Hibbs, *Quantum Mechanics and Path Integrals* (McGraw-Hill, New York, 1965).
24. C. Morette-DeWitt, A. Maheshwari, and B. Nerlin, *Phys. Rep.* **50**, 255 (1979); M. S. Marinov, *Phys. Rep.* **60**, 1 (1980).
25. C. N. Yang, *Phys. Rev. Lett.* **33**, 455 (1974).
26. L. H. Ford and A . Vilenkin, *J. Phys. A: Math. Gen.* **14**, 2353 (1981).
27. Y. Ne'eman and T. Regge, *Phys. Lett. B* **74**, 54 (1978); *Riv. Nuovo Cimento* **1**, #5 (series 3), 1 (1978); Y. Ne'eman, E. Takasugi, and J. Thierry-Mieg, *Phys. Rev. D* **22**, 2371 (1980).
28. J. Thierry-Mieg and Y. Ne'eman, *Ann. Phys. (N.Y.)* **123**, 247 (1979).
29. G. 't Hooft, *Nucl. Phys. B* **79**, 276 (1974); A. Polyakov, *JETP Lett.* **20**, 194 (1974); J. Arafune, P. G. D. Freund, and C. J. Goebel, *J. Math. Phys.* **16**, 433 (1975).
30. Y. Ne'eman, *Phys. Lett. B* **81**, 190 (1979); D. B. Fairlie, *Phys. Lett. B* **82**, 97 (1979); Y. Ne'eman and J. Thierry-Mieg, in *Differential Geometric Methods in Mathematical Physics*, P. L. Garcia, A. Perez-Rendon, and J. M. Souriau, eds., Springer-Verlag Lecture Notes in Math., Vol. 836 (Springer-Verlag, New York, 1980), pp. 318–348; *Proc. Natl. Acad. Sci. U.S.A.* **79**, 7068 (1982).
31. A. Polyakov, *Phys. Lett. B* **59**, 82 (1975); A. Belavin, A. Polyakov, A. Schwartz, and Y. Tyupkin, *Phys. Lett. B* **59**, 85 (1975).

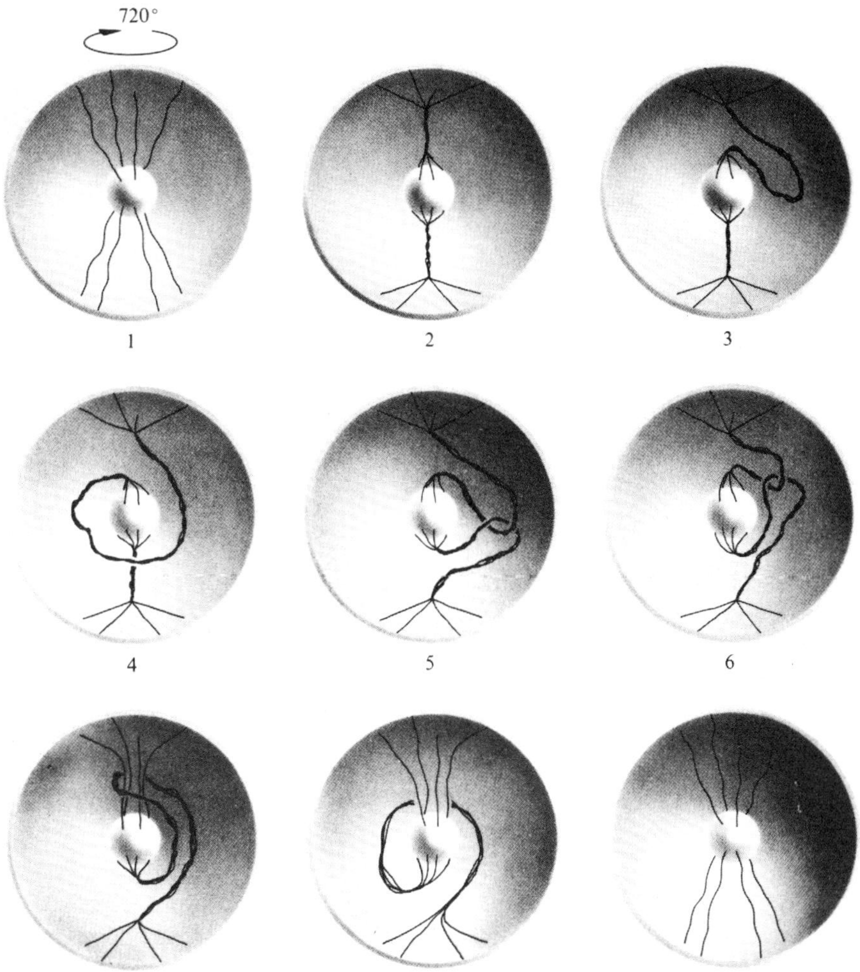


*Plate 31*

# Joint Wigner Distribution for Spin-1/2 Particles[1]

**Leon Cohen**[2] **and Marlan O. Scully**[3,4]

*Received July 19, 1985*

*The Wigner distribution has proven to be a useful tool in many quantum problems requiring a joint distribution of position and momentum. In the present paper we develop a joint distribution for spin components within the spirit of the Wigner distribution. This distribution provides an insight into the quantum theory of measurement. We also discuss how one may write joint distributions for two arbitrary noncommuting operators.*

## 1. INTRODUCTION

The Wigner distribution[(1)] has found numerous applications in a variety of fields. It is a function of position and momentum with the important property that it yields the correct quantum-mechanical marginal distributions

$$\int P(q, p)\, dp = |\psi(q)|^2 \tag{1}$$

$$\int P(q, p)\, dq = |\phi(p)|^2 \tag{2}$$

where $\psi$ and $\phi$ are the position and momentum wave functions respectively. Although $P(q, p)$ cannot be considered a proper probability distribution, since it can become negative, it can nonetheless be manipulated and used as such with considerable profit. For a review of the Wigner dis-

[1] It is a pleasure to dedicate this paper tọ Professor John A. Wheeler—physicist, philosopher, poet—an inspiration to us all.

[2] Center for Advanced Studies, University of New Mexico, Albuquerque, New Mexico 87131, and Hunter College of the City University, New York, New York 10021.

[3] Center for Advanced Studies, University of New Mexico, Albuquerque, New Mexico 87131, and Max-Planck Institut fur Quantenoptik, D-8046 Garching bei Munchen, West Germany.

[4] Work supported by the Office of Naval Research.

tribution (see Ref. 2). There are an infinite number of distributions satisfying the marginal distributions, and a method for readily generating them has been given.[(3)]

A quantum-mechanical particle, as contrasted to a classical particle, requires the fundamental property of spin for its complete description. The aim of this paper is to show how a joint distribution for spin components may be written in the same spirit as the Wigner distribution. Preliminary investigations along these lines have recently been published.[(4)] Also, O'Connell and Wigner[5] have studied the inclusion of spin and statistics in the Wigner distribution. That interesting problem is different in scope and direction from the present work.

To this end we consider two spin components $\hat{\sigma}(\theta_1, \phi_1) \equiv \hat{\sigma}_1$ and $\hat{\sigma}(\theta_2, \phi_2) \equiv \hat{\sigma}_2$ pointing in the $\theta_1 \phi_1$ and $\theta_2 \phi_2$ directions respectively, as in Appendix A. What we seek is a joint distribution of the $c$ numbers $s_1$ and $s_2$ consistent with the proper quantum-mechanical marginal distributions. In particular, we want a joint distribution of the variables $s_1$ and $s_2$ such that

$$\int P(s_1, s_2)\, ds_2 = P(s_1) \tag{3}$$

$$\int P(s_1, s_2)\, ds_1 = P(s_2) \tag{4}$$

where $P(s_1)$ and $P(s_2)$ are given by Eq. (A.7) and (A.8).

In Section 2 we consider possible joint distributions which satisfy Eqs. (3) and (4) as generated via the characteristic-function approach. The case of symmetric ordering is found, in Section 3, to lead to a particularly simple result. The case of Weyl ordering is developed in Section 5. In Section 4 connection is made with recent work on the Einstein–Podolsky–Rosen–Bohm (EPRB) problem which was based on quantum distribution theory.

## 2. CHARACTERISTIC FUNCTIONS

A systematic approach for writing joint distributions may be achieved by using characteristic functions.[(6)] For a joint distribution of two variables these are defined as

$$M(\xi, \tau) = \langle e^{i\xi s_1 + i\tau s_2} \rangle = \int e^{i\xi s_1 + i\tau s_2} P(s_1, s_2)\, ds_1\, ds_2 \tag{5}$$

[5] We note, however, that quantum distribution theory has been extensively applied in studying the dynamics of spin-1/2 systemas as they appear in quantum optics. However, the goal and thrust of those investigations is very different from that of the present work. See, for example, Ref. 5.

A knowledge of $M(\xi, \tau)$ will give the distribution by Fourier inversion. $M(\xi, \tau)$ may be calculated by the quantum-mechanical rules of finding expectation values,

$$M(\xi, \tau) = \langle \psi | \hat{M} | \psi \rangle \tag{6}$$

where $\hat{M}(\xi, \tau)$ is the operator corresponding to the classical function

$$\hat{M}(\xi, \tau) \leftrightarrow e^{i\xi s_1 + i\tau s_2} \tag{7}$$

As is well known, different ordering rules yield different $M(\xi, \tau)$ and hence different distribution functions. We will consider here two possibilities, namely the spin analog of the symmetrization rule

$$\hat{M}_S \leftrightarrow \tfrac{1}{2}[e^{i\xi\hat{\sigma}_1}, e^{i\tau\hat{\sigma}_2}]_+ = \tfrac{1}{2}[e^{i\xi\hat{\sigma}_1} e^{i\tau\hat{\sigma}_2} + e^{i\tau\hat{\sigma}_2} e^{i\xi\hat{\sigma}_1}] \tag{8}$$

and the Weyl ordering

$$\hat{M}_W \leftrightarrow e^{i\xi\hat{\sigma}_1 + i\tau\hat{\sigma}_2} \tag{9}$$

Before proceeding to the actual calculations of the $M$'s we note some general features of characteristic functions. The marginal conditions can be expressed as requiring that

$$M(\xi, 0) = \langle e^{i\xi\hat{\sigma}_1} \rangle = \cos \xi + i \langle \hat{\sigma}_1 \rangle \sin \xi \tag{10}$$

$$M(0, \tau) = \langle e^{i\tau\hat{\sigma}_2} \rangle = \cos \tau + i \langle \hat{\sigma}_2 \rangle \sin \tau \tag{11}$$

Hence, to check whether a distribution function satisfies the marginal conditions one can check that the characteristic function satisfies Eq. (10) and (11).

Also, if we have a function that satisfies Eqs. (10) and (11), then an infinite number of characteristic function (and hence distribution functions) may be obtained by forming [3a]

$$M_{\text{new}}(\xi, \tau) = f(\xi, \tau)\, M(\xi, \tau) \tag{12}$$

where $f(\xi, \tau)$ is function such that

$$f(\xi, 0) = f(0, \tau) = 1 \tag{13}$$

Clearly, if $M(\xi, \tau)$ satisfies Eqs. (10) and (11), so does $M_{\text{new}}(\xi, \tau)$.

We should point out that although we have been calling the $M(\xi, \tau)$'s characteristic functions, they are not proper in that they do not necessarily yield a positive joint distribution.

The characteristic functions of the single variable given by Eqs. (10) or (11) are proper, since they are derivable from a positive distribution, namely Eq. (A.8).

## 3. CALCULATION OF $M_S$ AND ITS DISTRIBUTION

Using

$$e^{i\xi\hat{\sigma}_1} = \cos\xi + i\hat{\sigma}_1 \sin\xi \tag{14}$$

and similarly for $\hat{\sigma}_2$, we have that

$$\tfrac{1}{2}[e^{i\xi\hat{\sigma}_1}, e^{i\tau\hat{\sigma}_2}]_+ = \cos\xi\cos\tau + i\hat{\sigma}_1 \sin\xi\cos\tau + i\hat{\sigma}_2 \sin\tau\cos\xi \\ - \sin\xi\sin\tau\cos\delta \tag{15}$$

where we have set, as per Eq. (A.9),

$$\cos\delta = \tfrac{1}{2}[\hat{\sigma}_1\hat{\sigma}_2 + \hat{\sigma}_2\hat{\sigma}_1] \tag{16}$$

and where $\delta$ is the angle between $\hat{\sigma}_1$ and $\hat{\sigma}_2$. Therefore,

$$M_S(\xi, \tau) = \cos\xi\cos\tau + i\langle\hat{\sigma}_1\rangle \sin\xi\cos\tau \\ + i\langle\hat{\sigma}_1\rangle \sin\tau\cos\xi - \sin\xi\sin\tau\cos\delta \tag{17}$$

To obtain the distribution we take the Fourier inverse of $M_S\ (\xi, \tau)$

$$P_S(s_1, s_2) = \frac{1}{4\pi^2}\int M(\xi, \tau)\, e^{-i\xi s_1 - i\tau s_2}\, d\xi\, d\tau \tag{18}$$

which straightforwardly gives

$$P_S(s_1, s_2) = P_{++}\ \delta(s_1 - 1)\,\delta(s_2 - 1) + P_{+-}\ \delta(s_1 - 1)\,\delta(s_2 + 1) \\ +\ P_{-+}\ \delta(s + 1)\,\delta(s_2 - 1) + P_{--}\ \delta(s_1 + 1)\,\delta(s_2 + 1) \tag{19}$$

where

$$\begin{aligned} P_{++} &= \tfrac{1}{4}(1 + \langle\hat{\sigma}_1\rangle + \langle\hat{\sigma}_2\rangle + \cos\delta) \\ P_{+-} &= \tfrac{1}{4}(1 + \langle\hat{\sigma}_1\rangle - \langle\hat{\sigma}_2\rangle - \cos\delta) \\ P_{-+} &= \tfrac{1}{4}(1 - \langle\hat{\sigma}_1\rangle + \langle\hat{\sigma}_2\rangle - \cos\delta) \\ P_{--} &= \tfrac{1}{4}(1 - \langle\hat{\sigma}_1\rangle - \langle\hat{\sigma}_2\rangle + \cos\delta) \end{aligned} \tag{20}$$

An alternative derivation is given in Appendix B.

Finally we note that we may take $\phi = 0$ and write the quantities $P_{\pm\pm}$ as given by Eq. (20) in the simple form

$$P_{\pm\pm} = \tfrac{1}{2}\langle[\hat{\pi}_\pm(\theta_1), \hat{\pi}_\pm(\theta_2)]_+\rangle \tag{20a}$$

$$= \tfrac{1}{8}\langle[(1 \pm \hat{\sigma}_1(\theta_1)), (1 \pm \hat{\sigma}_2(\theta_2))]_+\rangle. \tag{20b}$$

This observation is germaine to the discussion of the EPRB problem in the next section.

## 4. EPRB VIA TWO-PARTICLE AND SINGLE-PARTICLE JOINT CORRELATION FUNCTIONS

In a recent work[(4)] the EPRB problem[(7)] was analyzed using quantum distribution theory.[6] There it was noted that the simultaneous passage of both partners (spins 1 and 2) comprising a state $|\psi\rangle$ through two Stern–Gerlach apparati (SGA) oriented at angles $\theta_1$ and $\theta_2$ to the vertical is given by

$$\mathscr{P}_2(\theta_1, \theta_2) = \langle\psi|\, \pi_\theta^{(1)} \pi_\theta^{(2)}\, |\psi\rangle \tag{21}$$

where

$$\pi_\theta^{(i)} = |\theta_i\rangle\langle\theta_i|, \qquad i = 1, 2 \tag{22}$$

That Eq. (21) is indeed this probability is easily seen by noting that

$$\mathscr{P}_2(\theta_1, \theta_2) = |[\langle\theta_1| \otimes \langle\theta_2|]\,|\psi\rangle|^2 = \langle\psi|\theta_1\rangle \otimes |\theta_1\rangle\langle\theta_1| \otimes \langle\theta_2|\psi\rangle \tag{23}$$

and combining Eq. (22) with Eq. (23) we immediately get Eq. (21). Furthermore, we have (see Appendix A)

$$\hat{\pi}_\theta^{(i)} = \tfrac{1}{2}(1 + \sigma_z^{(i)} \cos\theta_i + \sigma_x^{(i)} \sin\theta_i) \tag{24}$$

Inserting Eq. (24) into Eq. (21) (and taking $|\psi\rangle$ to be a singlet state) yields the usual result

$$\mathscr{P}_2(\theta_1, \theta_2) = \tfrac{1}{4}[1 - \cos(\theta_1 - \theta_2)] \tag{25}$$

Thus far everything is straightforward. The only (slightly) different twist to this treatment of the EPRB problem is the use of projection operators, $\pi(\theta)$, to calculate $\mathscr{P}_2(\theta_1, \theta_2)$. The motivation for taking the trouble to formulate the problem in this way was the connection with quantum distribution theory as developed in Ref. 4. In the present section we see that the simultaneous passage probability as given by Eq. (21) leads to an interesting connection with the *single-particle* analog of the EPRB argument.

The key point of the EPRB "paradox" is the extension of the *two-particle* singlet state logic to *one particle* via reasonable physical arguments, as in Refs. 7 and 9 and as follows. First we measure $\hat{\pi}_\theta^{(1)}$, which is to say $\hat{\sigma}_\theta^{(1)}$, on the 1st particle, and the same sort of measurement is made on particle 2 by a second SGA at $\theta_2$. Then it is argued that (due to the anticorrelation of

[6] The present spin-1/2 example is due to Bohm.[(8)]

the spins) we now "know" both $\langle\hat{\sigma}_{\theta_1}^{(1)}\rangle$ and $\langle\hat{\sigma}_{\theta_2}^{(1)}\rangle$ for spin 1. But since $\hat{\sigma}_x$ and $\hat{\sigma}_z$ are noncommuting operators we are in trouble with the tenets of quantum theory. This is the essence of the EPRB paradox.

But in the previous section we have developed the means for treating such problems, i.e., the quantum phase-space distributions $P(s_1, s_2)$. This is just the sort of tool we need to provide a proper quantum-mechanical answer to the present question, i.e.,

> "What is the probability that a single particle have simultaneous projections along $\theta_1$ and $\theta_2$ or, in technical jargon, what is the joint probability $P(\theta_1, \theta_2)$ for our spin?"

To this end, consider the *single-particle* expectation value given by Eq. (20a). There is an obvious similarity between Eqs. (20a) and (21), but whereas the two-particle probability (21) follows from basic quantum-mechanical arguments [Eqs. (22) and (23)], the logic leading to the single-spin joint distribution parameter (20a) does not allow an immediate interpretation in terms of simple probabilities. Nevertheless, the single-particle average (20a) turns out to have an interesting connection with two-particle average equations (21) and (25). To see this we write

$$\hat{\pi}_+(\theta_\alpha) = \tfrac{1}{2}(1 + \hat{\sigma}_z \cos\theta_\alpha + \hat{\sigma}_x \sin\theta_\alpha), \qquad \alpha = 1, 2 \tag{26a}$$

and upon inserting Eq. (26a) into Eq. (20a) we find (for $|\psi\rangle = |\uparrow\rangle$)

$$P_{++}(\theta_1, \theta_2) = \tfrac{1}{4}[1 + \cos\theta_1 + \cos\theta_2 + \cos(\theta_1 - \theta_2)] \tag{27a}$$

Now this corresponds to spin projections along $+\theta_1$ and $+\theta_2$. An exactly similar expression can be written for projections $-\theta_1$ and $-\theta_2$. In this regard Eq. (26a) is replaced by

$$\hat{\pi}_-(\theta_\alpha) = \tfrac{1}{2}(1 - \hat{\sigma}_z \cos\theta_\alpha - \hat{\sigma}_x \sin\theta_\alpha) \tag{26b}$$

and the analog of Eq. (27a) corresponding to spin along $-\theta_1$ and $-\theta_2$ is found to be

$$\begin{aligned} P_{--}(\theta_1, \theta_2) &= \langle\psi|\hat{\pi}_-(\theta_1)\,\hat{\pi}_-(\theta_1)|\psi\rangle \\ &= \tfrac{1}{4}[1 - \cos\theta_1 - \cos\theta_2 + \cos(\theta_1 - \theta_2)] \end{aligned} \tag{27b}$$

Finally, we note that, taking the incoherent average of Eqs. (27a) and (27b), we have

$$P(\theta_1, \theta_2) = \tfrac{1}{2}[P_{++}(\theta_1, \theta_2) + P_{--}(\theta_1, \theta_2)] = \tfrac{1}{4}[1 + \cos(\theta_1 - \theta_2)] \tag{28}$$

which is the same as the two-particle result (25) except for a minus sign. The sign difference arises because in the spin singlet case we are looking at *antiparallel* spins. This means that to compare Eqs. (25) and (28) we should replace $\theta_2 \to \theta_2 + \pi$ in Eq. (25).

Thus, we are left with the interesting result that the single-particle averages $P_{++}(\theta_1, \theta_2)$ and $P_{--}(\theta_1, \theta_2)$ regain the singlet two-particle result (25). That the EPRB singlet state implies an incoherent mixed-case density matrix for any one member of the singlet pair has been noted before,[9] and the present work reinforces that result.

## 5. CALCULATION OF $M_W(\xi, \tau)$ AND ITS DISTRIBUTION

To evaluate $M_W(\xi, \tau)$ we expand the exponential

$$e^{i\xi\hat{\sigma}_1 + i\tau\hat{\sigma}_2} = \sum_{n=0}^{\infty} \frac{(i\xi\hat{\sigma}_1 + i\tau\hat{\sigma}_2)^{2n}}{(2n)!} + \sum_{n=0}^{\infty} \frac{(i\xi\hat{\sigma}_1 + i\tau\hat{\sigma}_2)^{2n+1}}{(2n+1)!} \tag{29}$$

Now

$$(\xi\hat{\sigma}_1 + \tau\hat{\sigma}_2)^2 = \xi^2\hat{\sigma}_1^2 + \tau^2\sigma_2^2 + \xi\tau(\hat{\sigma}_1\hat{\sigma}_2 + \hat{\sigma}_1\hat{\sigma}_2) = \xi^2 + \tau^2 + 2\xi\tau\cos\delta \tag{30}$$

and hence

$$c^{i\xi\hat{\sigma}_1 + i\tau\hat{\sigma}_2} = \cos\chi + i(\xi\sigma_1 + \tau\sigma_2)\frac{\sin\chi}{\chi} \tag{31}$$

where

$$\chi = [\xi^2 + \tau^2 + 2\xi\tau\cos\delta]^{1/2} \tag{32}$$

and in the following we shall consider only the simple case of $\delta = \pi/2$. We therefore have for the characteristic function

$$M_W(\xi, \tau) = \cos\chi + i(\xi\langle\sigma_1\rangle + \tau\langle\sigma_2\rangle)\frac{\sin\chi}{\chi} \tag{33}$$

$$= \cos\chi - i\left(\langle\sigma_1\rangle\frac{\partial}{\partial\xi} + \langle\sigma_2\rangle\frac{\partial}{\partial\tau}\right)\cos\chi \tag{34}$$

To obtain the distribution associated with the above characteristic function we first observe that, unlike the previous distribution, this distribution will have a continuous range of values for the random variables $s_1$ and $s_2$. But of course the marginals are discrete, consisting of only two possibilities for $s_1$ and $s_2$. We should therefore expect the joint distribution to have rather peculiar properties, as indeed it cannot be a smooth continuous function if it is to have the above behavior.

Define $L(s_1, s_2)$ to be the Fourier transform of the first term of Eq. (34)

$$L(s_1, s_2) = \frac{1}{4\pi^2} \int e^{-i\xi s_1 - i\tau s_2} \cos\chi \, d\xi \, d\tau \tag{35}$$

The Fourier transform of the second term of Eq. (34) may be expressed in terms of $L(s_1, s_2)$

$$-\frac{i}{4\pi^2} \int \left\{ \left( \langle \sigma_1 \rangle \frac{\partial}{\partial \xi} + \langle \sigma_2 \rangle \frac{\partial}{\partial \tau} \right) \cos\chi \right\} e^{-i\xi s_1 - i\tau s_2} \, d\xi \, d\tau \tag{36}$$

$$= -\frac{i}{4\pi^2} \iint \left( \langle \hat{\sigma}_1 \rangle \frac{\partial}{\partial \xi} + \langle \hat{\sigma}_2 \rangle \frac{\partial}{\partial \tau} \right) e^{i\xi s_1' + i\tau s_2'} L(s_1', s_2')$$

$$e^{-i\xi s_1 - i\tau_2} \, ds_1' \, ds_2' \, d\xi \, d\tau \tag{37}$$

$$= (-i) \int [\langle \hat{\sigma}_1 \rangle is_1' + \langle \hat{\sigma}_2 \rangle is_2'] \, L(s_1', s_2')$$

$$\delta(s_1 - s_1') \, \delta(s_2 - s_2') \, ds_1' \, ds_2' \tag{38}$$

$$= (\langle \hat{\sigma}_1 \rangle s_1 + \langle \hat{\sigma}_2 \rangle s_2) \, L(s_1, s_2) \tag{39}$$

The joint probability is therefore

$$P_W(s_1, s_2) = L(s_1, s_2)\{1 + \langle \sigma_1 \rangle s_1 + \langle \sigma_2 \rangle s_2\} \tag{40}$$

We now proceed to evaluate $L(s_1, s_2)$ but, for simplicity, continue to consider only the simple case $\delta = \pi/2$. Making the transformation

$$\xi = r \cos\theta \tag{41a}$$

$$\tau = r \sin\theta \tag{41b}$$

$$s_1 = \rho \cos\phi \tag{42a}$$

$$s_2 = \rho \sin\phi \tag{42b}$$

we have for Eq. (35)

$$= \frac{1}{4\pi^2} \int_0^\infty \int_0^{2\pi} (\cos r) \, e^{-ir\rho \cos(\theta - \phi)} r \, dr \, d\theta \tag{43}$$

where $\theta$ is the usual step function. Using the fact that

$$2\pi J_0(a)=\int_0^{2\pi} e^{-ia\cos(\theta-\phi)}\,d\theta \tag{44}$$

we have

$$L(\rho)=\frac{1}{2\pi}\int_0^{\infty}(\cos r)\,rJ_0(\rho r)\,dr \tag{45}$$

This integral is improper but can be expressed in terms of delta functions. The following integral is proper:

$$\int_0^{\infty} J_0(\rho x)\sin bx\,dx=\frac{1}{\sqrt{b^2-\rho^2}}=\frac{\theta(\rho)\,\theta(b-\rho)}{\sqrt{b^2-\rho^2}},\qquad 0<\rho<b \tag{46}$$

If we take the integral given by Eq. (45) to be the derivative of the integral given by Eq. (46), we have that

$$L(\rho)=\frac{1}{2\pi}\frac{\partial}{\partial b}\int_0^{\infty} J_0(\rho r)\sin(br)\,dr\big|_{b=1} \tag{47}$$

$$=\frac{1}{2\pi}\frac{\partial}{\partial b}\frac{\theta(\rho)\,\theta(b-\rho)}{\sqrt{b^2-\rho^2}}\bigg|_{b=1} \tag{48}$$

$$=\frac{1}{2\pi}\theta(\rho)\left\{\frac{\delta(1-\rho)}{(1-\rho^2)^{1/2}}-\frac{\theta(1-\rho)}{(1-\rho^2)^{3/2}}\right\} \tag{49}$$

We now examine the possible range for $s_1$ and $s_2$. From Eq. (47) we see that

$$0\leqslant\rho\leqslant 1 \tag{50}$$

or

$$0\leqslant s_1^2+s_2^2\leqslant 1 \tag{51}$$

Therefore $s_1$ and $s_2$ may take *any* value between minus one and one, but are jointly constrained by Eq. (51).

That the marginals are satisfied may be seen by direct integration. Of course, this follows from Eq. (32) by setting each of the variables equal to zero, but as we have such a strange distribution it is of interest to prove it directly. Consider integration over $s_1$; from Eqs. (40) and (48) we have

$$P(s_1)=\int P_{\mathrm{W}}(s_1,s_2)\,ds_1$$

$$=\frac{1}{2\pi}\int_{-\sqrt{b^2-s_2^2}}^{\sqrt{b^2-s_2^2}}\frac{\partial}{\partial b}\theta(\rho)\frac{\theta(b-\rho)}{\sqrt{b^2-\rho^2}}(1+s_1\langle\rho_1\rangle+s_2\langle\sigma_2\rangle)\bigg|_{b=1} \tag{52}$$

since $\rho = \sqrt{s_1^2 + s_2^2}$. The integration of the middle term is zero (by symmetry) and we have

$$\frac{1}{2\pi}(1 + s_2\langle\hat{\sigma}_2\rangle)2\frac{\partial}{\partial b}\theta(b^2 - s_2^2)\int_0^{\sqrt{b^2 - s_2^2}}[b^2 - s_1^2 - s_2^2]^{-1/2}\,ds_1 \tag{53}$$

The integral is equal to $\pi/2$ and hence

$$P(s_2) = (1 + s_2\langle\hat{\sigma}_2\rangle)\,b\delta(b^2 - s_2^2)|_{b=1} \tag{54}$$

$$= (1 + s_2\langle\hat{\sigma}_2\rangle)\,\delta(1 - s_2^2) \tag{55}$$

But

$$\delta(s_2^2 - 1) = \tfrac{1}{2}[\delta(s_2 - 1) + \delta(s_2 + 1)] \tag{56}$$

and therefore we finally have

$$P(s_2) = \tfrac{1}{2}(1 + s_2\langle\hat{\sigma}_2\rangle)[\delta(s_2 - 1) + \delta(s_2 + 1)] \tag{57}$$

which is the correct quantum-mechanical result [see Eq. (A.8)].

## 6. CONCLUSION

In conclusion we mention that the methods we have used for obtaining joint distributions for spin operators have been generalized to arbitrary operators[10] and we briefly summarize the results. Consider any two operators $A$ and $B$ and a wave function $\psi(q)$ in the coordinate representation. We seek a joint distribution of the $c$ numbers $a$ and $b$ which satisfy the quantum mechanical marginals

$$\int P(a, b)\,db = |\phi^A(a)|^2 \tag{58}$$

$$\int P(a, b)\,da = |\phi^B(b)|^2 \tag{59}$$

where $\phi^A(a)$ and $\phi^B(b)$ are the wave functions in the $A$ and $B$ representations

$$\phi^A(a) = \int v_a^*(q)\,\psi(q)\,dq \tag{60}$$

$$\phi^B(b) = \int w_b^*(q)\,\psi(q)\,dq \tag{61}$$

and $v_a(q)$ and $w_b(q)$ are the eigenfunctions of $A$ and $B$ with eigenvalues $a$ and $b$, respectively. We assume continuous spectrums as generalization to the discrete case is simply achieved by replacing integrals by summations where appropriate.

As for the spin case, we write the characteristic function corresponding to the analog of Weyl ordering and symmetrization,

$$M_1^{ab}(\theta, \tau) = \langle e^{i\theta A + i\tau B} \rangle = \int \psi^*(q)\, e^{i\theta A + i\tau B}\, \psi(q)\, dq \tag{62}$$

$$M_2^{ab}(\theta, \tau) = \left\langle \frac{e^{i\tau B}\, e^{i\theta A} + e^{i\tau A}\, e^{i\theta B}}{2} \right\rangle \tag{63}$$

We also define

$$M_3^{ab}(\theta, \tau) = \langle e^{i\tau B}\, e^{i\theta A} \rangle = \int \psi^*(q)\, e^{i\tau B}\, e^{i\theta A}\, \psi(q)\, dq \tag{64}$$

Evaluation of $M_1^{ab}(\theta, \tau)$ can be achieved in terms of the eigenfunctions and eigenvalues of the hermitian operator $\theta A + \tau B$,

$$(\theta A + \tau B)\, u_\gamma(q) = \gamma u_\gamma(q) \tag{65}$$

By writing the wave function in Eq. (62) in terms of the eigenfunctions $u_\gamma(q)$,

$$\phi(\gamma) = \int u_\gamma^*(q)\, \psi(q)\, dq, \qquad \psi(q) = \int u_\gamma(q)\, \phi(\gamma)\, d\gamma \tag{66}$$

the characteristic function $M_1(\theta, \tau)$ evaluates to

$$M_1^{ab}(\theta, \tau) = \iint \psi^*(q')\, u_\gamma(q')\, e^{i\gamma}\, u_\gamma^*(q)\, \psi(q)\, dq'\, dq\, d\gamma \tag{67a}$$

$$M_1^{ab}(\theta, \tau) = \iint \psi^*(q') \langle u_\gamma(q) | e^{i\gamma} | u_\gamma(q') \rangle\, \psi(q)\, dq'\, dq \tag{67b}$$

$$= \int |\phi(\gamma)|^2\, e^{i\gamma}\, d\gamma \tag{68}$$

Using Eqs. (60) and (61) the characteristic function $M_3^{ab}(\theta, \tau)$ may be readily expressed in terms of the transformation matrix between the $A$ and $B$ representation

$$M_3^{ab}(\theta, \tau) = \iint \phi^{*B}(b)\, T_{ab} \phi^A(a)\, e^{i\tau b + i\theta a}\, da\, db \tag{69}$$

where

$$T_{ab} = \int w_b^*(q)\, v_a(q)\, dq \tag{70}$$

For each characteristic function the joint distribution is obtained by fourier inversion

$$P^{ab}(a, b) = \frac{1}{4\pi^2} \iint M^{ab}(\theta, \tau)\, e^{-i\theta a - i\tau b}\, d\theta\, d\tau \tag{71}$$

In particular,

$$P_1^{ab}(a, b) = \iint \psi^*(q')\, G(a, b; q, q')\, dq\, dq' \tag{72}$$

where

$$G(a, b; q, q') = \frac{1}{4\pi^2} \iiint u_\gamma^*(q)\, e^{i\gamma - i\theta a - i\tau b}\, u_\gamma(q')\, d\gamma\, d\theta\, d\tau \tag{73}$$

and

$$P_3^{ab}(a, b) = \phi^{*B}(b)\, T_{ab} \phi^A(a) \tag{74}$$

and the distribution corresponding to $M_3^{ab}(\theta, \tau)$ is

$$P_2^{ab}(a, b) = \tfrac{1}{2}(P_3^{ab}(a, b) + \text{c.c.}) \tag{75}$$

As an example consider the case of position and momentum, $A = \hat{q}$ and $B = \hat{p}$, and we take for $a$ and $b$ the $c$ numbers $q$ and $p$. The eigenfunction equation given by Eq. (65) is readily solved:

$$u_\gamma(q) = \frac{1}{\sqrt{2\pi h \tau}}\, e^{i[q\gamma - (1/2)\theta q^2]/h\tau} \tag{76}$$

Calculating $G(q, p; x, x')$ by using Eq. (73),

$$G(q, p; x, x') = \frac{1}{2\pi h}\, \delta\left(\frac{x' + x}{2} - q\right) e^{i(x' - x)p/h} \tag{77}$$

and substituting into Eq. (72) straightforwardly yields the Wigner distribution. For the distribution given by Eq. (74) or (75) we have that $\phi^A(a) = \psi(q)$ and $\phi^B(b)$ then becomes the momentum wave function. The transformation matrix between the position and momentum representation is of course equal to

$$T_{qp} = \frac{1}{\sqrt{2\pi h}} e^{-iqp/h} \tag{78}$$

The resulting distribution given by Eq. (74) is then the symmetrical distribution which has been extensively studied.[11]

There are an infinite number of joint distributions which satisfy the quantum mechanical marginals. They can be generated by taking any characteristic function $M^{ab}(\theta, \tau)$, say, and forming a general characteristic function

$$M_g^{ab}(\theta, \tau) = M^{ab}(\theta, \tau)\, f(\theta, \tau) \tag{79}$$

where $f(\theta, \tau)$ is any function satisfying $f(\theta, 0) = f(0, \tau) = 1$.

The joint distribution is then obtained from

$$P_g^{ab}(a, b) = \frac{1}{4\pi^2} \iint M^{ab}(\theta, \tau)\, f(\theta, \tau)\, e^{-i\theta a - i\tau b}\, d\theta\, d\tau \tag{80}$$

By choosing different functions for $f(\theta, \tau)$ one may generate an infinite number of distributions satisfying the proper marginals.

The above distributions are bilinear in the wave function and will, in general, not be positive for all values of $a$ and $b$. However, there exist manifestly positive joint distributions. The case of position and momentum is readily generalized to two arbitrary operators,[12]

$$P_p^{ab}(a, b) = |\phi^A(a)|^2 |\phi^B(b)|^2 \{1 + cp(U, V)\} \tag{81}$$

where $p(U, V)$ is a function of two variables $U$ and $V$ which in turn depend on $a$ and $b$

$$U(a) = \int_{-\infty}^{a} |\phi^A(a')|^2\, da' \tag{82}$$

$$V(b) = \int_{-\infty}^{b} |\phi^B(b')|^2\, db' \tag{83}$$

The function $p(U, V)$ may be any function that satisfies

$$\int_0^1 p(U, V)\, dU = 0 \tag{84}$$

$$\int_0^1 p(U, V)\, dV = 0 \tag{85}$$

and the constant $c$ is chosen in the range $-1/l_2 \leqslant c \leqslant 1/l_1$, where $-l_1$ and $l_2$ are the absolute minimum and maximum of $p(U, V)$ in the unit square.

## APPENDIX A: SOME SPIN-1/2 ALGEBRA

We have need of the operator $\hat{\sigma}(\theta, \phi)$ corresponding to spin states pointing along the unit vector in the $\theta, \phi$ direction. This is given by

$$\hat{\sigma}(\theta, \phi) = e^{-i(\phi/2)\hat{\sigma}_z}[e^{-i(\theta/2)\hat{\sigma}_y}\,\hat{\sigma}_z\, e^{i(\theta/2)\hat{\sigma}_y}]\, e^{i(\phi/2)\hat{\sigma}_z} \tag{A.1}$$

where

$$\hat{\sigma}_z = \begin{pmatrix} 1 & 0 \\ 0 & -1 \end{pmatrix}, \quad \hat{\sigma}_x = \begin{pmatrix} 0 & 1 \\ 1 & 0 \end{pmatrix}, \quad \hat{\sigma}_y = \begin{pmatrix} 0 & -i \\ i & 0 \end{pmatrix} \tag{A.2}$$

The expression (A.1) may be simplified to yield

$$\hat{\sigma}(\theta, \phi) = \begin{bmatrix} \cos\theta & \sin\theta\, e^{-i\phi} \\ \sin\theta\, e^{i\phi} & \cos\theta \end{bmatrix} \tag{A.3}$$

and for $\phi = 0$ this is just

$$\hat{\sigma}(\theta) = \hat{\sigma}_z \cos\theta + \hat{\sigma}_x \sin\theta \tag{A.4}$$

The projection operators ($\phi = 0$)

$$\hat{\pi}_{\pm}(\theta) = |\pm\theta\rangle\langle\pm\theta| \tag{A.5}$$

may now be written in terms of $\hat{\sigma}(\theta)$ as

$$\hat{\pi}_{\pm}(\theta) = \tfrac{1}{2}(1 \pm \hat{\sigma}(\theta)) \tag{A.6}$$

The probabilities of spin in the $(+)$ or $(-)$ $\theta$ direction are then given by

$$P_{\pm} = \langle \hat{\pi}_{\pm}(\theta) \rangle = (1 \pm \langle \hat{\sigma}(\theta) \rangle) \tag{A.7}$$

which we may write in terms of the $c$-number spin variable $s$ as

$$\begin{aligned} P(s) &= \tfrac{1}{2}(1 + \langle \hat{\sigma} \rangle \delta_{s,1} + (1 - \langle \hat{\sigma} \rangle)\, \delta_{s,-1} \\ &= \tfrac{1}{2}(1 + s\langle \hat{\sigma} \rangle)[\delta(s-1) + \delta(s+1)] \end{aligned} \tag{A.8}$$

Finally we note that

$$\tfrac{1}{2}[\hat{\sigma}_1 \hat{\sigma}_2 + \hat{\sigma}_2 \hat{\sigma}_1] = \cos(\theta_1 - \theta_2) \tag{A.9}$$

as may easily be seen by setting $\phi = 0$ and doing the matrix multiplication using Eq. (A.3).

## APPENDIX B: ALTERNATIVE DERIVATION OF $P_S(s_1, s_2)$

We now give an alternative derivation of the symmetrical distribution. If we assume that there will be only four possible joint outcomes for $s_1$ and $s_2$, then to fix these four probabilities (assumed unknown now) four independent equations are neded. One equation is that of normalization:

$$P_{++} + P_{+-} + P_{-+} + P_{--} = 1 \tag{B.1}$$

Now, four equations can be written for the marginal conditions

$$P_{++} + P_{+-} = \tfrac{1}{2}(1 + \langle \hat{\sigma}_1 \rangle) \tag{B.2a}$$

$$P_{-+} + P_{--} = \tfrac{1}{2}(1 - \langle \hat{\sigma}_1 \rangle) \tag{B.2b}$$

$$P_{++} + P_{-+} = \tfrac{1}{2}(1 + \langle \hat{\sigma}_2 \rangle) \tag{B.2c}$$

$$P_{+-} + P_{--} = \tfrac{1}{2}(1 - \langle \hat{\sigma}_2 \rangle) \tag{B.2d}$$

But these equations are not independent, since, for example, Eq. (B.2d) minus Eq. (B2b) gives the same result as Eq. (B.2a) minus Eq. (B.2c), namely

$$P_{+-} - P_{-+} = \tfrac{1}{2}(\langle \hat{\sigma}_1 \rangle - \langle \hat{\sigma}_2 \rangle) \tag{B.3}$$

Similarly, subtracting Eq. (B.2a) from Eq. (B.2d) yields the same result as Eq. (B.2c) minus Eq. (B.2b):

$$P_{+-} - P_{--} = \tfrac{1}{2}(\langle \hat{\sigma}_1 \rangle + \langle \hat{\sigma}_2 \rangle) \tag{B.4}$$

Hence, thus far we have three independent equations.

To obtain a fourth equation we consider the expectation value of the first mixed moment. If we assume that the expectation value $\langle s_1 s_2 \rangle$ is obtained by calculating the expectation value of

$$\tfrac{1}{2}[\hat{\sigma}_1 \hat{\sigma}_2 + \hat{\sigma}_2 \hat{\sigma}_1] \tag{B.5}$$

then we have, using Eq. (A.9), that

$$\langle s_1 s_2 \rangle = \cos \delta \tag{B.6}$$

But in general the expectation value of $s_1 s_2$ is

$$1 \cdot 1 P_{++} + 1 \cdot (-1) P_{+-} + (-1) \cdot 1 P_{-+} + (-1) \cdot (-1) P_{--} \tag{B.7}$$

Hence, the fourth equation is

$$P_{++} - P_{+-} - P_{-+} + P_{--} = \cos \delta \tag{B.8}$$

Equations (B.1), (B.3), (B.4), and (B.8) constitute four independent equations which can be solved for the $P$'s. The solutions are the same $P$'s as given by Eqs. (20)–(20b). We feel that this is an interesting derivation because it shows that the single correspondence given by Eq. (B.5) is sufficient to determine the joint distribution.

## REFERENCES


1. E. Wigner, *Phys. Rev.* **40**, 749 (1932).
2. M. Hillery, R. O'Connell, M. Scully, and E. Wigner, *Phys. Rep.* **106**, 121 (1984).
3. (a) L. Cohen, *J. Math. Phys.* **7**, 781 (1966); **17**, 1893 (1976); (b) L. Cohen and Y. I. Zaparovanny, *J. Math. Phys.* **21**, 794 (1980).
4. M. Scully, *Phys. Rev. D* **28**, 2477 (1983).
5. W. Louisell, *Statistical Properties of Radiation* (Wiley, New York, 1967); R. N. Balazs and B. K. Jennings, *Phys. Rep.* **104**, 347 (1984).
6. J. Moyal, *Proc. Cambridge Philos. Soc.* **45**, 499 (1978).
7. A. Einstein, B. Podolsky, and N. Rosen, *Phys. Rev.* **47**, 777 (1935).
8. D. Bohm, *Quantum Theory* (Prentice-Hall, Englewood Cliffs, New Jersey, 1951), p. 614.
9. C. Cantrell and M. O. Scully, *Phys. Rev. C* **43**, 499 (1978).
10. L. Cohen and M. O. Scully, "Wigner Distribution for Two Arbitrary Operators," to be published.
11. C. L. Mehta, *J. Math. Phys.* **5**, 677 (1964); H. Margenau and R. N. Hill, *Prog. Theor. Phys.* **26**, 722 (1961); O. von Roos, *Phys. Rev.* **119**, 1174 (1960); T. Takabayasi, *Prog. Theor. Phys.* **11**, 341 (1954).
12. L. Cohen and Y. I. Zaparovanny, *J. Math. Phys.* **21**, 794 (1980); L. Cohen, in *Frontiers of Nonequilibrium Statistical Physics*, G. T. Moore and M. O. Scully, eds. (Plenum Press, New York, in press).

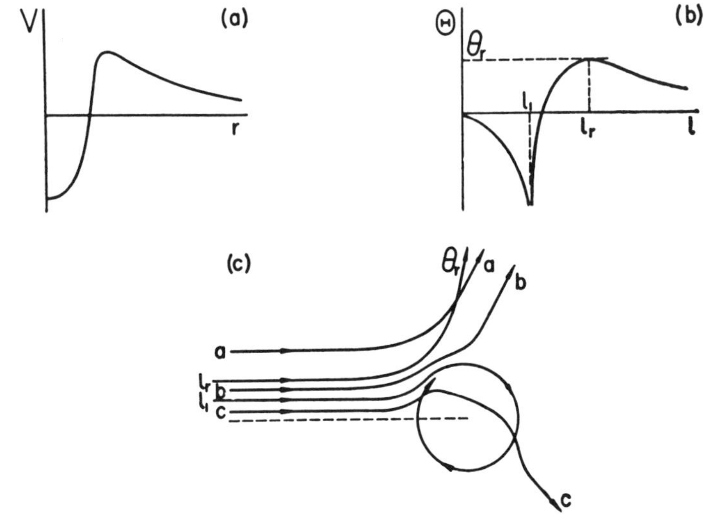


*Plate 32*

# Wedges I

**Cécile DeWitt-Morette,**[1,2,3] **Stephen G. Low,**[1,4] **Lawrence S. Schulman,**[2,5] **and Anwar Y. Shiekh**[6,7]

*Received September 30, 1985*

*The wedge problem, that is, the propagation of radiation or particles in the presence of a wedge, is examined in different contexts. Generally, the paper follows the historical order from Sommerfeld's early work to recent stochastic results—hindsights and new results being woven in as appropriate. In each context, identifying the relevant mathematical problem has been the key to the solution. Thus each section can be given both a physics and a mathematics title: Section 2: diffraction by reflecting wedge; boundary value problem of differential equations; solutions defined on mutiply connected spaces. Section 3: geometrical theory of diffraction; identification of function spaces. Section 4: path integral solutions; path integration on multiply connected spaces; asymptotics on the boundaries of function spaces. Section 5: probing the shape of the wedge and the roughness of its surface; stochastic calculus. Several propagators and Green functions are given explicitly, some old ones and some new ones. They include the knife-edge propagator for Dirichlet and Neumann boundary conditions, the absorbing knife edge propagator, the wedge propagators, the propagator for a free particle on a μ-sheeted Riemann surface, the Dirichlet and the Neumann wedge Green function.*

*To John Archibald Wheeler:* Recent interest in the wedge problem (Schulman, 1982; Shiekh, 1985) has led us to review several of its aspects from Sommerfeld's calculation of diffraction of light by a wedge in 1896 (Sommerfeld, 1896) to current investigations of stochastic diffusions when

[1] Department of Physics and Center for Relativity, University of Texas, Austin, Texas 78712.
[2] Supported in part by NSF grant PHY84-04931.
[3] Supported in part by SERC grant GR/D 15911.
[4] Supported in part by a NSERC (Canada) Fellowship.
[5] Department of Physics, Technion, Haifa, Israel, and Department of Physics, Clarkson University, Potsdam, New York 13676.
[6] Blackett Laboratory, Imperial College, London SW7 2BZ.
[7] Supported in part by SERC grant B/83301669.

sharp obstructions are present. The various facets of this problem touch upon your many interests and there is nobody more appropriate than you to whom we would like to dedicate this paper. There is also no occasion more appropriate than this Festschrift to offer it to you: In 1938 Sommerfeld celebrated his seventieth birthday, and for this occasion Pauli (1938) dedicated to his "old teacher"[8] a paper on wedges.[9]

It seems impossible to treat a subject with so many ramifications in a few pages. But we have all seen your colorful blackboards summarizing questions and answers of multifaceted issues; and we have all heard you saying at the end of a seminar, "Tell us in one sentence what we have learned today." Here are, in black and white, a few sentences on the wedge.

## 1. STATING THE WEDGE PROBLEM

Given a wedge, and a source of radiation or a source of particles, compute the intensity of the radiation or the density of the particles detected on the other side of the wedge. Solve (first) the problem in a plane, since many wedge problems in $\mathbb{R}^3$ can be reduced to problems in $\mathbb{R}^2$.

*Notation.* Points in $\mathbb{R}^3$ are given in cylindrical coordinates. The edge of the wedge is the $z$-axis. The angle of physical interest, i.e., the external angle for a solid wedge, is $\theta$; if $\theta = 2\pi$ the wedge is a half plane barrier, or as it is so often called, a knife edge; rational wedges refer to wedges of angle $\theta = \mu\pi/\nu$, where $\mu$ and $\nu$ are positive integers. The source is located at $a$ with coordinates $(R', \phi')$ and the detector is located at $b$ with coordinates $(R, \phi)$. (See Fig. 1.)

The Green function $G$ of the elliptic operator $\Delta + k^2$, where $\mathbf{k}$ is the wave vector of a monochromatic incident plane wave, satisfies $(\Delta + k^2)\, G(x, a) = -\delta(x-a)$. The elementary kernel of the Schrödinger equation for propagation from $a$ to $b$ in time $t$ is labelled $K(b, t; a)$; it satisfies $(\partial/\partial t - i\hbar\Delta/2M)\, K(x, t; a) = \delta(t)\, \delta(x-a)$. Subscripts $e$ and $w$ on $G$ and $K$ denote the edge and wedge cases and superscripts + and − refer, respectively, to the Neuman and Dirichlet boundary conditions. The definitions of special functions are the ones used by Gradshteyn and Ryzhik (1980), and by Abramowitz and Stegun (1965). The Bessel function

[8] Not "old" then; this is a term used by Pauli, with affection, many years later, in a letter sent to Cecile Morette and Bryce DeWitt on the occasion of their marriage.

[9] Rubinowicz (1938) also writing at this occasion refers to Sommerfeld's 1896 paper as his "first great scientific achievement." Indeed, some years before, Lord Rayleigh in the article on "Wave Theory" in the *Encyclopaedia Britannica* had stated that the mathematical difficulties were so formidable that no successful attempt had yet been made to solve this problem.

$K_\nu(z)$ of imaginary argument is given in terms of the Bessel functions of the first kind $J_\nu(z)$ by

$$K_\nu(-ix) = \pi \exp(\tfrac{1}{2}i\pi\nu)\,\frac{J_{-\nu}(x) - \exp(-i\pi\nu)\,J_\nu(x)}{2\sin(\nu\pi)}$$

The special functions used in the references are related to the ones we use as follows:

$$U_\nu(x)\ (\text{in Sommerfeld, 1896}) := \exp(-\tfrac{1}{2}i\pi\nu)\,K_\nu(-ix)$$

$$K_\nu(x)\ (\text{in Sommerfeld, 1896}) := -\frac{\pi}{2}\,Y_\nu(x)$$

In reviewing the problem, we generally follow historical order, but bring hindsight and describe new results as we proceed.

## 2. EXACT AND ASYMPTOTIC SOLUTIONS OF THE DIFFRACTION OF LIGHT BY A WEDGE

In 1896, A. Sommerfeld, then a *Privat-dozent*, solved the problem of the diffraction of light from a perfectly reflecting knife edge by recognizing that Fresnel diffraction is a well-defined mathematical boundary value problem,[10] and making a "decisive"[11] *ansatz* suggested by Riemann's solution for a similar problem.

Riemann had computed the potential $\chi$ of a point charge at $(R', \phi')$ outside a conducting wedge of external angle $\theta = \mu\pi/\nu$, illustrated in Fig. 1, as follows. The potential $\chi$ satisfies Laplace's equation

$$\Delta\chi(r, \psi) = 0 \qquad \text{at } (r, \psi) \neq (R', \phi') \tag{1}$$

and the boundary condition $\chi = 0$ on the surface of the wedge. The wedge problem for a monochromatic incident plane wave (**k** perpendicular to the edge of the wedge) reduces to the solution of the two-dimensional scalar wave equation (Helmholtz equation)

$$\Delta u + k^2 u = 0 \qquad \text{at } (r, \psi) \neq (R', \phi') \tag{2}$$

[10] Sommerfeld (1954), p. 249, referring to this 1896 paper "With this problem it was first demonstrated that Fresnel diffraction constitutes a well-defined mathematical boundary value problem (Fraunhofer diffraction cannot be treated directly by this method but only as a limiting case of Fresnel diffraction)."

[11] Pauli (1983), p. 925.

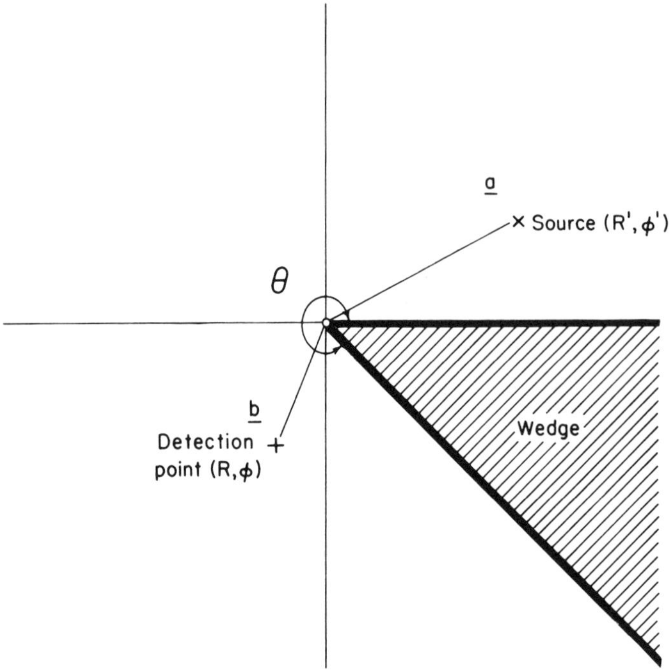


Fig. 1. Diffraction from a wedge.

together with either of the following boundary conditions:

$u = 0$ for a transverse magnetic wave (Dirichlet)
$\partial u/\partial n = 0$ for a transverse electric wave (Neumann)
where $n$ is the normal to the surface.

(1) This problem had been solved by the method of images for $\theta = \pi/\nu$ for integral $\nu$. The boundary conditions had been implemented through an algebraic sum of contributions from the source and its images placed symmetrically with respect to the wedge surfaces. When $\theta = \pi/\nu$ the images are all inside the wedge, which is not a region of physical interest. When $\theta \neq \pi/\nu$ some of the images (finitely many if $\theta$ is a rational multiple of $\pi$, infinitely many otherwise) are in the physical region outside the wedge. Riemann, in his doctoral thesis, developed a technique to overcome this difficulty in the rational case. In the complex plane a wedge of angle

$\theta = \mu\pi/v$ with $\mu$, $v$ integral may be mapped into a wedge of angle $\theta = \pi/v$ by the transformation

$$z \to Z := T(z) = z^{1/\mu}$$

The solution of the Laplace equation (1) in the $z$-variable is an analytic function

$$w(z) = u(z) + iv(z), \qquad z = x + iy$$

The solution in the $Z$-variable

$$W(Z) = U(Z) + iV(Z) = W(T(z)) = w(z)$$

is also an analytic function. The boundary conditions in the $Z$-variable are implemented by images inside the wedge, and the problem can be solved in the previously described manner. The transformation back to the $z$-variable yields a multivalued function $w(z)$. The values of physical interest can be obtained by restricting $z$ to $T^{-1}(Z)$ for $Z$ outside the wedge. Alternatively a many-valued function may be made single valued by considering it defined on a many-sheeted Riemann surface. In this picture the images appear on the lower folds of the many-sheeted Riemann surface. They are "hidden" and act only to implement the boundary conditions.

(2) Recognizing the mathematical similarity of the wedge problem in electrostatic theory and in diffraction theory, Sommerfeld looked for a many-valued solution of the Helmholtz equation (2) satisfying the appropriate boundary conditions, or "in the language introduced by Riemann,"[12] for a solution defined on a many-sheeted Riemann surface.

There are two classes of boundary conditions:

(i) The radiation condition at infinity[13], for which the field must behave like an outgoing spherical wave at infinity. This condition is satisfied[14] by the following Green's function for the Helmholtz equation. When $R < R'$

$$G(R, \phi; R', \phi') = \frac{1}{2\pi} \sum_{n=0}^{\infty} \gamma_n J_{n/\mu}(kR) \exp\left(\frac{-i\pi n}{2\mu}\right) K_{n/\mu}(-ikR') \times \cos\left(\frac{n}{\mu}(\phi - \phi')\right) \tag{3}$$

where $\gamma_0 = 1/\mu$ and $\gamma_n = 2/\mu$ for $n > 0$. If $R' < R$, interchange $R$ and $R'$.

[12] Sommerfeld (1954), p. 251.

[13] It seems that Sommerfeld was the first person to identify the radiation condition in scattering problems.

[14] Sommerfeld (1896), Eqs. (12) and (13), p. 356.

(ii) The wedge boundary condition (Dirichlet or Neumann) which is implemented by the contributions of images placed appropriately on a $\mu$-sheeted Riemann surface.

In his 1896 paper Sommerfeld fully developed only the solution for the knife edge but showed how his method can be used for an arbitrary wedge. In 1897, fully occupied with other work, Sommerfeld urged Carslaw to continue this investigation and guided him in the early stages of his work. Carslaw (1899) solved the Helmholtz equation (2) and the diffusion equation

$$\frac{\partial u}{\partial t} = \kappa \Delta u$$

with the knife-edge boundary in the form of a definite Fresnel integral on the real line instead of a series expansion.

$$K_e^{\mp} = \frac{1}{4\pi\kappa\sqrt{\pi t}} \left\{ \exp\left[ -\frac{r^2}{4\kappa t} \right] \int_{-\infty}^{(R'R/\kappa t)^{1/2} \cos \frac{1}{2}(\phi - \phi')} \exp[-\lambda^2]\, d\lambda \right.$$
$$\left. \mp \exp\left[ -\frac{r'^2}{4\kappa t} \right] \int_{-\infty}^{(R'R/\kappa t)^{1/2} \cos \frac{1}{2}(\phi + \phi')} \exp[-\lambda^2]\, d\lambda \right\} \tag{4}$$

where

$$r^2 := R^2 + R'^2 - 2R'R\cos(\phi - \phi'), \qquad r'^2 := R^2 + R'^2 - 2R'R\cos(\phi + \phi')$$

This classic solution has recently been obtained by path integral techniques independently by Schulman (1982, 1984), Wiegel *et al.*,[15] and Shiekh (1986).

Solutions for other geometries of the source and the wedge and simplified versions of the calculations can be found in Macdonald (1902) and Carslaw (1916–1919). We note here a solution (5) derived by Macdonald which has also been obtained recently by path integration (Shiekh, 1985). For a Dirichlet wedge, and $R < R'$

$$G_w^- = \frac{2}{\theta} \sum_{n=1}^{\infty} \exp\left( \frac{-in\pi^2}{2\theta} \right) J_{n\pi/\theta}(kR)\, K_{n\pi/\theta}(-ikR')$$
$$\times \sin\left( \frac{n\pi\phi}{\theta} \right) \sin\left( \frac{n\pi\phi'}{\theta} \right) \tag{5}$$

For $R > R'$ interchange $R$ and $R'$.

[15] Wiegel and Boersma (1983) were the first to notice that the path integral calculation gave a formula previously computed by Carslaw.

For a Neumann wedge with $R < R'$,

$$G_w^+ = \frac{1}{\theta}\left\{J_0(kR)\,K_0(-ikR') + 2\sum_{n=1}^{\infty} \exp\left(\frac{-in\pi^2}{2\theta}\right) J_{n\pi/\theta}(kR)\,K_{n\pi/\theta}(-ikR')\cos\left(\frac{n\pi\phi}{\theta}\right)\cos\left(\frac{n\pi\phi'}{\theta}\right)\right\} \tag{6}$$

For $R > R'$ interchange $R$ and $R'$.

The series in (3), (5), and (6) converge slowly for large arguments of the Bessel function. Asymptotic solutions were developed by Sommerfeld; they are valid only away from the boundary of the geometric shadow. Pauli developed alternative asymptotic solutions valid on the shadow boundary.

## 3. GEOMETRICAL THEORY OF DIFFRACTION. JWKB APPROXIMATIONS

(i) Keller, in 1953, extended the geometric theory of optics to include diffraction effects.[16] This makes it possible to solve problems such as the diffraction of waves by wedges at the JWKB level of approximation without having to solve the full partial differential equation. The power of the method comes from the fact that one can solve a wide class of problems at the JWKB level by solving a number of comparatively simple canonical problems in the immediate neighborhood of a diffracting edge or vertex. A justification for the localization of the problem will be presented in the section on stochastic diffusion where it will be shown to be a consequence of Kac's "principle of imperceptibility" (Kac, 1959).

The geometric theory of optics models the propagation of light by rays which satisfies Fermat's principle. [See, for instance, Baker and Copson (1950), Longhurst (1957), Felsen (1976), Kouyoumjiam (1975), and Wait (1960).] This principle states that the path $q$ taken by a ray extremizes the optical length $L$ which is a function(al) on the space $\mathscr{C}(a, b)$ of continuous paths $c$ from $a$ to $b$,

$$L\colon \mathscr{C}(a, b) \to \text{by } c \to \int_a^b n(c(s))\, ds \tag{7}$$

In this expression, the paths $c$ are the maps

$$c\colon \mathbb{R}_+ \to M\colon s \mapsto c(s)$$

[16] Keller's generalization also incorporated effects due to caustics and "creeping" or grazing rays. As only diffractional effects are considered here, the interested reader is referred to Keller's original work. The extensive literature that it has stimulated is reviewed in the work of Lewis *et al.* (1967) and Wait (1960).

where $s$ is the arc length of the path and $M$ is a smooth manifold (which may have a boundary) in which the rays propagate. In this discussion $M$ will be assumed to be three dimensional; $a, b \in M$. The index of refraction $n$ is a function on $M$.

The usual theory of geometrical optics considers only direct, reflected and refracted rays. Direct rays, if any, are *smooth* paths $q$ which extremize the optical length for paths $c$ in the function space $\mathscr{C}(a, b)$. Reflection and refraction occur when there are two-dimensional subspaces of $M$—called surfaces of discontinuity—on which $n$ is discontinuous or which form the boundary of $M$. For clarity, these surfaces of discontinuity will be assumed to be smooth except along "edges," one-dimensional subspaces of the surface along which the curvature of the surface is discontinuous. In turn, the edges will be assumed to be smooth except at "vertices" where the curvature of the edge is discontinuous.

The reflected and refracted rays are relative extrema of the optical length for paths $c$ restricted to the subspace $\mathscr{D}_\rho(a, b)$ of $\mathscr{C}(a, b)$ of paths which touch a point on the interior (i.e., the point is not on an edge) of a surface of discontinuity $\rho$ times.

Keller realized that this theory could be generalized to include diffracted rays (and, as noted in a previous footnote, grazing and caustics) by decomposing $\mathscr{C}(a, b)$ into the disjoint subspaces $\mathscr{D}_{\rho,\sigma,\tau}(a, b)$. The integers $\rho$, $\sigma$, $\tau$ label, respectively, the number of smooth arcs (which may be simply points) the curves $c$ have in the interior of the surfaces of discontinuity, on the edges and vertices.

Then, the variation problem of $L$ restricted to the subspace $\mathscr{D}_{\rho,\sigma,\tau}(a, b)$ yields the rays which have a total of $\rho$ reflections, refractions, and/or grazings along the surface, $\sigma$ diffractions by an edge, and $\tau$ diffractions by a vertex.

(ii) Congruence of rays normal to a given surface $S_0$ (possibly a point) can be used to construct approximate monochromatic solutions $u(x)\exp(-i\omega t)$ to the wave equation called JWKB approximations:

$$u(x) = \sum_i A(q_i(s)) \exp(ik\psi(q_i(s))), \qquad k = \omega/c \tag{8}$$

for all $q_i$ such that $q_i(s) = x$. The phase function $\psi(q(s))$ is the optical length along the ray $q$ from $S_0$ to $q(s)$. The surfaces of constant phase are the wave fronts. $\psi(x)$ is a possibly multivalued, real or complex solution of the eikonal equation. The amplitude $A(q_i(s))$ is a function of the Jacobi field along $q_i$ determined by the congruence of rays. The energy flux per unit area is proportional to $nA^2$, and the law of conservation of energy determines the amplitude. In the presence of caustics, which are characterized by Jacobi fields with vanishing boundary conditions, or in the presence of diffraction when one ray splits into infinitely many rays, the JWKB approximation takes a more complicated form. In this paper we consider only wedges. By analogy with the reflection and transmission coefficient which apportion the amplitude attached to an incident ray between its reflected and transmitted rays, Keller proposes to multiply the amplitude by an "edge diffraction" coefficient, or a "vertex diffraction" one as the case may be. Of interest for the wedge problem is the "straight edge diffraction" coefficient which has been computed by Keller as follows. Given an incident plane wave $u_i$ of amplitude $A_i$ propagating in a direction normal to the edge, with phase $\psi_i$ at the edge, the diffracted wave fronts are cylinders with the edge as their axis, and Keller writes the approximate field constructed from the diffracted rays

$$u(r, \psi) = \sum_i DA_i r^{-1/2} \exp(i(kr + \psi_i)) \tag{9}$$

The straight edge diffraction coefficient is determined by comparing (9) with the exact solution (5), (6) obtained by Sommerfeld—or rather with the asymptotic value of the exact solution for large values of $kr$. It is found to be

$$D = \frac{\exp(i\pi/4)\sin(\pi^2/\theta)}{\theta(2k/\pi)^{1/2}}\left[\left(\cos\frac{\pi^2}{\theta} - \cos\frac{\pi(\phi - \phi')}{\theta}\right)^{-1} \mp \left(\cos\frac{\pi^2}{\theta} - \cos\frac{\pi(\phi + \phi' - 4\pi)}{\theta}\right)^{-1}\right] \tag{10}$$

$\phi'$ and $\phi$ are the incident and diffracted ray angles given in Fig. 1 and $\theta$ is the exterior angle of the wedge.

Keller's theory then asserts that this coefficient may be used to solve a wide class of diffraction problems in which the surface of discontinuity locally looks like a wedge.

The question remains of solving these canonical problems without recourse to known exact solutions. Functional integration provides methods of directly obtaining approximate solutions of the form suggested by Keller. In certain cases, as we shall see in the next section, it does even more; it provides exact solutions to diffraction problems.

## 4. PATH INTEGRAL SOLUTIONS

Diffraction plays an important role in the experimental verification of quantum mechanics. It is therefore surprising that edges and wedges have not played a more central role in the theoretical development of quantum mechanics. Indeed, they first appear in papers whose primary purpose is classical electromagnetic wave diffraction (Buslaev, 1968; Keller and McLaughlin, 1975; Lee, 1978; Schulman, 1982, 1984; Knoll and Schaeffer, 1976; Crandall, 1983a, b; Low, 1984; Shiekh, 1985). They are based on functional integral techniques; some use them to solve a heat equation, others to solve a Schrödinger equation.

Path integral solutions of the wedge problem require use of the following tools:

(i) Computation of path integrals on multiply connected spaces. (Sect. 4.1)

(ii) Extension of the method of stationary phase to problems where the critical point of the action does not fall inside the domain of integration. (Sec. 4.2)

We present here two path integral calculations for the wedge problem in quantum mechanics: one (Shiekh, 1985; Sect. 4.1 consists mainly of unpublished results of Shiekh) uses linear combinations of explicit solutions of the Aharonov–Bohm effect (i.e., a path integral on multiply connected spaces) and the other (Schulman 1982, 1984) computes the propagator directly when a knife edge is present using a stationary phase method. Both solutions are exact.

(1) (i) In an Aharonov–Bohm configuration (Aharonov and Bohm, 1959), the probability amplitude $K$ for a particle of mass $M$ to go from $a$ to $b$ in time $t$ is

$$K(b, t; a; \alpha) = \sum_{m=-\infty}^{\infty} \exp(2\pi i m\alpha)\, \kappa_m \tag{11}$$

where $m$ is the number of times a path winds around the solenoid, and $\alpha$ is given by the flux in the solenoid, i.e., by the vector potential $A$ integrated along a loop around the solenoid.

$$\alpha = \frac{e}{h} \oint A \cdot dl \tag{12}$$

The partial amplitude $\kappa_m$ is the sum over all paths in the $m$-homotopy class (Schulman 1968, 1971, 1981; Laidlaw and DeWitt-Morette, 1971). It has been obtained independently by Edwards (1967), Berry (1980, see also 1972, 1981), Morandi and Menossi (1984), and Shiekh (1986). It is given by

$$\kappa_m = \frac{M}{2\pi i\hbar t} \exp\left(\frac{iM}{2\hbar t}(R'^2 + R^2)\right) \int_{-\infty}^{+\infty} d\lambda \exp\Big(i\lambda(\phi + 2\pi m)\Big)$$
$$\times I_{|\lambda|}\left(-\frac{i}{\hbar t} MR'R\right)$$

Using

$$\sum_{m=-\infty}^{\infty} \exp(im\theta) = 2\pi \sum_{n=-\infty}^{\infty} \delta(\theta + 2\pi n)$$

one obtains

$$K(\phi, R, t; \phi', R'; \alpha)$$

$$= \frac{M}{2\pi i\hbar t} \exp\left(\frac{iM}{2\hbar t}(R'^2 + R^2)\right)$$

$$\times \sum_{n=-\infty}^{\infty} \exp(-i(n+\alpha)(\phi - \phi'))(-i)^{|n+\alpha|} J_{|n+\alpha|}\left(\frac{1}{\hbar t} MR'R\right) \tag{13}$$

(ii) This propagator can readily be used to obtain the propagator $K_e$ for an absorbing (in a sense to be made precise shortly) knife edge. The Aharonov–Bohm propagator $K(b, t; a; \alpha)$ given in Eq. (13) is a function of the flux $\alpha$, and hence is a function of the equivalence class $[A]$ of gauge-related potentials satisfying (12). We can describe the Aharonov–Bohm effect in terms of any one member of this class and here we choose $A$ to be zero except along the half-plane barrier. The probability amplitude $K(b, t; a; \alpha)$ changes by a factor $\exp(2\pi i\alpha)$ when the detector located at $b$ crosses the knife edge. Thus one can construct $K_e$ (absorbing knife edge) by adding the contribution of $K(b, t; a; \frac{1}{2})$ and $K(b, t; a; 0)$; the latter kills the former on the far side of the half-plane barrier (away from the source). The barrier is absorbing, not in the sense that paths are destroyed at the half-plane barrier, but in the sense that their contribution interferes destructively on the far side.

$$K_{\text{edge}} \equiv K_e = \tfrac{1}{2}(K(\alpha = \tfrac{1}{2}) + K(\alpha = 0)) \tag{14}$$

Using the identity (Gradshteyn and Ryzhik, 1980)

$$\tfrac{1}{2}J_0(z) + \sum_{n=1}^{\infty} (-i)^{n/2} J_{n/2}(z)\cos(n\theta)$$

$$= (i\pi)^{-1/2} \exp(-iz\cos(2\theta)) \int_{-\infty}^{(2z)^{1/2}\cos\theta} \exp(is^2)\, ds \tag{15}$$

we obtain $K_e$ as a Fresnel integral:

$$K_e(\phi, R, t; \phi', R')$$

$$= \frac{M}{2i\pi\hbar t} \exp\left(i\frac{M}{2\hbar t}(R^2 + R'^2)\right)(i\pi)^{-1/2}$$

$$\times \exp\left(-i\frac{R'RM}{\hbar t}\cos(\phi - \phi')\right) \int_{-\infty}^{(2R'RM/\hbar t)^{1/2}\cos((\phi - \phi')/2)} \exp(is^2)\, ds \tag{16}$$

Since the Aharonov–Bohm propagator is a function only of the equivalence class $[A]$, any potential in this class gives the same result for the absorbing knife edge. It follows that the propagator is (to some extent) independent of the orientation and shape of the barrier.

(iii) We shall show that the propagator $K_e$ constructed from two different Aharonov–Bohm propagators is in fact a free propagator defined on a two-sheeted Riemann surface. It follows that it can be used to construct the reflecting knife edge propagators: $K_e^-$ for Dirichlet boundary conditions, $K_e^+$ for Neumann boundary conditions. Since $K_e$ is bivalued when $\phi - \phi'$ is not restricted to the interval $[0, 2\pi)$, it is natural to consider it as a single valued propagator on a two-sheeted Riemann surface. The "top" sheet of this Riemann surface may be identified with the physical space in which the source and detector are located. Thus, the knife edge may be modeled by a free particle on this surface, changing sheets where the knife edge would have been. The linear combination (14) is such that

$$\lim_{t \to 0^+} K_e(\phi, R, t; \phi', R') = \delta(R - R') \sum_{n=-\infty}^{\infty} \delta(\phi - \phi' - 4\pi n)$$

i.e., such that the source is on the physical sheet.

A perfectly reflecting knife edge is modeled by introducing an image of the source on the lower sheet and adding the contributions of the free propagators $K_e$, one with source $S_0$, the other with source $S_1$. To achieve Dirichlet boundary conditions ($K_e^- = 0$ on the knife edge) the two contributions must be of equal amplitude and opposite phase. On the other hand, to achieve Neumann conditions ($\partial K_e^+ / \partial n = 0$, $n$ being the normal to the surface) the two contributions must be of equal amplitude and equal phase.

$$K_e^{\mp}(\phi, R, t; \phi', R') = K_e(\phi, R, t; \phi', R') \mp K_e(\phi, R, t; -\phi', R') \qquad (17)$$

where $K_e$ is given by (16). This is precisely the result obtained by Schulman (1982) by an entirely different method (see Sect. 4.2). Schulman and C. M. Newman (private communication) have realized that (17) can be obtained from the Aharonov–Bohm propagator with the method of images. The derivation of (17) brings new insight to Wiegel and Boersma's derivation.

(iv) These results can be generalized to absorbing and reflecting wedges. It is clear mathematically what needs to be done. The fact that the free propagator on a Riemann surface can be obtained from a linear combination of different Aharonov–Bohm propagators [generalization of (14)] is not as intuitive as in the knife edge problem and is carried out in the Appendix.

First we determine the Riemann surface to be considered for computing propagators in the presence of a wedge of external angle $\theta - \mu\pi/\nu$. We need images obtained by reflecting the source in each boundary wall. However, the image which achieves proper boundary conditions on one wall disturbs the boundary conditions on the other one. So we add images of these images and repeat the process until the set of images closes on itself.

Images stand at $\psi = \pm\phi' \pm 2k\theta$ ($k$ an integer). For instance, the image $S_1$ is at $-\phi$, $\bar{S}_1$ is at $-\phi + 2\theta$, $S_2$ is at $\phi + 2\theta$, $\bar{S}_2$ is at $\phi - 2\theta$, etc. These are conveniently grouped as those obtained after an even number of reflections and those after an odd number

| Even | Odd | |
|---|---|---|
| $\psi = \phi' + 2\pi k \dfrac{\mu}{\nu}$, | $\psi = -\phi' + 2\pi k \dfrac{\mu}{\nu}$, | $k = 1, \ldots, \nu$ |

This set has $2\nu$ elements and exists upon a $\mu$-sheeted Riemann surface.

On a $\mu$-sheeted Riemann surface, paths must loop a multiple of $\mu$ times to go from the source to the detector which are both on the top (physical) sheet. We can repeat the calculation (Shiekh, 1986) leading from (11) to (13) when $\alpha = 0$ for paths which loop around a multiple of $\mu$ times.

$$K_\mu = \sum_{m=-\infty}^{\infty} \kappa_{\mu m}$$

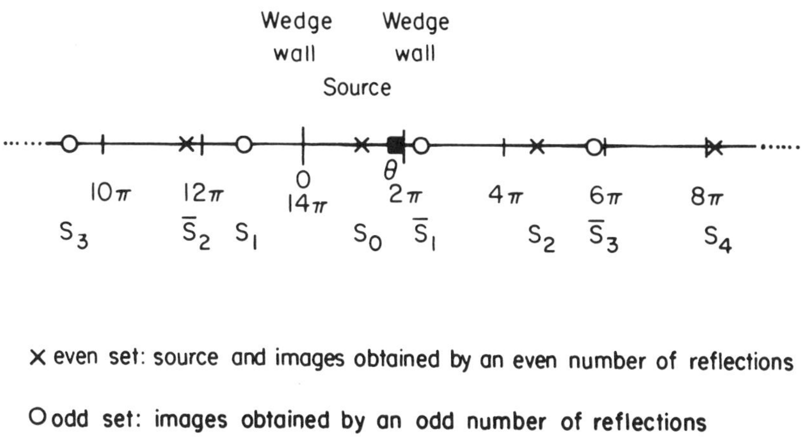


Fig. 2. Image positions on the $\mu$-sheeted Riemannian surface for a wedge at external angle $\theta = 7\pi/4$ and source at $\phi' = 7\pi/6$. In this figure $S_1$ is the reflection of $S_0$ by the wall at $\psi = 0$, $S_1$ is the reflection of $S_1$ by the wall at $\psi = \theta$, $S_2$ is the reflection of $S_1$ by the wall at $\psi = \theta$, $S_2$ is the reflection of $S_1$ by the wall at $\psi = 0$, and so forth.

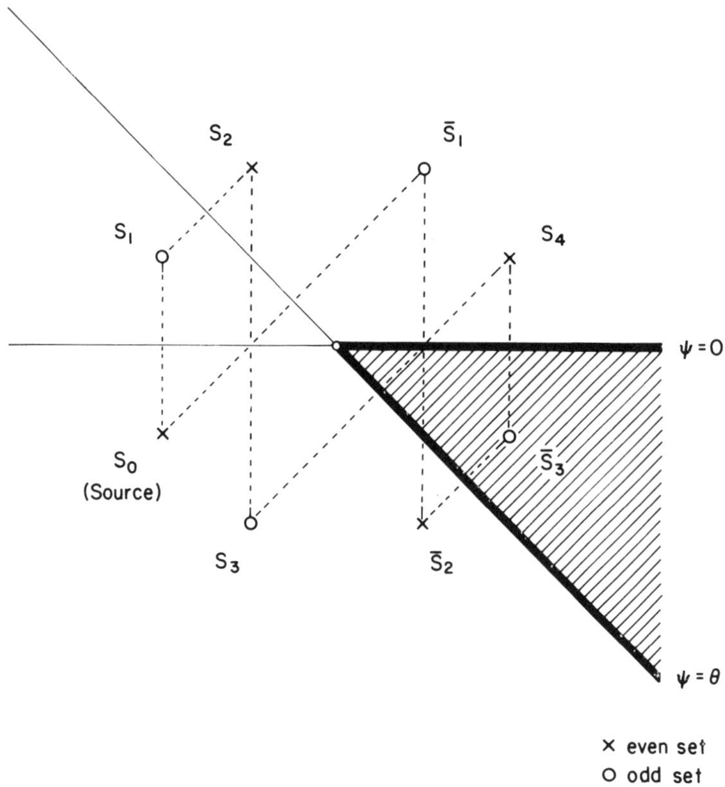


Fig. 3. The use of images to implement boundary conditions.

and obtain for the free propagator on a $\mu$-sheeted Riemann surface

$$K_\mu(\phi, R, t; \phi', R') = \frac{1}{\mu}\frac{M}{2\pi i\hbar t}\exp\left[\frac{iM}{2\hbar t}(R^2+R'^2)\right]$$
$$\times \sum_{n=-\infty}^{\infty} \exp\left[-i\frac{n}{\mu}(\phi-\phi')\right](-i)^{|n/\mu|}J_{|n/\mu|}\left(\frac{MR'R}{\hbar t}\right) \tag{18}$$

With the source and all its images having the same strength, the Dirichlet boundary conditions are achieved by running the odd and even set defined above with opposite phase; the Neumann conditions are achieved by running them in phase. The propagator $K^-$ which satisfies the Dirichlet boundary conditions and the propagator $K^+$ which satisfies the Neumann ones are given by

$$K_{\text{wedge}}^{\mp} \equiv K_w^{\mp} = \frac{M}{2\pi i\hbar t} \exp\left[\frac{iM}{2\hbar t}(R^2 + R'^2)\right] \frac{\pi}{\theta}$$
$$\times \left\{J_0\left(\frac{MR'R}{\hbar t}\right) + 2 \sum_{n=1}^{\infty} \exp\left(\frac{-in\pi^2}{2\theta}\right) J_{n\pi/\theta}\left(\frac{MR'R}{\hbar t}\right)\right.$$
$$\left.\times \cos\left(\frac{n\pi}{\theta}(\phi - \phi')\right)\right\} \mp \{\phi' \to -\phi'\}$$
$$= \frac{M}{2\pi i\hbar t} \exp\left[\frac{iM}{2\hbar t}(R^2 + R'^2)\right] \frac{\pi}{\theta}$$
$$\times \left\{J_0\left(\frac{MR'R}{\hbar t}\right)(1 \mp 1) + 2 \sum_{n=1}^{\infty} \exp\left(\frac{-in\pi^2}{2\theta}\right) J_{n\pi/\theta}\left(\frac{MR'R}{\hbar t}\right)\right.$$
$$\left.\times \cos\left(\frac{n\pi}{\theta}(\phi - \phi')\right) \mp \cos\left(\frac{n\pi}{\theta}(\phi + \phi')\right)\right\}, \qquad \text{where } \theta = \frac{\pi\mu}{\nu} \tag{19}$$

The perfectly absorbing wedge (in the sense given for the absorbing knife edge) can be obtained by superposing these two solutions. The absorbing edge and wedge propagators are not solutions of boundary value problems of partial differential equations because they are the sum of terms (14) which satisfy differential equations for different values of $\alpha$. This may explain Sommerfeld's scanty remarks on the absorbing wedge.

As remarked after Eqs. (5) and (6) for the wedge problem in electromagnetic diffraction, the series in (19) converge slowly for large arguments of the Bessel functions. These series are equal to finite sums which are preferable for numerical calculations.

$$\frac{1}{2} J_0(z) + \sum_{n=1}^{\infty} \exp\left(-in\frac{\pi\nu}{2\mu}\right) J_{n\nu/\mu}(z) \cos\left(n\frac{\nu}{\mu}\phi\right)$$
$$= \frac{1}{2\nu} \sum_{k=1}^{\nu} \left[\exp(-iz\cos(\phi + 2\pi k\mu/\nu))\left(-i\sum_{n=1}^{\mu-1} \cos\left[\left(\frac{\mu-n}{\mu}\right)(\phi + 2\pi k\frac{\mu}{\nu}\right)\right]\right.$$
$$\times \int_0^z \left(\exp\left(+in\frac{\pi}{2\mu}\right) J_{-n/\mu}(u) - \exp\left(-in\frac{\pi}{2\mu}\right) J_{n/\mu}(u)\right)$$
$$\left.\times \exp(+iu\cos(\phi + 2\pi k\mu/\nu))\, du + 1\right)\Bigg] \tag{20}$$

This equation is derived in the Appendix. It generalizes for rational wedges Eq. (15) used for edges where $\nu = 1$ and $\mu = 2$. The right-hand side of (20) is, however, cumbersome analytically and we retain the left-hand side for analytic manipulations.

We recall the key formulas which relate time-dependent Schrödinger kernels $K$ and monochromatic solutions $G$ of the wave equation with a point source so that we can relate the formulas in this section to formulas in Sect. 2.

The time-dependent Schrödinger equation for the kernel $K$

$$i\hbar \frac{\partial}{\partial t} K - \frac{(i\hbar)^2}{2M} \Delta K = i\hbar\, \delta(x - x_0)\, \delta(t)$$

can be converted to the Helmholtz equation with a point source

$$-k^2 G - \Delta G = \delta(x - x_0)$$

using the Fourier transform pair

$$G = \frac{i\hbar}{2M} \int_{-\infty}^{+\infty} K \exp(i\omega t)\, d\tau$$

and

$$K = \frac{1}{2\pi} \frac{2M}{i\hbar} \int_{-\infty}^{+\infty} G \exp(-i\omega t)\, d\omega, \qquad \omega = \hbar k^2/2M$$

The time-independent massless kernel $G_w$ may then be obtained by Fourier transform of the time dependent nonrelativistic kernel $K_w$

$$G = \frac{i\hbar}{2M} \int_{-\infty}^{+\infty} K \exp(i\omega t)\, dt = \frac{i\hbar}{2M} \int_{0}^{\infty} K \exp(i\omega t)\, dt$$

since $K = 0$ for $t < 0$. This transforms our nonrelativistic scalar kernel $K_w$ on a $\mu$-sheeted Riemannian surface into

$$G_w = \frac{1}{2\pi} \frac{1}{\mu} \sum_{n=-\infty}^{\infty} \exp\left( -\frac{in}{\mu} (\phi - \phi') \right) (-i)^{|n/\mu|} J_{|n/\mu|}(kR)\, K_{|n/\mu|}(-ikR') \qquad (21)$$

when $R < R'$; while for $R > R'$ interchange $R$ and $R'$. The following identity has been used in computing (21):

$$\int_0^\infty \frac{1}{\chi} \exp(-\alpha\chi - \beta/\chi)\, J_\nu(\gamma/\chi)\, d\chi$$
$$= 2J_\nu\{(2\alpha[(\beta^2 + \gamma^2)^{1/2} - \beta])^{1/2}\}\, K_\nu\{(2\alpha[(\beta^2 + \gamma^2)^{1/2} + \beta])^{1/2}\}$$

where Re $\alpha > 0$; Re $\beta > 0$; $\gamma > 0$.

The conditions on $\alpha$ and $\beta$ make the integrals finite and eliminate solutions with waves incoming from infinity. Equation (21) agrees with Sommerfeld's solution.

Images may then be introduced as before to implement boundary conditions. For $R < R'$, the expression is

$$G_w^{\mp} = \frac{1}{2\theta}\left((1 \mp 1)\, J_0(kR)\, K_0(-ikR') + 2 \sum_{n=1}^{\infty} \exp\left(\frac{-in\pi^2}{2\theta}\right)\right.$$
$$\left. \times\, J_{n\pi/\theta}(kR)\, K_{n\pi/\theta}(-ikR') \left\{\cos\left(\frac{n\pi}{\theta}(\phi - \phi')\right) \mp \cos\left(\frac{n\pi}{\theta}(\phi + \phi')\right)\right\}\right) \tag{22}$$

while for $R > R'$ interchange $R$ and $R'$. This result agrees with Carslaw's (1919).

As remarked after Eqs. (5), (6), and (19), the series in (22) converges slowly for large arguments of the Bessel functions. One can reexpress (22) as finite sums by taking the Fourier transforms of the finite-term solutions for $K_w^{\mp}$ which are obtained by replacing the series in (19) by the finite sums given in (20).

In conclusion, the path integral calculation of the free propagator on a $\mu$-sheeted Riemannian surface (18) together with the method of images makes it possible to compute explicitly propagators with boundary conditions appropriate to the wedge problem (see also Goodman (1981)).

2. The knife edge problem in quantum mechanics was first solved (Schulman, 1982) using path integration and stationary phase methods. The result turned out to be an exact solution of the Schrödinger equation with (reflecting) knife edge boundary conditions. For this problem the method of stationary phase introduced in functional integration in 1951 (Morette, 1951) had to be generalized because the path which extremizes the action (i.e., the critical point of the action) is not inside the domain of integration but on its boundary.

(i) Stationary phase in functional integration is a natural generalization of stationary phase in ordinary integrals[17] which in its simplest form gives an asymptotic approximation for large $\lambda$ of integrals of the following form:

$$F(\lambda) = \int_X h(x) \exp(i\lambda f(x))\, d\mu(x) \tag{23}$$

where $h$ is a real-valued smooth function of *compact* support on the $n$-dimensional riemannian manifold $X$ with volume element $d\mu(x)$, and where the critical points of $f$ (i.e., the solutions of $f'(y) = 0$, $y \in X$) are assumed to

[17] See, for instance, Bleistein and Handelsman (1975) or Choquet-Bruhat *et al.* (1982), pp. 593–595.

be nondegenerate,[18] that is, the determinant of the Hessian $\partial^2 f/\partial y^i\,\partial y^j$ does not vanish. In this case,

$F(\lambda)=O(\lambda^{-N})$ for any $N$ if $f$ has no critical point on the support of $h$

$F(\lambda)=O(\lambda^{-n/2})$ if $f$ has a finite number of nondegenerate critical points on the support of $h$.

These results are modified when $h$ does not vanish on the boundary $\partial X$ of $X$. If there is no critical point of $f$ in $X$, then the leading term is given by boundary terms which are obtained by integration by parts. For example, for $n=1$ and $X=[a,b]$ the first integration by parts gives

$$F(k)=\frac{1}{i\lambda}\frac{h(x)}{f'(x)}\exp(ikf(x))\bigg|_a^b-\frac{1}{i\lambda}\int_a^b\left(\frac{h(x)}{f'(x)}\right)'\exp(i\lambda f(x))\,dx \tag{24}$$

After $N$ integrations by parts, the boundary terms consist of a polynomial in $(i\lambda)^{-1}$ of order $N$ and the remaining integral is of order $(\lambda)^{-N}$.

For $X$ an $n$-dimensional Riemannian manifold,[19] use the identity

$$\frac{1}{i\lambda}\,\boldsymbol{\omega}(x)[\exp(i\lambda f(x))]\equiv\exp(i\lambda f(x)) \tag{25}$$

where $\boldsymbol{\omega}(x)$ is the derivation at $x$ defined by

$$\boldsymbol{\omega}(x):=\frac{\mathbf{v}(x)}{|\mathbf{v}(x)|^2},\qquad \mathbf{v}(x)=g^{ij}(x)\frac{\partial f}{\partial x^i}\frac{\partial}{\partial x^j}=v^j\frac{\partial}{\partial x^j} \tag{26}$$

$\boldsymbol{\omega}$ is defined for all $x$ which are not critical points.

It follows that, if there is no critical point $y\in X$,

$$\begin{aligned}F(\lambda)&=\frac{1}{i\lambda}\int_X h(x)\,\boldsymbol{\omega}(x)[\exp(i\lambda f(x))]\,d\mu(x)\\&=\frac{1}{i\lambda}\int_{\partial X} h(x)\exp(i\lambda f(x))\,\boldsymbol{\omega}(x)\cdot d\boldsymbol{\sigma}(x)\\&\quad-\frac{1}{i\lambda}\int_X \operatorname{div}(h(x)\,\boldsymbol{\omega}(x))\exp(i\lambda f(x))\,d\mu(x)\end{aligned} \tag{27}$$

where $d\boldsymbol{\sigma}$ is the $(n-1)$-dimensional surface element on $\partial X$.

[18] Generalizations when the critical points are degenerate (DeWitt-Morette *et al.*, 1983; DeWitt-Morette and Nelson, 1984; DeWitt-Morette, 1984) give, in particular, rainbows and glories (Ford and Wheeler, 1959).

[19] To determine the dependence of $F$ on $\lambda$, it is sufficient to consider the case $X\subset\mathbb{R}^n$ and set $h(x)=1$. In that case one can use the identity

$$\frac{1}{i\lambda}\nabla\cdot(\exp(i\lambda f)\,\mathbf{u})=\exp(i\lambda f)+\frac{1}{i\lambda}\exp(i\lambda f)\,\nabla\cdot\mathbf{u}$$

with $\mathbf{u}\equiv\nabla f/|\nabla f|^2$. Then

$$F(\lambda)=\frac{1}{i\lambda}\int_{\partial X}d\boldsymbol{\sigma}\cdot\mathbf{u}\exp(i\lambda f)-\frac{1}{i\lambda}\int_X d^n x\exp(i\lambda f)\,\nabla\cdot\mathbf{u}$$

The dependence on $\lambda$ is the same as the one-dimensional integral (24), and repeated integration by parts will produce the same dependence on $\lambda$ as before. The leading term for large $\lambda$ is the boundary contribution of (27).

$$F(\lambda) \sim (i\lambda)^{-1} \int_{\partial X} h(x) \exp(i\lambda f(x))\, \boldsymbol{\omega}(x) \cdot d\boldsymbol{\sigma}(x) \tag{28}$$

Under appropriate regularity and compactness conditions on $X$, $f|_{\partial X}$ has at least one extremal value (i.e., it has at least one critical point), and the $n-1$ dimensional integral in (28) can be approximated by the stationary phase method. It contributes a term of order $\lambda^{-(n-1)}$ and therefore

$$F(\lambda) \simeq \lambda^{-1} \lambda^{-(1/2)(n-1)} \simeq \lambda^{-(1/2)(n+1)}$$

$f|_{\partial X}$ attains its extrema at points $y \in \partial X$ when

$$(\partial f|_X)(y)/\partial y^i = 0$$

i.e., when the components of grad $f$ in $T_y X$ vanish. This means that grad $f$ attains its extrema when grad $f(y)$ is normal to $\partial X$.

In conclusion, we have

(A) (nondegenerate) critical points[20] $h$ of compact support $F(\lambda) = O(\lambda^{-n/2})$

(B) no critical points $h \neq 0$ on the boundary $F(\lambda) = O(\lambda^{-n/2 - 1/2})$.

This means that for a regular compact boundary $\partial X$, the integral "loses" a power of $k$ when there is no critical point of $f$ in $X$.

(ii) We shall now investigate the stationary phase approximation of a path integral when there is no critical point of the action $S$ in the domain of integration. Consider the knife edge problem, $\theta = 2\pi$, in which an otherwise free particle cannot pass through a thin barrier along the positive $x$-axis (see Fig. 4). Consider an $N$-dimensional approximation to the path integral representation of the propagator of a particle of mass 1 in $\mathbb{R}^2$.

$$K^N(b, t; a) = \int_{\Omega^N} d\mathbf{x}_1 \cdots d\mathbf{x}_N \left( \sqrt{\frac{1}{2\pi i\hbar\varepsilon}} \right)^{2N} \exp\left( \frac{i}{\hbar} S^N(\mathbf{x}_1, \ldots, \mathbf{x}_N) \right) \tag{29}$$

where $\varepsilon = t/N$, $\mathbf{x}_0 = a$, $\mathbf{x}_{N+1} = b$, and

$$S^N = \sum_{j=0}^{N} |\mathbf{x}_{j+1} - \mathbf{x}_j|^2 / 2\varepsilon \tag{30}$$

[20] If, but only if, the $n$-dimensional integral is separable, can one speak of "critical points in a lower-dimensional subspace of $X$." Otherwise a critical point is defined by $\partial f(y)/\partial y^j = 0$ for all $j$. The statement $\partial f(y)/\partial y^j = 0$ for some $j$ is not coordinate independent.

The domain of integration $\Omega^N$ consists of all sequences of points $\{\mathbf{x}_1,..., \mathbf{x}_N\}$ such that the broken line from $a$ to $b$ defined by these points does not cross the barrier. Convergence of $K^N$ to the propagator is guaranteed by the Trotter product formula (Schulman 1981). A critical point of $S^N$ is a solution of the discretized Euler–Lagrange equation

$$\mathbf{x}_{j+1} - 2\mathbf{x}_j + \mathbf{x}_{j-1} = 0, \qquad \forall j$$

i.e., a sequence of points in $\mathbb{R}^2$ which are aligned. If $a$ and $b$ are such that $|\phi' - \phi| > \pi$, then the critical point of $S^N$ is not in $\Omega^N$, and we expect the leading term[21] of $K^N$ for small $\hbar$ to come from the boundary $\partial\Omega^N$ of $\Omega^N$; but $\partial\Omega^N$, or its limit $\partial\Omega$ when $N = \infty$, is a more complex concept than the boundary of $X$. A possible approach for investigating this issue is indicated briefly after Eq. (38).

(iii) Although the role of the boundary $\partial\Omega^N$ is not yet well understood in the stationary-phase approximation of path integrals whose phase has no critical points, the propagator $K(b, t; a)$ for the knife edge has been computed directly by stationary phase methods (Schulman, 1984) and the result is fortuitously exact (Schulman, 1982). Here is an outline of the calculation.

$$K(b, t; a) = \int d^2c\, K(b, t - t_c; c)\, K(c, t_c; a) \tag{31}$$

where the arbitrary intermediate time $t_c$ is chosen to be the time when a free particle going from $(a, t = 0)$ to the origin $O$ to $(b, t)$ reaches $O$. When $c$ is visible both from $a$ and $b$ and located as in Fig. 4, we have direct contributions

$$K_{\text{Direct}}(b, t - t_c; c) = (2\pi i\hbar(t - t_c))^{-1} \exp(i\, |b - c|^2/2\hbar(t - t_c)) \tag{32}$$

$$K_{\text{Direct}}(c, t_c; a) = (2\pi i\hbar t_c)^{-1} \exp(i\, |c - a|^2/2\hbar t_c) \tag{33}$$

and one reflected contribution

$$K_{\text{Reflected}}(c, t_c; a) = (2\pi i\hbar t_c)^{-1} \exp(i\, |c' - a|^2/2\hbar t_c) \tag{34}$$

[21] Note in (29) that the $\hbar$ dependence of $K^N$ comes both from the normalizaing factors and from the phase, in contrast to the $\lambda$ dependence of $F$ in (23) which comes only from the phase. Counting powers of $h$ is facilitated by writing path integrals not formally as $\int \mathscr{D}x \exp(iS/\hbar)$ but as prodistributions (DeWitt-Morette 1972, 1974; DeWitt-Morette *et al.*, 1979).

$$\int_\Omega dw(\omega) \exp\left( -\frac{i}{\hbar} \int V(\omega(s))\, ds \right)$$

where $w$ is the complex gaussian whose covariance equals $(i\hbar/m)$ times the Wiener (1929) covariance. Alternatively one can count the "relevant powers of $\hbar$" by comparing a propagator with its free counterpart.

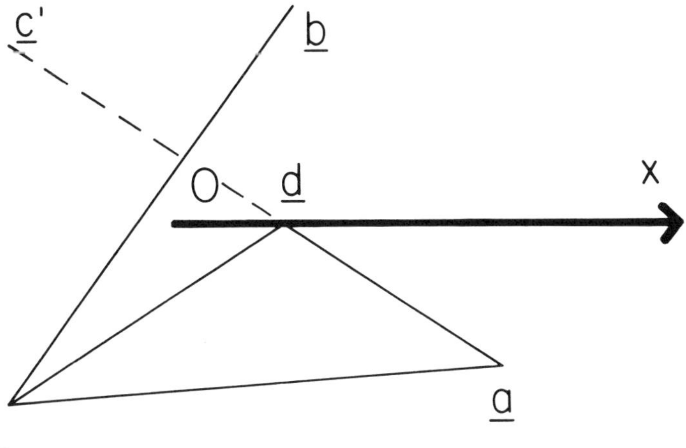


Fig. 4. For this geometry, the classical path contributions to the propagator are *acb* and *adcb*.

where $c'$ is symmetric to $c$ with respect to the barrier plane. It is understood that each term appears only in its classically allowed region. Inserting (32)–(34) into (31) gives

$$K^{\mp}(b, t; a) = K_{\mathrm{DD}}(b, t; a) \mp K_{\mathrm{RD}}(b, t; a)$$

where

$$K_{\mathrm{DD}}(b, t; a) = \int d^2c K_{\mathrm{D}}(b, t - t_c; c)\, K_{\mathrm{D}}(c, t_c; a)$$

$$K_{\mathrm{RD}}(b, t; a) = \int d^2c K_{\mathrm{D}}(b, t - t_c; c)\, K_{\mathrm{R}}(c, t_c; a)$$

$K^-(b, t; a)$ is the propagator for a reflecting knife edge with Dirichlet boundary condition, $K^+(b, t; a)$ for one with Neumann conditions. It is sufficient to compute $K_{\mathrm{DD}}$. After some algebra, we have

$$K_{\mathrm{DD}}(b, t; a) = K_0(2\pi i\hbar)^{-1}\gamma \int_0^\infty \rho\, d\rho \int_{\phi' - \pi}^{\phi + \pi} d\alpha \times \exp\left(\frac{i}{2\hbar}(\gamma\rho^2 - 2\rho v(\cos(\alpha - \phi') + \cos(\alpha - \phi)))\right) \quad (35)$$

where

$$\gamma = t_c^{-1} + (t - t_c)^{-1}$$

$$v = |\underline{a}0|/t_c = |0\underline{b}|/(t - t_c)$$

$$K_0 = (2\pi i\hbar t)^{-1} \exp\left(\frac{i}{2\hbar t}(|\underline{a}0|^2 + |0\underline{b}|^2)\right)$$

where $|\underline{a0}|$ and $|\underline{0b}|$ are the length of the line segments from $a$ to $O$ and $b$ to $O$ respectively in Fig. 4. The $\rho$ and $\alpha$ coordinates of the critical point of the phase in the integrand of (35) satisfy

$$\gamma\rho - v(\cos(\alpha-\phi')+\cos(\alpha-\phi))=0 \tag{36}$$

$$\sin(\alpha-\phi')+\sin(\alpha-\phi)=0 \tag{37}$$

The relevant solution of (37) is $\alpha=\frac{1}{2}(\phi+\phi')$ so that (36) yields

$$\rho=\frac{2v}{\gamma}\cos(\tfrac{1}{2}(\phi'-\phi))$$

$\rho$ being positive, the coupled equations (36) and (37) have a solution if $\phi'-\phi<\pi$, i.e., if there is a straight line from $a$ to $b$ in the space $\Omega$ of allowed paths. If there is, the stationary phase approximation of (35) is the free propagator $K_0$ in $\mathbb{R}^2$ proportional to $h^{-1}$. If not, the stationary phase approximation of (35) is dominated by the boundary terms, and, after some calculations (Schulman, 1984) one obtains

$$K^{\mp}(b,t;a)=K_0[\exp(-im^2)\,h(-m)\mp\exp(-in^2)\,h(-n)] \tag{38}$$

where

$$\begin{aligned} m&=(2/\hbar\gamma)^{1/2}\sin\omega_2, \qquad & n&=(2/\hbar\gamma)\,v\sin\omega_1\\ \omega_2&=\tfrac{1}{2}(\phi'-\phi)-\frac{\pi}{2}, \qquad & \omega_1&=\tfrac{1}{2}(\phi'+\phi)\\ h(m)&=(\pi)^{-1/2}\exp(-i\pi/4)\int_{-\infty}^{m}\exp(it^2)\,dt \end{aligned} \tag{39}$$

Note that for $m>0$, $h(-m)$ is of order unity, while for $m<0$ the stationary point in the integral defining $h$ is outside the range of integration. Thus for $m<0$, $h=O(1/m)=O(\hbar^{1/2})$.

The path integral calculation of the knife edge problem reduces to a two-dimensional integral (31) because the path integral is a cylindrical one. The relationship between a cylindrical path integral and a finite-dimensional integral (DeWitt-Morette, 1976; DeWitt-Morette *et al.* 1979) suggests a method for investigating the boundary $\partial\Omega$ of the domain of integration and its contribution to the path integral. But the work remains to be done. The propagators $K^{\mp}$ are identical with the ones obtained by Carslaw in 1899. The wedge problem, a great problem of the nineteenth century, continues to serve as a source of inspiration today—and Feynman's path integrals developed when he was a graduate student with John A. Wheeler brings new insight to this classic problem. It shifts the focus from partial differential equations with boundary values to integrals over restricted spaces of paths. It brings stochastic calculus (averages over continuous paths) and differential calculus together. A path integral solves a partial differential equation with given boundary conditions. But it does more:

(i) It incorporates the boundary conditions in the definition of its domain of integration. In other words, boundary conditions are not, as in differential calculus, an *additional* requirement satisfied by a solution of a partial differential equation whose compatibility has to be checked.

(ii) It often can answer global questions more readily than a differential equation because the domain of integration consists of paths which take their values in the configuration space of the system and probes its global properties. A partial differential equation states only local relationships between a function and its derivatives.

## 5. STOCHASTIC DIFFUSIONS

Although solutions of the Schrödinger equation can be obtained by analytic continuation of solutions of the heat equation, their asymptotics are quite different. The reason that analytic continuation of the full solutions does not imply analytic continuation of their asymptotics for small diffusion coefficient $D$ or small values of $\hbar$ is that these equations are singular in the limits $D=0$ and $\hbar=0$ respectively. This issue requires further study, but two simple remarks can readily be made.

(i) In a Schrödinger JWKB propagator the paths $\xi_0$ which minimize tha action $S(\xi_0)$ play no more an important role in asymptotic estimates than other extrema, whereas the minima and maxima of the "action" in a diffusion equation play different roles.

(ii) In a Schrödinger JWKB propagator the phase factor $\exp(iS(\xi_0)/\hbar)$ plays no role in asymptotic estimate probabilities; the relevant term is the Van Vleck determinant and the caustics, if any, dominate the answer. In the asymptotics of a diffusion equation the exponential term $\exp(-D^{-1}S(\xi_0))$ dominates the answer. A very instructive example has been analyzed by Michael Berry.[22] It arises from contemplating the question of how male moths use their sense of smell to find their way to females. Female moths emit pheromones which propagate by diffusion aided by convection. The concentration $C(\mathbf{r}, t)$ of particles with diffusion constant $D$ in a wind of velocity $\mathbf{V}(\mathbf{r}, t)$ satisfies the equation

$$\frac{\partial C(\mathbf{r}, t)}{\partial t} = D\Delta C(\mathbf{r}, t) - \mathbf{V}(\mathbf{r}, t)\cdot C(\mathbf{r}, t) \tag{40}$$

[22] Berry (to appear). An example which deserves more than the brief summary given here. These paragraphs owe much to remarks made by Michael Berry during his visit to Austin.

In the limit $D=0$, the concentration $C(\mathbf{r}, t)$ vanishes everywhere except along the paths where the action $S(\xi_0)$ vanishes. The quantum mechanical counterpart of this equation is

$$i\hbar \frac{\partial \psi}{\partial t} = H(\mathbf{r}, -i\hbar\nabla)\,\psi \tag{41}$$

where the classical Hamiltonian $H = p^2 + \mathbf{V}\cdot\mathbf{p}$. The velocity field $\mathbf{V}$ is assumed to represent an incompressible flow, $\nabla\cdot\mathbf{V}=0$, and there is no need to symmetrize the $\mathbf{V}\cdot\mathbf{\nabla}$ term in the quantum Hamiltonian. Equation (41) is the equation for a particle in a magnetic field $\mathbf{B} = -\frac{1}{2}\nabla\times\mathbf{V}$ and an electric field $\mathbf{E} = \frac{1}{2}(\mathbf{V}\cdot\nabla\mathbf{V} + \mathbf{V}\times(\nabla+\mathbf{V}) + \partial\mathbf{V}/\partial t)$. Its short-wave limit is dominated by caustics of the trajectories generated by the classical hamiltonian.[23]

It should also be noted that for the Schrödinger equation semiclassical asymptotics are not necessarily identical to short-time asymptotics. For instance the JWKB propagator for a particle in a Riemannian manifold is not equal to the short-time propagator.[24] A more intuitive example is the propagator between two nearby points which can be connected by multiple geodesics. All geodesics contribute to the JWKB propagator, but only the shortest one contributes to the short-time propagator.

In spite of the differences between the Schrödinger and the heat equations, techniques and insight developed in stochastic calculus can sometimes be transferred[25] to Feynman path integrals, and we shall indicate some results in the theory of stochastic diffusion of interest for the wedge problem: namely Molchanov's results, the stochastic definitions of regular and irregular boundary points, and a Feynman–Kac formula for solutions of Dirichlet problems. Applications of these results to wedge problems are under consideration.

1. The wedge problem in diffusion was first studied by Molchanov (1975) who investigated, in particular, diffusion on a riemannian manifold $M$ which is a nonconvex polygon (see Fig. 5).[26] He computed the new effects which occur when the geodesic $\gamma$ from the source at $a$ to the detector at $b$ has points in common with the boundary $\partial M$. He worked out in detail

[23] Berry (1986).

[24] DeWitt-Morette *et al.* (1980), Eqs. (11) and (8) slightly modified to yield the position-to-position transition.

[25] DeWitt-Morette *et al.* (1979).

[26] We refer the reader to Ikeda (1982, 1985) for diffusion in the presence of smooth obstructions—as opposed to polygonal obstructions. One could argue that a change of scale makes a rounded wedge look like a sharp wedge. This is indeed correct when dealing with exact solutions, but not so when dealing with asymptotics: one cannot always interchange the semiclassical limit and the change of scale which makes a round wedge sharp.

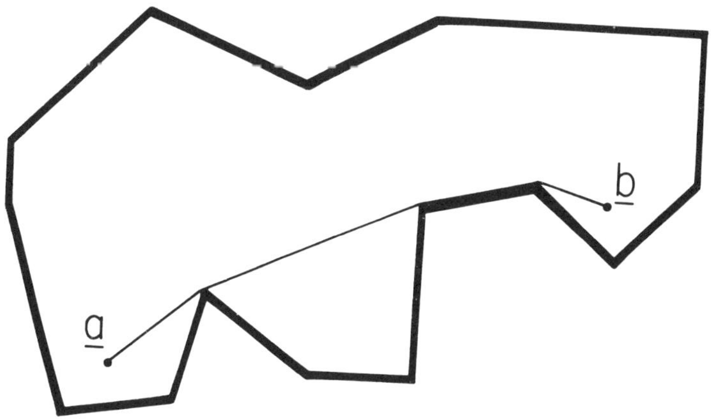


Fig. 5. A geodesic $\gamma$ from $a$ to $b$ in a nonconvex polygonal domain $M$.

diffusion on two-dimensional spaces, but his method should carry over to higher-dimensional spaces.

It is possible to recast the wedge problem as a problem of diffusion in a nongeodesically convex Riemannian manifold $M$ with boundary $\partial M$ because the boundary has an effect[27] on the diffusion *only* at the points of contact of the geodesic $\gamma$ with the boundary $\partial M$.

Kac called this situation the "principle of imperceptibility" of the boundary: Only if the geodesic *touches* the boundary do new effects appear; proximity is not enough. These new effects are the diffraction effects. That they are determined entirely by the points of contact justifies Keller's geometric theory of diffraction.

By computing upper and lower bounds of probabilities, Molchanov established the following results for $M$ two dimensional.

(i) If the polygonal geodesic $\gamma$ has $\rho$ sides and $\tau$ points in common with the boundary and if $\gamma$ has a break at the end points of each arc and at each vertex, then the transition density $p(t, a, b)$ for diffusion from $a$ to $b$ in time $t$ is

$$p(t, a, b) = O(t^{k-1} \exp(-L(\gamma)^2/2t)) \tag{42}$$

where $k = 2\rho + \frac{1}{2}\tau$ and $L(\gamma)$ is the length of $\gamma \in \mathscr{D}_{\rho,0,\tau}(a, b)$.

Note that $k$ does not depend on the length of the sides $\rho$ common to the geodesic and the boundary but only on their number. Note also that four vertices contribute the same amount to $k$, as one arc, however long. Molchanov calls the dependence of $k$ on $\rho$ and $\tau$ the "inertia" of Brownian

[27] See Molchanov (1975), Theorem 4.1, p. 24 for a precise statement.

motion: it is "harder" for a Brownian particle to go around $\tau > 4$ isolated vertices than to creep along one entire side of $\partial M$.

(ii) The angle of the boundary at a vertex does not affect the leading term of the transition probability expanded in powers of $t$, provided the part of $\partial M$ lying in some $\varepsilon$-neighborhood of the vertex can be approximated (in the sense of inclusion) by two angles (see Fig. 6).

2. Lebesgue (1924) discovered that the classical Dirichlet problem is not well posed in $d > 2$ dimensions if the boundary does not satisfy some regularity condition. Indeed consider a $d \geqslant 3$ dimensional ball and push a sharp thorn (called noadays a "Lebesgue thorn") into its side, place a heater inside the ball at the tip $O$ of the thorn, and maintain the walls at zero temperature away from $O$ (see Fig. 7).

As time goes to infinite, the temperature $u$ inside the ball should solve Dirichlet's problem

$$\Delta u(a) = 0 \qquad \text{for} \quad a \in M$$

$$\lim_{a=b} u(a) = f(b) \qquad \text{for} \quad b \in \partial M, \text{ here } f(0) = 1 \qquad \text{and} \qquad f(b \neq 0) = 0$$

But it can be shown that, if the thorn is sharp enough,

$$\lim_{a=0} u(a) < 1, \text{ hence } \lim_{a=0} u(a) \neq f(0)$$

A person inside the ball will be cold, no matter how close to the heater.

The definition of a regular boundary point in differential calculus is very complex,[28] but has a simple, equivalent, expression in stochastic calculus,[29] in terms of the first time a Brownian particle on the boundary

[28] Lebesgue (1924).

[29] See for instance Durett (1984), pp. 248–251, for an introduction to regular and irregular boundaries and Itô and McKean (1985), pp. 261–264, for the Dirichlet problem in stochastic calculus, and p. 259 for Wiener's necessary and sufficient condition for a boundary point to be regular. Other useful references are Barber and Ninham (1970); Gikhman and Skorokhod (1969); and Port and Stone (1978).

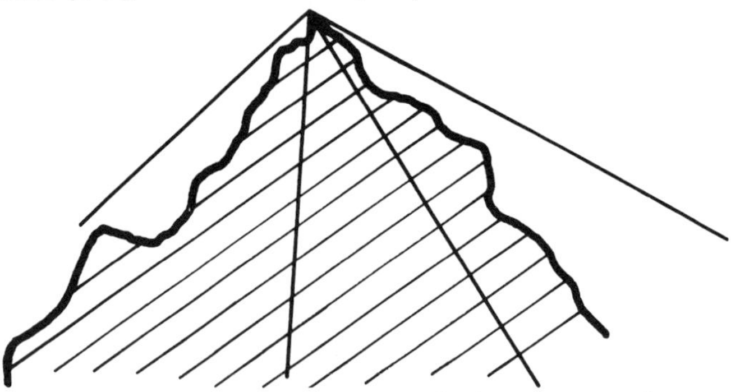

Fig. 6. An irregular wedge whose tip fits between two angles.

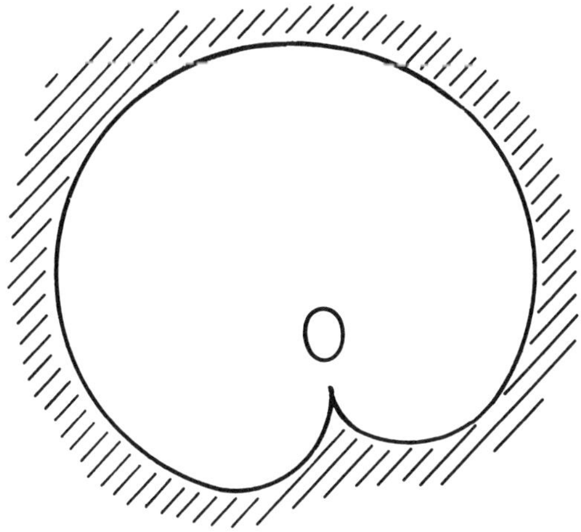

Fig. 7. A Lebesgue thorn.

$\partial M$ at time 0 leaves the domain $M$. The first exit time of a Brownian path $B(t, \omega)$, where $\omega$ is a point in a probability space $\Omega$ which labels the path, is naturally

$$\theta(\omega) = \inf\{t; B(t, \omega) \notin M\} \tag{43}$$

A point $b \in \partial M$ is said to be regular if the first exit time of a Brownian particle at $b$ at time 0 vanishes with probability one, i.e., if

$$P_b(\theta = 0) := \int_{\chi_{\theta = 0}} dw(\omega) = 1 \tag{44}$$

where

$$\chi_{\theta = 0} = \{\omega; \theta(\omega) = 0\}$$

and $w$ is the Wiener measure in the probability space $\Omega$.

We shall construct a Lebesgue's thorn of variable "sharpness" so that we can state when it is "sharp enough" for its tip to be an irregular point. Consider a $d$ ($d \geqslant 3$) dimensional block of side $[-1, 1]$ with a thorn removed (see Fig. 8).

$$M := [-1, 1]^d - \text{thorn}$$

$$\text{thorn} := \bigcup_{n=0}^{\infty} n\text{-block}$$

$$n\text{-block} := [-2^{-n}, -2^{-n-1}] \times [-a_n, a_n]^{d-1}$$

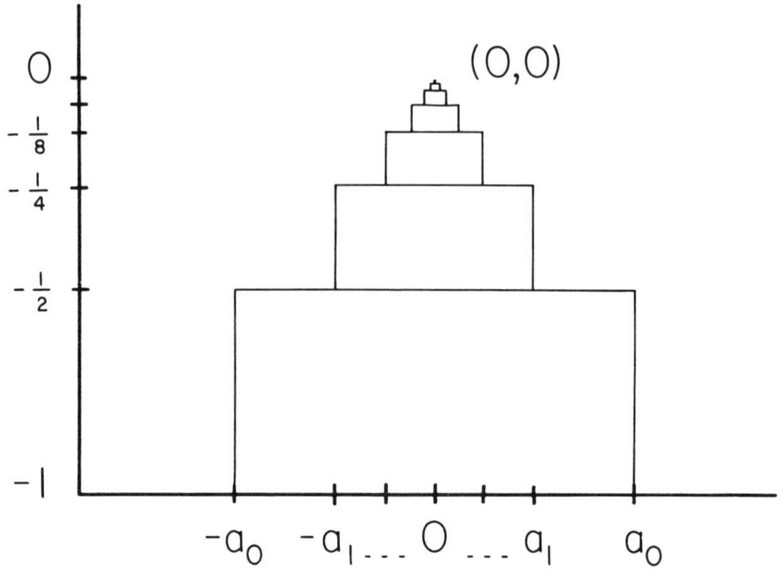


Fig. 8. Projection of a Lebesgue's thorn on two dimensions.

The origin is an irregular point $P_0(\theta > 0) > \frac{1}{2}$ if the sequence $\{a_n\}$ tends to zero sufficiently rapidly. Poincaré's (sufficient) "cone condition" says that 0 is regular if

$$\lim_{n=\infty} \inf 2^n a_n > 0$$

Wiener's necessary and sufficient condition says that 0 is irregular

$$\text{in } d=3, \qquad \text{if and only if } -\infty \leqslant \sum_{n=1} \log(2^n a_n) < \infty$$

$$\text{in } d \geqslant 4, \qquad \text{if and only if } \sum_{n=1}^{\infty} (2^n a_n)^{d-3} < \infty$$

3. Continuous solutions of the Dirichlet problem in a domain $M$ with regular boundary $\partial M$

$$\begin{cases} \dfrac{1}{2} \Delta u + Vu = 0 \\ \qquad u|_{\partial M} = \phi \end{cases} \tag{45}$$

can be expressed, under fairly general assumptions for $\phi$, as a path integral,[29] namely

$$u(x)=\int_{\Omega} dw(\omega)\exp\left(\int_0^{\theta(\omega)} V(B(t,\omega))\,dt\right)\phi(B(\theta,\omega))$$

$$\equiv \mathbb{E}_x\left(\exp\left(\int_0^{\theta} V(B(t))\,dt\right)\phi(B(\theta))\right) \tag{46}$$

where $w$ is the Wiener measure on $\Omega$, $B(\cdot,\omega)$ is a brownian path originating at $x$ at time $t=0$, $\theta$ is the first exit time (42), and $\mathbb{E}_x$ is defined by the path integral over $\Omega$.

The path integral solution of the Dirichlet problem bears some resemblance to the Feynman–Kac solution of the diffusion equation

$$\begin{cases} \dfrac{\partial K}{\partial t}=\dfrac{1}{2}\Delta K+VK \\ K(x,0)=\phi(x) \end{cases}$$

which, written with the above notation, is

$$K(x,t)=\mathbb{E}_x\left(\exp\left(\int_0^t\left(V(B(s))\,ds\right)\phi(B(t))\right)\right) \tag{47}$$

The important difference between (46) and (47) is that the limit of integration $\theta$ is path dependent; computing explicitly (46) is thus much more difficult than computing (47). At present (46) has been used primarily to obtain bounds. Recalling that time-dependent solutions for monoenergetic waves are readily obtained from time-independent Helmholtz equations, Eq. (46) offers an interesting approach to investigating diffusion with boundary.

## 6. CONCLUSION

The propagators and the Green functions presented in this paper are the knife edge propagator for Dirichlet and Neumann boundary conditions, (4), (38), and (17), the absorbing knife edge propagator (16), the

[30] See, for instance, Friedman (1976) Vol I, p. 145, for analytical details. Note also the stochastic solution of Poisson's equation:

$$u(x)=\mathbb{E}_x\left(\phi(B\theta)+\int_0^{\theta} c(B(s))\,ds\right) \text{ satisfies } 1/2\Delta u+c=0,\ u|_{\partial M}=\phi.$$

wedge propagators (19). The propagator for a free particle on a $\mu$-sheeted Riemann surface (18), and the Dirichlet and the Neumann wedge Green function (5), (6), and (22).

Each section presents the wedge problem in a particular context. In each case, identifying the relevant mathematical problem is the key to the solution; we supplement Sect. 1 with mentions of the corresponding mathematical problems.

2. Diffraction by reflecting wedge. Boundary value problem of differential equation. Solutions defined on multiply connected spaces.

3. Geometrical theory of diffraction. Identification of function spaces.

4. Path integral solutions. Path integration on multiply connected spaces. Asymptotics on the boundaries of function spaces.

5. Probing the shape of the wedge and the roughness of its surface. Stochastic calculus.

Problems 2 and 3 are by now well understood; progress is expected in problems 4 and 5. We would be delighted to write *Wedges II* for the next Wheeler *Festschrift*.

## ACKNOWLEDGMENTS

Michael Berry's visit was delightful and very instructive. We are grateful to Wolfgang Schleich for a critical reading of Sommerfelds's 1896 paper. We thank the Physics Department of Imperial College for having made possible A.S.'s visit to Austin.

## APPENDIX TO SECTION 4.1

1. Free propagator on a $\mu$-sheeted Riemann surface constructed as a linear combination of Aharonov–Bohm propagators for different fluxes (generalization of 4.4).

As discussed previously, our use of images leads to the consideration of a $\mu$-sheeted Riemannian surface. Only solenoids whose strength ($\alpha$) are a multiple of $1/\mu$ exist upon such a surface. Combine these such that the source lies only within the physical sheet by requiring that

$$\lim_{t \to 0^+} K = \delta(R - R') \sum_{n=-\infty}^{\infty} \delta((\phi - \phi') - 2\pi\mu n) \qquad \text{(A.1)}$$

Starting from the arbitrary linear combination

$$\begin{aligned} K &= \frac{M}{2\pi i\hbar t} \exp\left[\frac{iM}{2\hbar t}(R^2+R'^2)\right] \sum_{n=-\infty}^{\infty} a_n \exp\left[-\frac{in}{\mu}(\phi-\phi')\right] \\ &\quad \times (-i)^{|n/\mu|} J_{|n/\mu|}\left(\frac{MR'R}{\hbar t}\right) \\ &= \frac{M}{2\pi i\hbar t} \exp\left[\frac{iM}{2\hbar t}(R^2+R'^2)\right] \sum_{n=-\infty}^{\infty} a_n \exp\left[-\frac{in}{\mu}(\phi-\phi')\right] \\ &\quad \times I_{|n/\mu|}\left(\frac{-iMR'R}{\hbar t}\right) \end{aligned} \tag{A.2}$$

and noting that

$$I_\nu(z)_{|z|\to\infty} \to \frac{\exp(z)}{\sqrt{2\pi z}} \; |\arg z| < \frac{\pi}{2}$$

(regularization being implicitly used throughout), this may be cast in the form

$$\begin{aligned} \lim_{t\to 0^+} K &= \lim_{t\to 0^+} \frac{M}{2\pi i\hbar t} \exp\left[\frac{iM}{2\hbar t}(R^2+R'^2)\right] \\ &\quad \times \sqrt{\frac{i\hbar t}{2\pi R'RM}} \exp\left[-\frac{iR'RM}{\hbar t}\right] \sum_{n=-\infty}^{\infty} a_n \exp\left[-\frac{in}{\mu}(\phi-\phi')\right] \end{aligned}$$

The angular factor is required to be a series of delta functions. The $a_n$'s are in fact nothing but the Fourier coefficients. Fourier analysis yields the Poisson sum

$$\sum_{m=-\infty}^{\infty} \exp(im\theta) = 2\pi \sum_{n=-\infty}^{\infty} \delta(\theta+2\pi n)$$

from which one can deduce that the $a_n$'s are equal and so can be denoted simply as $a$, leading to

$$\lim_{t\to 0^+} K = \lim_{t\to 0^+} a\mu \sqrt{\frac{M}{2\pi i\hbar R'Rt}} \exp\left[\frac{iM}{2\hbar t}(R-R')^2\right] \sum_{n=-\infty}^{\infty} (\delta(\phi-\phi')-2\pi\mu n)$$

An inspection of

$$\lim_{t\to 0^+} \int_0^\infty \sqrt{\frac{M}{2\pi i\hbar R'Rt}} \exp\left[\frac{iM}{2\hbar t}(R-R')^2\right] R\, dR \tag{A.3}$$

reveals that the integrand contributes only in the neighborhood of $R=R'$. Therefore, (A.3) reduces to

$$\sqrt{\frac{M}{2\pi i\hbar t}}\int_{-\infty}^{\infty}\exp\left[\frac{iM}{2\hbar t}(R-R')^2\right]dR=1$$

and so behaves as $\delta(R-R')$. Hence

$$\lim_{t\to 0^+} K=a\mu\,\delta(R-R')\sum_{n=-\infty}^{\infty}\delta((\phi-\phi')-2\pi\mu n)$$

and so $a=1/\mu$ by using a comparison with A.1. This yields the complete kernel

$$K=\frac{1}{\mu}\frac{M}{2\pi i\hbar t}\exp\left[\frac{iM}{2\hbar t}(R^2+R'^2)\right]$$
$$\times\sum_{n=-\infty}^{\infty}\exp\left[-i\frac{n}{\mu}(\phi-\phi')\right](-i)^{|n/\mu|}J_{|n/\mu|}\left(\frac{MR'R}{\hbar t}\right)$$

from which we proceed as previously.

2. Proof of Eq. (20) [generalization of (15)]
To evaluate a finite sum form for

$$S(z,\phi)=\tfrac{1}{2}J_0(z)+\sum_{n=1}^{\infty}\exp\left[-i\frac{n}{2}\frac{\pi}{\mu}\right]J_{n/\mu}(z)\cos\left(\frac{n}{\mu}\phi\right)$$

where $\mu$ is a positive integer.

The aproach taken is to develop a partial differential equation (w.r.t. $z$) satisfied by $S$, which is then integrated to yield an integral form for $S$.

Look at

$$S_\phi:=\tfrac{1}{2}J_0+\sum_{n=1}^{\infty}\exp\left[i\frac{n\pi}{2\mu}\right]J_{n/\mu}\exp\left[-i\frac{n}{\mu}\phi\right],\qquad S=\tfrac{1}{2}(S_\phi+S_{-\phi})$$

$$\frac{\partial S_\phi}{\partial z}=-\tfrac{1}{2}J_1+\sum_{n=1}^{\infty}\exp\left[-i\frac{n\pi}{2\mu}\right]\tfrac{1}{2}(J_{(n/\mu)-1}-J_{(n/\mu)+1})\exp\left[i\frac{n}{\mu}\phi\right]$$

having used

$$J'_\nu=\tfrac{1}{2}(J_{\nu-1}-J_{\nu+1}),\qquad \nu\neq 0,\qquad J'_0=-J_1$$

So,

$$\frac{\partial S_\phi}{\partial z}=-\tfrac{1}{2}J_1+\tfrac{1}{2}\left\{\sum_{n=1}^{\infty}\exp\left[-i\frac{n\pi}{2\mu}\right]J_{(n/\mu)-1}\exp\left[i\frac{n}{\mu}\phi\right]\right.$$
$$\left.-\sum_{n=1}^{\infty}\exp\left[-i\frac{n\pi}{2\mu}\right]J_{(n/\mu)+1}\exp\left[i\frac{n}{\mu}\phi\right]\right\}$$

Let $n \to n+\mu$ in the first summation and $n \to n-\mu$ in the second:

$$\frac{\partial S_\phi}{\partial z} = -\tfrac{1}{2}J_1 + \tfrac{1}{2}\left\{-i\exp[i\phi]\sum_{n=1-\mu}^{\infty}\exp\left[-i\frac{n\pi}{2\mu}\right]J_{n/\mu}\exp\left[i\frac{n}{\mu}\phi\right]\right.$$
$$\left. - i\exp[-i\phi]\sum_{n=1+\mu}^{\infty}\exp\left[-i\frac{n\pi}{2\mu}\right]J_{n/\mu}\exp\left[i\frac{n}{\mu}\phi\right]\right\}$$

Complete the summations from 1 to $\infty$:

$$\frac{\partial S_\phi}{\partial z} = -\tfrac{1}{2}J_1 - \frac{i}{2}\exp[i\phi]\sum_{n=1-\mu}^{0}\exp\left[-i\frac{n\pi}{2\mu}\right]J_{n/\mu}\exp\left[i\frac{n}{\mu}\phi\right]$$
$$+\frac{i}{2}\exp[-i\phi]\sum_{n=1}^{\mu}\exp\left[-i\frac{n\pi}{2\mu}\right]J_{n/\mu}\exp\left[i\frac{n}{\mu}\phi\right]$$
$$-i\cos(\phi)\sum_{n=1}^{\infty}\exp\left[-i\frac{n\pi}{2\mu}\right]J_{n/\mu}\exp\left[i\frac{n}{\mu}\phi\right]$$

$$\frac{\partial S_\phi}{\partial z} = -\tfrac{1}{2}J_1 - \frac{i}{2}\exp[i\phi]\left\{J_0 + \sum_{n=1-\mu}^{-1}\exp\left[-i\frac{n\pi}{2\mu}\right]J_{n/\mu}\exp\left[i\frac{n}{\mu}\phi\right]\right\}$$
$$+\frac{i}{2}\exp[-i\phi]\left\{-iJ_1\exp[i\phi] + \sum_{n=1}^{\mu-1}\exp\left[-i\frac{n\pi}{2\mu}\right]J_{n/\mu}\exp\left[i\frac{n}{\mu}\phi\right]\right\}$$
$$-i\cos(\phi)\sum_{n=1}^{\infty}\exp\left[-i\frac{n\pi}{2\mu}\right]J_{n/\mu}\exp\left[i\frac{n}{\mu}\phi\right]$$

Let $n \to -n$ in the first summation:

$$\frac{\partial S_\phi}{\partial z} = -\frac{i}{2}\exp[i\phi]\,J_0 - \frac{i}{2}\sum_{n=1}^{\mu-1}\exp\left[+i\frac{n\pi}{2\mu}\right]J_{-n/\mu}\exp\left[i\frac{\mu-n}{\mu}\phi\right]$$
$$+\frac{i}{2}\sum_{n=1}^{\mu-1}\exp\left[-i\frac{n\pi}{2\mu}\right]J_{n/\mu}\exp\left[-i\frac{\mu-n}{\mu}\phi\right]$$
$$-i\cos(\phi)\sum_{n=1}^{\infty}\exp\left[-i\frac{n\pi}{2\mu}\right]J_{n/\mu}\exp\left[i\frac{n}{\mu}\phi\right]$$

Recall that

$$\frac{\partial S}{\partial z} = \frac{1}{2}\frac{\partial}{\partial z}(S_\phi + S_{-\phi})$$
$$= -\frac{i}{2}\cos(\phi)J_0 - \frac{i}{2}\sum_{n=1}^{\mu-1}\left\{\exp\left[+i\frac{n\pi}{2\mu}\right]J_{-n/\mu} - \exp\left[-i\frac{n\pi}{2\mu}\right]J_{n/\mu}\right\}\cos\left(\frac{\mu-n}{\mu}\phi\right)$$
$$-i\cos(\phi)\sum_{n=1}^{\infty}\exp\left[-i\frac{n\pi}{2\mu}\right]J_{n/\mu}\cos\left(\frac{n}{\mu}\phi\right)$$

This leads to the finite-term partial differential equation for $S$

$$\frac{\partial S}{\partial z}(z,\phi) = -\frac{i}{2}\sum_{n=1}^{\mu-1}\left\{\exp\left[+i\frac{n\pi}{2\mu}\right]J_{-n/\mu}(z) - \exp\left[-i\frac{n\pi}{2\mu}\right]J_{n/\mu}(z)\right\} \times \cos\left(\frac{\mu-n}{\mu}\phi\right) - i\cos(\phi)\,S(z,\phi)$$

which is of the form

$$\frac{\partial S}{\partial z}(z,\phi) = p(z,\phi)\,S(z,\phi) + q(z,\phi)$$

and has the solution

$$S(z,\phi) = \exp\left[\int p(z,\phi)\,dz\right]\left\{\int^{z} q(u,\phi)\exp\left[-\int^{u} p(v,\phi)\,dv\right]du + f(\phi)\right\}$$

where $f$ is an arbitrary function of $\phi$. Hence,

$$S(z,\phi) = \exp[-iz\cos(\phi)]\left\{-\frac{i}{2}\sum_{n=1}^{\mu-1}\cos\left(\frac{\mu-n}{\mu}\phi\right) \times \int_0^z\left(\exp\left[+i\frac{n\pi}{2\mu}\right]J_{-n/\mu}(u) - \exp\left[-i\frac{n\pi}{2\mu}\right]J_{n/\mu}(u)\right) \times \exp[+iu\cos(\phi)]\,du + f(\phi)\right\}$$

So,

$$\tfrac{1}{2}J_0(z) + \sum_{n=1}^{\infty}\exp\left[-i\frac{n\pi}{2\mu}\right]J_{n/\mu}(z)\cos\left(\frac{n}{\mu}\phi\right) = \exp[-iz\cos(\phi)]\left\{-\frac{i}{2}\sum_{n=1}^{\mu-1}\cos\left(\frac{\mu-n}{\mu}\phi\right)\int_0^z\left(\exp\left[+i\frac{n\pi}{2\mu}\right]J_{-n/\mu}(u) - \exp\left[-i\frac{n\pi}{2\mu}\right]J_{n/\mu}(\mu)\right)\exp[+iu\cos(\phi)]\,du + f(\phi)\right\}\text{iME}$$

Put $z = 0$ to find $f(\phi)$. This gives $f(\phi) = 1/2$; having used the fact that, we have

$$J_\nu(0) = 0, \qquad \nu > 0, \qquad J_0(0) = 1$$

Finally, the identity is obtained as

$$\tfrac{1}{2}J_0(z)+\sum_{n=1}^{\infty}\exp\left[-i\frac{n\pi}{2\mu}\right]J_{n/\mu}(z)\cos\left(\frac{n}{\mu}\phi\right)$$

$$=\exp[-iz\cos(\phi)]\left\{-\frac{i}{2}\sum_{n=1}^{\mu-1}\cos\left(\frac{\mu-n}{\mu}\phi\right)\int_0^z\left(\exp\left[+i\frac{n\pi}{2\mu}\right]J_{-n/\mu}(u)\right.\right.$$

$$\left.\left.-\exp\left[-i\frac{n\pi}{2\mu}\right]J_{n/\mu}(u)\right)\exp[+iu\cos(\phi)]\,du+\frac{1}{2}\right\}$$

This reduces to the known result for $\mu=2$ [Gradshteyn and Ryzhik (1980), p. 973, 8.511(5)], namely,

$$\tfrac{1}{2}J_0(z)+\sum_{n=1}^{\infty}\exp\left[-in\frac{\pi}{4}\right]J_{n/2}(z)\cos\left(n\frac{\phi}{2}\right)=(i/\pi)^{-1/2}\exp[iz\cos(\phi)]$$

$$\times\int_{-\infty}^{\sqrt{(2z)}\cos(\phi/2)}\exp(+it^2)\,dt$$

on using

$$J_{-1/2}(z)=\sqrt{(2/\pi z)}\cos(z),\qquad J_{1/2}(z)=\sqrt{(2/\pi z)}\sin(z)$$

the equality

$$\frac{1}{2}=(i/\pi)^{-1/2}\int_{-\infty}^{0}\exp(-it^2)\,dt$$

and the variable change

$$t=\sqrt{(2u)}\cos(\phi/2)$$

One may further generalize this identity by adding together this expression for angles given by

$$\phi+2\pi k\frac{\mu}{\nu},\qquad k=1,\ldots,\nu \text{ where } \nu \text{ is a positive integer}$$

The left-hand side of the identity developed becomes

$$\frac{1}{2}\sum_{k=1}^{\nu}\left\{J_0+\sum_{n=1}^{\infty}\exp\left[-i\frac{n\pi}{2\mu}\right]J_{n/\mu}\left(\exp\left[-i\frac{n}{\mu}\left(\phi+2\pi k\frac{\mu}{\nu}\right)\right]\right.\right.$$

$$\left.\left.+\exp\left[+i\frac{n}{\mu}\left(\phi+2\pi k\frac{\mu}{\nu}\right)\right]\right)\right\}$$

But

$$\sum_{k=1}^{\nu}\exp\left[-i2\pi k\frac{n}{\nu}\right]\quad\begin{cases}=\nu & \text{if } n \text{ is a multiple of } \nu\\ =0 & \text{if not}\end{cases}$$

Which is most easily seen by noticing that this is the sum of evenly spaced unit vectors in the complex plane.

So the left-hand side becomes

$$\nu\left\{\tfrac{1}{2}J_0+\sum_{n=1}^{\infty}\exp\left[-in\frac{\pi\nu}{2\mu}\right]J_{n\nu/\mu}\cos\left(n\frac{\nu}{\mu}\phi\right)\right\}$$

and one obtains the generalization of the first identity developed:

$$\tfrac{1}{2}J_0(z)+\sum_{n=1}^{\infty}\exp\left[-in\frac{\pi\nu}{2\mu}\right]J_{n\nu/\mu}(z)\cos\left(n\frac{\nu}{\mu}\phi\right)$$

$$=\frac{1}{2\nu}\sum_{k=1}^{\nu}\left\{\exp\left[-iz\cos\left(\phi+2\pi k\frac{\mu}{\nu}\right)\right]\left\{-i\sum_{n=1}^{\mu-1}\cos\left[\left(\frac{\mu-n}{\mu}\right)\left(\phi+2\pi k\frac{\mu}{\nu}\right)\right]\right.\right.$$

$$\times\int_0^z\left(\exp\left[+i\frac{n\pi}{2\mu}\right]J_{-n/\mu}(u)-\exp\left[-i\frac{n\pi}{2\mu}\right]J_{n/\mu}(u)\right)$$

$$\left.\left.\times\exp\left[+iu\cos\left(\phi+2\pi k\frac{\mu}{\nu}\right)\right]du+1\right\}\right\}$$

These results are useful in that they convert an infinite series which suffers from slow convergence (for large Bessel function argument) into an expression with a finite number of terms (for $\mu$, $\nu$ finite).

## REFERENCES


Abramowitz, M., and Stegun, I. (1965), *Handbook of Mathematical Functions* (Dover, New York).

Aharonov, Y., and Bohm, D. (1959), "Significance of Electromagnetic Potentials in the Quantum Theory," *Phys. Rev.* **115**, 485.

Baker, Bevan B., and Copson, E. T. (1950), *The Mathematical Theory of Huygens' Principle*, and edn. (Clarendon Press, Oxford).

Barber, Michael N., and Ninham, B. W. (1970), *Random 2nd Restricted Walks* (Gordon and Breach, New York).

Berry, M. V. (1980), "Exact Aharonov–Bohm Wave Function Obtained by Applying Dirac's Magnetic Phase Factor," *Eur. J. Phys.* **1**, 240–244.

Berry, M. V. (1981), "Singularities in Waves and Rays," in *Les Houches Session XXXV: Physics of Defects*, R. Baham *et al.*, eds. (North-Holland, Amsterdam), pp. 456–543.

Berry, M. V., "Scaling and Non-Gaussian Fluctuation in the Catastrophe Theory of Waves," *Prometheus* (UNESCO publication) (to appear).

Berry, M. V. and Mount, K. E. (1972), "Semiclassical Approximations in Wave Mechanics," *Rep. Prog. Phys.* **35**, 315–397.

Berry, M. V., and Upstill, C. (1980), "Catastrophe Optics: Morphologies of Caustics and Their Diffraction Patterns," in *Progress in Optics XVIII*, E. Wolf, ed. (North-Holland, Amsterdam), pp. 258–346.

Berry, M. V., Chambers, R. G., Lange, M. D., Upstill, C., and Walmsley, J. C. (1980), "Wavefront dislocations in the Aharonov–Bohm effect and its water wave analogue," *Eur. J. Phys.* **1**, 154–162.

Bleistein, N., and Handelsman, R. A. (1975), *Asymptotic Expansion of Integrals* (Holt, Rinehart and Winston, New York).

Boersma, J., and Wiegel, F. W. (1983), "Asymptotic Expansions for a Remarkable Class of Random Walks," *Physica A* **122**, 334–344.

Buslaev, V. S., (1968) "Continuum Integrals and the Asymptotic Behaviour of the Solutions of Parabolic Equations as $t \to 0$. Applications to diffraction," in *Topics in Mathematical Physics*, Vol. 2, M. Sh. Birman, ed. (Consultants Bureau, New York), pp. 67–86.

Carslaw, H. S. (1899), "Some Multiform Solutions of the Partial Differential Equation of Physical Mathematics and Their Applications," *Proc. London Math. Soc.* **30**, 121–163.

Carslaw, H. S. (1916–1919), "Diffraction of Waves by a Wedge of Any Angle," *Proc. London Math. Soc.* **18**, 281–306.

Choquet-Bruhat, Y., DeWitt-Morette, C., and Dillard-Bleick, M. (1982) *Analysis, Manifolds and Physics*, revised edn. (North-Holland, New York).

Crandall, R. E. (1983a), "Exact Propagator for Motion Confined to a Sector," *J. Phys. A* **16**, 513–519.

Crandall, R. E. (1983b), "Exact Propagator for Reflectionless Potentials," *J. Phys. A* **16**, 3005–3011.

DeWitt, Cecile Morette (1972), "Feynman Path's Integral" *Commun. Math. Phys.* **28**, 47–67.

DeWitt-Morette, C. (1974), "Feynman Path Integrals: I. Linear and Affine Techniques; II. The Feynman–Green Function," *Commun. Math. Phys.* **37**, 63–81.

Dewitt-Morette, C. (1976), "The Semiclassical Expansion" *Ann. Phys.* **97**, 367–399; **101**, 683.

DeWitt-Morette, C. (1984), "Feynman Path Integrals. From the Prodistribution Definition to the Calculation of Glory Scattering," in *Stochastic Methods and Computer Techniques in Quantum Dynamics*, H. Mitter and L. Pittner, eds., *Acta Phys. Aust. Supp.* **26**, 101–170.

DeWitt-Morette, C., and Nelson, B. (1984), "Glories—and Other Degenerate Points of the Action," *Phys. Rev. D* **29** 8, 1663–1668.

DeWitt-Morette, C., Maheshwari, A., and Nelson, B. (1979), "Path Integration in Non-Relativistic Quantum Mechanics," *Phys. Rep.* **50** S, 256–372.

DeWitt-Morette, C., Elworthy, K. D., Nelson, B. L., and Sammelman, G. S. (1980), "A stochastic Scheme for Constructing Solutions of the Schrödinger Equations," *Ann. Inst. Henri Poincaré* **32**, 327–341.

Durrett, Richard (1984), *Brownian Motion and Martingales in Analysis* (Wadsworth Advanced Books and Software, Belmont, California).

Edwards, S. F. (1967), "Statistical Mechanics with Topological Constraints: I," *Proc. Phys. Soc.* **91**, 513–519.

Felsen, L. B. (1976), "Propagation and Diffraction of Transient Fields in Non-Dispersive and Dispersive Media," in *Transient Electromagnetic Fields*, L. B. Felsen, ed. (Springer-Verlag, New York), pp. 2–176.

Ford, Kenneth W., and Wheeler, John A. (1959a), "Semiclassical Description of Scattering," *Ann. Phys. (N.Y.)* **7**, 259–286.

Ford, Kenneth W., and Wheeler, John A. (1959b), "Application of Semiclassical Scattering Analysis," *Ann. Phys. (N.Y.)* **7**, 287–322.

Friedman, Avner (1976), *Stochastic Differential Equations and Applications*, Vols. 1 and 2 (Academic Press, New York).

Gikhman, I. I., and Skorokhod, A. V. (1969), *Introduction to the Theory of Random Processes* (Saunders, Philadelphia).

Goodman, Mark (1981), "Path Integral Solution to the Infinite Square Well," *Am. J. Phys.* **49**, 843–847.

Gradshteyn, I. S., and Ryzhik, I. M. (1980), *Tables of Integrals, Series, and Products* (Academic Press, New York).

Ikeda, Nobuyuki (1982), "On the Asymptotic Behavior of the Fundamental Solution of the Heat Equation on Certain Manifolds," in *Proceedings, Taniguchi Symposium, Kantata*, pp. 169–195.

Ikeda, Nobuyuki (1985), Lectures given at the University of Warwick (unpublished).

Itô, Kiyosi, and McKean, Henry P. Jr., (1965), *Diffusion Processes and Their Sample Paths* (Springer-Verlag, New York).

Kac, M. (1959), *Probability and Related Topics in Physical Sciences*, (Interscience, New York).

Keller, J. B. (1953–1958), "A Geometric Theory of Diffraction," in *Calculus of Variations and Its Applications* (Proceedings of a Symposium in Applied Mathematics), L. M. Grarves, ed. (McGraw-Hill, New York).

Keller, J. B. (1960–1962), "Geometrical Theory of Diffraction," *J. Opt. Soc. Am.* **S2**, 2, 116–130.

Keller, J. B., and McLaughlin, D. W. (1975), "The Feynman Integral," *Am. Math. Mon.* **82**, 451–465.

Knoll, J., and Schaeffer, R. (1976), "Semiclassical Scattering Theory with Complex Trajectories: I. Elastic Waves," *Ann. Phys. (N.Y.)* **97**, 307–366.

Kouyoumjiam, R. G. (1975), "The Geometric Theory of Diffraction and Its Application," in *Numerical and Asymptotic Techniques in Electromagnetics*, R. Mittra, ed., Springer-Verlag Topics in Applied Physics Vol. 3 (Springer-Verlag, New York), pp. 165–215.

Laidlaw, Michael G. G., and DeWitt-Morette, Cecile (1971), "Feynman Functional Integrals for Systems of Indistinguishable Particles," *Phys. Rev. D* **36**, 1375–1378.

Lebesgue, H. (1924), "Conditions de Regularité, Conditions D'irregularité, Conditions D'impossibilité Dans le Problème de Dirichlet," *C.R. Acad. Sci. (Paris)*, 349–354.

Lee, S. W. (1978), "Path Integrals for Solving Some Electromagnetic Edge Diffractions Problems," *J. Math. Phys.* **19**, 1414–1422.

Lewis, R. M., and Boersma, J. (1969), "Uniform Asymptotic Theory of Edge Diffraction," *J. Math. Phys.* **10**, 2291–2305.

Lewis, R. M., Bleistein, N., and Ludwig, D. (1967), "Uniform Asymptotic Theory of Creeping Waves," *Comm. Pure Appl. Math.* **20**, 295–328.

Longhurst, R. S. (1957), *Geometric and Physical Optics* (Longmans, Green and Co., London), p. 439.

Low, S. G. (1984), "The Knife Edge Problem," in *Stochastic Methods and Computer Techniques in Quantum Dynamics*, H. Mitter and L. Pittner, eds., *Acta Phys. Austriaca, Suppl.* XXVI 171–184.

Macdonald, H. M. (1902), *Electric Waves*, Adams prize essay in the University of Cambridge (Cambridge University Press, Cambridge), pp. 186–198.

Molchanov, S. A. (1975), "Diffusion Processes and Riemannian Geometry," *Russ. Math Surv.* **30**, 1–63 (from *Usp. Mat. Nauk* **30**, 3–59).

Morandi, G. and Menossi, E. (1984), "Path Integrals in Multiply Connected Spaces and the Aharanov–Bohm Effect," *Eur. J. Phys.* **5**, 49–58.

Morette, C. (1951), "On the Definition and Approximation of Feynman's Path Integrals," *Phys. Rev.* **81**, 848–852.

Olariu, S., and Popescu, I. Lovitzu (1985), "The Quantum Effects of Electromagnetic Fluxes," *Rev. Mod. Phys.* **57**, 339–436.

Pauli, W. (1938), "On Asymptotic Series for Functions in the Theory of Diffraction of Light," *Phys. Rev.* **54**, 924–931.

Port, Sidney C., and Stone, Charles J. (1978), *Brownian Motion and Classical Potential Theory* (Academic Press, New York).

Rubinowicz, A. (1938), "On the Anomalous Propagation of Phase in the Focus," *Phys. Rev.* **54**, 931–936.

Schulman, L. S. (1968), "A Path Integral for Spin," *Phys. Rev.* **176**, 1558–1569.

Schulman, L. S. (1971), "Approximate Topologies," *J. Math. Phys.* **12**, 304–308.

Schulman, L. S. (1981), *Techniques and Applications of Path Integration* (Wiley, New York).

Schulman, L. S. (1982), "Exact Time-Dependent Green's Function for the Half-Plane Barrier," *Phys. Rev. Lett.* **49**, 599–601.

Schulman, L. S. (1982–1984), "Ray Optics for Diffraction: A Useful Paradox in a Path Integral Context," in *The Wave-Particle Dualism* (Proceedings of a conference in honor of Louis de Broglie, Perugia, 1982), S. Diner, D. Fargue, G. Lochak, and F. Selleri, eds. (D. Reidel, Dordrecht).

Shiekh, A. Y. (1985). Private communication, now incorporated in section 4.1.

Shiekh, A. Y. (1986), "Calculation of Nonrelativistic Scattering of Charged Scalar Particles from a Classical Aharanov–Bohm Solenoid without an Explicit Gauge Choice," *Ann. Phys. (N.Y.)* **166**, 299–321.

Sommerfeld, A. (1896), "Mathematische Theorie der Diffraction," *Math. Ann.* **47**, 317–374.

Sommerfeld, A. (1954), *Optics* (Academic Press, New York), pp. 247–272.

Wait, J. R. (1960), "Diffraction Corrections to the Geometrical Optics of Low-Frequency Propagation," in *Electromagnetic Wave Propagation*, M. Desirant and J. L. Michels, eds. (Academic Press, New York).

Wiener, N. (1929), "The Dirichlet problem," *J. Math. Phys.* **3**, 127–146.

Wiegel, F. W., and Boersma, J. (1983), "The Green Function for the Half-plane Barrier: Derivation from Polymer Entanglement Probabilities," *Physica A* **122**, pp. 325–333.

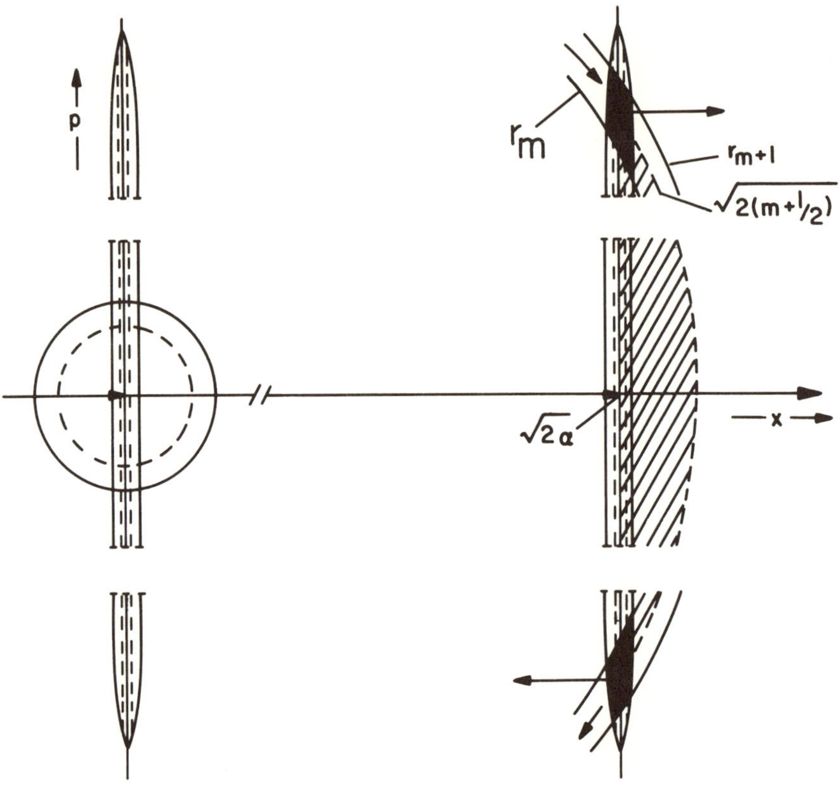


*Plate 33*

# The Measurement of Quantum Noise Reduction in Squeezed States

W. G. Unruh[1]

*Received September 16, 1985*

*The use of a photomultiplier to measure the reduction in counting statistics noise (from the usual $N^{1/2}$ form) which is expected to occur for squeezed coherent light is shown to lead to no reduction or to an increase of the noise unless the number of photons in a fundamental measuring time of the photomultiplier is very large. This fundamental time is estimated to be less than $10^{-13}$ sec for ordinary detectors. A technique for using squeezed states for determining the time a photomultiplier takes to detect the presence of a photon is presented.*

## 1. INTRODUCTION

John Wheeler has spent much of his time in recent years thinking about the implications which quantum measurement theory has on our understanding of the universe. The field of quantum measurement theory has long been regarded as an unwanted stepchild of physics, as something which philosophers worried about, but which has no place in real physics. This attitude is changing, and one of the reasons is that we are realizing that the field can help give insights even into the problems of mundane physics. In one area in particular, that of the design of detectors which will operate in the quantum regime such as in gravity wave detection, the old questions are resurfacing. One of these old questions that this work touches on is the question of when a quantum measurement can be said to have actually taken place. In addition, this is a problem in which the time dependence of the measurement process plays an important role.

Recently, great interest has been shown in the use of "squeezed states" of light in order to reduce the counting statistics noise in devices like laser

[1] Department of Physics, University of British Columbia, Vancouver, B.C., Canada V6T 2A6.

interferometers. This recent interest arose in the attempts to improve the noise performance of laser interferometric detectors for gravitational radiation, but has spread to other areas of quantum optics as well.[1] In the simplest case, the promise of the technique is that one could reduce the counting statistics noise (shot noise) to levels well below the standard $N^{1/2}$ (where $N$ is the total photon count in some time interval) noise level. For reasonable levels of $N$, such a reduction could be quite significant.

The problem which I considered was "If the number count fluctuation is less than $N^{1/2}$, what would one see if one made the time resolution of the photomultiplier sufficiently fine that one could see the arrival of the individual photons?." The only way I could imagine the reduction in shot noise was if the photons would be seen to arrive almost equally spaced in time. In order to test this hypothesis, I had a student, K. Coughlin, calculate the arrival time correlations for photons in a squeezed state, into a model photon detector. The hypothesis was not borne out. Instead we found that as the time resolution of the detector was made better and better, the noise began to get worse at a certain time resolution, so that long before one could see the arrival of individual photons, the shot noise had become much worse than $N^{1/2}$. The reduction in shot noise by the use of squeezed states depended crucially on the behavior of the detector used. For a detector with a good fundamental time resolution, the squeezing of the state reduced the shot noise only if the number of photons in the resolution time of the detector was greater than about $e^{6r}$, where $r$ is the usual squeezing parameter.

I will present the results of this calculation here, using a simple model for the photon field and for the behavior of the photodetector. The results should, however, be independent of these models.

## 2. MODEL PHOTON FIELD AND DETECTOR

In this investigation, it will be crucial to look at the photon field as a multimode (continuum) system.[2] Many people have regarded the field as a single simple harmonic oscillator (a single mode). Although questions of time resolution can be answered in the single mode model if one is very careful, the analysis is full of pitfalls. Furthermore, the multimode model used here is sufficiently simple that it leads to little additional computation over the single mode models. On the other hand, the full structure of the true electromagnetic field is not needed, and so I will use as my model for the photon field, a scalar field propagating in one spatial dimension. One can think of this as a model for one of the polarization states of the electromagnetic field, with the spatial variations in the direction perpendicular to the propagation direction ignored.

The equation of motion for this scalar field, $\phi$, is given by

$$\left(\frac{\partial^2}{\partial t^2}-\frac{\partial^2}{\partial x^2}\right)\phi=0$$

In the usual way, I expand the quantum field operator, $\phi$, in normal modes, with the usual creation and annihilation operators, $a_\omega^\dagger$ and $a_\omega$, as

$$\phi=\sum_\lambda\int(a_{\omega\lambda}\,e^{-i\omega(t-\lambda\kappa)}+a_{\omega\lambda}^\dagger\,e^{i\omega(t-\lambda\kappa)})\,d\omega/(2\pi\omega)^{1/2}$$

where $\lambda$ is $\pm 1$ and designates whether the mode is travelling to the right or the left.

I will place the field into the multimode squeezed state $|\Phi, r\rangle$ defined by

$$(\cosh r\,a_{\Omega-\nu}+\sinh r\,a_{\Omega+\nu}^\dagger)|\Phi, r\rangle=e^r\Phi\,\delta(\nu)|\Phi, r\rangle \tag{1}$$

with $\Phi$ real. I will restricted myself to $\lambda=1$ in all of the following. This state corresponds to a squeezed state, with a mean value for the field of $\Phi$ at frequency $\Omega$ and with squeezing parameter $r$. Equation (1) will be assumed to be valid for all values of $\Omega\pm\nu$ which lie within the bandwidth of the detector.

I will model the detector by assuming[2] that the detector responds to the quadratic field operator given by

$$N(t)=\int\phi(t+\tau)\,\phi(t-\tau)\,h(\tau)\,d\tau$$

The usual number current operator for the scalar field is obtained if I choose $h(\tau)=\delta'(\tau)/2\pi i$. In this case, I will choose $h$ so that its Fourier transform is peaked around the positive frequency $2\Omega$ and has a width of $\Delta$ around $2\Omega$. If we average this operator over a time period which is much longer than $1/\Delta$, we will obtain the number of particles whose frequencies lie within the bandwidth defined by the fourier transform of $h$.

[2] The details of the assumption about the exact form of the response of the detector are not important to the result. For example, one could instead use the operator $N(t)$ defined by

$$A(t)=\int h'(t-t')\,\Phi(t')\,dt'$$

$$N(t)=A(t)^\dagger A(t)$$

where the Fourier transform of $h'$ is concentrated over a width $\Delta$ around $+\Omega$. As long as the averaging time $T$ is long compared with $1/\Delta$, the results will be the same.

## 3. NOISE FOR SQUEEZED STATE

Let us now calculate the fluctuations in the expected photon number over an averaging time long compared with $1/\Delta$, the response time of the detector. I will define

$$N_i = \int N(t)\, w(t-t_i)\, dt$$

where $w(t)$ is a function of unit amplitude over a time interval $T$ centered around $t=0$, and falls to zero outside of this interval. The successive times $t_i$ are separated by time intervals of the order of $T$ so that $\int w(t-t_i)\, w(t-t_j) \sim T\delta_{ij}$. I therefore want to look at the terms $\Delta N_i^2 = \langle N_i^2\rangle - \langle N_i\rangle^2$ which represent the fluctuations in the expected values of $N_i$.

One finds

$$N_i = \int_{\omega,\omega'>0} a^\dagger_\omega a_{\omega'}\, W(\omega-\omega')\, H(\omega+\omega')\, e^{i(\omega-\omega')t_i} d\omega\, d\omega'/2\pi(\omega\omega')^{1/2}$$

where $W$ and $H$ are the Fourier transforms of $w$ and $h$ respectively. I have used the fact that $W$ is nonzero only for small arguement ($\omega \ll 1/T$), while $H$ is nonzero only when its arguement is near $2\Omega$.

Using the definition of the state given in Eq. (1), the expectation value of $N_i$ is found, after some calculations, to be given by

$$\langle N_i\rangle = \left(\Phi^2 H(2\Omega)/\Omega + \sinh^2 r \int H(2\omega)\, d\omega/\omega\right) W(0)/2\pi$$

To calculate the fluctuations, $\Delta N_i^2$, define the operators $b_{\Omega+\nu}$ by

$$b_{\Omega+\nu} = \cosh r\, a_{\Omega+\nu} - \sinh r\, a^\dagger_{\Omega-\nu} - \Phi e^r\, \delta(\nu)$$

so that Eq. (1) defining the state becomes

$$b_{\Omega+\nu}\, |\Phi, r\rangle = 0$$

for all $\nu$ sufficiently near 0 (i.e., for $\nu$ less than a few times $\Delta$). Expanding the $a_\omega$ in terms of the $b$ operators, one finds that

$$\Delta N_i^2 = \left\{\Phi^2 \left(e^{-2r}\left(e^{-r}\int H^2(2\Omega+2\nu)\,|W(\nu)|^2\, d\nu\right)\right.\right.$$

$$+ \sinh^2 r \cosh^2 r \int |W(2\Omega-\omega-\omega')|^2 [H^2(2\Omega-\omega+\omega')$$

$$\left. + H(2\Omega-\omega+\omega')\, H(2\Omega+\omega'+\omega)]\, d\omega\, d\omega'\right\} \Big/ (2\pi\Omega)^2$$

In order to understand this expression, we can make some approximations. Since by assumption, $1/T$ is much less than $\Delta$, the integral in the first terms may be approximated by

$$\int H^2(2\Omega + 2\nu)\,|W(\nu)|^2\,d\nu = H^2(2\Omega)\int |W(\nu)|^2\,d\nu$$

$$= H^2(2\Omega)\,2\pi\int |w(t)|^2\,dt$$

$$= 2\pi H^2(2\Omega)\,T$$

If one changes integration variables to $\xi = \omega - \omega'$ and $\zeta = 2\Omega - (\omega + \omega')$ in the second term, these integral terms become

$$2\int |W(\xi)|^2 d\xi \int H(2\Omega - \zeta)[H(2\Omega - \zeta) + H(2\Omega + \zeta)]\,d\zeta$$

$$= 4\pi T 2 H^2(2\Omega)\Delta$$

where I have assumed that $H$ is essentially a smoothly peaked real function around $2\Omega$. Finally, looking at $\langle N_i \rangle$, we have

$$\langle N_i \rangle = (\Phi^2 + \sinh^2 r\,\Delta/2)\,H(2\Omega)\,W(0)/4\pi\Omega$$

Since $w(t)$ is assumed to be equal to 1 over a time interval of order $T$, one has that $W(0) = T$. Furthermore, if $r = 0$, $\Phi^2/2\pi$ is the density of particles per unit time at frequency $\Omega$, so in order that the detection probability for a particle at frequency $\Omega$ be unity (i.e., unit quantum efficience), we must take $H(2\Omega)/2\Omega$ to be 1. We can now write the expression for $\Delta N_i^2$ as

$$\Delta N_i^2 = \langle N_i \rangle\,e^{-2r} - e^{-2r}\sinh^2 r\,\Delta T/4\pi$$
$$+ \sinh^2 r\cosh^2 r\,\Delta T/2\pi$$

Let me now interpret this expression. The first term is the reduction in the counting noise that one expects from the squeezing of the state. For large $r$, however, the other terms represent an increase in the noise, an increase which will dominate if

$$\langle N_i \rangle/\Delta T < \sim e^{6r}$$

But, $\langle N_i \rangle/T$ is just then mean number of particles per unit time, while $1/\Delta$ is of the order of the response time of the detector. The expression $\langle N_i \rangle/T\Delta$ is thus the mean number of particles expected in a time period corresponding to the response time of the detector. Note that this is much less than the number, $\langle N_i \rangle$, expected in the *averaging* time $T$. The lesson is that only if the mean number of particles in the beam in one response time of the detector is much greater than $e^{6r}$ will the squeezing reduce the noise.

This gives us a technique for measuring this fundamental response time for a detector. The noise is a nontrivial function of both the squeezing parameter and the intensity of quanta in the beam. For a given squeezing (i.e., a given value of $r$), the noise fluctuations are essentially independent of the intensity of the beam until a certain threshold intensity. That threshold is determined by $1/\Delta$, the response time of the detector, and by $r$. A measurement of the noise as a function of intensity for a given value of $r$ will therefore give the value of $\Delta$ for that particular detector.

It may seem that there is an easy way of improving the performance of a detector. The quantity $\Delta$ corresponds roughly to the bandwidth of the detector. It would seem easy to make the bandwidth small by putting a narrowpass filter in front of the detector. The detector will then detect only those incoming particles whose frequency lies within the narrow bandwidth. This, however, does not work. At those frequencies at which the detector does not have unit transmissivity, it will act to mix the incoming squeezed state with ordinary vacuum fluctuations which will worsen, rather than improve, the noise performance. That is, $r$ will effectively become a function of $\nu$ falling off to 0 outside the bandwidth of the filter. If the filter bandwidth is smaller than that of the detector, this will degrade the performance of the detector.

We can now apply these results to a photomultiplier. The fundamental response time of the photomultiplier is rather difficult to estimate. It is not the usual resolution time (say, picoseconds for a very fast photomultiplier, or much longer for slower ones), as this is determined by things like the residual capacitances in the photomultiplier. This time is closer to the averaging time $T$ in the above analysis than it is to $1/\Delta$. A better estimate would be the inverse bandwidth of the photomultiplier. This bandwidth will probably give an underestimate however, since there may be a number of physical processes which determine the bandwidth, each of which has a much smaller bandwidth. However, the bandwidths of these individual processes are probably not too much smaller than the total bandwidth, or one would see very fine structure in the respose curve of the photomultiplier. I will thus take $\Delta$ of the order of $10^{13}$ Hz. The above analysis then implies that the number of photons hitting the detector in a time period of the order of $10^{-13}$ sec must be much greater than unity (say, 10). For optical photons, this corresponds to an energy flux of order $10^{-5}$ watts.

The only way to alleviate this high power requirement is to find a highly resonant type detector, for which the value of $\Delta$ is much smaller than $10^{13}$ Hz. On the other hand, if one had a detector which could respond accurately to such high power densities, squeezing the incoming light would allow one to measure this fundamental detector response time. For a given input flux of quanta, one could look at the dependence of the noise on the squeezing parameter, which I will take as $e^{2r}$. The noise will drop initially, as the squeezing parameter is increased. When it has attained a value of about $(\langle N_i \rangle / T\Delta)^{1/3}$, the noise would begin to increase again rapidly. Since the photon flux $\langle N_i \rangle / T$ is known, one would thus measure $\Delta$, the fundamental response bandwidth of the detector.

This illustrates the noncommutativity on measurement and averaging in quantum mechanics. If one employ a detector which is capable of determining the arrival time of a photon to a time interval of order $\tau$, but one averages the results over some long time interval $T$, this is not equivalent to using a detector whose response time is of order $T$. Forgetting that one (or rather that one's detector) has made a series of measurements does not eliminate the quantum noise associated with that series of measurements.

## ACKNOWLEDGMENTS

I would like to thank K. Coughlin for doing the initial calculations which led to this work, especially for not getting the answer I told her she should get. This work was partially supported by the National Science and Engineering Research Council of Canada, by a UBC senior Killam fellowship, and by an NSERC Steacie Fellowship, for which I am very grateful.

## REFERENCES

1. C. M. Caves, *Phys. Rev. D* **23**, 1693 (1981); H. P. Yuen, *Phys. Rev. A* **13**, 2226 (1976).
2. W. G. Unruh, "Quantum Noise in the Interferometer Detector," in *Quantum Optics, Experimental Gravitation, and Measurement Theory*, P. Meystre and M. O. Scully, eds. (Plenum Press, New York, 1983); C. M. Caves, *Phys. Rev. D* **26**, 1817 (1982).

*Plate 34*

# Quantum Mechanics without Probability Amplitudes

**William K. Wootters**[1]

*Received September 11, 1985*

*First steps are taken toward a formulation of quantum mechanics which avoids the use of probability amplitudes and is expressed entirely in terms of observable probabilities. Quantum states are represented not by state vectors or density matrices but by "probability tables," which contain only the probabilities of the outcomes of certain special measurements. The rule for computing transition probabilities, normally given by the squared modulus of the inner product of two state vectors, is re-expressed in terms of probability tables. The new version of the rule is surprisingly simple, especially when one considers that the notion of complex phases, so crucial in the evaluation of inner products, is entirely absent from the representation of states used here.*

## 1. INTRODUCTION

What is the origin of the structure of quantum mechanics? This is one of a number of deep questions John Wheeler has brought to our attention in recent years.[(1)] It can also be phrased this way: Why is nature so constructed that we find it convenient to describe it in terms of complex probability amplitudes? Perhaps some insight can be gained into this question if we consider what quantum mechanics looks like when it is *not* expressed in terms of probability amplitudes. If, as Professor Wheeler has argued, the origin of quantum mechanics' structure is to be sought in a theory of observation and observers and meaning, then we would do well to focus our attention not on amplitudes but on quantities which are more directly observable. My aim in doing the work presented here has been to express as much as possible of the physical content of quantum mechanics purely

[1] Department of Physics and Astronomy, Williams College, Williamstown, Massachusetts 01267.

in terms of *probabilities*, which are directly observable, rather than in terms of amplitudes. This aim is only partially fulfilled in the present paper: We express some but not all of the essential features of quantum mechanics in terms of probabilities. But this is only intended to be a beginning, and there is no obvious barrier in the path.

It is obviously *possible* to devise a formulation of quantum mechanics without probability amplitudes. One is never forced to use any quantities in one's theory other than the raw results of measurements. However, there is no reason to expect such a formulation to be anything other than extremely ugly. After all, probability amplitudes were invented for a reason. They are not as directly observable as probabilities, but they make the theory simple. I hope to demonstrate here that one *can* construct a reasonably pretty formulation using only probabilities. It may not be quite as simple as the usual formulation, but it is not much more complicated.

The key element in this no-amplitude formulation is the existence of what can be called "mutually unbiased" measurements. Loosely speaking, two measurements are mutually unbiased, in the sense used here, if they are as noncommutative as two measurements can possibly be. For example, in the case of a spin-1/2 particle, the measurements $S_x$ and $S_y$ are mutually unbiased: if the outcome of the measurement $S_x$ can be predicted with certainty, then the two possible outcomes of $S_y$ are equally likely.

In order to construct the kind of formulation of quantum mechanics we want, we first need a way of expressing any quantum state in terms of probabilities. For any given system, we will select once and for all a particular set of special "reference measurements" in terms of which all the states will be defined. For each state, we imagine setting up a table as follows: each column of the table corresponds to one of the special reference measurements, and the entries in a given column are the probabilities of the outcomes of that particular measurement. Such a table of probabilities will be sufficient to define the state. Exactly how the reference measurements are chosen will be explained in detail later; in the simplest cases they are a set of mutually unbiased measurements. For example, we will take $S_x$, $S_y$, and $S_z$ to be a set of reference measurements for spin-1/2 particles.

We next need a way of computing the probability of a transition from one state to another when an appropriate measurement is performed. In the usual formulation this probability, for pure states, is equal to the squared modulus of the inner product of the two states vectors. In our formulation we have no state vectors but only probability tables. One might think it would be difficult to extract from these tables information which normally comes from the inner product, since the latter depends crucially on the complex phases of the state vectors' components. Surprisingly, it turns out not to be particularly difficult. Equally surprisingly, one finds that the amount of work involved in computing these transition probabilities depends on whether the number of outcomes of a complete measurement on the system is prime or composite.

One would like to go further and find out how to express in terms of probability tables the ordinary Schrödinger evolution of a quantum state, which is normally given by a family of unitary operators. This work has not yet been done. Also, we restrict our attention here to systems with finite-dimensional state spaces, although I think an extension to the infinite-dimensional case ought to be possible.

In Sect. 2 we first ask how many mutually unbiased measurements—that is, measurements all unbiased with respect to each other—we would *like* to be able to find for an $N$-dimensional state space, and we see that there do exist that many if $N$ is prime. We discuss the representation of states by probability tables in Sec. 3, and the computation of transition probabilities in Sect. 4. Finally, in Sect. 5 we mention some of the mathematical coincidences that make our formulation possible. In order to save space, I will not include the proofs of all the results presented here.

## 2. MUTUALLY UNBIASED MEASUREMENTS

The concept of mutually unbiased measurements can be motivated by the following practical question: What is the best way to ascertain the quantum state of a given ensemble? The general problem of state determination has been treated in Refs. 2 and 3. The idea of using mutually unbiased measurements has been developed by Ivanovic.[3]

Consider, for example, an ensemble of neutral spin-1/2 particles. We do not know the state of the ensemble—it may be pure or mixed—and would like to find it by making a series of measurements on the ensemble. We are interested only in the *spin* state and not the translational state; so we are looking for a $2 \times 2$ density matrix. Suppose we begin by measuring the $z$-component of spin of a large number of the particles. We find that 60% of them yield the value $+1/2$ and 40% yield the value $-1/2$. So far we do not have enough information to determine the state. It takes three independent real numbers to define a $2 \times 2$ density matrix, and so far we have obtained only one, namely, the probability of the outcome $S_z = +1/2$. To get the second number we need to make a different measurement, say $S_y$, on another subensemble, and to get the third number we need to make a third measurement, say $S_x$, on a third subensemble. Once these measurements have been made, we will have enough information to determine the state.

It is helpful to analyze this spin-1/2 problem in more detail. A general $2\times 2$ density matrix can be written in the form

$$\rho = b|s\rangle\langle s| + (1-b)I/2 \tag{1}$$

where $I$ is the identity matrix, $|s\rangle$ is a pure spin state, and $b$ is a real number such that $0\leqslant b\leqslant 1$. The quantity $b$ can be thought of as the degree of purity of the state. Each state $\rho$ can be represented by a point in the unit sphere in three dimensions: the distance of the point from the center is given by $b$, and the direction from the center is given by the unique unit vector $\mathbf{n}$ such that $|s\rangle$ is an eigenstate of $\mathbf{n}\cdot\boldsymbol{\sigma}$ with eigenvalue $+1$. If a particle in the state $\rho$ is subjected to a measurement of, say, $S_z$ (where the $z$-axis is taken to be vertical), then the probability of the outcome $S_z = +1/2$ turns out to be equal to the height of the point representing $\rho$, measured from the level of the bottom of the sphere and measured in units of the sphere's diameter. Thus, if we perform the measurement $S_z$ on a large subensemble and find that two-thirds of the particles yield the outcome $S_z = +1/2$, we can conclude that the ensemble's "state-point" lies on a horizontal plane two-thirds of the way from the bottom of the sphere to the top.

This way of representing states makes it easy to see how we can determine the state of an ensemble from three different measurements performed on subensembles. Each measurement restricts $\rho$ to a plane slice of the sphere. As long as the three axes along which the spin component is measured are not coplanar, the intersection of the three planes gives us the point representing the actual state of the ensemble.

Thus, almost any choice of the three measurements will allow a determination of the ensemble's state. However, if there is any *error* in our estimates of the probabilities—there will always be statistical error if the ensemble is finite—then not all triples of measurements are equally good. For example, if the three axes are not coplanar but make very small angles with each other, then each measurement gives practically the same information; the state-point is pinned down well in one dimension but hardly at all in the other two dimensions. Without going into a formal argument—because this is not the main point of the paper—I would like to suggest that it is best, in some reasonable sense of "best," to choose three axes that are mutually perpendicular. In that way each different measurement gives us as much new information as possible. Thus, the set of measurements consisting of $S_x$, $S_y$, and $S_z$ is, I think, an optimal set for determining the density matrix.

Let us now try to generalize these ideas to systems whose pure-state space has not just two dimensions but $N$ dimensions. In that case the most general state is given by an $N \times N$ density matrix, which requires $N^2 - 1$ independent real numbers for its specification. Any single complete measurement—we will consider here only complete measurements, i.e., without degeneracy—has $N$ possible outcomes, so it can gives us $N-1$ independent probabilities. Therefore, just on the basis of counting the number of variables, we expect that the number of different measurements we need in order to determine the density matrix completely is $(N^2-1)/(N-1)$, which equals $N+1$. For example, when $N$ equals 2, as for a spin-1/2 particle, we need to make three different measurements, as we have already seen.

We are now faced with the question: What is the best choice of the $N+1$ measurements? As in the spin-1/2 case, we would like the measurements to be as different from each other as possible. I conjecture that the kind of difference we would most desire is expressed by the property of being *mutually unbiased*, which we now define. Let $B$ and $B'$ be two orthonormal bases for the $N$-dimensional state space. We will say these two bases are mutually unbiased if for every vector $|v\rangle$ in $B$ and every vector $|v'\rangle$ in $B'$, $|\langle v | v' \rangle|^2 = 1/N$. Two *measurements* are defined to be mutually unbiased if the bases composed of their eigenstates are mutually unbiased.

We would thus like to find, if possible, $N+1$ bases for an $N$-dimensional state space, all unbiased with respect to each other. If such bases can be found, then the measurements to which they correspond will be as different from each other as possible, and will constitute, I think, an optimal set of measurements for ascertaining the state of an ensemble. (I assume that every basis does correspond to a realizable measurement; i.e., there is no superselection rule.) In two dimensions, we have seen that it is possible to find $N+1=3$ mutually unbiased bases. Here they are explicitly, in a particular representation:

$$B^0 = \left\{ \begin{pmatrix} 1 \\ 0 \end{pmatrix}, \begin{pmatrix} 0 \\ 1 \end{pmatrix} \right\}, \qquad B^1 = \left\{ \begin{pmatrix} 1/\sqrt{2} \\ 1/\sqrt{2} \end{pmatrix}, \begin{pmatrix} 1/\sqrt{2} \\ -1/\sqrt{2} \end{pmatrix} \right\},$$

$$B^2 = \left\{ \begin{pmatrix} 1/\sqrt{2} \\ i/\sqrt{2} \end{pmatrix}, \begin{pmatrix} 1/\sqrt{2} \\ -i/\sqrt{2} \end{pmatrix} \right\} \tag{2}$$

We now ask: Is it possible to find $N+1$ mutually unbiased bases in $N$ dimensions?

I do not know the complete answer to this question. I do know that if $N$ is a *prime* number, the answer is yes. The existence of the desired bases in that case has been demonstrated by Ivanovic[3] by explicit construction. Let $B_l^{nm}$ be the $l$th component of the $m$th vector in the $n$th basis. (Here $l = 1,\dots, N$; $m = 1,\dots, N$; and $n = 0,\dots, N$.) For prime values of $N$ greater than

2, one can show by direct calculation that the following bases are all unbiased with respect to each other:

$$B_l^{0m} = \delta_{ml}$$

$$B_l^{Nm} = \frac{1}{\sqrt{N}} \exp\left[\frac{2\pi i}{N} ml\right] \tag{3}$$

$$B_l^{nm} = \frac{1}{\sqrt{N}} \exp\left[\frac{2\pi i}{N} n(m+l)^2\right], \qquad \text{for} \quad n = 1,\dots, N-1$$

This is obviously not the only possible set of $N+1$ mutually unbiased bases. One can, of course, freely change the ordering of the vectors in each basis or the labelling of the bases themselves, and one can perform a simultaneous rotation on all the bases. Nevertheless, for the rest of this paper we will work only with the particular set of bases given above. This is done in the same spirit as choosing a coordinate system.

Here are some interesting mathematical questions concerning mutually unbiased bases to which I do not know the answers. (1) When $N$ is prime, can one find *more* than $N+1$ mutually unbiased bases? I suspect not. (2) For a general composite $N$, what is the maximum number of mutually unbiased bases one can find? On the basis of trial and error, I would guess that the maximum number is less than $N+1$ for all composite values of $N$, i.e., that the desired bases can be found if and only if $N$ is prime. For our purpose, it turns out that the mutually unbiased bases for prime $N$ are all we will need.

Each of the above bases is supposed to correspond to a realizable measurement. It would be interesting to try to figure out how one might actually go about performing these measurements. In the case of spin-1/2 particles, it is easy: one measures $S_x$, $S_y$, and $S_z$ by appropriately rotating the spin of each particle before it enters the inhomogeneous field of a Stern–Gerlach magnet. For systems with larger $N$ the problem is considerably more difficult, and I have not yet found any reasonably practical solution.

## 3. THE REPRESENTATION OF QUANTUM STATES IN TERMS OF PROBABILITIES

### 3.1. Prime-Dimensional State Space

Let $\rho$ be an $N \times N$ density matrix where $N$ is prime. A system in the state $\rho$, when subjected to the $n$th measurement of Eq. (2) or (3), will yield the $m$th outcome with probability

$$p_{nm} = \sum_{k,l=1}^{N} B_k^{nm*} \rho_{kl} B_l^{nm} \tag{4}$$

From the set of all these $p_{nm}$'s, with $n = 0, \ldots, N$ and $m = 1, \ldots, N$, one can reconstruct the density matrix $\rho$ as follows:

$$\begin{aligned} \rho_{kk} &= p_{0k} \\ \rho_{kl} &= \sum_{n=1}^{N} \sum_{m=1}^{N} p_{nm} B_k^{nm} B_l^{nm*}, \qquad \text{for} \quad k \neq l \end{aligned} \tag{5}$$

Equation (5) can be verified by substituting for the $p_{nm}$'s the expressions given in Eq. (4) and for the $B^{nm}$'s the expressions given in Eqs. (2) and (3).

Thus, for prime $N$, we can take the measurements defined by Eqs. (2) and (3) as our reference measurements, with respect to which any quantum state can be represented uniquely by a table of probabilities. Equations (4) and (5) provide a way of going back and forth between the representation of a state by a density matrix and its representation by a probability table. To give an example, in the case $N = 3$, the state

$$\rho = \begin{pmatrix} 1 & 0 & 0 \\ 0 & 0 & 0 \\ 0 & 0 & 0 \end{pmatrix} \tag{6}$$

is represented by the table

| | | measurement | | | |
|---|---|---|---|---|---|
| | | 0 | 1 | 2 | 3 |
| | 1 | 1 | 1/3 | 1/3 | 1/3 |
| outcome | 2 | 0 | 1/3 | 1/3 | 1/3 |
| | 3 | 0 | 1/3 | 1/3 | 1/3 |

(7)

Notice that the numbers in each column add to 1, as they must since they are the probabilities of the three possible outcomes of a given reference measurement. Notice also that because this particular state is an eigenstate of measurement #0, it has equal probabilities of all the possible outcomes of all the other measurements; this follows from the fact that the four reference measurements are mutually unbiased.

Does every conceivable probability table correspond to a possible quantum state? The answer is no. There are certain restrictions that every realizable probability table must satisfy. Are these restrictions equalities or inequalities? Answer: The only equalities that must be satisfied are that the numbers in each column add to 1. Otherwise the restrictions are inequalities. These inequalities can be summarized in one requirement: The probability table $p$ must be such that the matrix $\rho$ computed from it via Eq. (5) has no negative eigenvalues.

It would be good to find a more direct statement of the inequalities which does not involve $\rho$. As a first step in this effort, we state here *one* of the required inequalities: the sum of the squares of all the elements of the table must be less than or equal to 2. This fact follows from the requirement of non-negative eigenvalues, but by itself it is not enough to guarantee the non-negativity. Incidentally, one can show that for any probability table which does correspond to an actual state, the sum of the squares *equals* 2 if and only if the state is pure.

### 3.2. Composite-Dimensional State Space

When $N$ is not prime, the construction given in Eq. (3) does not yield a set of mutually unbiased bases. We therefore need to generalize our means of choosing a set of reference measurements.

Our approach will be to treat every system having a composite-dimensional state space as if it were a composite system. That is, only systems with prime $N$ will be treated as truly fundamental and indivisible. It may well be that nature is actually structured in that way. Our approach is certainly most appealing if that is the case, and it does not seem improbable that it is. However, even if there do exist fundamental systems with composite $N$, it is still possible to *treat* them as if they had components. In the case of a spin-3/2 particle, for example, the particle's four-dimensional state space could be factorized into two two-dimensional factor spaces, such that the operator "sign of $S_z$" acts only on one of these spaces, and the operator "magnitude of $S_z$" acts only on the other. These two smaller spaces could be thought of as belonging to two subsystems, which together constitute the original spin-3/2 particle. In this spirit, we will regard any system for which $N$ has $f$ prime factors as a composite system consisting of $f$ primitive subsystems.

The reference measurements for composite systems are chosen as follows. Imagine performing on every primitive subsystem a reference measurement for that subsystem, chosen from the set specified in Sect. 3.1. The performance of all these measurements, i.e., one on each subsystem, is counted as a reference measurement on the whole system, and every reference measurement is constructed in this way. Thus, if $N = N_1 N_2 \cdots N_f$, where each $N_i$ is prime, the total number of reference measurements on the whole system is $(N_1+1)(N_2+1)\cdots(N_f+1)$, since there are $N_i+1$ reference measurements for the $i$th subsystem.

Consider for example a system consisting of an electron and a proton; as always we consider only the spin state. For this system we have

$N_1 = N_2 = 2$ and $N = 4$. A typical reference measurement would consist of performing a measurement of, say, $S_y$ on the electron and $S_x$ on the proton. There are altogether nine different reference measurements, each of which has four possible outcomes, so we can imagine constructing a probability table having nine columns and four rows. As we will see shortly, such a table contains a complete description of the system's state.

For a general composite system, each of the reference measurements has eigenstates of the form

$$v_{l_1 \cdots l_f} = B_{l_1}^{n_1 m_1} B_{l_2}^{n_2 m_2} \cdots B_{l_f}^{n_f m_f} \tag{8}$$

Here each $n_i$ has a value from 0 to $N_i$, and the ordered set $(n_1,..., n_f)$ labels the measurement: it says we are performing the $n_1$th reference measurement on the first subsystem, the $n_2$th on the second, and so on. Each $m_i$ has a value from 1 to $N$, and the ordered set $(m_1,..., m_f)$ labels the particular outcome, or particular eigenvector, of the measurement. The $l_i$'s label the components of this eigenvector. The components of the density matrix of the composite system can be labelled as

$$\rho_{k_1 \cdots k_f; l_1 \cdots l_f}, \qquad \text{where} \quad k_i, l_i = 1,..., N_i \tag{9}$$

When the measurement $(n_1,..., n_f)$ is performed on the state $\rho$, the probability of the outcome $(m_1,..., m_f)$ is

$$p_{n_1 \cdots n_f; m_1 \cdots m_f} = \sum_{k_1,..., k_f} \sum_{l_1,..., l_f} (B_{k_1}^{n_1 m_1} \cdots B_{k_f}^{n_f m_f})^* \rho_{k_1 \cdots k_f; l_1 \cdots l_f} (B_{l_1}^{n_1 m_1} \cdots B_{l_f}^{n_f m_f}) \tag{10}$$

Equation (10) tells us how to go from the density matrix of a state to its probability table. As in the case of prime $N$, we would like to have a prescription for going the other direction, i.e., for recovering the density matrix $\rho$ from the $p$'s. The prescription turns out to be very similar to that given in Eq. (5):

$$\rho_{k_1 \cdots k_f; l_1 \cdots l_f} = \sum_{n_1=1}^{N_1} \cdots \sum_{n_f=1}^{N_f} \sum_{m_1=1}^{N_1} \cdots \sum_{m_f=1}^{N_f} p_{n_1 \cdots n_f; m_1 \cdots m_f} \times (B_{k_1}^{n_1 m_1} \cdots B_{k_f}^{n_f m_f})(B_{l_1}^{n_1 m_1} \cdots B_{l_f}^{n_f m_f})^* \tag{11}$$

if $k_i \neq l_i$ for each $i = 1,..., f$. For each value of $i$ for which $k_i = l_i$, one eliminates the sum over $n_i$ and replaces $n_i$ in the summand by zero. Equation (11) is our guarantee that the probability table contains all the information one needs to determine the state.

Again we need to address the question: what restrictions must one impose on a probability table in order to ensure that it corresponds to a genuine quantum state? As before, there will be some equalities and some inequalities. Here are the equalities that must be satisfied:

(i) The probabilities of the outcomes of reference measurements performed on any *subset* of the $f$ primitive subsystems cannot depend on the choice of measurements on the *other* subsystems. For example, the reduced probability table obtained by summing $p_{n_1 \cdots n_f; m_1 \cdots m_f}$ over $m_i$ cannot depend on $n_i$.

(ii) For each measurement $(n_1, ..., n_f)$, the sum over *all* possible outcomes, $\sum_{m_1,...,m_f} p_{n_1 \cdots n_f; m_1 \cdots m_f}$, must equal 1.

One can show quite generally, i.e., not necessarily in the context of quantum mechanics, that a probability table satisfying the above two restrictions contains exactly $N^2 - 1$ independent real numbers. This is precisely the number needed to specify a general quantum state.

There are still some *inequalities* that any legitimate probability table must satisfy. As before, these can be summarized by saying that the matrix $\rho$ defined by Eq. (11) must have no negative eigenvalues.

### 3.3. Systems of Identical Particles

When a system contains identical particles, it is no longer true that one needs $N^2 - 1$ numbers to specify a state of the system. Consider, for example, a system of two electrons. If they were not identical, the spin state of this system would be described by an ordinary $4 \times 4$ density matrix. However, because of the indistinguishability, odd and even spin states must be correlated to spatial states of different parity, so they cannot be coherently superposed. This means that the density matrix for the spin, written in the usual singlet-triplet basis, must be block diagonal, containing a $3 \times 3$ block and a $1 \times 1$ block. (We do not consider here the description of the spatial part of the state.) The number of independent numbers in such a density matrix is not $4^2 - 1 = 15$, but only 9.

In such a case, our description of states in terms of probability tables is modified only slightly. We require the probability table to be invariant under the interchange of identical particles; in other respects the description is the same as before. For example, in the two-electron case, we require that $p_{n_1 n_2; m_1 m_2} = p_{n_2 n_1; m_2 m_1}$; one can convince oneself that this restriction does indeed cut the number of independent variables from 15 to 9. Quite generally, one can show that for any system containing identical particles, every unsymmetrized but otherwise legitimate probability table yields, upon symmetrization, a table corresponding to a legitimate symmetrized density matrix. This would obviously be true if we were requiring that *no* measurement show any difference among the indistinguishable particles. But we are only requiring that none of the *reference* measurements show any difference among the particles. This restriction turns out to be enough.

## 4. CALCULATION OF TRANSITION PROBABILITIES

Suppose a system in state $S$ is subjected to a measurement one of whose eigenstates is a different state $S'$. What is the probability $P(S \to S')$ that the outcome corresponding to $S'$ will occur? This probability is normally obtained by taking the trace of $\rho\rho'$, where $\rho$ is the density matrix of the original state $S$ and $\rho'$ is the density matrix of the pure state $S'$. How do we express this same transition probability in terms of probability tables?

If the system is primitive, the answer could hardly be simpler. Let $p_{nm}$ and $p'_{nm}$ be the probability tables for the two states $S$ and $S'$. The transition probability $P(S \to S')$ turns out to be given by

$$P(S \to S') = \left( \sum_{n=0}^{N} \sum_{m=1}^{N} p_{nm} p'_{nm} \right) - 1 \tag{12}$$

The remarkable simplicity of this result is the main advantage of using *mutually unbiased* measurements to define states, rather than some equally complete but different set of measurements. The advantage is analogous to that gained by using an orthogonal coordinate system, as opposed to an oblique coordinate system, to analyze a problem in, say, classical mechanics.

For composite systems, the calculation of transition probabilities is somewhat more complicated, but it is still reasonably simple considering that we are working only with probabilities and not with amplitudes. To simplify the notation let us define the symbol $[p, p']$ to mean the sum of products of corresponding terms of the probability tables $p$ and $p'$. That is,

$$[p, p'] = \sum_{n_1=0}^{N_1} \cdots \sum_{n_f=0}^{N_f} \sum_{m_1=1}^{N_1} \cdots \sum_{m_f=1}^{N_f} p_{n_1 \cdots n_f; m_1 \cdots m_f} p'_{n_1 \cdots n_f; m_1 \cdots m_f} \tag{13}$$

The quantity $[p, p']$ has the following interpretation: it is the sum, over all reference measurements, of the probability that the primed and unprimed states will yield exactly the same outcome. We can call $[p, p']$ the "coincidence sum" for the two states.

Let us denote by ${}^i p$ the reduced probability table obtained from $p$ by summing over $m_i$. It gives the probabilities of the outcomes of all measurements which do not involve the $i$th primitive subsystem. For example, ${}^{1}p_{n_2 \cdots n_f; m_2 \cdots m_f} = \sum_{m_1} p_{n_1 \cdots n_f; m_1 \cdots m_f}$ is the probability of getting the outcome $(m_2,..., m_f)$ when the measurement $(n_2,..., n_f)$ is performed on subsystems 2 through $f$. As we have mentioned before, this probability is independent of what measurement is made on subsystem 1. Likewise, ${}^{ij}p$ is the table obtained from $p$ by summing over $m_i$ and $m_j$, and the analogous symbols with more indices are defined similarly.

The transition probability between two states $S$ and $S'$ of a general system, with probability tables $p$ and $p'$, turns out to be given by the following combination of coincidence sums:

$$P(S \to S') = \{p, p'\} \equiv [p, p'] - \sum_{i=1}^{f} [{}^{i}p, {}^{i}p'] + \sum_{\substack{i,j=1 \\ i<j}}^{f} [{}^{ij}p, {}^{ij}p'] - \cdots + (-1)^f \tag{14}$$

The last term comes from $(-1)^f [{}^{1,...,f}p, {}^{1,...,f}p']$, but in this expression each of the "tables" in the brackets consists of the single number 1, so the bracket also equals 1. Equation (14) is the main result of this paper. Notice that Eq. (12) is a special case of Eq. (14).

We have been thinking of the state $S'$ as a pure state, so that it could be an eigenstate of a measurement. It happens, not surprisingly, that $\{p, p'\}$ is always equal to the trace of $\rho\rho'$, even when both states are mixed.

## 5. DISCUSSION

We have seen how to represent a general quantum state (of a system with a finite-dimensional state space) by a "probability table," and how to compute the transistion probability between two states from their probability tables. As a by-product, we have obtained a prescription, Eq. (11), for finding the density matrix of an arbitrary system directly from measurable probabilities.

It is interesting to list some of the mathematical facts that have allowed us to write down at least this much of quantum mechanics purely in terms of probabilities.

(i) The number of real numbers needed to specify a quantum state, $N^2 - 1$, divided by the number of numbers obtainable from a given measurement, $N - 1$, happens always to be an integer. This means that of the data collected from the $N + 1$ reference measurements, $N$ being prime, every available number has to be used. There are no redundant variables in the data. Thus our probability tables were "just the right size" for specifying a state.

(ii) It happens that for an $N$-dimensional complex vector space, with $N$ prime, there *are* $N+1$ mutually unbiased bases, exactly as many as we needed.

(iii) For composite systems, it was not surprising that for every legitimate probability table, the results of measurements on *part* of the system did not depend on the choice of measurement performed on the rest of the system. This property is, after all, the reason we cannot use the violation of Bell's inequality to transmit information faster than light.[(4)] We noted that this requirement, along with the obvious requirement that the sum of the probabilities of all the outcomes of a measurement be equal to 1, reduced the number of independent numbers a table could contain down to $N^2-1$, which is precisely the number needed to specify a state. This agreement between "numbers obtainable" and "numbers needed" allowed us to apply to the treatment of *composite* systems the results we had obtained for primitive systems.

Let me pursue this last point a little further. Consider a hypothetical world where, as in the quantum mechanics of our world, one needs to perform several different measurements on an ensemble in order to determine its state. Let $g(N)$ be the number of different measurements needed if each measurement has $N$ possible outcomes. (For quantum mechanics, $g(N) = N+1$.) Suppose further that, as in quantum mechanics, none of the $(N-1)\,g(N)$ independent probabilities obtained from these measurements is redundant; it takes exactly that many numbers to determine the state.

In such a world, let $Z_1$ and $Z_2$ be two systems, having respectively $N_1$ and $N_2$ possible outcomes of a complete measurement. Let $M_1$ be a set of $g(N_1)$ measurements sufficient for the determination of the state of $Z_1$, and let $M_2$ be an analogous set for $Z_2$. Suppose we now *insist* that, as in quantum mechanics, one can obtain a complete determination of the state of the composite system $Z_1+Z_2$, with no redundant numbers,[2] by performing on an ensemble of such systems every possible combination of a measurement from $M_1$ and a measurement from $M_2$. Assuming as usual that the choice of measurement on one subsystem does not affect the other, one can show

[2] Given the fact that the choice of measurement on one subsystem cannot influence another subsystem—let us call this "the causality condition"—an experimenter using our method of determining an ensemble's state knows from the outset that not all the probabilities he obtains are going to be independent of each other. But this lack of independence, arising from the causality condition, is not what we are calling "redundancy." Rather, we imagine that the experimenter will select from his data a maximal set of numbers which *are* logically independent, even given the causality condition. If it turns out that, because of the laws of physics, some of *these* numbers do in fact depend on each other, then we call these numbers redundant.

that the number of independent probabilities obtainable from these measurements is

$$(N_1-1)\,g(N_1)+(N_2-1)\,g(N_2)+(N_1-1)(N_2-1)\,g(N_1)\,g(N_2) \quad (15)$$

On the other hand, according to our assumptions, the number of independent numbers needed to specify the state of $Z_1+Z_2$ is

$$(N_1N_2-1)\,g(N_1N_2) \quad (16)$$

Therefore, if the measurements are to give a complete and nonredundant characterization of the state, we must have

$$\begin{aligned}(N_1-1)\,g(N_1)+(N_2-1)\,g(N_2)+(N_1-1)(N_2-1)\,g(N_1)\,g(N_2)\\ =(N_1N_2-1)\,g(N_1N_2)\end{aligned} \quad (17)$$

In any world in which our assumptions are satisfied, the positive-integer-valued function $g(N)$ must satisfy Eq. (17) for all positive integers $N_1$ and $N_2$. We can think of Eq. (17) as follows: it is the condition that must be satisfied in order that *those measurements which are just sufficient for determining the states of the subsystems are, when performed jointly, also just sufficient for determining the state of the whole system.* In quantum mechanics, $g(N)=N+1$, which does satisfy Eq. (17). Another solution is $g(N)=1$, which corresponds to a world where on each system there is only one complete measurement that can be performed. Such a world could be called "classical," although it differs from the world of classical physics in that the outcomes are discrete. Any function $g(N)$ of the form $g(N)=\sum_{k=0}^{K} N^k$ is also a solution of Eq. (17). It would be interesting to know if there are any other solutions. In any case, the fact that the $g(N)$ for quantum mechanics does satisfy Eq. (17) made it possible for us to treat composite systems as we did.

In conclusion, we have here the beginning of a formulation of quantum mechanics expressed purely in terms of quantities which are in principle directly observable. The way in which states are represented is quite different from that of the usual formulation, and the rule for computing transition probabilities also looks quite different. But the physical content is the same. The question, "why complex amplitudes?," could therefore be rephrased to read, "why does the transition rule involve the particular combination of probabilities given by $\{p,p'\}$?" The restatement may or may not be easier to answer than the original question. It certainly gives us a new perspective on the problem.

## ACKNOWLEDGMENTS

I would like to thank Asher Peres and Bill Lenhart for a number of stimulating and enlightening conversations on this subject.

Also, on this special occasion, I would like to thank John Wheeler for what could be called a retarded influence and an advanced influence: from the past, much encouragement and support; from the future, the vision of an utterly simple but marvelously strange universe.

## REFERENCES

1. John A. Wheeler, "Genesis and Observership," in *Foundational Problems in the Special Sciences*, R. E. Butts and J. Hintikka, eds. (Reidel, Dordrecht, Holland, 1977); "Frontiers of Time," in *Problems in the Foundations of Physics (Proceedings of the International School of Physics "Enrico Fermi": course 72)* (Italian Physical Society, 1979); "Beyond the Black Hole," in *Some Strangeness in the Proportion*, Harry Woolf, ed. (Addison-Wesley, Reading, Massachusetts, 1980); *Am. J. Phys.* **51**, 398 (1983).
2. W. Gale, C. Guth, and G. T. Trammell, *Phys. Rev.* **165**, 1434 (1968); W. Band and J. L. Park, *Found. Phys.* **1**, 133 (1970); J. L. Park and W. Band, *Found. Phys.* **1**, 211 (1971); W. Band and J. L. Park, *Found. Phys.* **1**, 339 (1971); *Am. J. Phys.* **47**, 188 (1979); I. D. Ivanovic, *J. Math. Phys.* **24**, 1199 (1983).
3. I. D. Ivanovic, *J. Phys. A*: *Mat. Gen.* **14**, 3241 (1981).
4. John S. Bell, *Physics* **1**, 195 (1964); John F. Clauser, Michael A. Horne, Abner Shimony, and Richard A. Holt, *Phys. Rev. Lett.* **23**, 880 (1969); Stuart J. Freedman and John F. Clauser, *Phys. Rev. Lett.* **28**, 938 (1972); P. J. Bussey, *Phys. Lett. A* **90**, 9 (1982).

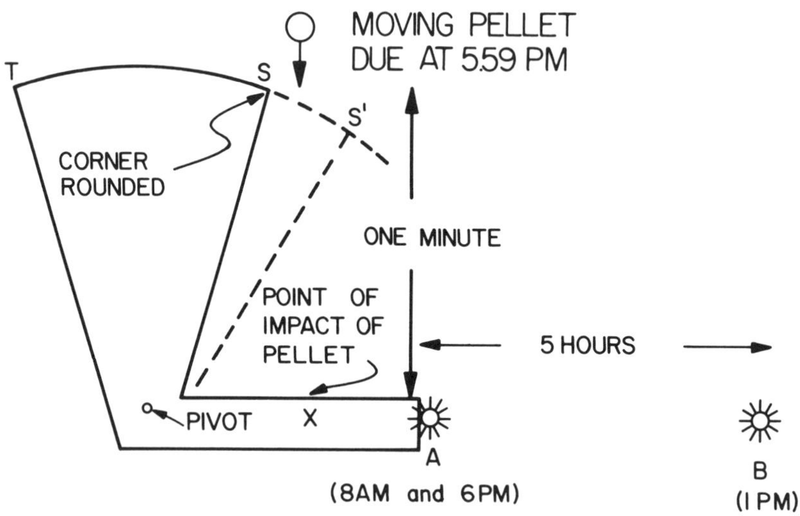
MOVING PELLET
DUE AT 5.59 PM
T
S
S'
CORNER
ROUNDED
ONE MINUTE
POINT OF
IMPACT OF
PELLET
5 HOURS
PIVOT
X
A
(8AM and 6PM)
B
(1 PM)

*Plate 35*

# Quantum Mechanical Computers[1]

**Richard P. Feynman**[2]

*Received March 15, 1985*

*The physical limitations, due to quantum mechanics, on the functioning of computers are analyzed.*

## 1. INTRODUCTION

This work is a part of an effort to analyze the physical limitations of computers due to the laws of physics. For example, Bennett[(1)] has made a careful study of the free energy dissipation that must accompany computation. He found it to be virtually zero. He suggested to me the question of the limitations due to quantum mechanics and the uncertainty principle. I have found that, aside from the obvious limitation to size if the working parts are to be made of atoms, there is no fundamental limit from these sources either.

We are here considering ideal machines; the effects of small imperfections will be considered later. This study is one of principle; our aim is to exhibit some Hamiltonian for a system which could serve as a computer. We are not concerned with whether we have the most efficient system, nor how we could best implement it.

Since the laws of quantum physics are reversible in time, we shall have to consider computing engines which obey such reversible laws. This problem already occurred to Bennett,[(1)] and to Fredkin and Toffoli,[(2)] and a great deal of thought has been given to it. Since it may not be familiar to

---

[1] *Editor's note*: This article, which is based on the author's plenary talk presented at the CLEO/IQEC Meeting in 1984, originally appeared in the February 1985 issue of *Optics News*. It is here reprinted with kind permission of Professor Feynman and *Optics News*.

[2] Department of Physics, California Institute of Technology, Pasadena, California 91125.

you here, I shall review this, and in doing so, take the opportunity to review, very briefly, the conclusions of Bennett,[(2)] for we shall confirm them all when we analyze our quantum system.

It is a result of computer science that a universal computer can be made by a suitably complex network of interconnected primitive elements. Following the usual classical analysis we can imagine the interconnections to be ideal wires carrying one of two standard voltages representing the local 1 and 0. We can take the primitive elements to be just two, NOT and AND (actually just the one element NAND = NOT AND suffices, for if one input is set at 1 the output is the NOT of the other input). They are symbolized in Fig. 1, with the logical values resulting on the outgoing wires, resulting from different combinations of input wires.

From a logical point of view, we must consider the wires in detail, for in other systems, and our quantum system in particular, we may not have wires as such. We see we really have two more logical primitives, FAN OUT when two wires are connected to one, and EXCHANGE, when wires are crossed. In the usual computer the NOT and NAND primitives are implemented by transistors, possibly as in Fig. 2.

What is the minimum free energy that must be expended to operate an ideal computer made of such primitives? Since, for example, when the AND operates the output line, $c'$ is being determined to be one of two values, no matter what it was before, the entropy change is ln 2 units. This represents a heat generation of $kT \ln 2$ at temperature $T$. For many years it was thought that this represented an absolute minimum to the quantity of heat per primitive step that had to be dissipated in making a calculation.

The question is academic at this time. In actual machines we are quite concerned with the heat dissipation question, but the transistor system used actually dissipates about $10^{10}kT$! As Bennett[(3)] has pointed out, this arises because to change a wire's voltage we dump it to ground through a resistance; and to build it up again we feed charge, again through a resistance, to the wire. It could be greatly reduced if energy could be stored in an inductance, or other reactive element.

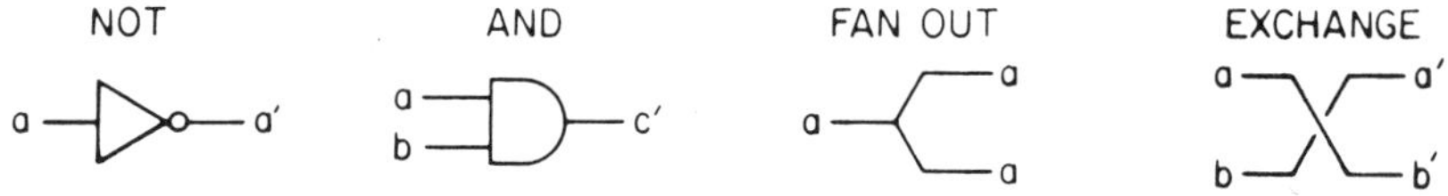


| a | a' |
|---|---|
| 0 | 1 |
| 1 | 0 |

| a | b | c' |
|---|---|---|
| 0 | 0 | 0 |
| 0 | 1 | 0 |
| 1 | 0 | 0 |
| 1 | 1 | 1 |

| a | b | a' | b' |
|---|---|---|---|
| 0 | 0 | 0 | 0 |
| 0 | 1 | 1 | 0 |
| 1 | 0 | 0 | 1 |
| 1 | 1 | 1 | 1 |

Fig. 1. Primitive elements.

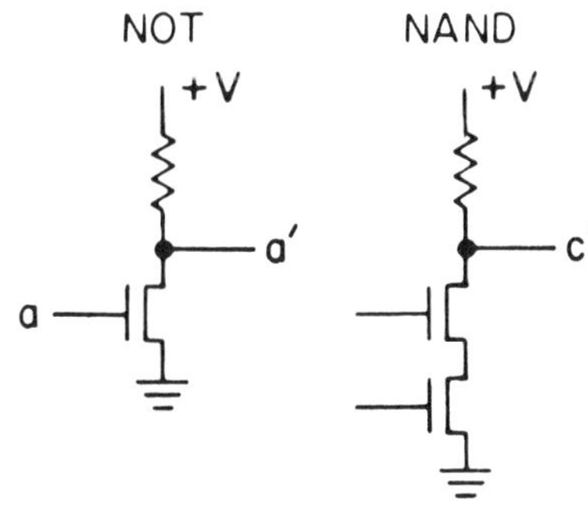


Fig. 2. Transistor circuits for NOT and NAND.

However, it is apparently very difficult to make inductive elements on silicon wafers with present techniques. Even Nature, in her DNA copying machine, dissipates about $100kT$ per bit copied. Being, at present, so very far from this $kT \ln 2$ figure, it seems ridiculous to argue that even this is too high and the minimum is really essentially zero. But, we are going to be even more ridiculous later and consider bits written on one atom instead of the present $10^{11}$ atoms. Such nonsense is very entertaining to professors like me. I hope you will find it interesting and entertaining also.

What Bennett pointed out was that this former limit was wrong because it is not necessary to use irreversible primitives. Calculations can be done with reversible machines containing only reversible primitives. If this is done the minimum free energy required is independent of the complexity or number of logical steps in the calculation. If anything, it is $kT$ per bit of the output answer.

But even this, which might be considered the free energy needed to clear the computer for further use, might also be considered as part of what you are going to do with the answer—the information in the result if you transmit it to another point. This is a limit only achieved ideally if you compute with a reversible computer at infinitesimal speed.

## 2. COMPUTATION WITH A REVERSIBLE MACHINE

We will now describe three reversible primitives that could be used to make a universal machine (Toffoli[(4)]). The first is the NOT which evidently loses no information, and is reversible, being reversed by acting again with NOT. Because the conventional symbol is not symmetrical we shall use an $X$ on the wire instead (see Fig. 3a).

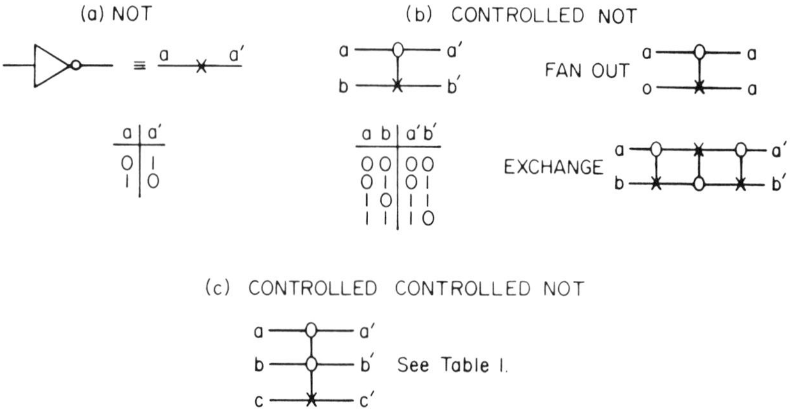


Fig. 3. Reversible primitives.

Next is what we shall call the CONTROLLED NOT (see Fig. 3b). There are two entering lines, $a$ and $b$, and two exiting lines, $a'$ and $b'$. The $a'$ is always the same as $a$, which is the control line. If the control is activated $a=1$ then the out $b'$ is the NOT of $b$. Otherwise $b$ is unchanged, $b'=b$. The table of values for input and output is given in Fig. 3. The action is reversed by simply repeating it.

The quantity $b'$ is really a symmetric function of $a$ and $b$ called XOR, the exclusive or; $a$ or $b$ but not both. It is likewise the sum modulo 2 of $a$ and $b$, and can be used to compare $a$ and $b$, giving a 1 as a signal that they are different. Please notice that this function XOR is itself not reversible. For example, if the value is zero we cannot tell whether it came from $(a, b)=(0, 0)$ or from $(1, 1)$ but we keep the other line $a'=a$ to resolve the ambiguity.

We will represent the CONTROLLED NOT by putting a 0 on the control wire, connected with a vertical line to an $X$ on the wire which is controlled.

This element can also supply us with FAN OUT, for if $b=0$ we see that $a$ is copied onto line $b'$. This COPY function will be important later on. It also supplies us with EXCHANGE, for three of them used

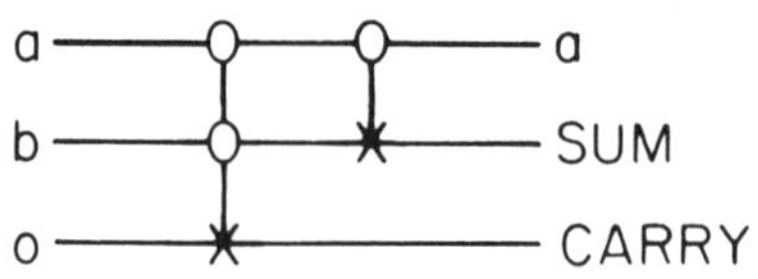


Fig. 4. Adder.

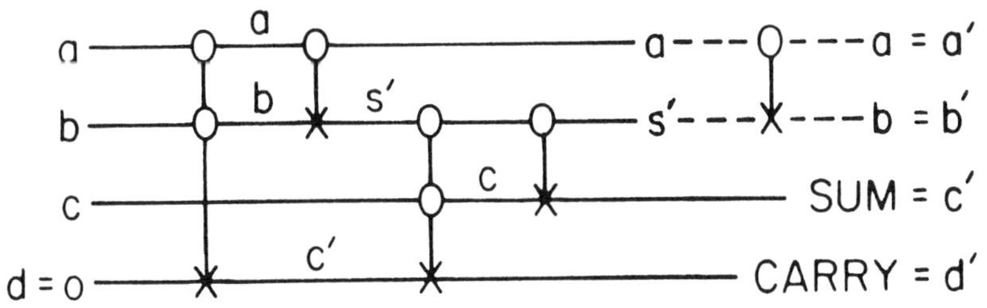


Fig. 5. Full adder.

successively on a pair of lines, but with alternate choice for control line, accomplishes an exchange of the information on the lines (Fig. 3b).

It turns out that combinations of just these two elements alone are insufficient to accomplish arbitrary logical functions. Some element involving three lines is necessary. We have chosen what we can call the CONTROLLED CONTROLLED NOT. Here (see Fig. 3c) we have two control lines $a$, $b$, which appear unchanged in the output and which change the third line $c$ to NOT $c$ only if both lines are activated ($a=1$ and $b=1$). Otherwise $c'=c$. If the third line input $c$ is set to 0, then evidently it becomes $1 (c'=1)$ only if both $a$ and $b$ are 1 and therefore supplies us with the AND function (see Table I).

Three combinations for $(a, b)$, namely $(0, 0)$, $(0, 1)$, and $(1, 0)$, all give the same value, 0, to the AND $(a, b)$ function so the ambiguity requires two bits to resolve it. These are kept in the lines $a$, $b$ in the output so the function can be reversed (by itself, in fact). The AND function is the carry bit for the sum of $a$ and $b$.

From these elements it is known that any logical circuit can be put together by using them in combination, and in fact, computer science

**Table I.**

| $a$ | $b$ | $c$ | $a'$ | $b'$ | $c'$ |
|---|---|---|---|---|---|
| 0 | 0 | 0 | 0 | 0 | 0 |
| 0 | 0 | 1 | 0 | 0 | 1 |
| 0 | 1 | 0 | 0 | 1 | 0 |
| 0 | 1 | 1 | 0 | 1 | 1 |
| 1 | 0 | 0 | 1 | 0 | 0 |
| 1 | 0 | 1 | 1 | 0 | 1 |
| 1 | 1 | 0 | 1 | 1 | 1 |
| 1 | 1 | 1 | 1 | 1 | 0 |

shows that a universal computer can be made. We will illustrate this by a little example. First, of course, as you see in Fig. 4, we can make an adder, by first using the CONTROLLED CONTROLLED NOT and then the CONTROLLED NOT in succession, to produce from $a$ and $b$ and 0, as input lines, the original $a$ on one line, the sum on the second line, and the carry on the third.

A more elaborate circuit is a full adder (see Fig. 5), which takes a carry $c$ (from some previous addition) and adds it to the two lines $a$ and $b$ and has an additional line $d$ with a 0 input. It requires four primitive elements to be put together. Besides this total sum, the total of the three, $a$, $b$, and $c$ and the carry, we obtain on the other two lines two pieces of information. One is the $a$ that we started with, and the other is some intermediary quantity that we calculated on route.

This is typical of these reversible systems; they produce not only what you want in output, but also a certain amount of garbage. In this particular case, and as it turns out in all cases, the garbage can be arranged to be, in fact, just the input, if we would just add the extra CONTROLLED NOT on the first two lines, as indicated by the dotted lines in Fig. 5; we see that the garbage would become $a$ and $b$, which were the inputs of at least two of the lines. (We know this circuit can be simplified but we do it this way for illustrative purposes.)

In this way, we can by various combinations produce a general logic unit that transforms $n$ bits to $n$ bits in a reversible manner. If the problem you are trying to do is itself reversible, then there might be no extra garbage, but in general, there are some extra lines needed to store up the information which you would need to be able to reverse the operation. In other words, we can make any function that the conventional system can, plus garbage. The garbage contains the information you need to reverse the process.

And how much garbage? It turns out in general that if the output data that you are looking for has $k$ bits, then starting with an input and $k$ bits containing 0, we can produce, as a result, just the input and the output and no further garbage. This is reversible because knowing the output and the input permits you, of course, to undo everything. This proposition is always reversible. The argument for this is illustrated in Fig. 6.

Suppose we began with a machine $M$, which, starting with an input, and some large number of 0's, produces the desired outut plus a certain amount of extra data which we call garbage. Now we have seen that the copy operation which can be done by a sequence of CONTROLLED NOT's is possible, so if we have originally an empty register, with the $k$ bits ready for the output, we can, after the processor $M$ has operated, copy the output from the $M$ onto this new register.

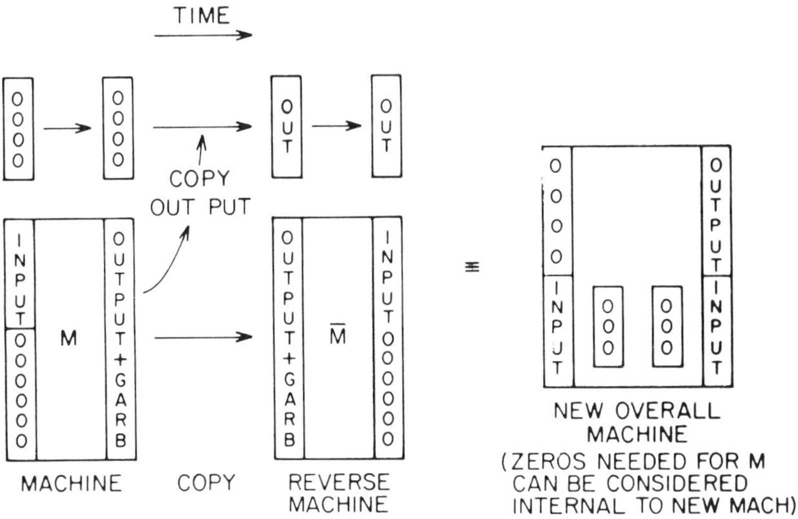


Fig. 6. Clearing garbage.

After that, we can build the opposite machine, the $M$ in reverse, the reverse machine, which would take this output of $M$ and garbage and turn it into the input and 0's. Thus, seen as an overall machine, we would have started with the $k$ 0's of the register for the output, and the input, and ended up with those $k$ 0's occupied by the output data, and repeat the inut as a final product. The number of 0's that was originally needed in the $M$ machine in order to hold the garbage is restored again to 0, and can be considered as internal wires inside the new complete machine ($M$, $\bar{M}$ and copy).

Overall, then, we have accomplished what we set out to do, and therefore garbage need never be any greater than a repetition of the input data.

## 3. A QUANTUM MECHANICAL COMPUTER

We now go on to consider how such a computer can also be built using the laws of quantum mechanics. We are going to write a Hamiltonian, for a system of interacting parts, which will behave in the same way as a large system in serving as a universal computer. Of course the large system also obeys quantum mechanics, but it is in interaction with the heat baths and other things that could make it effectively irreversible.

What we would like to do is make the computer as small and as simple as possible. Our Hamiltonian will describe in detail all the internal computing actions, but not, of course, those interactions with the exterior involved in entering the input (preparing the initial state) and reading the output.

How small can such a computer be? How small, for instance, can a number be? Of course a number can be represented by bits of 1's and 0's. What we are going to do is imagine that we have two-state systems, which we will call "atoms." An $n$ bit number is then represented by a state of a "register," a set of $n$ two-state systems.

Depending upon whether or not each atom is in one or another of its two states, which we call $|1\rangle$ and $|0\rangle$, we can of course, represent any number. And the number can be read out of such a register by determining, or measuring, in which state each of the atoms are at a given moment. Therefore one bit will be represented by a single atom being in one of two states, the states we will call $|1\rangle$ and $|0\rangle$.

What we will have to do then can be understood by considering an example; the example of a CONTROLLED CONTROLLED NOT. Let $G$ be some sort of an operation on three atoms $a$, $b$, and $c$, which converts the original state of $a$, $b$, and $c$ into a nex appropriate state, $a'$, $b'$, $c'$, so that the connection between $a'$, $b'$, and $c'$ and $a, b, c$, are just what we would have expected if $a$, $b$, and $c$ represented wires, and the $a'$, $b'$, and $c'$ were the output wires of a CONTROLLED CONTROLLED NOT.

It must be appreciated here that at the moment we are not trying to move the data from one position to another; we are just going to change it. Unlike the situation in the actual wired computer in which the voltages on one wire then go over to voltages on another, what we are specifically making is something simpler, that the three atoms are in some particular state, and that an operation is performed, which changes the state to new values, $a'$, $b'$, $c'$.

What we would have then is that the state, in the mathematical form $|a', b', c'\rangle$, is simply some operation $G$ operating on $|a, b, c\rangle$. In quantum mechanics, state changing operators are linear operators, and so we'll suppose that $G$ is linear. Therefore, $G$ is a matrix, and the matrix elements of $G$, $G_{a',b',c',a,b,c}$ are all 0 except those in Table I, which are of course 1.

This table is the same table that represents the truth value table for the CONTROLLED CONTROLLED NOT. It is apparent that the operation is reversible, and that can be represented by saying that $G^*G = 1$, where the * means Hermitian adjoint. That is to say, $G$ is a unitary matrix. (In fact $G$ is also a real matrix $G^* = G$, but that's only a special case.) To be more specific, we are going to write $A_{ab,c}$ for this special $G$. We shall use the same matrix $A$ with different numbers of subscripts to represent the other primitive elements.

To take a simple example, the NOT, which would be represented by $A_a$, is the simple matrix

$$\begin{bmatrix} 0 & 1 \\ 1 & 0 \end{bmatrix}$$

This is a $2 \times 2$ matrix and can be represented in many ways, in different notations, but the particular one we will use to define these is by the method of creation and annihilation operators. Consider operating in this case, on a single line $a$. In order to save alphabets, let us call $\underline{a}$ the matrix

$$\underline{a} = \begin{bmatrix} 0 & 1 \\ 0 & 0 \end{bmatrix}$$

which annihilates the 1 on atom $a$ and converts it to 0; $\underline{a}$ is an operator which converts the state $|1\rangle$ to $|0\rangle$. But, if the state of the atom were originally $|0\rangle$, the operator $\underline{a}$ produces the number 0. That is, it doesn't change the state, it simply produces the numerical value zero when operating on that state. The conjugate of this thing, of course, is

$$\underline{a^*} = \begin{bmatrix} 0 & 0 \\ 1 & 0 \end{bmatrix}$$

which creates, in the sense that operating on the 0 state, it turns it to the 1 state. In other words, it moves from $|0\rangle$ to $|1\rangle$. When operating on the $|1\rangle$ state there is no further state above that which you can create, and therefore it gives it the number zero. Every other operator $2 \times 2$ matrix can be represented in terms of these $\underline{a}$ and $\underline{a^*}$. For example, the product $\underline{a}^*\underline{a}$ is equal to the matrix

$$\underline{a^*a} = \begin{bmatrix} 1 & 0 \\ 0 & 0 \end{bmatrix}$$

which you might call $N_a$. It is 1 when the state is $|1\rangle$ and 0 when the state is $|0\rangle$. It gives the number that the state of the atom represents. Likewise the product

$$\underline{aa^*} = \begin{bmatrix} 0 & 0 \\ 0 & 1 \end{bmatrix}$$

is $1 - N_a$, and gives 0 for the up-state and 1 for the down-state. We'll use 1 to represent the diagonal matrix,

$$\begin{bmatrix} 1 & 0 \\ 0 & 1 \end{bmatrix}$$

As a consequence of all this, $\underline{aa^*} + \underline{a}^*\underline{a} = 1$.

It is evident then that our matrix for NOT, the operator that produces NOT, is $A_a = \underline{a} + \underline{a}^*$ and further of course, that's reversible, $A_a^* A_a = 1$, $A_a$ is unitary.

In the same way the matrix $A_{a,b}$ for the CONTROLLED NOT can be worked out. If you look at the table of values for CONTROLLED NOT you see that it can be written this way:

$$\underline{a}^*\underline{a}(\underline{b} + \underline{b}^*) + \underline{a}\underline{a}^*$$

In the first term, the $\underline{a}^*\underline{a}$ selects the condition that the line $a = 1$, in which case we want $\underline{b} + \underline{b}^*$ the NOT to apply to $b$. The second term selects the condition that the line $a$ is 0, in which case we want nothing to happen to $b$ and the unit matrix on the operators of $b$ is implied. This can also be written as $1 + \underline{a}^*\underline{a}(\underline{b} + \underline{b}^* - 1)$, the 1 representing all the lines coming through directly, but in the case that $a$ is 1, we would like to correct that by putting in a NOT instead of leaving the line $b$ unchanged.

The matrix for the CONTROLLED CONTROLLED NOT is

$$A_{ab,c} = 1 + \underline{a}^*\underline{a}\underline{b}^*\underline{b}(\underline{c} + \underline{c}^* - 1)$$

as, perhaps, you may be able to see.

The next question is what the matrix is for a general logic unit which consists of a sequence of these. As an example, we'll study the case of the full adder which we described before (see Fig. 5). Now we'll have, in the general case, four wires represented by $a$, $b$, $c$, and $d$; we don't necessarily have to have $d$ as 0 in all cases, and we would like to describe how the object operates in general (if $d$ is changed to 1 $d'$ is changed to its NOT). It produces new numbers $a'$, $b'$, $c'$, and $d'$, and we could imagine with our system that there are four atoms labeled $a, b, c, d$ in a state labeled $|a, b, c, d\rangle$ and that a matrix $M$ operates which changes these same four atoms so that they appear to be in the state $|a', b', c', d'\rangle$ which is appropriate for this logic unit. That is, if $|\psi_{\text{in}}\rangle$ represents the incoming state of the four bits, $M$ is a matrix which generates an outgoing state $|\psi_{\text{out}}\rangle = M|\psi_{\text{in}}\rangle$ for the four bits.

For example, if the input state were the state $|1, 0, 1, 0\rangle$, then, as we know, the output state should be $|1, 0, 0, 1\rangle$; the first two $a'$, $b'$ should be 1, 0 for those two first lines come streight through, and the last two $c'$, $d'$ should be 0, 1 because that represents the sum and carry of the first three, $a$, $b$, $c$, bits in the first input, as $d = 0$. Now the matrix $M$ for the adder can easily be seen as the result of five successive primitive operations, and therefore becomes the matrix product of the five successive matrices representing these primitive objects.

$$M = A_{a,b} A_{b,c} A_{bc,d} A_{a,b} A_{ab,d}$$

The first, which is the one written farthest to the right, is $A_{ab,d}$ for that represents the CONTROLLED CONTROLLED NOT in which $a$ and $b$ are the CONTROL lines, and the NOT appears on line $d$. By looking at the diagram in Fig. 5 we can immediately see what the remaining factors in the sequence represent. The last factor, for example, $A_{a,b}$, means that there's a CONTROLLED NOT with a CONTROL on line $a$ and NOT on line $b$. This matrix will have the unitary property $M^*M=1$ since all of the $A$'s out of which it is a product are unitary. That is to say, $M$ is a reversal operation, and $M^*$ is its inverse.

Our general problem, then, is this. Let $A_1, A_2, A_3$,,,,' $A_k$ be the succession of operations wanted, in some logical unit, to operate on $n$ lines. The $2^n \times 2^n$ matrix $M$ needed to accomplish the same goal is a product $A_k \cdots A_3 A_2 A_1$, where each $A$ is a simple matrix. How can we generate this $M$ in a physical way if we know how to make the simpler elements?

In general, in quantum mechanics, the outgoing state at time $t$ is $e^{iHt}\psi_{\text{in}}$, where $\psi_{\text{in}}$ is the input state, for a system with Hamiltonian $H$. To try to find, for a given special time $t$, the Hamiltonian which will produce $M=e^{iHt}$ when $M$ is such a product of noncommuting matrices, from some simple property of the matrices themselves, appears to be very difficult.

We realize however, that at any particular time, if we expand the $e^{iHt}$ out (as $1+iHt-H^2t^2/2- \cdots$) we'll find the operator $H$ operating an innumerable arbitrary number of times, once, twice, three times, and so forth, and the total state is generated by a superposition of these possibilities. This suggests that we can solve this problem of the composition of these $A$'s in the following way.

We add to the $n$ atoms, which are in our register, an entirely new set of $k+1$ atoms, which we'll call "program counter sites." Let us call $q_i$ and $q_i^*$ the annihilation and creation operators for the program site $i$ for $i=0$ to $k$. A good thing to think of, as an example, is an electron moving from one empty site to another. If the site $i$ is occupied by the electron, its state is $|1\rangle$, while if the site is empty, its state is $|0\rangle$.

We write, as our Hamiltonian

$$H = \sum_{i=0}^{k-1} q_{i+1}^* q_i A_{i+1} + \text{complex conjugate}$$

$$= q_1^* q_0 A_1 + q_2^* q_1 A_2 + q_3^* q_2 A_3 + \cdots + q_0^* q_1 A_1^*$$

$$+ q_1^* q_2 A_2^* + q_2^* q_3 A_3^* + \cdots$$

The first thing to notice is that if all the program sites are unoccupied, that is, all the program atoms are initially in the state 0, nothing happens because every term in the Hamiltonian starts with an annihilation operator and it gives 0 therefore.

The second thing we notice is that if only one or another of the program sites is occupied (in state $|1\rangle$), and the rest are not (state $|0\rangle$), then this is always true. In fact the number of program sites that are in state $|1\rangle$ is a conserved quantity. We will suppose that in the operation of this computer, either no sites are occupied (in which case nothing happens) or just one site is occupied. Two or more program sites are never both occupied during normal operation.

Let us start with an initial state where site 0 is occupied, is in the $|1\rangle$ state, and all the others are empty, $|0\rangle$ state. If later, at some time, the final site $k$ is found to be in the $|1\rangle$ state, (and therefore all the others in $|0\rangle$) then, we claim, the $n$ register has been multiplied by the matrix $M$, which is $A_k \cdots A_2 A_1$ as desired.

Let me explain how this works. Suppose that the register starts in any initial state, $\psi_{\text{in}}$, and that the site, 0, of the program counter is occupied. Then the only term in the entire Hamiltonian that can first operate, as the Hamiltonian operates in successive times, is the first term, $q_1^* q_0 A_1$. The $q_0$ will change site number 0 to an unoccupied site, while $q_1^*$ will change the site number 0 to an occupied site. Thus the term $q_1^* q_0$ is a term which simply moves the occupied site from the location 0 to the location 1. But this is multiplied by the matrix $A_1$ which operates only on the $n$ register atoms, and therefore multiplies the initial state of the $n$ register atoms by $A_1$.

Now, if the Hamiltonian happens to operate a second time, this first term will produce nothing because $q_0$ produces 0 on the number 0 site because it is now unoccupied. The term which can operate now is the second term, $q_2^* q_1 A_2$, for that can move the occupied point, which I shall call a "cursor." The cursor can move from site 1 to site 2 but the matrix $A_2$ now operates on the register; therefore the register has now got the matrix $A_2 A_1$ operating on it.

So, looking at the first line of the Hamiltonian, if that is all there was to it, as the Hamiltonian operates in successive orders, the cursor would move successively from 0 to $k$, and you would acquire, one after the other, operating on the $n$ register atoms, the matrices, $A$, in the order that we would like to construct the total $M$.

However, a Hamiltonian must be hermitian, and therefore the complex conjugate of all these operators must be present. Suppose that at a given stage, we have gotten the cursor on site number 2, and we have the matrix $A_2 A_1$ operating on the register. Now the $q_2$ which intends to move that occupation to a new position need not come from the first line, but may have come from the second line. It may have come, in fact, from

$q_1^* q_2 A_2^*$ which would move the cursor back from the position 2 to the position 1.

But note that when this happens, the operator $A_2^*$ operates on the register, and therefore the total operator on the register is $A_2^* A_2 A_1$ in this case. But $A_2^* A_2$ is 1 and therefore the operator is just $A_1$. Thus we see that when the cursor is returned to the position 1, the net result is that only the operator $A_1$ has really operated on the register. Thus it is that as the various terms of the Hamiltonian move the cursor forwards and backwards, the $A$'s accumulate, or are reduced out again.

At any stage, for example, if the cursor were up to the $j$ site, the matrices from $A_1$ to $A_j$ have operated in succession on the $n$ register. It does not matter whether or not the cursor on the $j$ site has arrived there, by going directly from 0 to $j$, or going further and returning, or going back and forth in any pattern whatsoever, as long as it finally arrived at the state $j$.

Therefore it is true that if the cursor is found at the site $k$, we have the net result for the $n$ register atoms that the matrix $M$ has operated on their initial state as we desired.

How then could we operate this computer? We begin by putting the input bits onto the register, and by putting the cursor to occupy the site 0. We then check at the site $k$, say, by scattering electrons, that the site $k$ is empty, or that the site $k$ has a cursor. The moment we find the cursor at site $k$ we remove the cursor so that it cannot return down the program line, and then we know that the register contains the output data. We can then measure it at our leisure. Of course, there are external things involved in making the measurements, and determining all of this, which are not part of our computer. Surely a computer has eventually to be in interaction with the external world, both for putting data in and for taking it out.

Mathematically it turns out that the propagation of the cursor up and down this program line is exactly the same as it would be if the operators $A$ were not in the Hamiltonian. In other words, it represents just the waves which are familiar from the propagation of the tight binding electrons or spin waves in one dimension, and are very well known. There are waves that travel up and down the line and you can have packets of waves and so forth.

We could improve the action of this computer and make it into a ballistic action in the following way: by making a line of sites in addition to the ones inside, that we are actually using for computing, a line of say, many sites, both before and after. It's just as though we had values of the index $i$ for $q_i$, which are less than 0 and greater than $k$, each of which has no matrix $A$, just a 1 multiplying there. Then we had have a longer spin chain, and we could have started, instead of putting a cursor exactly at the beginning site 0, by putting the cursor with different amplitudes on different sites representing an initial incoming spin wave, a wide packet of nearly definite momentum.

This spin wave would then go through the entire computer in a ballistic fashion and out the other end into the outside tail that we have added to the line of program sites, and there it would be easier to determine if it is present and to steer it away to some other place, and to capture the cursor. Thus, the logical unit can act in a ballistic way.

This is the essential point and indicates, at least to a computer scientist, that we could make a universal computer, because he knows if we can make any logical unit we can make a universal computer. That this could represent a universal computer for which composition of elements and branching can be done is not entirely obvious unless you have some experience, but I will discuss that to some further extent later.

## 4. IMPERFECTIONS AND IRREVERSIBLE FREE ENERGY LOSS

There are, however, a number of questions that we would like to discuss in more detail such as the question of imperfections.

There are many sources of imperfections in this machine, but the first one we would like to consider is the possibility that the coefficients in the couplings, along the program line, are not exactly equal. The line is so long that in a real calculation little irregularities would produce a small probability of scattering, and the waves would not travel exactly ballistically, but would go back and forth.

If the system, for example, is built so that these sites are built on a substrate of ordinary physical atoms, then the thermal vibrations of these atoms would change the couplings a little bit and generate imperfections. (We should even need such noise for with small fixed imperfections there are shallow trapping regions where the cursor may get caught.) Suppose then, that there is a certain probability, say $p$ per step of calculation (that is, per step of cursor motion, $i \to i+1$), for scattering the cursor momentum until it is randomized ($1/p$ is the transport mean free path). We will suppose that the $p$ is fairly small.

Then in a very long calculation, it might take a very long time for the wave to make its way out the other end, once started at the beginning —because it has to go back and forth so many times due to the scattering. What one then could do would be to pull the cursor along the program line with an external force. If the cursor is, for example, an electron moving from one vacant site to another, this would be just like an electric field trying to pull the electron along a wire, the resistance of which is generated by the imperfection or the probability of scattering. Under these circumstances we can calculate how much energy will be expended by this external force.

This analysis can be made very simply: it is an almost classical analysis of an electron with a mean free path. Every time the cursor is scattered, I am going to suppose it is randomly scattered forward and backward. In order for the machine to operate, of course, it must be moving forward at a higher probability than it is moving backward. When a scattering occurs therefore, the loss in entropy is the logarithm of the probability that the cursor is moving forward, divided by the probability the cursor was moving backward.

This can be approximated by (the probability forward − the probability backward)/(the probability forward + the probability backward). That was the entropy lost per scattering. More interesting is the entropy lost per net calculational step, which is, of course, simply $p$ times that number. We can rewrite the entropy cost per calculational step as

$$pv_D/v_R$$

where $v_D$ is the drift velocity of the cursor and $v_R$ its random velocity.

Or if you like, it is $p$ times the minimum time that the calculation could be done in (that is, if all the steps were always in the forward direction), divided by the actual time allowed.

The free energy loss per step then, is $kT \times p \times$ the minimum time that the calculation could be done, divided by the actual time that you allow yourself to do it. This is a formula that was first derived by Bennett. The factor $p$ is a coasting factor, to represent situations in which not every site scatters the cursor randomly, but it has only a small probability to be thus scattered.

It will be appreciated that the energy loss per step is not $kT$ but is that divided by two factors. One, $(1/p)$, measures how perfectly you can build the machine and the other is proportional to the length of time that you take to do the calculation. It is very much like a Carnot engine, in which in order to obtain reversibility, one must operate very slowly. For the ideal machine where $p$ is 0, or where you allow an infinite time, the mean energy loss can be 0.

The uncertainty principle, which usually relates some energy and time uncertainty, is not directly a limitation. What we have in our computer is a device for making a computation, but the time of arrival of the cursor and the measurement of the output register at the other end (in other words, the time it takes in which to complete the calculation) is not a define time. It's a question of probabilities, and so there is a considerable uncertainty in the time at which a calculation will be done.

There is no loss associated with the uncertainty of cursor energy; at least no loss depending on the number of calculational steps. Of course, if you want to do a ballistic calculation on a perfect machine, some energy would have to be put into the original wave, but that energy, of course, can be removed from the final wave when it comes out of the tail of the program line. All questions associated with the uncertainty of operators and the irreversibility of measurements are associated with the input and output functions.

No further limitations are generated by the quantum nature of the computer *per se*, nothing that is proportional to the number of computational steps.

In a machine such as this, there are very many other problems, due to imperfections. For example, in the registers for holding the data, there will be problems of cross-talk, interactions between one atom and another in that register, or interaction of the atoms in that register directly with things that are happening along the program line, that we did not exactly bargain for. In other words, there may be small terms in the Hamiltonian besides the ones we have written.

Until we propose a complete implementation of this, it is very difficult to analyze. At least some of these problems can be remedied in the usual way by techniques such as error correcting codes, and so forth, that have been studied in normal computers. But until we find a specific implementation for this computer, I do not know how to proceed to analyze these effects. However, it appears that they would be very important, in practice. This computer seems to be very delicate and these imperfections may produce considerable havoc.

The time needed to make a step of calculation depends on the strength or the energy of the interactions in the terms of the Hamiltonian. If each of the terms in the Hamiltonian is supposed to be of the order of 0.1 electron volts, then it appears that the time for the cursor to make each step, if done in a ballistic fashion, is of the order $6 \times 10^{-15}$ sec. This does not represent

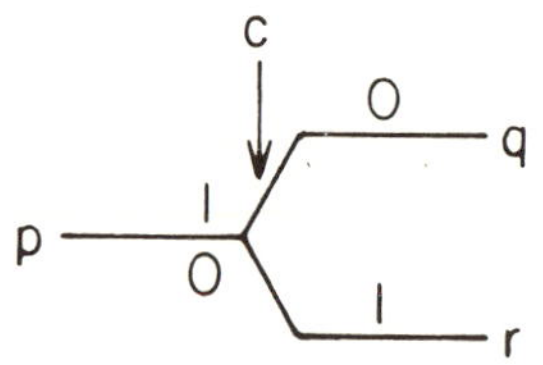


$$H = q^* c p + r^* c^* p + p^* c^* q + p^* c r$$

IF c = 1 GO p TO q AND PUT c = 0
IF c = 0 GO p TO r AND PUT c = 1
IF c = 1 GO r TO p AND PUT c = 0
IF c = 0 GO q TO p AND PUT c = 1

Fig. 7. Switch.

an enormous improvement, perhaps only about four orders of magnitude over the present values of the time delays in transistors, and is not much shorter than the very short times possible to achieve in many optical systems.

## 5. SIMPLIFYING THE IMPLEMENTATION

We have completed the job we set out to do—to find some quantum mechanical Hamiltonian of a system that could compute, and that is all that we need say. But it is of some interest to deal with some questions about simplifying the implementation. The Hamiltonian that we have written involves terms which can involve a special kind of interaction between five atoms. For example, three of them in the register, for a CONTROLLED CONTROLLED NOT, and two of them as the two adjacent sites in the program counter.

This may be rather complicated to arrange. The question is, can we do it with simpler parts. It turns out that we can indeed. We can do it so that in each interaction there are only three atoms. We are going to start with new primitive elements, instead of the ones we began with. We'll have the NOT all right, but we have in addition to that simply a "switch" (see also Priese[5]).

Supposing that we have a term, $q^*cp + r^*c^*p +$ its complex conjugate in the Hamiltonian (in all cases we'll use letters in the earlier part of the alphabet for register atoms and in the latter part of the alphabet for program sites). See Fig. 7. This is a switch in the sense that, if $c$ is originally in the $|1\rangle$ state, a cursor at $p$ will move to $q$, whereas if $c$ is in the $|0\rangle$ state, the cursor at $p$ will move to $r$.

During this operation the controlling atom $c$ changes its state. (It is possible also to write an expression in which the control atom does not change its state, such as $q^*c^*cp + r^*cc^*p$ and its complex conjugate but, there is no particular advantage or disadvantage to this, and we will take the simpler form.) The complex conjugate reverses this.

If, however, the cursor is at $q$ and $c$ is in the state $|1\rangle$ (or cursor at $r$, $c$ in $|0\rangle$), the $H$ gives 0, and the cursor gets reflected back. We shall build all our circuits and choose initial states so that this circumstance will not arise in normal operation, and the ideal ballistic mode will work.

With this switch we can do a number of things. For example, we could produce a CONTROLLED NOT as in Fig. 8. The switch $a$ controls the operation. Assume the cursor starts at $s$. If $a = 1$ the program cursor is carried along the top line, whereas if $a = 0$ it is carried along the bottom line, in either case terminating finally in the program site $t$.

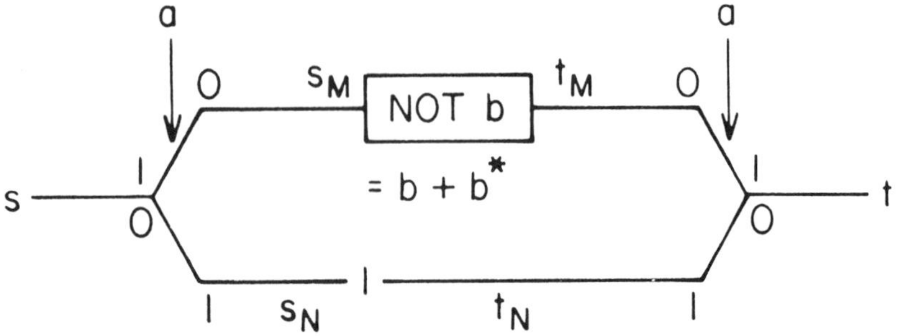


Fig. 8. CONTROLLED NOT by switches.

In these diagrams, horizontal or vertical lines will represent program atoms. The switches are represented by diagonal lines and in boxes we'll put the other matrices that operate on registers such as the NOT $b$. To be specific, the Hamiltonian for this little section of a CONTROLLED NOT, thinking of it as starting at $s$ and ending at $t$, is given below:

$$H_c(s, t) = s_M^* as + t^* a^* t_M + t_M^*(b + b^*)\, s_M + s_N^* a^* s$$
$$+ t^* a t_N + t_N^* s_N + \text{c.c}$$

(The c.c means to add the complex conjugate of all the previous terms.)

Although there seem to be two routes here which would possibly produce all kinds of complications characteristic of quantum mechanics, this is not so. If the entire computer system is started in a definite state for $a$ by the time the cursor reaches $s$, the atom $a$ is still in some definite state (although possibly different from its initial state due to previous computer operations on it). Thus only one of the two routes is taken. The expression may be simplified by omitting the $s_N^* t_N$ term and putting $t_N = s_N$.

One need not be concerned in that case, that one route is longer (two cursor sites) than the other (one cursor site) for again there is no interference. No scattering is produced in any case by the insertion into a chain of coupled sites, an extra piece of chain of any number of sites with the same mutual coupling between sites (analogous to matching impedances in transmission lines).

To study these things further, we think of putting pieces together. A piece (see Fig. 9) $M$ might be represented as a logical unit of interacting parts in which we only represent the first input cursor site as $s_M$ and the final one at the other end as $t_M$. All the rest of the program sites that are between $s_M$ and $t_M$ are considered internal parts of $M$, and $M$ contains its registers. Only $s_M$ and $t_M$ are sites that may be coupled externally.

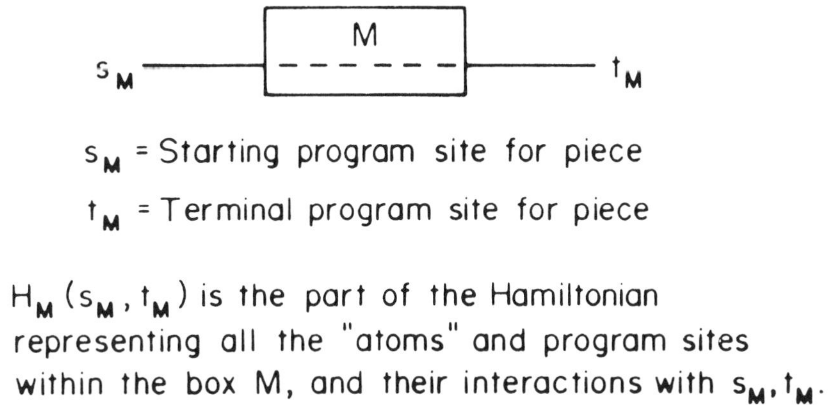


Fig. 9. One "piece."

The Hamiltonian for this subsection we'll call $H_M$ and we'll identify $s_M$ and $t_M$, as the name of the input and output program sites by writing $H_M(s_M, t_M)$. So therefore $H_M$ is that part of the Hamiltonian representing all the atoms in the box and their external start and terminator sites.

An especially important and interesting case to consider is when the input data (in the regular atoms) comes from one logical unit, and we would like to transfer it to another (see Fig. 10). Suppose that we imagine that the box $M$ starts with its input register with 0 and its output (which may be the same register) also with 0. Then we could use it in the following way. We could make a program line, let's say starting with $s'_M$ whose first job is to exchange the data in an external register which contains the input, with $M$'s input register which at the present time contains 0's.

Then the first step in our calculation, starting, say, at $s'_M$, would be to make an exchange with the register inside of $M$. That puts zero's into the original input register and puts the input where it belongs inside the box $M$. The cursor is now at $s_M$. (We have already explained how exchange can be made of controlled NOTs.) Then as the program goes from $s_M$ to $t_M$ we find the output now in the box $M$. Then the output register of $M$ is now cleared as we write the results into some new external register provided for that purpose, originally containing 0's. This we do from $t_M$ to $t'_M$ by exchanging data in the empty external register with the $M$'s output register.

We can now consider connecting such units in different ways. For example, the most obvious way is succession. If we want to do first $M$ and then $N$ we can connect the terminal side of one to the starting side of the other as in Fig. 11, to produce a new effective operator $K$, and the Hamiltonian then for $H_K$ is

$$H_K(s_K, t_K) = H_M(s_K, t) + H_N(t, t_K)$$

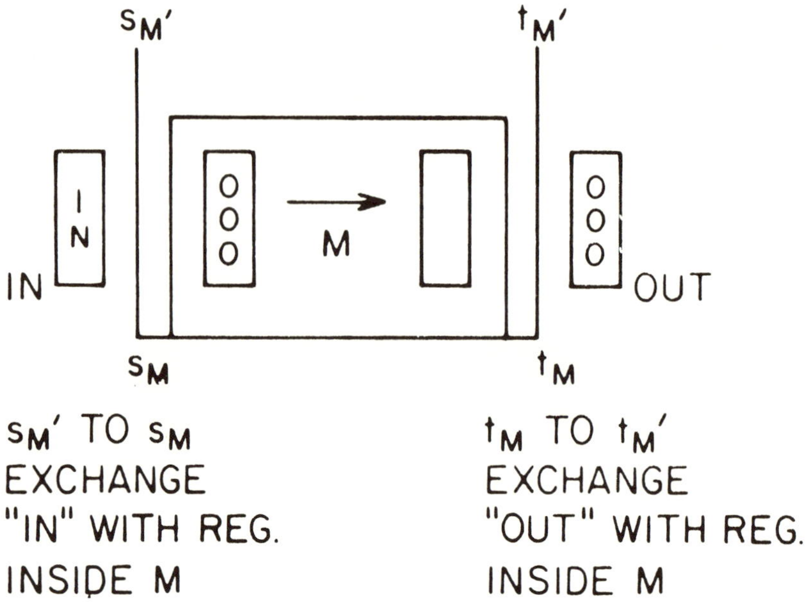


Fig. 10. Piece with external input and output.

The general conditional, if $a=1$ do $M$, but if $a=0$ do $N$, can be made, as in Fig. 12. For this

$$H_{\text{cond}}(s_c, t_c) = (s_M^* a s_c + t_c^* a^* t_M + s_N^* a^* s_c + t_c^* a t_N + \text{c.c.}) + H_M(s_M, t_M) + H_N(s_N, t_N)$$

The CONTROLLED NOT is the special case of this with $M = \text{NOT}\, b$ for which $H$ is

$$H_{\text{NOT}b}(s, t) = s^*(b + b^*)\, t + \text{c.c.}$$

and $N$ is no operation $s^*t$.

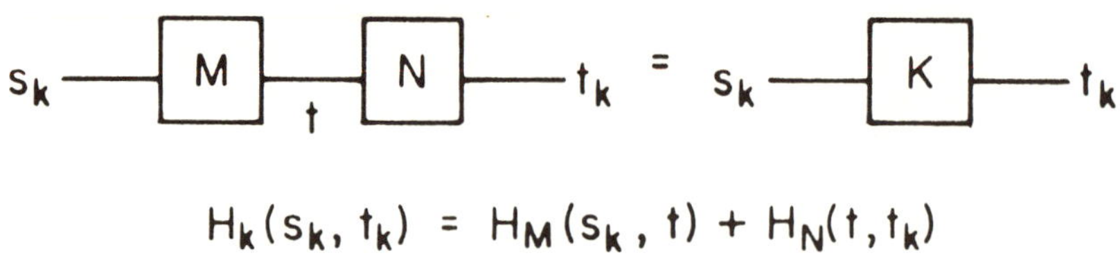


Fig. 11. Operations in succession.

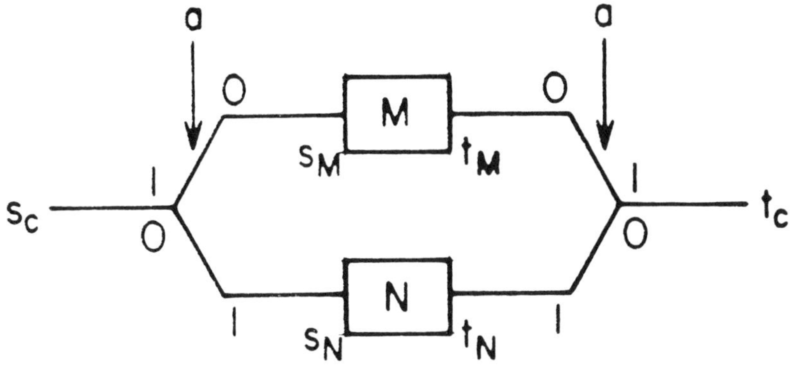


Fig. 12. Conditional if $a = 1$ then $M$, else $N$.

As another example, we can deal with a garbage clearer (previously described in Fig. 6) not by making two machines, a machine and its inverse, but by using the same machine and then sending the data back to the machine in the opposite direction, using our switch (see Fig. 13).

Suppose in this system we have a special flag which is originally always set to 0. We also suppose we have the input data in an external register, an empty external register available to hold the output, and the machine registers all empty (containing 0's). We come on the starting line $s$.

The first thing we do is to copy (using CONTROLLED NOT's) our external input into $M$. Then $M$ operates, and the cursor goes on the top line in our drawing. It copies the output out of $M$ into the external output register. $M$ now contains garbage. Next it changes $f$ to NOT $f$, comes down on the other line of the switch, backs out through $M$ clearing the garbage, and uncopies the input again.

When you copy data and do it again, you reduce one of the registers to 0, the register into which you coied the first time. After the coying, it goes out (since $f$ is now changed) on the other line where we restore $f$ to 0

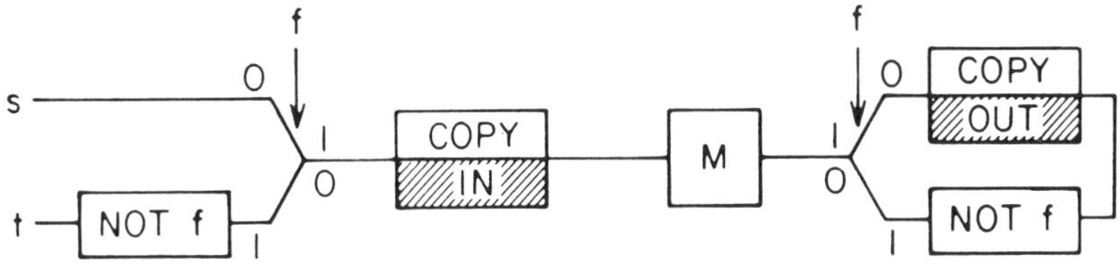


Fig. 13. Garbage clearer.

and come out at $t$. So between $s$ and $t$ we have a new piece of equipment, which has the following properties.

When its starts, we have, in a register called IN, the input data. In an external register which we call OUT, we have 0's. There is an internal flag set at 0, and the box, $M$, is empty of all data. At the termination of this, at $t$, the input register still contains the input data, and the output register contains the output of the effort of the operator $M$. $M$, however, is still empty, and the flag $f$ is reset to 0.

Also important in computer programs is the ability to use the same subroutine several times. Of course, from a logical point of view, that can be done by writing that bit of program over and over again, each time it is to be used, but in a practical computer, it is much better if we could build that section of the computer which does a particular operation, just once, and use that section again and again.

To show the possibilities, here, first just suppose we have an operation we simply wish to repeat twice in succession (see Fig. 14). We start at $s$ with the flag $a$ in the condition 0, and thus we come along the line, and the first thing that happens is we change the value of $a$. Next we do the operation $M$. Now, because we changed $a$, instead of coming out at the top line where we went in, we come out at the bottom line, which recirculates the program back into changing $a$ again; it restores it.

This time as we go through $M$, we come out and we have the $a$ to follow on the uper line, and thus come out at the terminal, $t$. The Hamiltonian for this is

$$H_{MM}(s, t) = (s_N^* a^* s + s_M^*(a^* + a)\, s_N + x^* a^* t_M + s_N^* a x + t^* a t_M + \text{c.c.}) + H_M(s_M, t_M)$$

Using this switching circuit a number of times, of course, we can repeat an operation several times. For example, using the same idea three

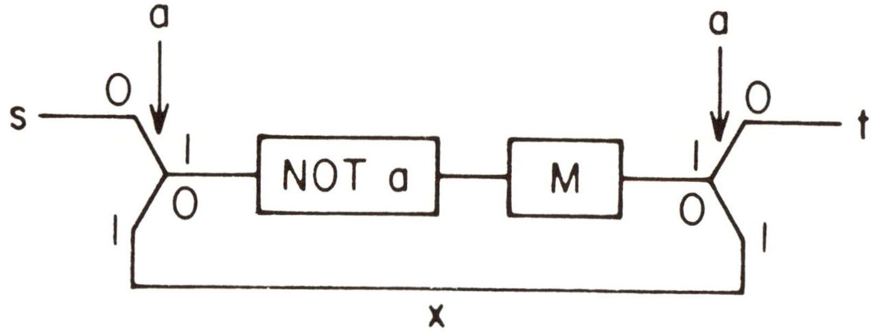


Fig. 14. Do $M$ twice.

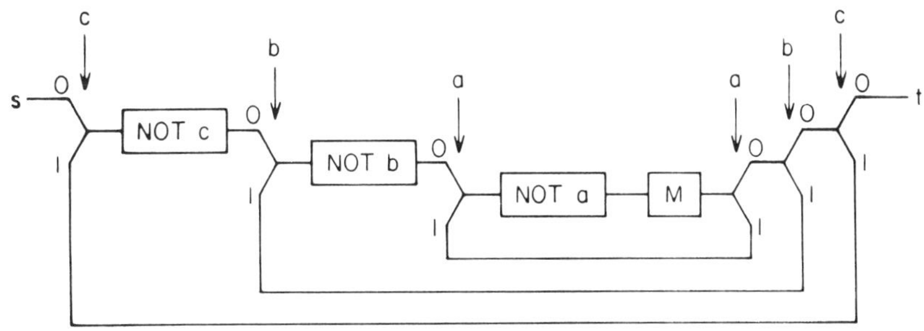


Fig. 15. Do $M$ eight times.

times in succession, a nested succession, we can do an operation eight times, by the apparatus indicated in Fig. 15. In order to do so, we have three flags, $a$, $b$, and $c$. It is necessary to have flags when operations are done again for the reason that we must keep track of how many times its done and where we are in the program or we'll never be able to reverse things.

A subroutine in a normal computer can be used and emptied and used again without any record being kept of what happened. But here we have to keep a record and we do that with flags, of exactly where we are in the cycle of the use of the subroutine. If the subroutine is called from a certain place and has to go back to some other place, and another time is called, its origin and final destination are different, we have to know and keep track of where it came from and where it's supposed to go individually in each case, so more data have to be kept. Using a subroutine over and over in a reversible machine is only slightly harder than in a general machine. All these considerations appear in papers by Fredkin, Toffoli, and Bennett.

It is clear by the use of this switch, and successive uses of such switches in trees, that we would be able to steer data to any oint in a memory. A memory would simply be a place where there are registers into which you could copy data and then return the program. The cursor will

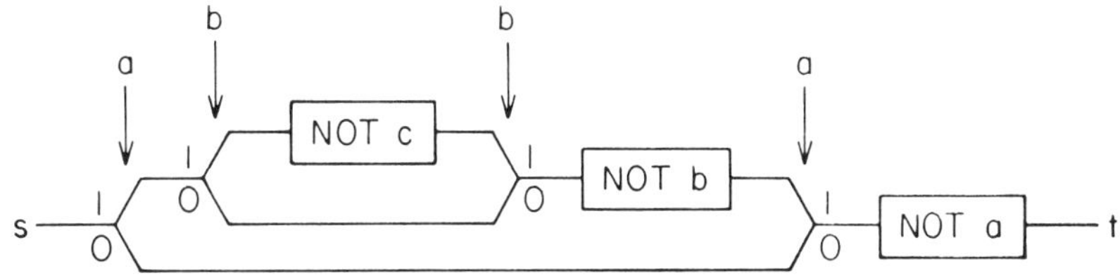


Fig. 16. Increment counter (3-bit).

have to follow the data along. I suppose there must be another set of tree switches set the oposite direction to carry the cursor out again, after copying the data so that the system remains reversible.

In Fig. 16 we show an incremental binary counter (of three bits $a$, $b$, $c$ with $c$ the most significant bit) which keeps track of how many net times the cursor has passed from $s$ to $t$. These few examples should be enough to show that indeed we can construct all computer functions with our SWITCH and NOT. We need not follows this in more detail.

## 6. CONCLUSIONS

It's clear from these examples that this quentum machine has not really used many of the specific qualities of the differential equations of quantum mechanics.

What we have done is only to try to imitate as closely as possible the digital machine of conventional sequential architecture. It is analogous to the use of transistors in conventional machines, where we do not properly use all the analog continuum of the behavior of transistors, but just try to run them as saturated on or off digital devices so the logical analysis of the system behavior is easier. Furthermore, the system is absolutely sequential—for example, even in the comparison (exclusive or) of two $k$ bit numbers, we must do each bit successively. What can be done, in these reversible quantum systems, to gain the speed available by concurrent operation has not been studied here.

Although, for theoretical and academic reasons, I have studied complete and reversible systems, if such tiny machines could become practical there is no reason why irreversible and entropy creating interactions cannot be made frequently during the course of operations of the machine.

For example, it might prove wise, in a long calculation, to ensure that the cursor has surely reached some oint and cannot be allowed to reverse again from there. Or, it may be found practical to connect irreversible memory storage (for items less frequently used) to reversible logic or short-term reversible storage registers, etc. Again, there is no reason we need to stick to chains of coupled sites for more distant communication where wires or light may be easier and faster.

At any rate, it seems that the laws of physics present no barrier to reducing the size of computers until bits are the size of atoms, and quantum behavior holds dominant sway.

## REFERENCES[3]

1. C. H. Bennett, "Logical Reversibility of Computation," *IBM J. Res. Dev.* **6**, 525–532 (1979).
2. E. Fredkin and T. Toffoli, "Conservative Logic," *Int. J. Theor. Phys.* **21**, 219–253 (1982).
3. C. H. Bennett, "Thermodynamics of Computation—A Review," *Int. J. Theor. Phys.* **21**, 905–940 (1982).
4. T. Toffoli, "Bicontinuous Extensions of Invertible Combinatorial Functions," *Math. Syst. Theory* **14**, 13–23 (1981).
5. L. Priese, "On a Simple Combinatorial Structure Sufficient for Sublying Nontrivial Self-Reproduction," *J. Cybern.* **6**, 101–137 (1976).

[3] I would like to thank T. Toffoli for his help with the references.

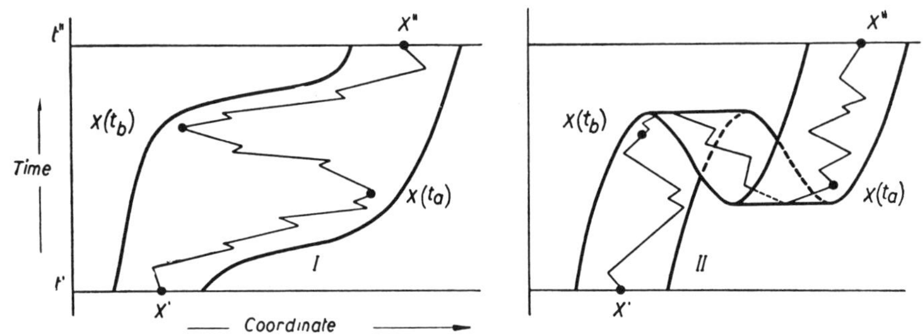


*Plate 36*

# Computability and Physical Theories

**Robert Geroch**[1] **and James B. Hartle**[2]

*Received August 22, 1985*

*The familiar theories of physics have the feature that the application of the theory to make predictions in specific circumstances can be done by means of an algorithm. We propose a more precise formulation of this feature—one based on the issue of whether or not the physically measurable numbers predicted by the theory are computable in the mathematical sense. Applying this formulation to one approach to a quantum theory of gravity, there are found indications that there may exist no such algorithms in this case. Finally, we discuss the issue of whether the existence of an algorithm to implement a theory should be adopted as a criterion for acceptable physical theories.*

"Can it then be that there is... something of use for unraveling the universe to be learned from the philosophy of computer design?"

—J. A. Wheeler[(1)]

## 1. INTRODUCTION

The process of making predictions in physics involves two steps. The first is the formulation of a theory covering a broad class of phenomena; the second, the application of that theory to the specific circumstances of interest. For making predictions of the motion of the planets of our solar system, for example, the first step is the formulation of Newton's theories of mechanics and gravitation; the second the solution of a system of many dozens of ordinary differential equations. These two steps are rather different in character. The first involves the selection, from a wide variety of phenomena, of those one judges to have a common cause, and then the for-

[1] Enrico Fermi Institute, Chicago, Illinois 60637.

[2] Department of Physics, University of California, Santa Barbara, California 93106.

mulation of a mechanism for how that cause is to operate. The test of a good theory is its simplicity and breadth. The second step involves the selection of a specific algorithm by which to extract from the theory its implications. The test of a good algorithm is its convenience.

What underlies our confidence that physics can be structured in this way? Perhaps it is primarily our experience with present-day physical theories. In both classical and quantum physics, the theory culminates in the formulation of differential equations, while the implementation of the theory consists of selecting some algorithm—of which there are many—for solving the equations. But not all theories of physics need be of this type. We may someday in physics be confronted with a situation in which the line between the theory and its implementation is not so sharp—in which the activity of applying the theory is not so different from that of finding the theory in the first place. Specifically, we may reach a point at which there exist no algorithms whatever for applying a theory mechanically to specific circumstances. Indeed, there are suggestions that, with quantum gravity, we may already be at this point.

John Wheeler has stressed on several occasions[1,2] the relation between physical laws, computation, and logic. Here, on the occasion of his 75th birthday, we discuss certain possible formulations of, and attitudes toward, these issues. Section 2 is a review of mathematics. We briefly summarize the notion of solvability, by computer, of a mathematical problem, and the closely related notion of a computable number. Section 3 involves physics. We propose, in parallel with the notion of a computable number in mathematics, that of a measurable number in a physical theory. The question of whether there exists an algorithm for implementing a theory may then be formulated more precisely as the question of whether the measurable numbers of the theory are computable. We argue that the measurable numbers are in fact computable in the familiar theories of physics, but there is no reason why this need be the case in order that a theory have predictive power. Indeed, in some recent formulations of quantum gravity as a sum over histories, there are candidates for numbers that are measurable but not computable. In Sec. 4, we speculate as to the consequences for physics should there be accepted a theory having measurable numbers that are not computable. We suggest that such a theory should be no more unsettling to physics than has the existence of well-posed problems unsolvable by any algorithm been to mathematics.

But, in the spirit of many of John Wheeler's papers, we aim more to raise issues and stimulate discussion than to state conclusions.

## 2. COMPUTABILITY

Computability theory deals with the issue of whether certain well-posed mathematical problems can be solved by means of a digital computer. The key result is a theorem to the effect that one particular problem cannot be solved in this way. From this theorem, one shows unsolvability of a variety of other problems, including that of finding rational approximations to certain real numbers. These are called the noncomputable numbers. We briefly summarize this subject below.[3]

We begin with a few ground rules. First, we wish to deal only with digital computers, and not with analog, for we are concerned for the moment only with mathematics, and not with physics. We are interested only in what is possible in principle with such computers, and not in such issues as efficiency or speed. Fix some finite character set, e.g., the standard one of 256 characters. By a *string*, we mean a finite sequence of characters. Thus, while each string is itself finite, there is no overall upper limit to the allowed lengths of strings. The computer is to have access to an infinite number of memory locations, each capable of holding a string. All these locations are to be blank initially, in order to avoid a situation in which "the answer to the problem" is already encoded, from the beginning, in the memory. The language of the computer is to consist of the specification of certain allowed strings, called the instructions in that language. Of course, not every possible string need be an instruction. A program in the language consists of a finite sequence of instructions. We may represent a program by placing its instructions one after another to form a single long string. Again, not every possible string need be a program. Finally, we demand of the language that it include at least the following three instructions: (i) an input instruction, which temporarily interrupts the running of the program while the user reads in, for possible later use in the program, any string, (ii) a print instruction, which causes the computer to display to the user any string produced by the program, and (iii) a halt instruction, which causes the program to cease execution.

The above is an outline of the types of languages we have in mind. In practice, there are numerous specific languages in which computer programs are commonly written, e.g., Fortran, Basic, Pascal, Cobol, and various machine languages. One might think that whether a problem can be solved by a computer program will depend on the language in which that program is to be written. But, at least for these specific languages, this is not the case: A problem that can be solved in any one of these languages

[3] For a more detailed and rigorous treatment of this material, see, for example, Ref. 3.

can be solved in any other. Consider, for example, Pascal and Basic. One could write a program in Basic that "imitates Pascal", i.e., that takes a sequence of Pascal instructions, determines for each what would be done in Pascal, and then does precisely that. Any problem solvable in Pascal is thus solvable in Basic, and similarly for other combinations of these standard languages.

Since, in the sense above, the language is irrelevant, let us fix once and for all a language, and call it "Fortran," with the understanding that this word can be regarded as standing for any of the others. What problems can be "solved" by a Fortran program? The idea is to specify in some detail what one wants one's program to do, and then to determine whether or not there exists such a program.

Consider, as an example, Fermat's conjecture: that there exist no positive integers $x$, $y$, $z$, and $n$, with $n \geqslant 3$, such that $x^n + y^n = z^n$. Does there exist a Fortran program that prints "true" if Fermat's conjecture is true, or "false" if it is false, and then halts? The answer to the question, stated in this way, is yes. Indeed, consider the program, with just two instructions, that merely prints "true" and halts; and a second program that prints "false" and halts. Since Fermat's conjecture is either true or false, one of these is the desired program. So, there does indeed *exist* a program that does what we have asked. Similarly, any well-posed "yes-or-no" problem is solvable. One might be tempted to reformulate the question so that its answer requires the specification of *which* is the desired program. But it turns out to be difficult to do this in a precise way.

Clearly, the more interesting problems are those that are actually a sequence of problems. For example: Does there exist a program having a single input instruction such that, when any integer $n$ is read in, the program prints "prime" or "not prime" according to whether that integer is prime or not, and then halts? The answer of course is yes: Most computer users have written such a program. So, in this sense, the problem of determining whether or not a number is prime is solvable. Similarly, there exists a program that continues to print out the successive digits in the decimal expansion of $\pi$, never halting.

As a final example, we return to Fermat's conjecture, but this time with a different question. Does there exist a Fortran program having a single input instruction such that, when any integer $n \geqslant 3$ is read in, the program prints "yes" or "no" according to whether or not there are positive integers $x$, $y$, and $z$ with $x^n + y^n = z^n$, and then halts? Since Fermat's conjecture is open, we certainly are not in possession of such a program. [It would not do, for example, to have one's program merely try various combinations of $x$, $y$, and $z$, for then, were $n$ such that there are no suitable combinations, the program would never halt.] Indeed, it is even unclear whether or not there *exists* such a program.[4] There does, for example, if Fermat's conjecture is true (namely, the program that always prints "no" and halts), or if the conjecture is false for all but a finite number of $n$ (namely, the program that compares the $n$ read in with this finite number of candidates). But what if the $n$'s for which the theorem fails are more "scattered"?

The key theorem in this subject displays an explicit example of a problem that cannot be solved by any program. It is convenient, for this discussion, to introduce the term *monoprogram* to mean a program having a single input instruction. The problem is to determine, given any monoprogram and the string we intend to read in for its input, whether that monoprogram will ultimately halt. In more detail, consider a program $X$, having just two input instructions and operating as follows. Let, for the first input instruction, there be read in the string representing a monoprogram $P$, and for the second, an arbitrary string $S$. Then $X$ itself must always halt, having printed "halts" or "does not halt" according to whether or not the monoprogram $P$, with string $S$ read in for *its* input, halts or not. [Should the first string read in to $X$ not represent any monoprogram, then $X$ can do anything.] It is instructive to attempt to write, say, for Fortran, such a program $X$. One first thinks of various simple arrangements by which $P$, with $S$ read in, might halt, and so writes $X$ to look for those arrangements. But one continually discovers ever more complicated ways by which $P$ might halt, and so the program $X$ becomes more and more complicated. Eventually, with no end in sight, one gives up in frustration.

There exists no such program $X$. The proof is simple, but subtle. Suppose, for contradiction, that there were such an $X$. Construct from it the following monoprogram $P$. When string $S$ is read in for its input, $P$ merely runs program $X$, reading in string $S$ for *both* of $X$'s inputs. Then $P$ waits patiently for $X$ to finish. If $X$ would have printed "halts," then $P$ goes into an infinite loop, never halting; while if $X$ would have printed "does not halt," then $P$ halts. Thus, given program $X$ one could, by merely changing a few of its instructions as indicated above, write this program $P$. Now run program $P$, but, for its input string, read in the string representing the program $P$ itself. Will $P$ so run halt or not halt? Either answer gives a contradiction. Suppose, for example, that $P$, with the string representing $P$ read in, halts. This means, from the way $P$ was constructed, that $X$, with the string representing $P$ read in for both its inputs, will print "does not halt." But this in turn means, from the way $X$ is supposed to operate, that

[4] For purposes of this illustration, we ignore the recent result that Fermat's conjecture holds for all but at most a finite member of $n$. See Ref. 4.

$P$, with the string representing $P$ read in, does not halt—a contradiction. Similarly for the alternative supposition. This contradiction establishes in turn that the desired program $X$ does not exist. This result is called the *halting theorem.*

Note that the halting theorem asserts only that there is no general algorithm (i.e., no program $X$) for deciding whether or not programs halt. There need not be any specific program such that the issue of whether or not *that* program halts is undecidable. Indeed, in the formal sense there cannot be, for whether a specific program halts is a "yes-or-no" question. Imagine that one's job consists of receiving from one's employer a monoprogram and a string, and then trying to decide whether or not that program, with that string read in, halts. The first few cases turn out to be quite easy. Then one day one receives a more difficult case; but, after a month's work, devises a most clever solution. This new method works for the next few cases. But eventually one will receive a case requiring a new, still more clever, solution. The halting theorem states in essence that this job will never become routine— that new and ever more clever solutions will always be required.

It is of interest to see the role played in the halting theorem by the language being used. Clearly, the particular language—provided only that it has the three basic instructions—is irrelevant. But it is crucial that the program $X$ be written in the *same* language as are the programs $P$ that $X$ tests. Consider, for example, the simplest possible language—that having just the three instructions "input," "print," and "halt." Then the halting theorem in this case asserts that no program $X$ written in this simple language can solve the halting problem for programs written in this same language. One sees directly that this is true, for programs in this simple language cannot "branch." Suppose, however, that one allows the program $X$ to be written in Fortran. There does exist a Fortran program $X$ that will solve the halting problem for programs $P$ written in our simple language. [Program $X$ merely checks whether or not program $P$ includes a "halt" instruction.] But, by the halting theorem, no Fortran $X$ will solve the halting problem for Fortran $P$'s. Next, imagine a language that is essentially Fortran, but having an instruction of the form "Do, for $I=1$ to $\infty$," with appropriate modifications to incorporate this into the rest of the language. One can write a program $X$ in this "super-Fortran" that solves the halting problem for ordinary Fortran programs $P$. [Program $X$ basically asks whether $P$ has halted after $I$ steps, for "$I=1$ to $\infty$."] But again no such $X$ will solve the halting problem for programs $P$ also written in "super-Fortran."

While one can imagine the action of "super-Fortran" language, it is much more difficult to see how one might build a physical computer that actually runs programs in this language. To execute "Do, for $I = 1$ to $\infty$" in the traditional way would require performing an infinite number of steps in finite time. Indeed, to go one step further, it is difficult to see how *any* language that could actually be run on a physical computer could do more than Fortran can do. The idea that there is no such language is called Church's thesis.[5] Thus, under Church's thesis, there is a universal notion of an "algorithm"—of "the most that can be done on a physical computer"—and this notion is correctly captured by standard computer languages. The role of the thesis, then, is to give technical results of the form "... cannot be solved by a Fortran program" the broader meaning "... cannot be solved by any algorithm."

Since we shall shortly be concerned with physics, it is convenient to reexpress the above in terms of the objects with which physics deals: numbers.

A real number $w$ is said to be *computable* if there exists a (Fortran) program with a single input instruction such that, when any positive rational number $\varepsilon$ is read in, the program prints some rational number $r$ with $|w - r| < \varepsilon$ and then halts. Thus, a number is computable if there is an algorithm that generates rational approximations to the number. [Rationals are used in the definition because they can be "finitely displayed."] So, for example, the number $\pi$ is computable. Indeed, it is easy to write out an appropriate program in detail, using, e.g., a power-series expansion for $\pi$. Similarly, $\pi^e \sin \sqrt{2}$—and, in fact, any other number one can think of offhand—is computable.

That there must exist noncomputable numbers is easily seen, for the cardinality of the set of Fortran programs (that of the integers) is smaller than the cardinality of the set of real numbers. In fact, one can give an "explicit example" of such a number. Fix once and for all an ordering (first, second, etc.) for pairs consisting of a monoprogram and a string. [One might, for example, first fix an ordering for the (finite) character set. Then apply dictionary ordering to those pairs for which the program and string combined have just one character, then to those having just two, etc.] Now consider the following number:

$$K = \sum_{n=1}^{\infty} \alpha_n 3^{-n} \tag{1}$$

where $\alpha_n$ is one if the $n$th monoprogram with string (in the ordering above) halts, and zero if it does not. This sum, since it necessarily converges, certainly defines a number—in fact, one between 0 and 1. We now claim that this number $K$ is not computable. Indeed, suppose that there did exist a

[5] For further discussion of the physical limitations on computing, see, e.g., Ref. 5.

program $P$ to compute it, as above. Then we would immediately have a solution to the halting problem: If you wish to determine whether the $n$th monoprogram with string halts, merely read in to $P$ the value $\varepsilon = 3^{-n-1}$, and, from the rational approximation to $K$ then printed by $P$, determine whether or not the appropriate term appears in the sum (1). It thus follows, in light of the halting theorem, that $K$ is not computable. More generally, this strategy can always be applied to make the transition between "whether there exists a program to answer a sequence of yes-or-no questions" and "whether a certain number is computable."

It is curious that, while the number $K$ above is not computable as we have defined that term, it is "determinable" in a somewhat weaker sense. We claim: There exists a program that successively prints out an increasing sequence of rational numbers, never halting, such that this sequence converges to $K$. Construct such a program as follows. Fix an integer, say 10. Let our program write out the first ten monoprograms and strings, run each of these programs for just ten steps, and then evaluate the sum of the first ten terms in (1), letting $\alpha_n$ be 1 if the corresponding program has halted in ten steps, zero otherwise. This is the first rational approximation. The successive approximations are obtained by increasing the integer, say, to 100, then 1,000, etc., and repeating the process. There clearly results an increasing sequence of rational numbers converging to $K$. Note that this program does less than would be required to show $K$ computable, for we do not know at any stage how close the "rational approximation" is to $K$. If, for example, the first monoprogram required $10^{17}$ steps before halting, then our sequence of rationals would "jump upward" by about 1/3 very far along in the sequence. Note also that there can exist no program that prints out a *decreasing* sequence of rationals converging to $K$, for the existence of programs yielding both decreasing and increasing sequences would imply that $K$ itself is computable.

We emphasize that computability is an attribute of the *number itself*, and not how that number is presented. Thus, for example, the number $K$ given by Eq. (1) must be irrational, for every rational number is computable. As a further illustration of this point, consider the number

$$M = \sum_{n=1}^{\infty} 3^{-n}/L(n) \qquad (2)$$

where $L(n)$ is the number of steps taken by the $n$th monoprogram and string before it halts (or "infinite," i.e., omit the corresponding term in (2), if the program does not halt). Is the number $M$ computable? One might think of writing, along the following lines, a program to provide rational approximations to $M$. The program would first determine how many terms in the sum (2) would be required to approximate $M$ to within the given allowed error $\varepsilon$, would then determine the corresponding $L(n)$, and would finally take the sum of those terms. But this strategy must fail, for, by the halting theorem, there is no program that will determine the $L(n)$. Yet, despite this failure, the number $M$ is actually computable! A program that shows this is the following. In order to approximate $M$ to within error, say, $\varepsilon = 1/100$, it suffices to deal only with the first ten terms in the sum (2), and, even for these, only either to determine $L(n)$ or else ensure that it exceeds 1,000. So, given $\varepsilon = 1/100$, our program merely runs the first ten monoprograms with strings for 1,000 steps each, letting $L(n)$ be infinite for any monoprogram that has not by then halted. Similarly for other $\varepsilon$. Thus, a number may appear to be noncomputable by virtue of the way it is presented, and yet turn out to be computable by appeal to some more sophisticated algorithm.

## 3. MEASURABILITY

Imagine a physical theory whose prediction, for the result of some particular experiment, is the noncomputable number $K$ of Eq. (1). The theory has thus made a perfectly definite prediction, for $K$ is merely some specific number. Imagine next that we wish to test this prediction. We do so in the usual way: Select some desired accuracy $\varepsilon$, carry out the experiment to within that accuracy, and then check to see whether the number that results from the experiment is within $\varepsilon$ of $K$. But, in actually carrying out this test, we would be confronted by the following difficulty: There exists no algorithm to determine whether a given number is within $\varepsilon$ of $K$—for this is the essence of $K$'s being noncomputable. Of course, it is nonetheless possible to test the theory—and to do so to arbitrary accuracy. The point is only that the test cannot be carried out mechanically: Each new level of accuracy will require new ideas for the testing procedure. We regard this as the prototype of the situation in which a theory has no algorithm for its implementation. In this section, we formulate this issue more precisely. In the following section, we discuss possible consequences for physics should such a theory come to be accepted.

The key step is the identification of the experimentally determinable numbers that emerge from the theory—what we shall call the measurable numbers of the theory. This having been done, the issue of whether a theory can be mechanically applied can be stated thus: Are all measurable numbers of the theory computable? It turns out that "measurable number" must be formulated with some care, in particular to ensure that we thus distinguish only the final physical predictions of the theory, not numbers from intermediate calculations. As a consequence, this term must be given a meaning slightly more restrictive than its common one.

Regard number $w$ as measurable if there exists a finite set of instructions for performing an experiment such that a technician, given an abundance of unprepared raw materials and an allowed error $\varepsilon$, is able by following those instructions to perform the experiment, yielding ultimately a rational number within $\varepsilon$ of $w$. "Measurable" is analogous to, although of course much less precise than, "computable." The technician is analogous to the computer, the instructions to the computer program, the "abundance of unprepared raw materials" to the infinite number of memory locations, initially blank. Indeed, one can think of the measurable numbers as those that are "computable" using an analog, rather than a digital, computer.[6] A few examples will illustrate what we have in mind.

The number $\pi$ is measurable. Suitable instructions for this case might direct that a metal disk be turned on a lathe, that measurements of its circumference and diameter be taken, and that their ratio be computed. Of course, the care with which the disk is to be constructed, and with which the measurements are to be taken, would be dictated by the allowed error $\varepsilon$.

The number $e^{1/2}$ is measurable. Suitable instructions might direct that a projectile be released in a viscous fluid, that measurements be taken of the distance $d_1$ the projectile has traveled when it slows to half its initial speed as well as the ultimate distance traveled $d_2$, and that the combination $d_2/(d_2 - d_1)$ be computed. Why would not a similar argument show that $e^K$, where $K$ is the number given by Eq. (1), is measurable? It is only necessary to replace in the instructions "half" by "a fraction $K$." But these modified instructions would not be regarded as acceptable, for, since $K$ is not computable, it would not be possible for the technician to perform the experiment by merely "following instructions."

The ratio of the energies of two given levels in the helium atom is measurable. Suitable instructions might direct that there be constructed helium atoms, and that the transition energies be measured with appropriate care.

The fine structure constant is measurable. Suitable instructions[7] might direct that there be measured the wavelengths of the photons emerging from positronium decay as well as that of the lowest energy photon that

[6] For other connections between digital computers and the natural "analog" world, see Ref. 6. The thesis raised in these papers is that limitations on the power of digital computers are universally applicable to their analog counterparts and so can be used to place limitations on the behavior of physical systems. We are not directly concerned with this issue here.

[7] See Ref. 7 for this and other methods of measuring $\alpha$.

will ionize hydrogen in its ground state, and that the square root of twice the ratio of these wavelengths be computed.

The meaning of "unprepared raw materials" is illustrated by the following example. Place two metal rods on a table: Do the obvious instructions show the ratio of their lengths to be measurable? We contend not, for two carefully calibrated rods are hardly "unprepared." Were we to allow such examples, then clearly any number would be measurable. Measurable numbers are to be those that flow from the operation of physical laws, not from mere initial conditions. The analogous requirement, in the computable case, is that all memory locations initially be blank. Otherwise, any number would be computable. The lathe used to turn the disk in the first example above does not constitute "prepared raw materials," for one could build a lathe from scratch. The Earth–Moon mass ratio would not be regarded as measurable, since the Earth–Moon system is "prepared." But we would regard as measurable the proton–electron mass ratio.

All numbers in physics are essentially dimensionless, for even for example an elapsed time reported in seconds merely gives the ratio between that elapsed time and the elapsed time that is the standard second. But, for the issue of measurability, the choice of standard is important. If, for example, the standard second is 1/86,400 of the mean solar day, then instructions reporting times in these units would not be acceptable, for the experiment involves "prepared" raw materials. But this problem does not arise if the standard second is the time of 9, 192, 631, 700 vibrations of a certain hyperfine transition in $Cs^{133}$. It is curious that most such "unprepared" standards arise via quantum mechanics.

It is intended that "measurable" be an attribute of the number itself, and not of how that number is presented. Thus, a given number is either measurable or it is not, the instructions merely serving (as does the computer program in the computable case) to show it measurable. Consider, for example, instructions that direct the technician to work with an unstable system, so the number finally reported varies wildly from one run to the next. There is no number that these instructions show to be measurable. This is analogous to a Fortran program that prints rationals that vary wildly from one $\varepsilon$ to the next. There is no number that this program shows to be computable.

The notion "measurable" involves a mix of natural phenomena and the theory by which we describe those phenomena. Imagine that one had access to experiments in the physical world, but lacked any physical theory whatsoever. Then *no* number $w$ could be shown to be measurable, for, to demonstrate experimentally that a given instruction set shows $w$ measurable would require repeating the experiment an infinite number of times, for a succession of $\varepsilon$'s approaching zero. One could not even demonstrate that a given instruction set shows measurability of any number at all, for it could turn out that, as $\varepsilon$ is made smaller, the resulting sequence of experimentally determined rationals simply fails to converge. It is only a *theory* that can guarantee otherwise. The situation is analogous to that of trying to demonstrate that a given Fortran program shows some number to be computable. There is no general algorithm for deciding this. In particular, it would not do merely to run the program for a few selected values of $\varepsilon$.

So, "The number $w$ is measurable." is basically a statement about the theory within which we are working, and only indirectly—only to the extent that the theory correctly describes the physical world—a statement about natural phenomena. Thus, in the example of measurement of $\pi$, the theory is that of classical continuous elastic solids in Euclidean space. Classical continuous elastic solids is required to ensure that arbitrarily accurate measurements can, in principle, be made on the disk (without interference, for example, from quantum mechanics or atomic structure). Euclidean geometry ensures that the ratio obtained will be independent of disk size, and also that $\pi$ is actually the number to which the approximations converge. In the example of $e^{1/2}$, the theory is that of viscous fluids, and in particular its prediction of exponential decay. In the example of the helium atom, the theory is quantum mechanics; in that of the fine-structure constant, quantum electrodynamics.

Every computable number is measurable. This is easy to see: Let the instructions direct that the raw materials be assembled into a computer, and that a certain Fortran program—one specified in the instructions—be run on that computer. That is, every digital computer is at heart an analog computer.

We now ask whether, conversely, every measurable number is computable—or, in more detail, whether current physical theories are such that their measurable numbers are computable. This question must be asked with care. Consider, for example, the fine-structure constant $\alpha^{-1}$. This number is measurable in quantum electrodynamics, as discussed above. Is $\alpha^{-1}$ computable? The problem is that in quantum electrodynamics the actual value of $\alpha^{-1}$ is not specified by the theory, but rather is left to be determined by experiment. There is no "number," mathematically specified, whose computability can be investigated. There is no meaning to be attached to the statement "$\alpha^{-1}$ is computable." [Of course, were quantum electrodynamics reformulated with $\alpha^{-1}$ specified, say to be 137.036 (exactly!), then the computability of the fine-structure constant *would* be meaningful.] We are thus led to divide the measurable numbers into two classes: those assigned specific values by the theory, and those not. For example, $\pi$, as measured by the disk, and $e^{1/2}$, as measured by the viscous fluid, are clearly assigned values by the theories. The example of the helium atom is more subtle. If the theory is that of a helium atom with infinite-mass, pointlike nucleus and no relativistic effects, then the ratio of energy levels is assigned a specific value by that theory. But a finite mass brings in the electron–nucleon mass ratio, and relativistic effects the fine-structure constant, neither of which is assigned a specific value in conventional theory. Of course, specific values may be assigned in some future theory.

We suggest that the familiar theories of physics are such that all the measurable numbers specified by these theories are in fact computable. The following example, representing a "typical" theory whose mechanism is a differential equation, will support this suggestion. Let the state of the physical system be characterized by a single variable $x$, evolving with time according to the equation[8]

$$\frac{dx}{dt} = 5(x^2 - t^2)\sin(xt) \tag{3}$$

Let it be the case that a technician, given any allowed error $\varepsilon$, is able to (i) set the system initially in state $x$ within $\varepsilon$ of any given rational number, and (ii) wait for time $t$ within $\varepsilon$ of any given rational number, and then measure to within $\varepsilon$ the final state of the system. Then under these ground rules the measurable numbers of the theory are clearly the following: all numbers of the form $x(t_0)$, where $t_0$ is any computable number and $x(t)$ is the solution of Eq. (3) with initial condition any computable $x_0$. Are all of these numbers computable? This is a purely mathematical question, involving the character of the solutions of Eq. (3). It turns out that all these numbers are indeed computable, as one sees by noting that one could integrate Eq. (3) numerically on a computer. Thus, at least in this model theory, the measurable numbers are computable. Note that, were the coefficient 5 on the right in Eq. (3) replaced by a constant $\lambda$ not further specified, then the measurable numbers would not be specified by the theory; and were it replaced by the noncomputable number of Eq. (1), then the measurable numbers would not be computable. In a more realistic treatment, the issue of what the technician can set and what he can measure would be determined by the interactions in the theory. We note, finally, that a similar argument applies to field theories governed by hyperbolic equations,[9]

[8] The right side of Eq. (3) could as well be any smooth "explicit" function—or even some not so explicit. We here merely select one as an example in order to avoid the issue of what is the class of allowed functions.

[9] For examples of lack of computability in solutions of differential equations, see Ref. 8.

although the technician must for such examples be given considerable power to control initial conditions.

Thus, we argue, the conventional theories of physics do have the property that all measurable numbers specified by these theories are computable. Here, then, is at least one sense in which it is true that these familiar theories can be applied by means of algorithms.

There is a recent example[9] in which this issue comes very much to the fore. Consider one version of quantum gravity for closed cosmologies, formulated as a sum over histories. An observable in the theory is a functional, $A[\mathscr{G}]$, of geometries $\mathscr{G}$ on compact 4-manifolds. The expectation value of such an observable in the analog of the ground state is given, formally, by

$$\langle A \rangle = \frac{\sum A[\mathscr{G}] \exp(-I[\mathscr{G}])}{\sum \exp(-I[\mathscr{G}])} \tag{4}$$

where $I[\mathscr{G}]$ is the Euclidean gravitational action. The sums in Eq. (4) are over all compact 4-geometries—including in particular all possible topologies for the 4-manifolds. Now fix a "reasonable" observable $A$. Is the number given by Eq. (4) measurable? Is it computable?

As we have seen, even for simple theories the issue of whether a given number is measurable can be subtle. But this theory is anything but simple, for the theory purports to provide the quantum description of the entire Universe. What instructions would allow a technician to "measure" $\langle A \rangle$? Our understanding of the physics of quantum gravity is not yet to the point at which we can answer this question. Nonetheless, one expects that there will be *some* measurable numbers in the final theory, and such expectation values would seem to be good candidates.

Is $\langle A \rangle$ computable? Unfortunately, the mathematics of quantum gravity has not yet been developed to the point that we have a complete mathematical formulation of what Eq. (4) is to mean. But one possible meaning that has been proposed involves use of Regge calculus. Fix an integer $n$. Consider a simplicial 4-manifold with exactly $n$ vertices. That is, take $n$ points as vertices, and specify which pairs of vertices are to be connected by an edge, which triples of adjoining edges are to be connected by a face, which quadruples of adjoining faces are to be connected by a tetrahedron, and which quintuplets of adjoining tetrahedra are to be connected by a 4-simplex—all such that the resulting space is a topological 4-manifold. Next, specify numerical values for the lengths of all the edges of this simplicial 4-manifold. The result is a simplicial 4-geometry $\mathscr{G}$. We can now evaluate $A[\mathscr{G}]$ and $I[\mathscr{G}]$, and thus one term in each of the sums in Eq. (4). To evaluate the sums themselves, we proceed as follows: first keep the simplicial 4-manifold fixed and integrate over the values of the edge lengths; and then sum the result over all simplicial 4-manifolds with exactly $n$ vertices. There results the $n$-vertex simplicial approximation $\langle A \rangle$, a number we denote $\langle A \rangle_n$. Finally, take the limit of the $\langle A \rangle_n$ as $n$ goes to infinity—assuming this limit exists! The result is our candidate for the expectation value $\langle A \rangle$.

This particular meaning for Eq. (4) has the following curious feature. Imagine an attempt to evaluate the $\langle A \rangle_n$ by computer. One would select and read in some value for the integer $n$. The computer would then consider $n$ vertices, and try all combinations of which vertices are to be connected by edges, which edges by faces, etc. From the resulting list of simplicial complexes, the computer would then eliminate all those that are not 4-manifolds. Then, from the resulting list of simplicial 4-manifolds, the computer would eliminate all duplications—cases in which there is included more than one simplicial representation of a single 4-manifold. The computer would then perform, for each simplicial 4-manifold in this reduced list, the integrations over edge lengths. Finally, the computer would take the sums over the 4-manifolds. The result would be reported as $\langle A \rangle_n$. The curious feature is that there exists no computer that can carry out this task! The key problem is with the elimination of duplications. It is known that the issue of whether two simplicial 4-manifolds are topologically identical is undecidable. The theorem, in more detail, is the following.[10] Let simplicial 4-manifolds be specified by giving the number $n$ of vertices, which vertices are to be connected by edges, which edges by faces, etc. This information could be encoded, for example, into a string. Then: There exists no Fortran program that will accept as inputs two such strings, ultimately print "same manifold" or "different manifolds" according to whether the 4-manifolds represented by the strings are topologically the same or different, and then halt. Further, even the elimination of those simplicial complexes that are not manifolds may be problematic. It appears likely that the question of whether a simplicial complex is a manifold is decidable in dimension four, but undecidable in higher dimensions.[(10)]

All this is not to say that quantum gravity will admit measurable numbers that are not computable. The mere fact that one particular strategy for writing a program to compute the $\langle A \rangle_n$ fails does not mean that the $\langle A \rangle_n$ are noncomputable. We have seen an example in Eq. (2). Further, even should the $\langle A \rangle_n$ turn out to be noncomputable, this does not guarantee that their limit, $\langle A \rangle$, is also noncomputable. It is easy to write down examples of sequences of noncomputable numbers having computable limits. [The noncomputable numbers are dense in the reals.] It could also

[10] For a review of these results, see Ref. 10.

turn out that the meaning ultimately assigned to Eq. (4) is different from that indicated here. For example, one might simply adopt the rule that duplications of 4-manifolds are not to be excluded in the sums; or, alternatively, the rule that the sums are not to be over different 4-manifolds. It could turn out that the numbers ultimately measurable in the theory are quite different from these expectation values. Or, finally, what is ultimately quantum gravity may bear little resemblance to a sum over simplicial histories.

Yet quantum gravity does seem to be a serious candidate for a physical theory for whose application there is no algorithm.[11]

## 4. DISCUSSION

What must be the character of a physical theory in order that it be regarded as satisfactory? Surely, it must include some mechanism for making quantitative predictions—for producing, under specified experimental arrangements, quantities subject to experimental verification. Otherwise, we would hardly have a "theory." But this is by no means the only criterion. The familiar theories of physics possess, in addition to predictive power, such features as simplicity, generality, and elegance. Historically, these criteria have not always been incidental, for they have at times played a central role in the search for physical laws. General covariance is an example—one for which, remarkably, the "criterion" turned out to be empty!

We here raise the possibility of another type of criterion: that the theory be such that its predictions can be extracted not merely "in principle, possibly by ever higher levels of skill and sophistication," but also "in practice, mechanically." The criterion, in the present more precise version, is that all the measurable numbers specified by the theory be computable numbers. At issue is the status of this criterion.

First note that it may be difficult to decide whether or not a given theory satisfies this criterion. One mathematical formulation of the theory may provide no algorithm for implementing the theory, and yet another, equivalent, formulation does. The physical issue of whether a number is measurable, as well as the mathematical issue of whether it is computable, can be a subtle matter. Nevertheless, it should be far more straightforward to decide whether a theory satisfies this criterion than, for example, to decide whether a theory is "elegant." It is at least more a question of fact than of taste.

[11] For discussion and other examples of undecidable problems in physics and mathematical physics, see Refs. 11 and 12.

In present physical theories, the computability of the measurable numbers is manifest, so this issue is rarely raised. But now the issue may be forced upon us, by quantum gravity. There are indications that this theory may have measurable numbers that are not computable, at least if its formulation as a sum over simplicial histories is a good guide.

What, then, would be the consequences for physics should there be accepted a theory whose measurable numbers are not computable? Fix some experimental arrangement, and attempt to extract from such a theory its prediction for that experiment. To predict to within, say, 10%, one manipulates the mathematics of the theory for a while, arriving eventually at the predicted number. To predict to within 1%, it will be necessary to work much harder, and perhaps to invent a novel way to carry out certain mathematical steps. To predict to 0.1% would require still more new ideas. The process would be similar to that of trying to compute rational approximations to the noncomputable number given by Eq. (1). The theory certainly "makes definite predictions," in the sense that predictions are always in principle available. It is just that ever increasing degrees of sophistication would be necessary to extract those predictions. The prediction process would never become routine.

Clearly, this would not be a particularly desirable state of affairs. But it would seem to be merely an inconvenience—far from a disaster for physics.

The status of the other criteria for physical theories arose, not through general arguments, but rather through experience with specific examples. Whether physical theories will be required to possess some algorithm by which they can be implemented will presumably be settled by the same process.

## ACKNOWLEDGMENT

Many colleagues gave helpful advice on a preliminary version of this paper. We are grateful in particular to B. DeWitt, J. Friedman, B. Horne, G. Horowitz, W. Kohn, A. Komar, G. Kreisel, D. Scalapino, R. Sorkin, M. Srednicki, and S. Wolfram. This work was supported in part by the NSF.

## REFERENCES

1. J. A. Wheeler, *Int. J. Theor. Phys.* **21**, 557 (1982).
2. C. W. Misner, K. S. Thorne, and J. A. Wheeler, *Gravitation* (Freeman, San Francisco, 1970).

3. A. Yasuhara, *Recursive Function Theory and Logic* (Academic Press, New York, 1971).
4. G. Faltings, *Invent. Math.* **73**, 349 (1983).
5. *Int. J. Theor. Phys.* **21**, Nos. 3–4 (1982).
6. S. Wolfram, *Phys. Rev. Lett.* **54**, 735 (1985); *Sci. Am.* **251**, 188 (Sept. 1984).
7. B. N. Taylor, W. H. Parker, and D. N. Langenberg, *Rev. Mod. Phys.* **41**, 375 (1969).
8. M. B. Pour-El and I. Richards, *Ann. Math. Logic* **17**, 61 (1979); *Adv. Math.* **39**, 215 (1981).
9. J. B. Hartle, to appear.
10. W. Haken, in *Word Problems*, W. W. Boone, F. B. Cannonito, and R. C. Lyndon, eds. (North-Holland, Amsterdam, 1973).
11. A. Komar, *Phys. Rev. B* **133**, 542 (1964).
12. G. Kreisel, *Synthese* **29**, 11 (1974).

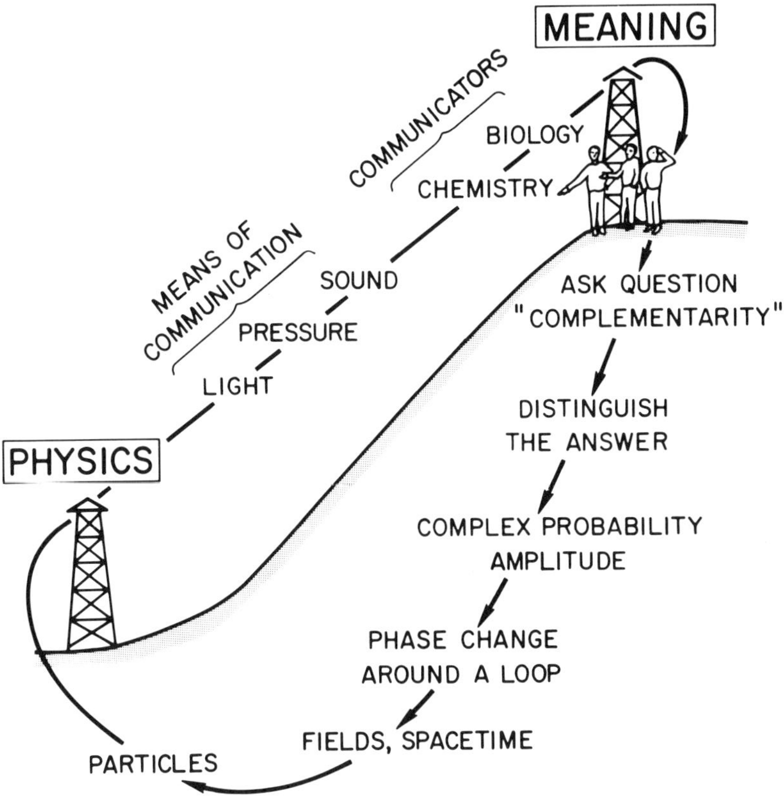
MEANING
COMMUNICATORS
BIOLOGY
CHEMISTRY
MEANS OF COMMUNICATION
SOUND
PRESSURE
LIGHT
PHYSICS
ASK QUESTION "COMPLEMENTARITY"
DISTINGUISH THE ANSWER
COMPLEX PROBABILITY AMPLITUDE
PHASE CHANGE AROUND A LOOP
FIELDS, SPACETIME
PARTICLES

*Plate 37*

# Computation and Physics: Wheeler's Meaning Circuit?

**Rolf Landauer**[1]

*Received September 23, 1985*

*Computation is a physical process, inevitably utilizing physical degrees of freedom. Computation, therefore, is restricted by the laws of physics and also by the construction materials and operating environments available in our actual universe. These restrictions have been investigated for a quarter century. A discussion of this field is provided, too concise to be a review, but only intended as a guide to the literature. Physical law, in turn, consists of algorithms for information processing. Therefore, the ultimate form of physical laws must be consistent with the restrictions on the physical executability of algorithms, which is in turn dependent on physical law.*

## 1. INTRODUCTION

The concern with the *ultimate* physical limitations of the computational process has received limited attention, but has had a sufficient diversity of contributions so that in a short paper we cannot really review these. We will only attempt to provide a road map to some of the existing work, and add a number of observations. We particularly stress the fact that not only does physics determine what computers can do, but what computers can do, in turn, will define the ultimate nature of the laws of physics. After all, the laws of physics are algorithms for the processing of information, and will be futile, unless these algorithms are implementable in our universe, with its laws and resources. "What computers can do," refers not to the foreseeable technological future, but rather to that which can be done in principle with the resources available in our actual physical universe. John

[1] IBM Thomas J. Watson Research Center, P. O. Box 218, Yorktown Heights, New York 10598.

Wheeler[1] has pointed us toward Era III of physics, in which, "...we have to seek nothing less than the foundation of physical law itself." Wheeler contrasts this to Era II, which commenced with Newton, where the laws of physics were to be discovered, but their origin was not a fit subject for discussion. Wheeler has given us a specific proposal for Era III, the "meaning circuit."[1,2] In this, physics gives rise to all the phenomena whose study has usually occupied us. These phenomena, in turn, give rise to biology, communicators, and their questions, and these lead to the laws of physics. We will be talking about a similar circuit, but without stressing the role of communicators. Our circuit may, or may not, be related to Wheeler's meaning circuit. We, therefore, need a more general label for a situation in which the laws of physics are part of a "circuit," i.e., defined in a self-consistent way—situations in which physical law not only explains the universe, but where the universe in turn determines the nature of physical law. We will invoke the label "Wheeler circuit," for this purpose.

Our Wheeler circuit is shown in Fig. 1. The right-side arrow shows that computation, or information handling, is tied to physical degrees of

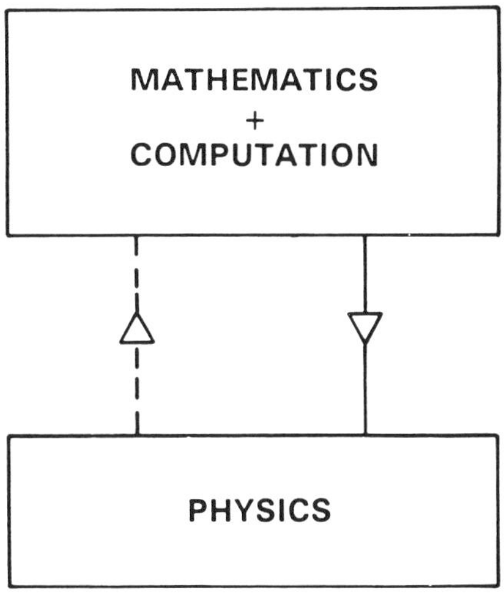


Fig. 1. Information handling, shown at the top, is dependent on that which the real physical universe permits, through available "parts," and through the laws of physics. This dependency is indicated by the right arrow. But the laws of physics are directions for processing information and, therefore, dependent on executable algorithms. This is indicated by the dotted arrow at left.

freedom, and therefore to the laws of physics. Most of the papers in our field are limited to that side. The arrow on the left side is the return portion of our Wheeler circuit. The laws of physics, as algorithms for information manipulation, are in turn limited to that which the universe permits. If we do not have enough degrees of freedom to be tied together into an unlimited memory, we cannot distinguish $\pi$ from a near neighbor. In the next section we review our limited understanding of the right side of Fig. 1, subsequently returning to the left side.

## 2. PHYSICAL LIMITS OF COMPUTATION

In our new and emerging field it is easier to ask questions, then it is to answer them. Some typical questions follow:

1. How much energy is required by an elementary step in a computation? How is this tied to $kT$ and to the uncertainty principle?
2. Can computation be made immune to noise?
3. How much of the universe can be tied into an effective computing structure? How large a memory is available to us?
4. Will the inevitable deterioration effects present in computing machinery limit the length of a computation? Or is it always possible to transfer the calculation, in time, to a more massive, more failure immune, machine?

Our progress, so far, is essentially limited to questions 1 and 2 above, and we, here, give an outline of that work. Other questions, which inevitably arise, relate to the minimal scale of time and of spatial extension, required for the execution of a logic function. There is no reason, in principle, why logic functions have to be tied to solid-state electronics. If we are clever enough we can use the internal degrees of freedom of the nucleus, or other fundamental particle interactions, to do this. Therefore, until physics has resolved the fundamental graininess of the universe, this question will remain unanswered.

## 3. REVERSIBLE COMPUTATION

Discarding information, during the computational process, requires energy dissipation. Computation, however, can be carried out in such a way that every step along the way is a one-to-one mapping and preserves information.[3,4] This, in turn, permits physical devices which can be reversed, i.e., devices which can run backwards to undo a logic operation which has been accomplished. Computers which operate this way are called "reversible" computers, and have been described in a number of papers. At this point, we cite only a sampling of this literature.[3–9] Reversible computers are assumed to be noisy, to have frictional forces proportional to their velocity, and are allowed to have some "manufacturing" errors; their parts do not have to be perfect and can be allowed to have limited errors.[10] Reversible computers dissipate an energy, per step, in proportion to the velocity with which the computation is executed, and have no minimal, unavoidable, dissipation per step. Reversible computers can also be made immune to noise-induced errors, not only if operated slowly, and not only to thermal equilibrium noise, but also to other forms of noise *in the intentionally used degrees of freedom.* The italicized phrase is intended to exclude fluctuations which cause equiment deterioration.

We now have about half a dozen reversible computer embodiments. The exact count is a matter of taste depending on the criteria of acceptability and on the criterion for labeling related proposals as distinct. A list of these proposals is contained in Ref. 9. Proposals also exist for reversible computers[11,12] which are dissipationless. In these conservative systems the computation proceeds with the kinetic energy with which the computation is launched. The proposal of Ref. 11, which came earlier, is better known, and physically more appealing, is illustrated in Fig. 2. The proposal of Ref. 12, however, has the advantage that the particles in its memory are

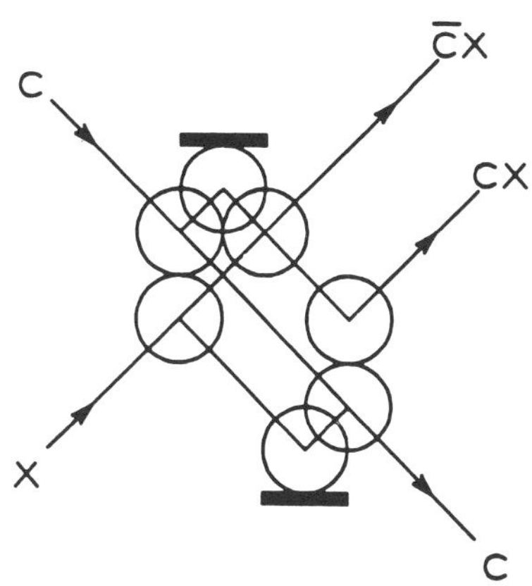


Fig. 2. Logic can be carried out by the collision of hard balls, with the presence or absence of a ball denoting 1 or 0, respectively. Reflecting mirrors then redirect the balls toward further interactions.

not moving, and have no kinetic energy. In contrast, the proposal of Ref. 11 requires an allowance, at the beginning of a calculation, for the maximum kinetic energy required, by storage, during the course of the program. The main problem, however, of both proposals, comes from their need for perfection. Thus, for example, an incorrectly tilted mirror, in Fig. 2, will misdirect a particle and cause it to deviate from its intended direction. In subsequent particle collisions, the angle of deviation will be amplified and, in addition, cause an error to be passed on to the other collision partner. Thus, dissipationless computation is not, in any serious sense, realizable. Throwing away information requires dissipation, and correcting an errant trajectory requires us to throw away information about the error. For example, a misaligned mirror in Fig. 2 leaves its mark on the trajectory, and error correction (within some qualifications taken up in the Appendix) throws away that information. In view of this, is there any real purpose in the discussion of "ballistic" computation, as conservative computation has come to be called? One purpose will become evident in the subsequent discussion of quantum mechanical computation. Let us, here, take up another reason. Assume that our machinery is relatively perfect, i.e., not far from its ideal specifications. This requirement for perfection includes, in the example of Fig. 2, not only proper mirror alignment, but also requires the initial conditions for the position and velocity of the colliding balls to be close to the specifications for that. The requirement for "close to perfection" also requires that the internal degrees of freedom of the components (mirrors and balls) be inactive, i.e., causing little or no damping, and introducing no fluctuations. Now, if we have such a system which is close to perfection, it can go through a number of successive logic operations before we have to invoke error correction.[13] Error correction consists of machinery which takes a ball, still sufficiently close to its intended trajectory so that the correct trajectory is recognizable, and restores the position (and perhaps velocity) to that of the intended trajectory. Such restandardization is done in ordinary transistor logic circuitry at every stage, driving the signal voltage, for example, to two extreme limits, 0 and $B+$. During error correction, the phase space of the billiard ball is reduced, and the associated entropy reduction must show up as an entropy increase elsewhere, i.e., as energy dissipation. If we use such a plan to restore the signals to their intended value, and do so only after a number of logic operations, then each independent source of error will have propagated on to affect a number of later logic variables. The error that will have to be corrected, therefore, is a cumulative error, and not just the error which (to use the specific vocabulary, appropriate to Fig. 2) first arose in the history of the particle being reset. The set of errors which exist, at a given time, among the various balls, or logic variables, do, of course, have correlations. Thus the phase space, associated with the errors, is much less than if the errors were really independent. Without the availability, however, of an independent record, external to the machine, of the nature and source of the errors, we cannot make use of this correlation, and must reset the different logic variables independently of each other. Indeed, it has been shown by Zurek,[14] that in the billiard ball computer the effective error phase space doubles, approximately, at every collision. Whether other forms of dissipationless computation behave equally poorly is still unknown. What has been gained through ballistic computation?[13] We now have computation which no longer has arbitrarily small dissipation. On the other hand, the dissipation is independent of computational velocity, as long as the system, except for the resetting apparatus, is really frictionless. To achieve this velocity independence we must also assume that the size of the required relative phase space reduction, at each resetting event, is independent of the computational velocity.

Thus, imperfect ballistic computers offer us a dissipation related to the error size, rather than to computational velocity. We stress, however, that this velocity independence, which permits *fast* computation with limited dissipation, is simply the result of an idealization: the complete absence of friction, except in error correction. On the other hand, the dissipative, noisy, and reversible computers really can be designed so that they operate in a fashion which is independent of small deviations from their ideal specification; in that case, no unwarranted idealization is involved.

In recent years the field has gone beyond the discussion of classical conservative computers, to the treatment of Hamiltonian systems which can carry out computation, and can be treated quantum mechanically. The realization that this can be done stems from the work of Paul Benioff,[15] and has been extended by Deutsch,[16] Feynman,[17] Peres,[18] and Zurek.[14] In all of these quantum schemes the quantum mechanical basis states are the obvious logic states of the computer, that is, the set of logically interesting and accessible information states. The computation then consists of a unitary mapping of these states into each other, as a computer would. A Hermitian Hamiltonian operator is constructed, acting on the interacting information-bearing binary degrees of freedom, which, in turn, can be considered to be spins. No other degrees of freedom are described. Thus, there is no dissipation, thermal reservoir, or source of fluctuations. The bits evolve, in time, as we would like to see them do in a computer. Benioff's original work provided two versions; one of these utilized a periodically time-dependent or "clocked" Hamiltonian. The other, time-independent Hamiltonian, may be considered to be more like a table-lookup machine than a computer. This time-independent Hamiltonian has to anticipate the complete trajectory of successive states for every possible program.[18] Feynman went beyond this to provide a time-independent Hamiltonian whose "design complexity," i.e., difficulty in specifying the Hamiltonian, seems just like that of an ordinary computer. In Benioff's time-independent Hamiltonian the unitary operator specifying the time evolution in a clock cycle *is* simple, it is only the Hamiltonian which is not simple. Stressing that distinction implies that we design or manufacture Hamiltonians, rather than unitary time evolution operators. Neither view, however, really seems correct. Design and manufacture deal with the assembly of unstructured parts, unrelated to the overall computational progress. That is not, obviously, equivalent to the specification of a Hamiltonian.

Deutsch[16] has pointed to the fact that we can have an initial state, which is quantum mechanical superposition of basis states corresponding to ordinary computer states. Such a superposition is of interest in its own right, and permits a computational mode not available to ordinary classical computers. Strict quantum mechanical coherence must, however, be maintained throughout the computation, if we wish to make use of this novel feature. Zurek[14] has pointed to the fact that quantum mechanical systems, with a discrete spectrum, do not allow the equivalent of a slightly misdirected billiard ball; therefore they are less susceptible to error. Peres,[18] elaborating on a brief comment by Zurek,[14] points to the fact that in a system in which a computation is done simultaneously on several computers, comparison can be used to detect errors, and to reset an erroneous bit. This requires a heat dissipation proportional to the error rate.

The quantum mechanical computers, as already stressed, produce the desired time evolution of interacting bits, without any other hardware, such as frames, cables, or even some sort of rigging to hold the information-bearing spins in their proper place. Furthermore, the spins must be much like the fictional spins in an Ising model; only selected spins should interact. Real particles, of course, do not behave like Ising spins. Real particles have an interaction that falls off with distance, but is not all that selective, and is not limited to nearest neighbors or next nearest neighbors. In the appendix we discuss some additional details related to quantum mechanical dissipationless computers.

It seems unlikely that the real universe permits machines with an infinite memory. Finite machines have a limited set of possible computational paths; all of these can, in principle, be checked out by the designer. Therefore, they may seem like uninteresting table-lookup devices. But *real* computers *are* finite machines, and nevertheless have a utility far beyond that of table-lookup devices. Finite machines can do a great many tasks not understood in advance by their designer; finite machines are more than table lookup devices. We need to characterize finite machines by a figure of merit that characterizes the universe of calculations they can do, compared to the complexity of the machine's own design. Unfortunately, such an approach does not yet exist.

## 4. OVERVIEW OF LIMITS AND IMPLICATIONS

The primary thrust of the detailed work in our field has been the elimination of limits which might have appeared as natural. $kT$ is *not* a minimal energy required per logic step; the uncertainty principle does *not* impose additional energy dissipation requirements. This does not mean that there are no limits; the real limits are likely to occur in connection with questions 3 and 4 out of our earlier list of four representative questions, in Section 2. These questions deal with the availability of parts in our universe. In fact, all of the existing classical reversible computer models already involve idealizations in that regard, e.g., the "hard" parts invoked by Bennett's springless mechanical Turing machine.[4] (The quantum models, which permit the free invention of Hamiltonians, hide that sort of question completely.) Furthermore, even the existing absence of limits represents a weak assertion. We can spend a lot less than $kT$ per computational step, if done slowly enough. But "slowly enough" may be too slow to be satisfactory—too slow, perhaps, to do prediction, or too slow, perhaps, before the computer falls apart; or too slow for the lifetime of the universe.

We have all been indoctrinated by the mathematician's sense of values. Given $\varepsilon$, $\exists N$ such that .... Now we ask whether these implied mathematical operations are, in fact, really executable. Are there enough degrees of freedom available, that can be tied into a computer, to permit us to express $\pi$ to any required precision? The real world has equipment deterioration problems. Is it reasonable to assume that these can be offset to any desired degree?

Before quantum mechanics, physics had an idealized view of the measurement process. Nature was there, and independent of our attempt to probe it. Our probing was limited only by the skill and dedication we wanted to invest in this process. We learned, eventually, that measurement really was a more integral part of nature and of physical law. Wheeler urges us to expand our notion of measurement to include comunication and the posing of questions. We, instead, emphasize that calculation, just like measurement, is not an idealizable and separable entity, but an integral part of the fabric of physics and physical law. Once we admit that the ordinary processes of mathematical analysis are in trouble, then, of course, the laws of physics expressed in terms of such operations are also in difficulty. In a universe which allows only limited precision, we cannot sensibly invoke a Schrödinger equation, Newton's laws, or any of the ordinary machinery of theoretical physics. Thus, in the long run, we need a self-consistent theory. Its information handling operations must satisfy the laws of physics and must be realizable with the parts available in the universe. On the other hand, the laws of physics, which these information handling operations must obey, must themselves be expressed in terms of executable operations. We are very far from such a goal.

My viewpoint, originally articulated in 1967,[(19)] can be traced back to P. W. Bridgman, Nobel laureate, high-pressure physicist, and author of a number of discussions of "operationalism." Bridgman, in 1934, published a paper: *A Physicist's Second Reaction to Mengenlehre*, in Vol. II of *Scripta Mathematica* (p. 3). This journal was published at Yeshiva University, and oriented to fundamental questions in mathematics. The paper starts with a remarkable qualifying note by the editor: "The *Scripta Mathematica* takes pleasure in publishing this interesting article by Professor P. W. Bridgman. As in the case of all articles, however, it is understood that this does not necessarily represent the views of the editors." Bridgman was, by that time, already well known, and science is normally full of divergent viewpoints. Fifty years later, we can only guess that the unusual editorial qualification represented the editor's desperation. He may have been hesitant to turn down a famous scientist and, at the same time, was afraid of the wrath of his readers. Bridgman's paper presents his attempt to wrestle with the paradoxes of set theory. He objects to self-referential definitions and asks for mathematical procedures in which each successive step can be carried out directly and unambiguously. In a remarkable coincidence(?) he uses the word *program* for such a succession of instructions. Bridgman would have been entirely satisfied by a restriction of mathematics to that which can be done on a Turing machine, which came along a few years later. Bridgman did not go on to ask about the *physical realizability* of these operations, as we have done. One cannot help but guess that if Bridgman had ever returned to the subject, e.g., if he had been exposed to a discussion of Turing machines, he would have asked: Can we really assume an unlimited supply of tape? Does the universe have enough degrees of freedom for that? Can we keep the degrees of freedom in place long enough? If we pile up enough equipment to do very precise and very reliable calculation, will that pile of equipment not have its own physical effect on its surroundings?

Mathematicians can, of course, devise closed formal systems, which contain abstract images of our ordinary analytical operations. Such systems are, however, exactly what they claim to be: closed formal systems. They have no relationship to the real world, and have no utility in following the motion of particles, or even in balancing bank accounts. Unfortunately, the recent surge of interest in computational limits has produced little motion toward the viewpoint just discussed. The fact that computer operations are physical events has, of course, seen widespread acceptance, e.g., in Wolfram's discussions.[20] But it is the appreciation for the "return circuit" that is still missing, except perhaps in the work of Hans Bremermann.[21] A very different approach is that of Fredkin[22] and Zuse,[23] who believe that the world *is* a computer, specifically a cellular automaton, with interlinked logic units acting on a minute scale of time and space. This is, in a sense, opposite to the notions we have put forth. We assert that the execution of algorithms is limited by our ability to harness the sparsely distributed existing degrees of freedom of the universe. Fredkin and Zuse, instead, fill the universe with a phenomenal amount of digital equipment.

There are, of course, others who are concerned with the fact that the real universe is not really characterized by a continuum. T. D. Lee[24] has proposed a "discrete mechanics." Feynman has stated[25] "... *everything* that happens in a finite volume of space and time would have to be exactly analyzable with a finite number of logical operations. The present theory of physics is not that way, apparently. It allows space to go down into infinitesimal distances, wavelengths to get infinitely great, terms to be summed in infinite order, and so forth; and therefore, if this proposition is right, physical law is wrong." That statement clearly has a resemblance to our exposition. Ross[26] has invoked a model of the real number line based on probabilistic considerations. None of these, however, seem to go as far as we do, and point out that the correct physical law must be based on algorithms whose nature is still to be found, rather than on classical mathematical analysis.

## ACKNOWLEDGMENTS

Conversations with D. Deutsch and A. Peres have helped to develop my viewpoint.

## APPENDIX

We relegate to this appendix some comments related to the details of the quantum mechanical computer descriptions. The comments are somewhat incidental, and fragmented. Some point to minor limitations in papers which, in other aspects, are pioneering. We include comments, here, which may require the reader to turn to the publication under discussion.

Let me start with my own mistake. In Ref. 12 I drew an analogy between a quantum mechanical computer moving through a succession of different physical states, and an electron in a one-dimensional disordered potential, suggesting that the resulting reflections, occurring within the computation, might "localize" the computation.[27,28] After all, different steps in the computation are associated with differing patterns of change. In some steps many bits change, in others only a few. For example, the number of Fredkin gate switches moved into the crossed track position, in the scheme described in Ref. 3, depends on the information content. On the other hand, in the billiard ball scheme,[11] the kinetic energy really is constant. Nevertheless, even there, the pattern of evolution of the billiard ball positions, per computational step, varies a great deal, and does not look much like motion through a periodic structure. The number of balls in collision varies from step to step. We do not, of course, have a developed quantum picture of the billiard ball computer, and, therefore, cannot really follow up on this periodicity question. Feynman's more general discussion, however, shows that if the computational Hamiltonian is perfect,[25] reflections need not happen. Feynman's Hamiltonian describes a system in which the cursor, or clock, motion controls the logic process, but the logic process has no back effect on the cursor motion. Such a system can be realized in a form very similar to the "Bit Space Pipe Computer" of Ref. 10. A system of this sort can be visualized as a set of planes (or higher dimensional spaces), one per cursor index. Each of these planes (or subspaces) contains a position and potential well for each possible computational state. Every state is then connected by a passageway, or tunneling channel, to its proper subsequent state, in the next plane. All tunneling channels are adjusted to provide the same degree of coupling. Such a computer is, however, unsatisfactory. The defect does not arise from a superficial resemblance to a table-lookup device; its design complexity need not be all that horrendous. The nature of the "tunnels," after all, can be independent of most of the data content in a particular state. But the number of tunnels, or channels, and therefore the number of parts, will be totally unreasonable.

Feynman's own discussion[25] of the imperfect computer, in turn, is not very explicit about the distinction between static imperfections, i.e., manufacturing defects on the one hand, and thermally excited vibrations, on the other. His discussion of entropy changes is tailored to the latter. Static defects do not affect the entropy, as might be constructed from Feynman's discussion; random static defects, in the cursor chain, produce localization, just as they do in the case of an electron in a disordered medium.[27,28] In that case, a small driving force, without the assistance of thermal vibrations, will not produce a computational drift velocity.

Zurek[14] points out that quantized systems, if we limit ourselves to discrete spectra, cannot have the equivalent of a slightly misdirected billiard ball. Classical systems readily permit a large probability for the occurrence of a very small error. Quantized systems, instead, favor errors which have the form of a small admixture of the wrong state, i.e., a small probability of a large error. This distinction seems, however, to be based more on the difference between discrete and continuous systems, than on the distribution between classical and quantal systems. In describing quantum computers we have felt free to invent arbitrary Hamiltonians, without the need to explain how the physical universe permits the construction of these Hamiltonians. Classical mechanics is accompanied by a tradition which does not encourage discussion of phase space mappings in such a sweeping way; we feel the need to draw pictures of the pieces, as if we were writing a patent disclosure. We do not, however, really need to allow, in Fig. 2, for a continuity of directions. We can idealize to a system in which the balls move along one-dimensional tracks and, occasionally, reach a junction where the interaction influences the choice of track. As far as the possibility for an incorrect velocity magnitude goes, do we really need to allow an arbitrary choice? Can we not idealize to a system which allows only one value for that?

The real world is not classical, it is quantum mechanical. If we are going to invent fictitious classical laws of motion, do they have to be Newton's? Can we not, instead, go directly to a discretized version of mechanics, as is done, for example by Greenspan?[29] Margolus[30] has investigated reversible cellular automata which simulate the dynamics of the billiard ball computer of Fig. 2. His cellular automata patterns map deterministically into each other, in a 1:1 fashion. Is that not a form of classical mechanics?

Consider the use of triple modular redundancy, abbreviated as TMR,[31] suggested by Zurek,[14] and described in detail by Peres,[18] to correct errors. In this scheme, all steps are carried out in triplicate (or higher odd multiples). After a suitable number of steps, results are compared. If two out of the three modules agree, the third processor is reset to agree with the other two. Resetting the error, in a digital system, without a continuous spectrum, need not involve any erasure; the erroneous value can be stored. During the reverse computation, however, when we restore the initial state, this can cause problems, and it is not clearly a worthwhile complication.

Let us now consider explicitly the case where, in a particular section of the computation, we generate a small admixture of an erroneous state, with a probability $\alpha \ll 1$, for the erroneous state. This error can affect many logic variables, at the next majority polling comparison in the TMR scheme.

These will not be independent errors; the errors will all occur in the same comparison module, if they occur at all. The quantum mechanical computer is really the ultimate EPR experiment in which the effects of one stage fan out and propagate to any other stages.[2] When an error does occur, however, as already stressed in an earlier section, we have no ready way to anticipate the correlation between errors in the variety of later logic variables. If we throw away the errors, we will be forced to treat the errors in the different logic variables as independent, and a separate energy dissipation penalty will be associated with each logic variable that has to be reset.

Errors in a computer can arise as a result of

1. Improper program loading.
2. Permanent hardware defects.
3. Hardware fluctuations.

If we reverse the computation, after completion, without any readout, and without any attempt at error correction, then errors of types 1 and 2 above would not prevent a return to the initial state. If, however, temporary fluctuations are present, we cannot expect a satisfactory reversal. Consider case 2, and a computational state

$$\psi = \gamma\psi_c + \beta\psi_e$$

with $|\gamma| \sim 1$, $|\beta| \ll 1$, $|\gamma|^2 + |\beta|^2 = 1$. $\psi_c$ is the desired computational state, and $\psi_e$ an erroneous digital state. The admixture of $\psi_e$ does not, by itself, prevent a satisfactory reversal of the computation, after completion. If, however, through error correction, or readout, we project onto one of the above states, $\psi_c$, or $\psi_e$, then we can no longer count on a completely satisfactory reversal. Error correction, however, during reversal, can lead us back to the desired initial state, with high probability.

If errors occur during the computation then, even in the presence of error correction methods, eventually an uncorrectable error will occur, when two out of the three modules are faulty. In the reversal process we cannot return to the original state, and the error will, very likely, have propagated so as to cause a great many erroneous bits in the supposed initial state. Thus, after enough computer programs have been run and reversed, but before a permanent error has had a high probability of occurrence, we will have to reset the computer's initial state. Resetting a bit to a known state, which it has reached with high probability, can be done with an energy expenditure small compared to $kT$, only requiring an energy expenditure proportional to the error probability.

[2] Implicit in Ref. 14, but also underlined by W. H. Zurek in private conversation.

## REFERENCES

1. J. A. Wheeler, in *Problems in Theoretical Physics*, A. Giovanni, F. Mancini, and M. Marinaro, eds. (University of Salerno Press, 1984), p. 121; J. Gliedman, *Sci. Digest* **92**(10), 34 (1984).
2. J. A. Wheeler, "Physics as Meaning Circuit: Three Problems," in *Frontiers of Non-equilibrium Statistical Physics*, G. T. Moore and M. O. Scully, eds. (Plenum Press, New York, in press).
3. R. Landauer, *Ber. Bunsenges.* **80**, 1048 (1976).
4. C. H. Bennett, *IBM J. Res. Dev.* **17**, 525 (1973).
5. C. H. Bennett, *Int. J. Theor. Phys.* **21**, 905 (1982).
6. R. Landauer, in *Computer Culture, The Scientific Intellectual and Social Impact of the Computer*, H. R. Pagels, ed., *Ann. N.Y. Acad. Sci.* **426**, 161 (1984).
7. R. Landauer and C. H. Bennett, *Sci. Am.* **253**(1), 48 (1985).
8. J. A. Wheeler and W. H. Zurek, *Quantum Theory and Measurement* (Princeton University Press, Princeton, 1983), pp. 782–785.
9. R. Landauer, in *Der Informationsbegriff in Technik und Wissenschaft*, O. G. Folberth and C. Hackl, eds. (Oldenbourg, Munich, 1986), p. 139.
10. R. Landauer and M. Büttiker, *Phys. Scri.* **T9**, 155 (1985).
11. E. Fredkin and T. Toffoli, *Int. J. Theor. Phys.* **21**, 219 (1982).
12. R. Landauer, *Int. J. Theor. Phys.* **21**, 283 (1982).
13. T. Toffoli, *Phys. Rev. Lett.* **53**, 1204 (1984).
14. W. H. Zurek, *Phys. Rev. Lett.* **53**, 391 (1984).
15. P. Benioff, *J. Stat. Phys.* **29**, 515 (1982); *Phys. Rev. Lett.* **48**, 1581 (1982).
16. D. Deutsch, *Proc. R. Soc. London A* **400**, 97 (1985).
17. R. P. Feynman, *Opt. News* **11**, 11 (1985); reprinted in *Found. Phys.* **16**, 507 (1986).
18. A. Peres, "Reversible Logic and Quantum Computers," to be published.
19. R. Landauer, *IEEE Spectrum* **4**, 105 (1967).
20. S. Wolfram, *Phys. Rev. Lett.* **54**, 735 (1985).
21. H. J. Bremermann, in *The Encyclopedia of Ignorance*, R. Duncan and M. Weston-Smith, eds. (Pergamon, New York, 1977).
22. T. Wright, *The Sciences* **25**(1), 7 (1985).
23. K. Zuse, *Int. J. Theor. Phys.* **21**, 589 (1982).
24. T. D. Lee, manuscript contribution to symposium, *Old and New Problems in Fundamental Physics*, held in honor of G. C. Wick (Pisa, Italy, October 25, 1984); R. Friedberg and T. D. Lee, *Nucl. Phys.* **B225**, 1 (1983).
25. R. P. Feynman, *Int. J. Theor. Phys.* **21**, 467 (1982).
26. D. K. Ross, *Int. J. Theor. Phys.* **23**, 1207 (1984).
27. *Localization, Interaction and Transport Phenomena in Impure Metals*, G. Bergmann, Y. Bruynseraede, and B. Kramer, eds. (Springer, Heidelberg, 1985).
28. D. J. Thouless, *Physica* **126B**, 18 (1984).
29. D. Greenspan, *Int. J. Theor. Phys.* **21**, 505 (1982).
30. N. Margolus, *Physica* **10D**, 81 (1984).
31. D. A. Rennels, *IEEE Trans. Comput.* **C-33**, 1116 (1984).

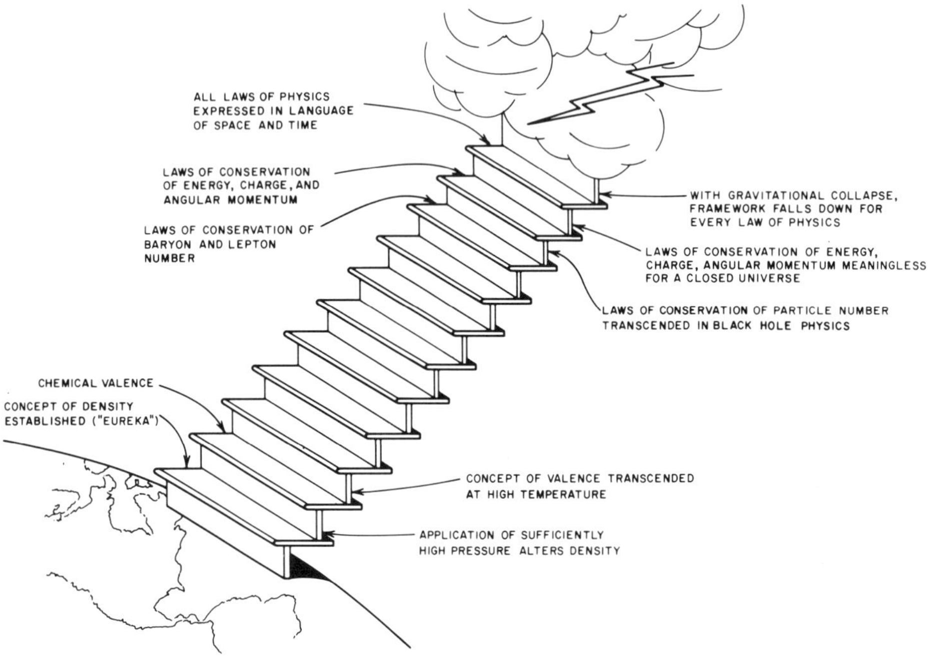

ALL LAWS OF PHYSICS EXPRESSED IN LANGUAGE OF SPACE AND TIME
LAWS OF CONSERVATION OF ENERGY, CHARGE, AND ANGULAR MOMENTUM
LAWS OF CONSERVATION OF BARYON AND LEPTON NUMBER
WITH GRAVITATIONAL COLLAPSE, FRAMEWORK FALLS DOWN FOR EVERY LAW OF PHYSICS
LAWS OF CONSERVATION OF ENERGY, CHARGE, ANGULAR MOMENTUM MEANINGLESS FOR A CLOSED UNIVERSE
LAWS OF CONSERVATION OF PARTICLE NUMBER TRANSCENDED IN BLACK HOLE PHYSICS
CHEMICAL VALENCE
CONCEPT OF DENSITY ESTABLISHED ("EUREKA")
CONCEPT OF VALENCE TRANSCENDED AT HIGH TEMPERATURE
APPLICATION OF SUFFICIENTLY HIGH PRESSURE ALTERS DENSITY


*Plate 38*

*Foundations of Physics, Vol. 16, No. 6, 1986*

# On Wheeler's Notion of "Law without Law" in Physics[1]

**David Deutsch**[2]

*Received August 6, 1985*

*Wheeler's idea that physical "laws" would not appear in a truly fundamental description of nature is critically examined.*

Is it possible that there is an ultimate law of physics, a principle $\mathscr{P}$ from which follows everything that is knowable about the material world? If so, what can we already infer or postulate about the form that this principle takes? Wheeler's idea of "law without law" is an attempt to begin to answer these questions.

If there were no all-explanatory physical principle $\mathscr{P}$ approachable by the methods of science, this would presumably mean that there exist aspects of the natural world that are fundamentally inaccessible to science. This would run directly counter to rationalism and to our view of physics as the universal science, which have hitherto been the driving forces behind progress in the subject and which we should be extremely reluctant to abandon.

But if $\mathscr{P}$ were itself a law of physics, then the problem of $\mathscr{P}$'s own origin—why that particular principle holds in nature rather than some other—would be forever insoluble. And hence $\mathscr{P}$ would not be all-explanatory within physics. So, paradoxically $\mathscr{P}$, the ultimate principle of physics, cannot be a "law" (of physics). Hence the expression "law without law."

[1] In honor of J. A. Wheeler's 75th birthday.

[2] Department of Astrophysics, Oxford University, South Parks Road, Oxford OX1 3RQ, England, and Center for Theoretical Physics, University of Texas at Austin, Austin, Texas 78712.

In what follows, I shall be criticizing Wheeler's actual proposals for a "law without law." But however flawed they may be, the motivating argument that I have just summarized, that $\mathscr{P}$ must exist but cannot be a physical law, seems to me to be very compelling and I shall not quarrel with it. Indeed, at the end of this article I make some suggestions of my own about what $\mathscr{P}$ must look like.

There seems to be a consensus among physicists that physics is now, for the first time, approaching the point where its fundamental theories will be truly universal in their intended scope. We hope soon to be constructing theories, not about this or that special type of physical system, as has always been the case in the past, but about the whole of material reality. Furthermore, many believe (see, e.g., Ref. 1) that nothing qualitatively new lies between us and such a unified theory of nature. There are problems, but it is widely expected that their solutions will take the same *form* as our existing theories.

Certainly quantum theory, the general framework within which all other physical theories are currently formulated, has no known experimental counterexample and no serious rival theory. Within the quantum framework, a physical theory amounts to the specification of an action functional and the quantum field configurations that constitute its domain. Thus the expectation is that we shall soon disover the ultimate "fundamental fields," their invariance properties, and their action functional. The principle of the stationarity of this action functional, together with the principles of quantum theory, would be the ultimate dynamical "laws of nature."

This consensus has been repeatedly attacked by Wheeler.[2,3]. He does not necessarily deny that we are now close to constructing universal *dynamical* theories. But he argues that by their very nature dynamical laws cannot be fundamental. The idea of an immutable dynamical law, far from being a feature we should expect in an "ultimate" description of the world, is in fact only characteristic of the temporary, incomplete descriptions we have had to resort to along the way.

For dynamical laws, at least as we have known them, are always expressible as conservation laws, or symmetry or invariance principles. And Wheeler argues that symmetry and invariance principles are always indicative of a level of structure *below* the one being described. Yet at the same time they mask the nature of this structure. For example, according to the ideal gas law

$$\frac{pV}{nT} = R$$

$n$ moles of gas at constant volume $V$ exert a pressure $p$ on their container proportional to the temperature $T$. Superficially this law gives no clue that there is a level of description of gases deeper than that via "pressure" and "temperature." But, many years after the ideal gas law was first postulated, its derivation from kinetic theory gave us a compelling token of the existence of an underlying structure, i.e., of the existence of molecules. Even that told us very little indeed about the molecules themselves. We can learn a little more about molecular and atomic structure by measuring small deviations from the ideal gas law, but this is an uphill struggle. The very concepts of "pressure" and "temperature" are fundamentally misleading. Nature at the molecular level cannot even be described in terms of them. Thus the complex form of the underlying theory is masked, rather than revealed, by the very "law" which is the token of its presence.

An invariance principle asserts that some physical quantity is *constant*, or is *independent* of certain variable parameters. For example, the gas law says that $R$ is independent of $p$, $V$, $T$, and $n$. This "immutability" worries Wheeler who points out that many of the quantities thought immutable in the past have later turned out to be very mutable under suitably extreme conditions. (And perhaps we should also bear in mind that this is not the first time in history that physicists have thought they were close to an all-explanatory theory.) Of course we must expect obsolete physical theories to contain obsolete invariance principles, so the fact that our old theories were wrong lends no credence to the proposition that our new ones will be too! But Wheeler is interested not so much in the demise *per se* of certain invariance principles, but in the manner of their demise of late. In the old days, a theory was abandoned when it turned out to be false; i.e., when it was refuted in favor of a truer theory, of which it became a limiting case. Ohm's law is a favorite elementary example, but for more recent cases one might cite the law of conservation of mass, which is violated in the special theory of relativity, or Euclid's parallel axiom which is violated in the general theory of relativity. But recently, some invariance principles have not really been *refuted* by the discovery of exceptions, or conflicts with better theories, but rather what Wheeler calls "*transcended*" by our increasing understanding of existing theories. A theory is transcended when it is found to be not so much less *true*, as less *fundamental*, than had previously been thought. The conservation law for baryon number, for example, was transcended when it was discovered that a black hole evaporating via the Hawking process emits equal numbers of baryons and antibaryons, even if it was originally formed entirely out of baryons. This would happen even if the divergence of the baryon current was everywhere zero in space-time. The success of the "big bang" cosmological model, and the singularity theorems of Penrose and Hawking (see Ref. 4) have led us to consider seriously the possibility that *all* our conservation laws must be transcended sooner or later, for at a space-time singularity the passage of time itself is not "immutable": time ends (or begins).

Perhaps we are indeed approaching the final truth of *dynamics*, only to find that dynamics is not, after all, fundamental in physics. What is fundamental, then? It must be conceded there are some physical problems which there is no reason to expect would be solved even if we knew the ultimate dynamical laws. These may be classified into two broad categories:

1. How is the actual solution of the dynamical equations distinguished from the infinity of possible ones?
2. Whet underlying principles require the dynamical laws to take this, and not some other form, and what principles unify the dynamical laws, the quantum framework, and the solution of (1)?

Prominently among problems of type (1) I would list

The origin of thermodynamic phenomena such as obedience to the second law of thermodynamics.

The nature of the quantum state of the universe (see, e.g., Refs. 5–7).

Determination of the boundary conditions and initial conditions on quantum fields.

And among problems of type (2):

The "factor ordering problem" in quantum theory, or the problem of finding a consistent operator formalism to replace the stopgap of "quantization."[(8)]

The relationship between physics and the theory of computation.[(9,10)]

Such problems have traditionally been regarded as secondary in importance, or even outside the scope of physics altogether. Wheeler[(3)] looks forward to a "third era" of physics in which they will be primary.

But Wheeler has intentions even more radical than this. It is really not just dynamics that he wants to regard as ultimately ephemeral, but the whole notion of a physical law. For suppose we were successful in solving all the type (1) and type (2) problems. Would the solution be a physical law or principle? Wheeler would say "no."

Our dilemma in contemplating the possibility of an all-explanatory physical theory $\mathscr{P}$ can be summarized thus: *Is* $\mathscr{P}$ *an analytic or a synthetic proposition*? An analytic proposition is one whose truth or falsity is a consequence of logic alone. A synthetic proposition is a meaningful proposition that is not analytic. Physical theories as we know them are always synthetic, because they must make testable assertions. An analytic proposition (that is true) makes no assertion about the world. It is empty. Thus, if $\mathscr{P}$ is to have testable theories as its consequences, it must itself be synthetic. On the other hand, if $\mathscr{P}$ is synthetic, it must contain some "arbitrary" postulates not deducible from anything more fundamental, so how can it be "all-explanatory"? How can it explain why these particular postulates, and not some others, are true in nature.

In the past, Wheeler and others have attempted to escape from this dilemma by seeking some fundamental mathematical structure with which the physical world could be identified, so that physical laws would have the status of theorems. Thus he[(11)] suggested that underlying geometry there might be a "pregeometry" with the structure of symbolic logic. Similarly Penrose[(12)] has suggested that "We [may all be] composed of abstract mathematics." Maybe so, but I believe that such approaches do not really address the problem of what $\mathscr{P}$ may be. For one cannot derive something synthetic (physics) from something analytic (mathematics). Some of the physical laws may be "theorems." But there will always be others which are "axioms." The entire physical content of such theories would lie in the method of identification of the physical world with the abstract mathematical structure.

Another possibility suggested by Wheeler is that underlying the physical laws there is no structure at all, just "chaos." In his words "There is no law except the law that there is no law." He had in mind the emergence of approximate lawlike behavior, such as the ideal gas law, from an underlying chaotic complexity like that of the motion of myriads of molecules. But again this proposal fails to address the problem. Lawlike behavior could never emerge from something that obeyed no laws. The motion of molecules is not lawless, it is *random*. And randomness implies a probability distribution function, and that function, or the principle from which it may be derived, is the missing underlying law.

Wheeler has recognized this deficiency in his original "order from chaos" idea, and has now modified it to

"Chaos" + "*regulating principle*" → approximate "laws"

In the case of kinetic theory, the "regulating principle" is the "principle of equal *a priori* probability" of statistical mechanics. But in making a fundamental "regulating principle" the new object of his search, Wheeler *ipso facto* abandons the search for "law without law." For a regulating principle is a law, and his modified proposal can be rewritten in the uninspiring form

"Stochastic laws" → approximate "deterministic laws"

Yet another possibility that Wheeler has investigated from time to time (e.g., Refs. 2 and 13) is that the underlying reality is composed entirely out of "individual acts of observation" (or "consciousness" or "measurement").

This sort of proposal has a long history, though not so much in physics as in antirealist philosophies of science such as subjectivism, instrumentalism, and logical positivism. I refer the reader to Popper's[14] refutations of these. Admittedly the fact that a proposed new physical theory is "philosophically untenable" should never deter a physicist: It is more likely that philosophy is wrong than physics. But I believe[15] that Wheeler's physical motivation in this case is also misguided, deriving as it does from an erroneous interpretation of quantum theory.

Nevertheless this last proposal, though hopeless in itself, has one salvageable feature which I believe really is a clue to the nature of $\mathscr{P}$: Instead of trying futilely to get "something for nothing," a synthetic proposition from an analytic, it places at the foundation of physics assertions which are synthetic (e.g. that there are "acts of observation" in the world), but which *have to be postulated anyway* for some reason outside physics.

What are these "synthetic propositions which have to be postulated anyway"? The idea that there are such propositions and that physical theories can be deduced from them is not new. Wheeler's appeal to the existence of acts of observation is also Descartes' ultimate source of justification of beliefs "*cogito ergo sum.*" And Kant made a whole collection of "synthetic *a prioris*," which were synthetic propositions that he believed were necessary for thought.

However, modern rationalist epistemology gets along without ever "justifying" beliefs; no method of justification has any authority. Nor need we nowadays accept that any synthetic proposition is "necessary for thought." After all, Kant included Euclidian geometry among his "synthetic *a prioris*," yet no physicist nowadays believes that Euclidian geometry is a true theory, let alone "necessary."

We cannot hope to find any synthetic proposition that can be proved or established in any way as necessarily true, or even probable. What we can do, however, is to examine more closely and systematically what synthetic propositions we are in fact, for whatever reason, assuming *a priori* (by which I now mean merely "prior to any physical principle"). For although there is no synthetic proposition so self-evident that we *must* assume it true, it is equally the case that we cannot make progress without making some *a priori* assumptions, even though as rationalists we should do this tentatively and always be ready to criticize and replace any one of them.

One thing that we always tacitly assume *a priori* in the search for any physical theory is that the physical process of that theory becoming known and expressed is not in itself forbidden by the theory. No physical principle that we can know can itself forbid our knowing it. That *every* physical principle must satisfy this highly restrictive property is a synthetic *a priori* proposition, not because it is *necessarily* true, but because we cannot help assuming it to be true in seeking to know the principle.

Very similar arguments motivate the "anthropic principle" in its various forms.[16] For example, it has been argued that the values of certain fundamental physical constants may be inferred with some precision from the very fact that someone exists who has asked what their values are.

It is hard to make watertight inferences from the anthropic principle because they always depend on assumptions about what physical conditions are necessary for the emergence of "someone who asks." However, the *a priori* meta-principle that "the theory we are constructing cannot say that we shall not construct it" can be expressed without anthropic references:

> *Successive approximations to every physical principle can be encoded in the states of realizable physical systems.*

Notice that this can be postulated independently of the existence of an all-explanatory principle $\mathscr{P}$. But if $\mathscr{P}$ exists, this meta-principle is surely part of it.

Another related set of hypotheses which has been formulated recently deal with the role of computation and computability in physics. Tipler (Ref. 16; cf. also my "law of increase of $Q$-logical depth," Ref. 9) has derived physical, in principle testable consequences from the postulate that the amount of information that can be processed in the lifetime of the universe be unbounded. Several authors have proposed (e.g., Ref. 10), for various reasons, that the laws of physics must be "computable." And I have proposed[9] that the Church–Turing hypothesis[17] be reformulated as the following physical principle:

> *Every physical system can be perfectly simulated by a universal model computing machine operating by finite means.*

In a broad sense what links all these ideas is that they attempt to derive physical consequences from postulates outside physics, specifically from epistemological assumptions, preferably ones we were "going to make anyway." If this turns out to be successful as a research program, another way of looking at it would be that epistemology is now being incorporated into physics, just as the scope of physics has historically been successively extended to take over other sciences such as astronomy, chemistry, geometry, and most recently computer science.

## ACKNOWLEDGMENT

This work was supported in part by NSF grant No. PHY7826592.

## REFERENCES

1. S. W. Hawking, "Is the End in Sight for Theoretical Physics?" Inaugural address as Lucasian Professor, Cambridge University, 1979.
2. J. A. Wheeler, in *Relativity, Quanta, and Cosmology*, F. de Finis, ed. (Harcourt-Brace-Jovanovich, New York, 1979).
3. J. A. Wheeler, *Am. J. Phys.* **51**(5), 398 (1983).
4. S. W. Hawking and G. F. R. Ellis, *The Large-Scale Structure of Space-Time*" (Cambridge University Press, Cambridge, 1973).
5. B. S. DeWitt, *Phys. Rev.* **160**, 1113 (1967).
6. S. W. Hawking, *Nucl. Phys.* **B239**, 257 (1984).
7. D. M. Page, in *Quantum Concepts in Space and Time*, R. Penrose and C. Isham, eds. (Oxford University Press, Oxford, 1986).
8. D. Deutsch, in *Quantum Theory of Gravity, Essays in Honor of the 60th Birthday of Bryce DeWitt, S.* Christensen, ed. (Adam Hilger, Bristol, 1984).
9. D. Deutsch, "Quantum Theory, the Church–Turing Principle, and the Universal Quantum Computer," *Proc. R. Soc. London A* **400**, 97 (1985).
10. T. Toffoli, *J. Comput. Sys. Sci.* **15**, 213 (1977).
11. C. W. Misner, K. S. Thorne, and J. A. Wheeler, *Gravitation* (W. H. Freeman, San Francisco, 1973), p. 1211.
13. J. A. Wheeler, in *Proceedings of the Nato Advanced Study Institute Workshop on Frontiers of Nonequilibrium Physics*, 1984 (Plenum, New York, 1985).
14. K. R. Popper, *Objective Knowledge* (Oxford University Press, Oxford, 1972).
15. D. Deutsch, *Int. J. Theor. Phys.* **24**, 11 (1985).
16. J. Barrow and F. Tipler, *The Anthropic Cosmological Principle* (Oxford University Press, Oxford, 1986).
17. D. R. Hofstädter, *Gödel, Escher, Bach, an Eternal Golden Braid* (Random House, New York, 1979).

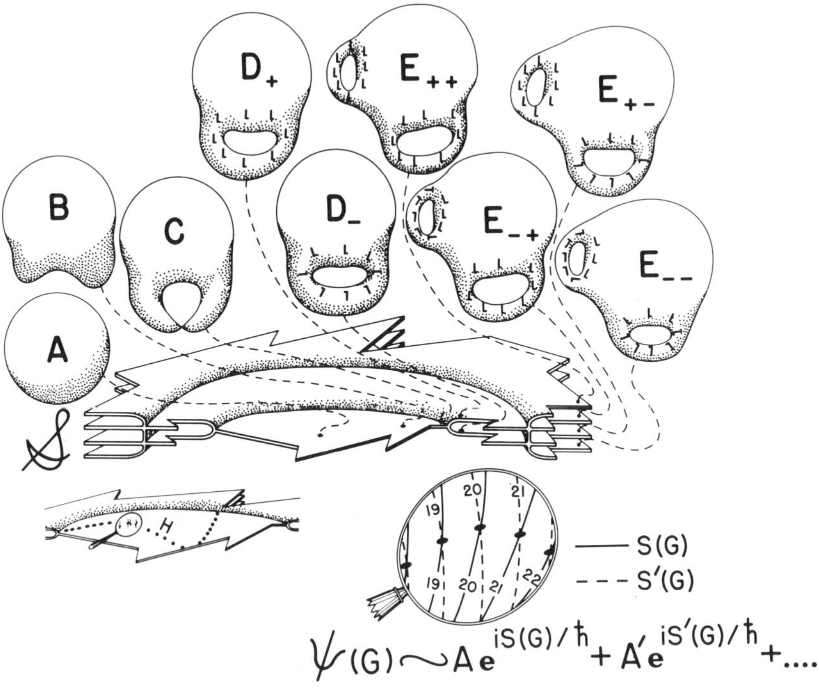


*Plate 39*

# Existence of "Free Will" as a Problem of Physics[1]

**Asher Peres**[2]

*Received June 17, 1985*

*The proof of Bell's inequality is based on the assumption that distant observers can freely and independently choose their experiments. As Bell's inequality is* experimentally *violated, it appears that distant physical systems may behave as a single, nonlocal, indivisible entity. This apparent contradiction is resolved. It is shown that the "free will" assumption is, under usual circumstances, an excellent approximation.*

I have set before you life and death, blessing and cursing: therefore choose life....

—*Deuteronomy* XXX, 19

## 1. BELL'S THEOREM

Consider two observers equipped with various measuring apparatuses. These observers are very distant from each other, so that *no communication* between them is possible while they are doing their experiments. For example, they may perform these experiments at times and places bearing a space-like relationship, as in Fig. 1. Nevertheless, the observers are able, at a later stage, to *compare* their results by sending *messages* to one another. Moreover, they may prearrange their experiments so that they observe *correlated* physical systems, such as particles emitted by a single source, somewhere in the common past of *both* observers (in the intersection of their past light-cones). We now make the following assumptions[(1,2)]:

(i) The first observer has the *choice* of measuring a dynamical variable $A$, with possible results $a = \pm 1$, *or* a dynamical variable

[1] Dedicated to Prof. John A. Wheeler on the occasion of his 75th birthday.

[2] Department of Physics, Technion-Israel Institute of Technology, 32 000 Haifa, Israel.

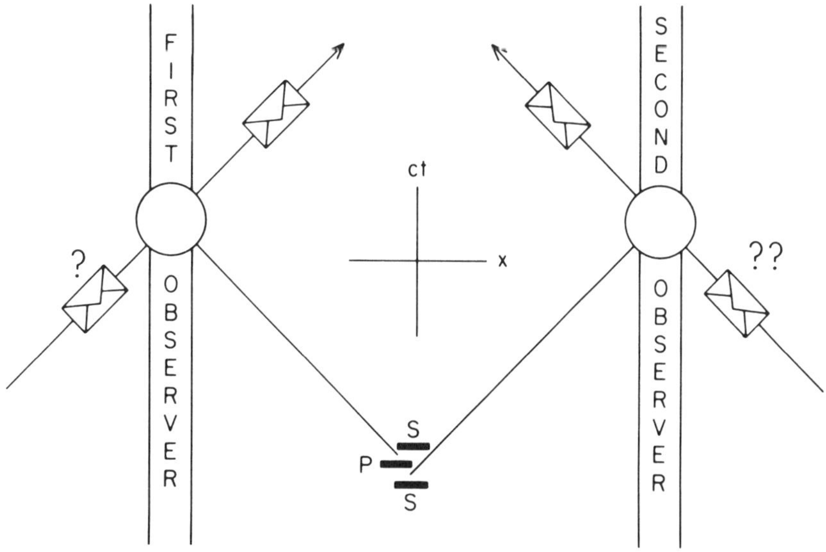


Fig. 1. Space-time diagram showing the world tubes of two observers (assumed mutually at rest), the trajectories of two photons emitted by a single atom (SPS cascade), and messages emitted by the observers in order to compare their results. The observers may select their experiments by "free will" or by means of random messages received from disconnected parts of the Universe (as in Fig. 3, below).

$C$, with possible results $c = \pm 1$. Likewise, the second observer has the *choice* of measuring a dynamical variable $B$, with possible results $b = \pm 1$, *or* a dynamical variable $D$, with possible results $d = \pm 1$.

(ii) Each one of these possible outcomes ($a$, $b$, $c$, and $d$) is *unpredictable* by the observer, but nevertheless *deterministic*.

(iii) Any outcome obtained by *one* of the observers does not depend on the *choice* of the experiment which is performed (was performed, will be performed) by the *other* observer.

Let us carefully examine these three assumptions:

(i) Each observer *can select* one of two possible *macroscopic procedures*, such as orienting a Stern–Gerlach magnet along the $x$ direction *or* the $y$ direction. These procedures are *mutually exclusive*, but each one is possible and is *freely* chosen by the observer. If we do not want to involve human free will, these procedures can be selected by automated instruments involving random elements. For example, we may have two telescopes pointing at opposite sides of the sky, and counting whether an odd or even number of photons arrive in a predetermined time interval.

Each one of these possible *macroscopic procedures* is called, euphemistically, the "measurement of a dynamical variable," (such as spin, polarization, etc.). These terms should *not* be understood in their *classical meanings*. Nowhere in this paper is it assumed that there exist dynamical variables having numerical values which are originally unknown to the observers, and then are revealed to them by means of measurements.[3] It is only assumed that some macroscopic procedures—called "measurements"—have various macroscopically distinct outcomes, which are conventionally labeled as $a = \pm 1$, etc. An example is shown in Fig. 2.

(ii) The second assumption, which I call *cryptodeterminism*, must be scrutinized very carefully. Consider a classical pinball machine. The trajectory of each ball is perfectly deterministic. Yet, it is utterly unpredictable. This system is *chaotic*[4]: tiny changes in the initial data, or tiny external perturbations, cause huge changes in the final outcome. Only statistical averages can be reliably computed. The same situation prevails in the classical kinetic theory of gases. In principle, all motions are perfectly deterministic. However, because of incomplete knowledge of all initial data, only statistical predictions can be made.

Likewise, the outcomes of quantum measurements are deemed "random" and only their statistical averages are predictable. Are they *really* random? This point of view is supported by many authors. For example, Landé[5] asserts that "the random distribution is a physical reality, and the determinism which only looks random is a purely academic construction." Then, Landé adds the following explanation: "A distribution of effects

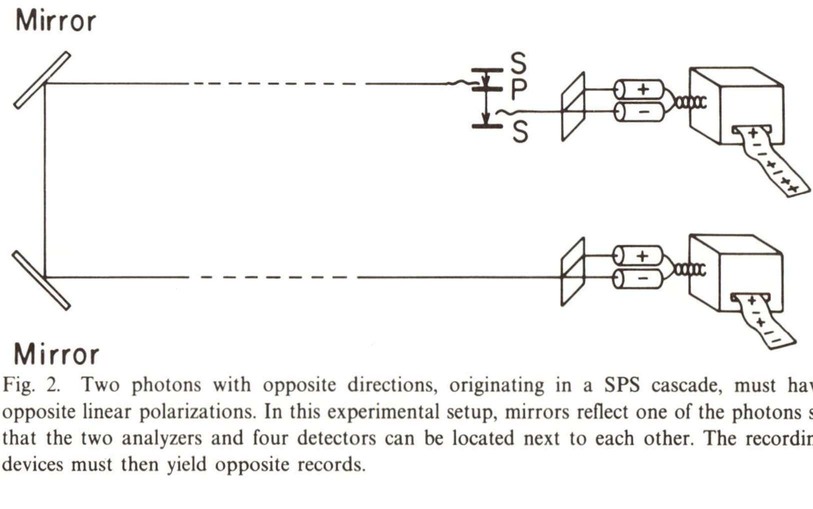


Fig. 2. Two photons with opposite directions, originating in a SPS cascade, must have opposite linear polarizations. In this experimental setup, mirrors reflect one of the photons so that the two analyzers and four detectors can be located next to each other. The recording devices must then yield opposite records.

satisfying the laws of error theory requires, just from the determinist's viewpoint, a corresponding distribution of causes at an earlier time and from there back to a still earlier time. A program of giving a strictly deterministic theory of statistically distributed events leads nowhere." Why nowhere? In the worst case, it may lead in a direction which Landé might, perhaps, have found hazardous. This *is* indeed the direction in which I intend to lead the reader of this essay.

Consider the experiment shown in Fig. 2, whereby one observes the polarizations of photon pairs originating in a SPS cascade. (Another possibility is a $2S-1S$ transition[6] which is a second-order two-photon decay process.) Each recording device yields a sequence of $+1$ and $-1$ which *appears* perfectly random: This sequence would pass every known test for randomness.[7] Yet, if one of the two sequences is known, the other one can be immediately guessed—it is the opposite of the first one. In particular, if the delay in observing one photon is long enough, one can predict with certainty the outcome of a measurement of its polarization from the result of a measurement already performed on the other photon. It thus appears that the outcome of such a polarization measurement is not at all random. It is perfectly determined, although the whole sequence may *appear* random if we lack *relevant information.*

This suggest the possible existence of *additional variables* (the so-called "hidden variables") which would be the "relevant information" allowing to one predict with certainty the outcome of any experiment. If we have no knowledge at all about these additional variables, we can still compute *averages* over their possible values. These averages ought to coincide with those obtained from quantum theory. Bell's celebrated paper[1] was formulated in a language appropriate to a hidden variable theory (without specific details of the theory). Here, I shall involve no theory at all and stay on strictly phenomenological grounds.

One may feel somewhat uneasy about assumption (ii) because it seems to contradict the spirit, if not the letter, of assumption (i). Observers, after all, are made of ordinary matter and are subject to the ordinary laws of physics.[2] Thus, why should observers be endowed with "free will" while the observed objects behave deterministically? Can't an observer be observed too? These difficult questions may be avoided by using, instead of living observers, some automated devices such as telescopes pointing toward opposite sides of the sky, as explained above. Assumption (i) may then be replaced by:

(i′) The *choices* of the experiments (whether to measure $A$ or $C$, and whether to measure $B$ or $D$) can be performed in a way involving *no a priori correlations.*

In other words, *there are neither physical laws nor initial conditions* imposing that if one measurement is, say, $A$ (rather than $C$) then the other measurement must be $B$, or must be $D$. It seems quite obvious that there is no such physical law, which is valid for *any* initial conditions. On the other hand, the hypothesis that we are not subject to special initial conditions, imposing a unique solution for each pair of measurements, is a matter of faith. One could also imagine that there is a worldwide conspiracy whereby everything is predetermined.

(iii) This third assumption is called *locality*, or *separability*. It should be clearly understood that the *result* obtained by each observer may well be correlated to the *result* obtained by the other one, if the particles which they test were correlated in the past (see, e.g., Fig. 2). However, it is reasonable to assume[8] that if the first observer measures $A$ and obtains $a = \pm 1$, say, this result cannot depend on whether the second observer decides (has decided, will decide) to measure $B$ rather than $D$.

It should be noted that assumption (iii) is *logically meaningless* unless we accept assumption (ii). Indeed, when we formulate (iii), we rely on counterfactual reasoning. We consider the results of two *possible but mutually exclusive* measurements.[9] This has a meaning only if these results are definite, even though only one of them can actually be known.

Having explained all these premises, it is easy now to prove Bell's theorem. Consider a sequence of pairs of measurements, as discussed above. For the $j$th pair, the outcomes are $a_j$, $b_j$, $c_j$, and $d_j$. These numbers are unknown, but they can be only $\pm 1$, so that

$$a_j b_j + a_j d_j + c_j b_j - c_j d_j \equiv \pm 2 \tag{1}$$

Taking an average over all values of $j$ (the serial number $j$ effectively plays the same role as Bell's hidden variables), we obtain

$$|\langle ab \rangle + \langle ad \rangle + \langle cb \rangle - \langle cd \rangle| \leqslant 2 \tag{2}$$

Now, although all the numbers $a_j$, $b_j$, $c_j$, and $d_j$ cannot be simultaneously known, *average values* such as $\langle ab \rangle$, namely the correlation of $A$ and $B$, *can* be measured experimentally. There have been many such experiments, the most famous of which, due to Aspect *et al.*,[10] is schematically shown in Fig. 3. The experimental result clearly *violates* inequality (2). Therefore, at least one of the assumptions (i′) or (ii) or (iii) must be wrong.

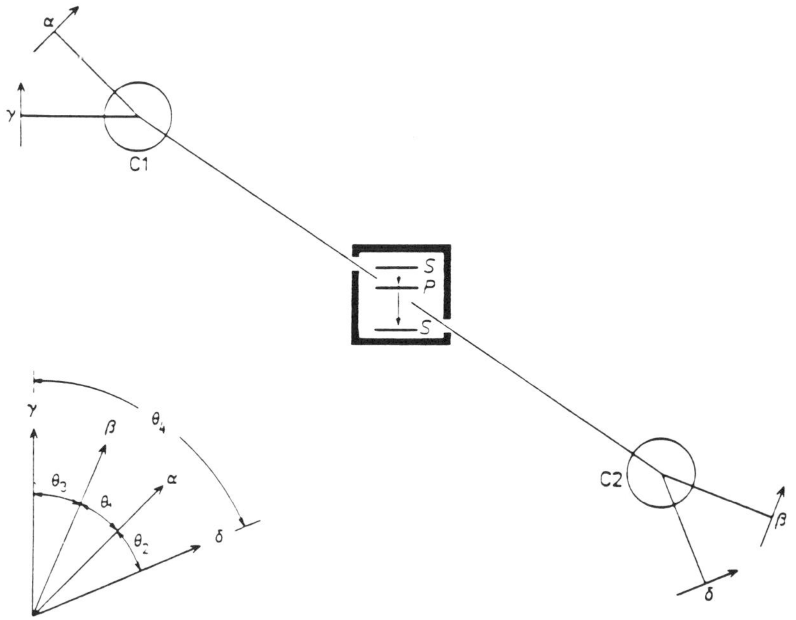


Fig. 3. A theorist's view of Aspect's experiment. Two photons with opposite polarizations pass through random optical commutators C1 and C2, toward polarization analyzers oriented along the directions $\alpha$, $\beta$, $\gamma$, and $\delta$. The diagram in the left corner shows the set of angles (three times $\pi/8$) giving the maximum violation of Bell's inequality.

## 2. ABANDON SEPARABILITY?

One may be reluctant to abandon (i) or (i′). As stated by Bell,[11] "separate parts of the world would be deeply and conspiratorially entangled, and our apparent free will would be entangled with them."

One is even more reluctant to abandon (ii) for which some experimental support is suggested, if not formally proved, by setups such as the one described in Fig. 2. Anyway, as explained above, if one repeals (ii), then (iii) becomes meaningless.

On the other hand, there is neither direct nor indirect experimental support for (iii). The only thing which we know for sure is that distant actions cannot have an immediate (or retroactive!) influence on results of *classical* measurements, or on quantum measurements which happen to be quasi-classical because they are predictable with 100% certainty, or on predictable quantum *average* values such as $\langle ab \rangle$, etc. None of these conflict with inequality (2). In particular, classical correlations always satisfy this inequality.[12]

We are thus led to abandon locality and consider the *pair* of photons in Aspect's experiment as a *single, nonlocal, indivisible entity.*[3,13] This may appear strange at first, but this is not essentially different from considering a *single* photon as a nonlocal, indivisible object: This is indeed how we explain that a single photon can pass through two widely separated slits and interfere with itself. A single photon may even originate in two different lasers.[14] (Actually, the situation is even more curious. In the Aspect experiment, the photons are distinguishable because they have different energies, and *almost* everything happens as if there were two distinct particles with reasonably well-defined positions and momenta at each instant. Only the *polarizations* of the photons are inseparably entangled.)

We have now reconciled all our knowledge with the violation of Eqs. (1) and (2). There are *no* physical quantities such as $a_j$: One must distinguish $a_{jB}$ from $a_{jD}$ (namely, the values obtained for $a_j$ if the second observer measures $B$ or $D$, respectively). One may even estimate how often $a_{jB} \neq a_{jD}$: this must happen in more than 41% of cases.[15]

However, once we accept that, doubts about (i) or (i′) appear again, as new Hydra's heads! Why can't our two observers too be considered as a single, nonlocal, indivisible entity? Their past histories are undoubtedly correlated, since they agreed to collaborate in some joint experiment. Why do we believe that each observer has a free will, although he or she consists of atoms subject to deterministic (or cryptodeterministic) laws?

Consider again the situation illustrated in Fig. 2. Each one of the records appears "random" (as if the corresponding apparatus had "free will") in the *absence of information* on the other record. That information is essential, because the two photons have *correlated histories.* They are, as already explained, a single, indivisible, nonlocal object. If we ignore that correlation and make no use of the corresponding information, then *every* outcome looks random. Likewise, when people are considered individually, and all their past interactions with other people or things are ignored, they appear to behave randomly—to have "free will."

Is our apparent free will limited by some unknown correlations? Let us again look at Fig. 2. Randomness was not completely eliminated by correlating the two sequences of results. While we can predict the second sequence after looking at the first one, we cannot predict the first one. Some randomness remains. Could we add information and reduce this randomness again? To do this, we would need to trace back the history of the atom which emitted the SPS cascade. If it was emitted by a coherent process, it might be possible to correlate these two photons with other observable phenomena. On the other hand, if the excitation process was thermal, then further correlations are *practically* lost. Turning now our attention to considerably more complex systems, such as human beings, it is obvious that any EPR[8] correlations must be immediately destroyed in a maze of irreversible processes. *Each one* of us thus behaves unpredictably, as if endowed with free will. This is why words such as "each one" can legitimately be used when we describe the behavior of people.

It *is* an excellent approximation to use pronouns such as "he" and "she" in everyday's life. Lawyers who prepare contracts *can* consider the various signatories as distinct entities. Indeed, these signatories are very complicated systems, which cannot be described completely. We know that actually they are only parts of an interconnected universe, but in our *approximate* description, they *appear* to be distinct and isolated. They behave "randomly." They have free will.

In our daily work as physicists, we are compelled to use only *incomplete* information on the world, because we cannot know everything. The methodology which we use in physics is the following. We divide the world in three parts, which may be called the *system*, the *observer* (or the observing apparatus), and the *rest of the world* (that is, most of it!) which is then considered as irrelevant. We assume that the system under observation is perfectly isolated from everything else, except from the observer testing it, if and when it is tested. This makes things *appear* simple. *This* is what makes physics an exact science. For example, we can compute the properties of the hydrogen atom to umpteen decimal places, because when we do these calculations there is only one hydrogen atom in our conceptual world.

We work as if the world could be dissected into small independent pieces. This is an illusion, because, as Bell's theorem shows, the entire world is interdependent. But there is no other way of working. The practical question with which we are faced is the following: given a finite amount of information, what are the possible outcomes of an ill-defined experiment which we prepare? The answer must be probabilistic, by the very nature of the problem. One may only speculate whether, in a complete description of the whole Universe, including our brains, determinism would be restored.

I don't know the answer to this question. Only God can know that, if He or She exists. However, since this is an essay on science, and not on theology, I shall now turn my attention to a more practical problem.

## 3. INFORMATION TRANSFER

Assuming that measurements are instantaneous (that is, very brief), can we use these nonlocal properties to transfer information *instantaneously* between distant observers? If we could, Einstein's theory of relativity would be in jeopardy.

There have been some attempts in this direction[16,17] and also some attempts to prove that this is impossible.[18–20] Why did the question arise? After all, nonlocal correlations do exist in classical physics. A classical analogue of the Aspect experiment can easily be imagined as follows: A bomb, initially at rest, explodes into two fragments carrying opposite angular momenta. Two observers measure the projections of these angular momenta along arbitrary directions and find them correlated.[12] Obviously, there is nothing in this scenario allowing one observer to send a signal to the other. Both observers only *receive* information from a common source.

How does this differ from the EPR effect? The point is that the angular momenta of the bomb fragments are believed to have objective values, even before they have been measured. On the other hand, the second photon in Fig. 2 *acquires* a definite polarization as a result of the "collapse of the wavefunction" resulting from the measurement of the polarization of the first photon. If this collapse is naively considered as a real physical phenomenon—which is nonsense[3,13,21]—there is here some new feature which was absent in classical physics.

Can it be used to convey information? For example, can the second observer guess what was the orientation of the apparatus used by the first one? Any proofs purporting to show that this is impossible, by using the Schrödinger equation (or other features of quantum theory), are not acceptable, because the "collapse" itself blatantly violates Schrödinger's equation. Thus, no proof of nonexistence has been proved to exist, as yet.

To prove impossibility, we must *postulate* (in agreement with quantum theory, but without relying on it explicitly) that

(iv) no measurement of a physical system can give us information about its state prior to the measurement, unless some information was already available beforehand.[22,23]

For example, if we know that a photon is certainly polarized along the $x$ direction or the $y$ direction, but we do not know which one of the two, we can ascertain this point by a "nondemolition measurement" that leaves the initial state unchanged. However, if we know *nothing* about the initial state, there is no way to discover it by experimenting on a *single* photon.[24] The most we could learn from such a measurement is that the initial state was not in the orthogonal complement of the final one. Postulate (iv) must be satisfied by any theory claiming that a physical system has a well-defined state, which can change abruptly when a measurement is performed. Otherwise—that is, if (iv) is not valid—it would be possible to convey information instantaneously by measuring correlated systems.

There is one more point in need of clarification. I have not yet properly defined what is information. What is a message? Intuitively, it is a relationship between a *decision* and an *observation*. When I write a letter, I can decide what to write—this is my "free will." But then, my friend who receives the letter cannot decide what to read. He, or she, can only passively observe the contents of the letter. (Notice that my friend and I are tacitly assumed to be distinct entities!) I have thereby defined a message—a relationship between a decision and an observation—but then there is the next hurdle: What is a decision? What is an observation? These obviously are physical processes, but how do we distinguish between them by some objective criterion, without using pronouns such as "I" or "he" or "she"?

My belief is that the distinction between decisions and observations is quantitative, rather than qualitative. A decision is influenced by too many factors to be predictable. It is like the evolution of a chaotic system which, although deterministic, is not computable.[(4)] Decisions thus *appear* completely random and this is why we claim that we have free will. An observation too is controlled by many factors, but in a way which is conspicuously deterministic, or at least follows simple probability rules. The distinction between past and future, cause and effect, thus seems reducible to the interdependence of events involving chaotic or regular dynamics.[(25,26)]

How can these qualitative notions be recast into a quantitative, dynamical theory? Consider a model consisting of two physical systems, with dynamical variables collectively called $R$ and $C$, respectively. These two systems are connected by a chain of Hookean oscillators—a "field"—having, in the long wavelength limit, a well-defined transition time $T$. Assume that the internal dynamics of $R$ and $C$ are regular and chaotic, respectively. Then, if the Hookean oscillators are at rest at some time $t = t_0$, or have only random (uncorrelated) motions at that time, one obtains, for $t > t_0$, non-vanishing correlations of the type $\langle R(t+T)\, C(t)\rangle$, and *only* of that type. The chaotic system can influence the regular one, with a delay $T$, but there are *no easily analyzable* effects propagating from the regular system (the "receiver") toward the chaotic one (the "emitter").

Likewise, for $t < t_0$, one has nonvanishing correlations $\langle R(t-T)\, C(t)\rangle$. In that case, the quiescent state assumed at $t = t_0$ (all the field variables at rest or, at least, uncorrelated) results from the *advanced* potential of the emitter. The problem therefore is time-symmetric[(27)]—it could not be otherwise. The arrow of time cannot result from dynamics alone, if each elementary step is reversible. It necessitates *two* separate assumptions: One is that there is a privileged time $t_0$ at which the field is quiescent. The other is that the field connects two very different systems, one of which is chaotic and the other regular, acting as the emitter and receiver, respectively.

## ACKNOWLEDGMENTS

It is a pleasure to dedicate this essay to John A. Wheeler, with whom these ideas were discussed during my visits to the University of Texas. However, Professor Wheeler bears no responsibility for any errors, misconceptions, or other fantasies in this paper.

Some parts of this article were written in Copenhagen. There could not be a more appropriate location for that. I am very grateful to NORDITA for hospitality.

I am also grateful for the hospitality of the Center for Philosophy and History of Science at Boston University and especially to Prof. Abner Shimony for many helpful comments.

Work done at Technion was supported by the Gerard Swope Fund, the New York Metropolitan Research Fund, and the Fund for Encouragement of Research.

## REFERENCES

1. J. S. Bell, *Physics* **1**, 195 (1964).
2. A. Peres and W. H. Zurek, *Am. J. Phys.* **50**, 807 (1982).
3. A. Peres, *Found. Phys.* **14**, 1131 (1984).
4. G. Iooss, R. H. G. Helleman, and R. Stora, eds., *Chaotic Behavior of Deterministic Systems* (North-Holland, Amsterdam, 1983).
5. A. Landé, *Foundations of Quantum Theory* (Yale University Press, New Haven, 1955), p. 4.
6. W. Perrie, A. J. Duncan, H. J. Beyer, and H. Kleinpoppen, *Phys. Rev. Lett.* **54**, 1790 (1985).
7. F. James, *Rep. Prog. Phys.* **43**, 1145 (1980).
8. A. Einstein, B. Podolsky, and N. Rosen, *Phys. Rev.* **47**, 777 (1935).
9. N. Bohr, *Phys. Rev.* **48**, 696 (1935).
10. A. Aspect, J. Dalibard, and G. Roger, *Phys. Rev. Lett.* **49**, 1804 (1982).
11. J. S. Bell, *J. Phys. C* **2**, 41 (1980).
12. A. Peres, *Am. J. Phys.* **46**, 745 (1978).
13. A. Peres, *Am. J. Phys.* **52**, 644 (1984).
14. R. L. Pfleegor and L. Mandel, *Phys. Rev.* **159**, 1084 (1967).
15. S. J. Feingold and A. Peres, *J. Phys. A* **13**, 3187 (1980).
16. N. Cufaro-Petroni, A. Garuccio, F. Selleri, and J. P. Vigier, *C. R. Acad. Sci. Paris B* **290**, 11 (1980).
17. N. Herbert, *Found. Phys.* **12**, 1171 (1982).
18. G. C. Ghirardi, A. Rimini, and T. Weber, *Lett. Nuovo Cimento* **27**, 293 (1980).
19. D. N. Page, *Phys. Lett. A* **91**, 57 (1982).
20. A. Shimony, in *Proc. International Symposium on Foundations of Quantum Mechanics* (Phys. Soc. Japan, Tokyo, 1984).
21. A. Peres, "When is a quantum measurement?", *Am. J. Phys.* **54** (1986), in press.

22. D. Finkelstein, in *Paradigms and Paradoxes*, R. C. Colodny, ed. (Univ. Pittsburgh Press, 1971), Vol. V; reprinted in *Logico-Algebraic Approach to Quantum Mechanics*, C. A. Hooker, ed. (Reidel, Dordrecht, 1975), Vol. II, pp. 141–160.
23. H. Paul, *Am. J. Phys.* **53**, 318 (1985).
24. A. Peres, *Am. J. Phys.* **43**, 1015 (1975).
25. A. Peres and L. S. Schulman, *Int. J. Theor. Phys.* **6**, 377 (1972).
26. L. Gatlin, *Int. J. Theor. Phys.* **19**, 25 (1980).
27. J. A. Wheeler and R. P. Feynman, *Rev. Mod. Phys.* **21**, 425 (1949).

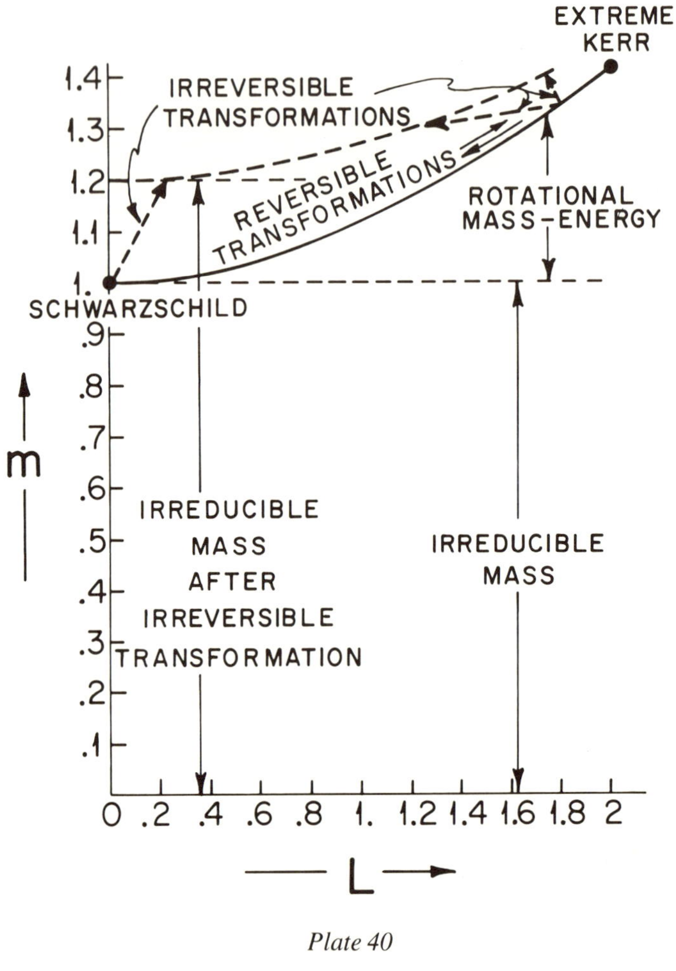
EXTREME
KERR
IRREVERSIBLE
TRANSFORMATIONS
REVERSIBLE
TRANSFORMATIONS
ROTATIONAL
MASS-ENERGY
SCHWARZSCHILD
IRREDUCIBLE
MASS
AFTER
IRREVERSIBLE
TRANSFORMATION
IRREDUCIBLE
MASS
m
L
1.4
1.3
1.2
1.1
1.
.9
.8
.7
.6
.5
.4
.3
.2
.1
0 .2 .4 .6 .8 1. 1.2 1.4 1.6 1.8 2

*Plate 40*

*Foundations of Physics, Vol. 16, No. 6, 1986*

# On the Nature and Origin of Complexity in Discrete, Homogeneous, Locally-Interacting Systems

**Charles H. Bennett**[1]

*Received March 12, 1986*

*The observed complexity of nature is often attributed to an intrinsic propensity of matter to self-organize under certain (e.g., dissipative) conditions. In order better to understand and test this vague thesis, we define complexity as "logical depth," a notion based on algorithmic information and computational time complexity. Informally, logical depth is the number of steps in the deductive or causal path connecting a thing with its plausible origin. We then assess the effects of dissipation, noise, and spatial and other symmetries of the initial conditions and equations of motion on the asymptotic complexity-generating abilities of statistical–mechanical model systems. We concentrate on discrete, spatially-homogeneous, locally-interacting systems such as kinetic Ising models and cellular automata.*

An important but subtle question is "why is the world complicated?" A widely accepted answer today is the doctrine of self-organization, which holds that, although nature obeys simple laws, these may lead in time to the spontaneous generation of complex structures. Self-organization is not seen as happening always and everywhere, but only under certain conditions. The most commonly noted of these is dissipation: the removal or dilution of entropy produced by a system so that it can go on producing more entropy. On the earth dissipation is induced chiefly by the temperature difference between the hot sun and cold night sky; on a cosmological scale, it is induced by the expansion of the universe. Converse to the idea that dissipation begets self-organization is the view that nothing very interesting is likely to happen in a world at thermal equilibrium,

[1] IBM Research, Yorktown Heights, New York 10598.

reflected in the epithet "heat death." In the modern world view, dissipation has taken over one of the functions formerly performed by God: It makes matter transcend the clod-like nature it would manifest at equilibrium, and behave instead in dramatic and unforseen ways, molding itself, for example, into thunderstorms, people, and umbrellas.

One would like to understand whether self-organization is an asymptotically qualitative phenomenon like phase transitions. In other words, is there a reasonable model universe in which complexity, appropriately defined, not only increases, but increases without bound in the limit of infinite space and time? A positive answer to this question would not explain the natural history of our particular finite world, but would suggest that its quantitative complexity can legitimately be viewed as an approximation to a well-defined qualitative property of infinite systems. On the other hand, a negative answer would suggest that our world should be compared to a simple chemical reaction–diffusion system that self-organizes on a macroscopic, but still finite, scale. A thorough understanding of the physical prerequisites for qualitative self-organization will ultimately shed some light on the difficult issue of the extent to which our world's observed complexity is conditioned by the posterior existence of sentient observers.

Although the question of qualitative self-organization ought ultimately to be posed for more general models, here we will mainly consider cellular automata in the broad sense of lattice models with finitely many states per site, that evolve deterministically or probabilistically, in continuous or discrete time, according to a uniform local law. We do not consider spatially inhomogeneous systems such as spin glasses because, in all known cases, the inhomogeneity is not a fundamental physical feature, but represents quenched-in (and in principle temporary) disorder generated by an underlying homogeneous dynamical process.

Among the most realistic and least computer-like of cellular automata are probabilistic systems in which all local transition probabilities are positive. Since all states are mutually accessible, these systems, if finite, are ergodic, in the stochastic sense of relaxing to a unique stationary distribution; however, infinite systems of the same sort need not be ergodic (e.g. the Ising model at low temperature in zero field).

Such a system is considered reversible if its matrix of transition probabilities is of the "microscopically reversible" form $DSD^{-1}$, where $D$ is diagonal and $S$ symmetric. In that case a movie of the system at equilibrium would look the same shown forwards as backwards, and the stationary distribution can be represented (exactly for asynchronous automata, approximately for synchronous) as the Boltzmann exponential of a locally additive potential. On the other hand, if the local transition probabilities are not microscopically reversible, the stationary macrostate is dissipative (corresponding physically to a system whose environment continually removes entropy from it), a movie of the system would not look the same forwards as backwards, and the distribution of microstates in general cannot be approximated by the exponential of any locally additive potential. Asynchronous reversible probabilistic cellular automata, otherwise known as generalized kinetic Ising models, are widely studied in statistical mechanics.

At the other extreme are deterministic cellular automata. The classical examples of these automata (e.g., Conway's game of "life") had noninvertible transition rules and had no particular connection to physics, but recently there has been considerable interest in reversible deterministic automata, whose dynamics, though discrete, is physically realistic in that it has locally-additive conserved quantities analogous to energy or momentum (Hardy, *et al.*,[7] Pomeau,[9] Margolus,[8] Vichniac[12]). Many of these systems have the paradoxical property of behaving like a computer for special initial conditions, but like a traditional statistical-mechanical model (e.g., a gas or Ising model) for generic initial conditions.

An important qualitative proerty related to subjective complexity is computational universality, a computationally universal system being one that can be programmed through its initial conditions to simulate any digital computation. For example, the computational universality of the Conway game implies that one can find an initial configuration that will evolve so as to turn a certain site on if and only if white has a winning strategy at chess, another initial configuration that will do so if and only if the millionth decimal digit of $\pi$ is a 7, and so on.

The property of computational universality was originally demonstrated for systems such as Turing machines and irreversible deterministic cellular automata having little resemblance to the systems ordinarily studied in mechanics and statistical mechanics. Later some reversible, deterministic systems (e.g., the hard shere gas with appropriate initial and boundary conditions (Fredkin and Toffoli[4]), and Margolus' reversible billiard ball cellular automaton (Margolus[8]) which discretely models this gas) have been shown to be computationally universal. Very recently (Gacs,[5] Gacs and Reif[6]) certain irreversible, noisy systems (synchronous probabilistic cellular automata in one and three dimensins with all local transition probabilities positive) have been shown to be universal, and there appears to be a good liklihood of generalizing these results to analogous asynchronous automata. Computational universality therefore now appears to be a property that physically realistic models can have; moreover, if a system does have this property, it is by definition capable of behavior as complex as any that can be digitally simulated.

Unfortunately, all these models appear to require special initial conditions to cause them to embark on a meaningful computation. Can a similar model be made to generate states of unbounded complexity from simple, uniform initial conditions? Indeed, how should the complexity of a state be defined? We digress to answer this question.

Computational universality by itself is an unsuitable criterion of complexity for our puroses because it is a functional property of systems rather than a structural property of states. Thus it makes no distinction among:

1) trivial states, that are subjectively simple and will remain so under the universal system's dynamics;

2) "ambitious" states that are simple now, but will be transformed into organized states by the system's dynamics;

3) "logically deep" states that contain evidence that a nontrivial computation has already been performed to generate them.

The notion of logical depth, which we identify with the subjective complexity or organization of a state, can be formally defined by a computerized version of the Occam's razor paradigm, which exhorts us to explain each phenomenon by thc most cconomical hypothesis able to explain it at all. In the computerized version, the phenomenon $x$ to be explained, appropriately digitized as a binary string, is regarded as the output of a universal computer, and the most economical hypothesis to explain $x$ is identified with the "minimal program" to compute $x$, i.e., the shortest (in number of bits) binary input which causes the computer to compute exactly $x$ as output. It is natural then to regard this computation of $x$ from its minial program as the intrinsically most plausible causal history of $x$; and hence the run time of this computation measures the logical depth, or intrinsic complexity, of $x$. A more careful development of this notion (Bennett[(1)]) taking appropriately weighted account of near-minimal programs as well as the minial one, yields a complexity measure that is reasonably robust and machine-independent. This may at first seem surprising, but it is the result of an essential property of universal computers, their ability to stimulate one another. This ability means that an object that is shallow (rapidly coputable by a program of near-minimal size) on one computer will not be very deep on another, since programs for the first computer can be translated to run on the second, increasing their size by at most an additive constant, and their run time by at most a polynomial. Similar arguments show that deterministic computations cannot quickly transform shallow objects into deep ones, while probabilistic computations can do so only with low probability, as it were by accident. On the other hand, depth can be reliably produced by slow computations, notably by a version of Cantor's diagonal argument, in which the computer first generates a list of strings quickly computable by small programs, then outputs the first string not on the list.

Logical depth is thus an asymptotically absolute property: A deep structure is one that cannot plausibly have been generated quickly either in the system containing it or in any other system whose dynamics can be efficiently digitally simulated. Logical depth has the right intuitive properties for a complexity measure: subjectively trivial structures (such as a string of zeros, or a random string of zeros and ones produced by coin tossing) are shallow; on the other hand, deep structures can be produced slowly, though not quickly, by a universal computer. It may at first seem surprising that random objects are shallow, but a typical random output $x$ typically cannot be produced at all by a small program, and so a "print program," containing a *verbatim* description of the desired output, is fast-running and of near-minimal size. Armed with the notion of depth, we can say that a system self-organizes if it produces, with high probability, increasingly deep local configurations in the course of its evolution, configurations that could not have been produced quickly by any other evolution with a simple initial condition and simple laws of motion.

Returning to the question of self-organization from a uniform initial state, an immediate problem is to break translational symmetry. This can be done by a probabilistic system, or by a deterministic system with a uniformly random initial state, but there remains the problem (evident in simulations of "life" from a "soup" initial condition), that whatever nontrivial structure is formed in one place will also be formed in infinitely many other places, so some conflict-resolution is necessary if any of the structures are to grow indefinitely in size and complexity.

A simple example of conflict resolution is illustrated in Fig. 1, which represents the history of a one-dimensional chain of sites each of which may contain an east-walker or a west-walker. Initially, at about half the sites in the system, an egg hatches, producing an east-walker and a west-walker who begin walking apart from each other with velocity $\frac{1}{2}$. The region behind and between such a pair of walkers constitutes a "domain," and each domain increases in width linearly with time as long as it does not encounter other domains. When domains meet, the colliding walkers have a duel, and one of them at random survives and continues walking, thereby enlarging his domain at the expense of the other. The numerical evidence suggests that, at any time $t$, at least $\frac{1}{4}$ of the sites in an infinite chain would belong to domains of width $t$, which has been growing since time 0.

Now imagine that each initial hatching of an egg initiates its own coy of a nontrivial, depth-producing computation, carried out by other

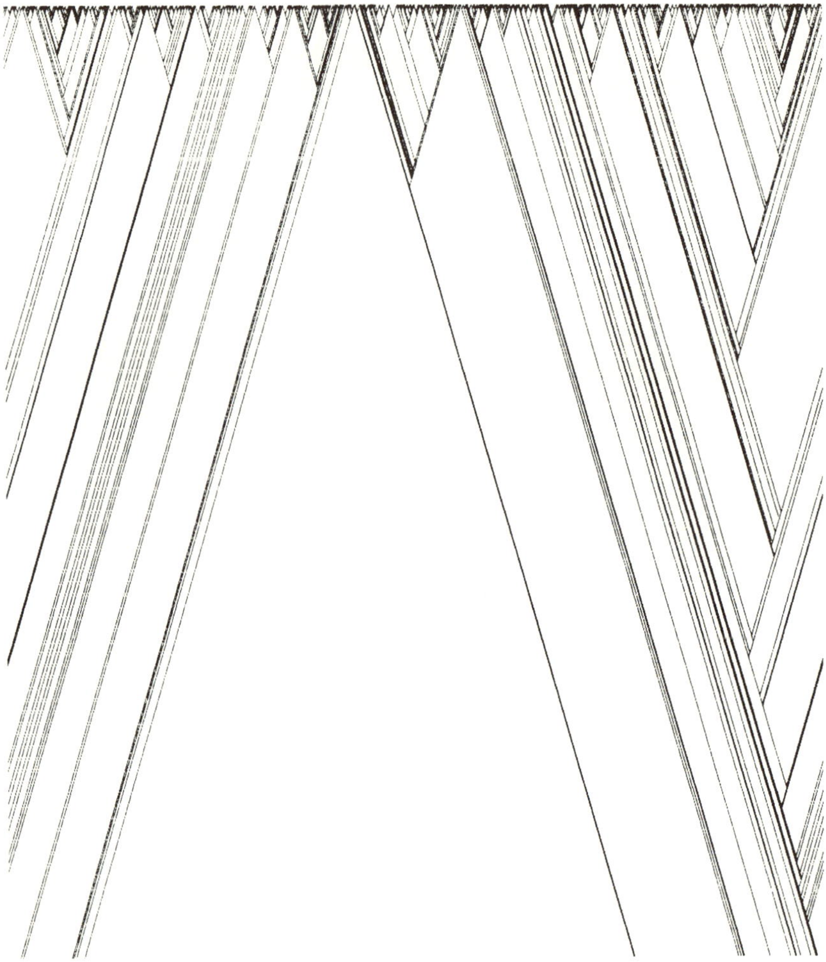

Fig. 1. Time history (time increases downward) of conflict resolution between domains growing in an initially homogeneous medium.

variables at each site not shown in the figure (universal one-dimensional deterministic cellular automata are well known), and that this computation continues unimpeded as long as its host domain continues to grow. On the other hand, when a domain is conquered, its space is made available for the computation of the conquering domain. Under these conditions, the system as a whole will self-organize in the desired manner, a finite fraction of sites at each time *t* belonging to coherent computations of age *t*.

This system is unsatisfying chiefly because it depends on exact left–right symmetry of the probabilistic duel outcome. If walkers of one sign were more likely to win, then almost all walkers of the other sign would be eliminated, and after this time almost all the territory would belong to nongrowing domains of some typical characteristic size and corresondingly bounded depth.

A more convincing example of self-organization would be a probabilistic system that was robust with respect to small perturbations both of the initial condition and transition probabilities. The one-dimensional (Gacs[5]) and three-dimensional (Gacs and Reif[6]) universal error-correcting automata mentioned earlier are robust with respect to perturbations of the transition probabilities, but they require special initial conditions.

In the case of self-organization from a uniform initial state, it is not clear how much robustness with respect to transition probabilities can be achieved, since there may be a conflict between the need to amplify fluctuations that break translational symmetry, while at the same time suppressing fluctuations that would lead to errors in the computation.

At this point we may again ask why there seems to be a qualitative difference in the complexity of reversible and dissipative systems. The answer to this question is not fully understood.

Systems such as the Gacs or Gacs–Reif automaton that perform reliable computations in the presence of noise require the active use of redundancy to correct errors after they occur. Such error-correction, being a many-to-one mapping of the computer's logical state, decreases the entropy of the computer and so must increase the entropy of the environment. Hence it can only be performed in a dissipative system. As yet no general theory exists of the dissipation–reliability–noise tradeoff in such systems. The analysis of the thermodynamic cost of error correction is complicated by the fact that a noisy system's tendency to make errors is itself a thermodynamic driving force that can be partially harnessed to do useful work (perhaps correcting other errors).

Another important difference between reversible and irreversible systems is that only the former are subject to the Gibbs phase rule. In other words, for a reversible system at a generic point in the parameter space of its Hamiltonian, one phase has lowest free energy and will nucleate and grow, making the system ergodic and thus incapable of universal computation (a system that forgets its initial conditions certainly cannot be programmed by them). Irreversibility removes this limitation by making the ordinary free energy of a $d$-dimensional model ill-defined, while the $d+1$-dimensional free energy (defined in terms of the stationary distribution of histories) still exists, but depends on the local transition probabilities in a degenerate manner, so that it remains identically zero as the transition probabilities are varied (Domany and Kinzel,[3] Toom,[11]

Bennett and Grinstein[2]). Thus nonergodic behavior of reversible systems is the exception, occurring only at special points in the phase diagram, while the same behavior can occur at generic points in the phase diagram of irreversible systems.

What is not entirely clear is why in nature the Gibbs phase rule enforces as much simplicity as it does, the unique stable phase of an equilibrium system typically being either spatially homogeneous (e.g., liquid) or periodic with a fairly short period (crystal). By contrast, Hamiltonian lattice models have been designed (Radin[10]) in which the lowest-energy phase is quasiperiodic rather than periodic. The scarcity or absence of these phases in nature is still not fully understood.

## ACKNOWLEDGMENTS

I wish to thank Peter Gacs, Geoff Grinstein, Norman Margolus, Charles Radin, and Tom Toffoli for numerous helpful discussions on recent aspects of this work.

## REFERENCES

1. C. H. Bennett, in *Emerging Syntheses in Science* (Santa Fe Institute, Santa Fe, New Mexico, 1985), David Pines, ed., pp. 297–313. See also unpublished manuscript "On the Logical Depth of Sequences and their Reducibilities to Random Sequences" (1982). The first published description of logical depth is by G. Chaitin, *I.B.M. J. Res. Develop.* **21**, 350–359 (1977).
2. C. H. Bennett and G. Grinstein, *Phys. Rev. Lett.* **55**, 657–660 (1985).
3. E. Domany and W. Kinzel, *Phys. Rev. Lett.* **53**, 311 (1985).
4. E. Fredkin and T. Toffoli, *Internat. J. Theor. Phys.* **21**, 219 (1982).
5. P. Gacs, Technical Report No. 132, Computer Science Dept. University of Rochester, 1983, to appear in *J. Comp. Sys. Sc.*
6. P. Gacs and J. Reif, *Proceedings of the 17th ACM Symposium on the Theory of Computing*, 388–395 (1985).
7. J. Hardy, O. de Pazzis, and Y. Pomeau, *J. Math. Phys.* **14**, 470 (1973).
8. N. Margolus, *Physica* **10D**, 81–95 (1984).
9. Y. Pomeau, *J. Phys.* **17**, 12 (1984).
10. C. Radin, *J. Math. Phys.* **26**, 1342 (1985).
11. A. L. Toom, in *Advances in Probability 6* (Multicomponent Systems), R. L. Dobrushin, ed. (Dekker, New York, 1980), pp. 549–575.
12. G. Vichniac, *Physica* **10D**, 96 (1984).

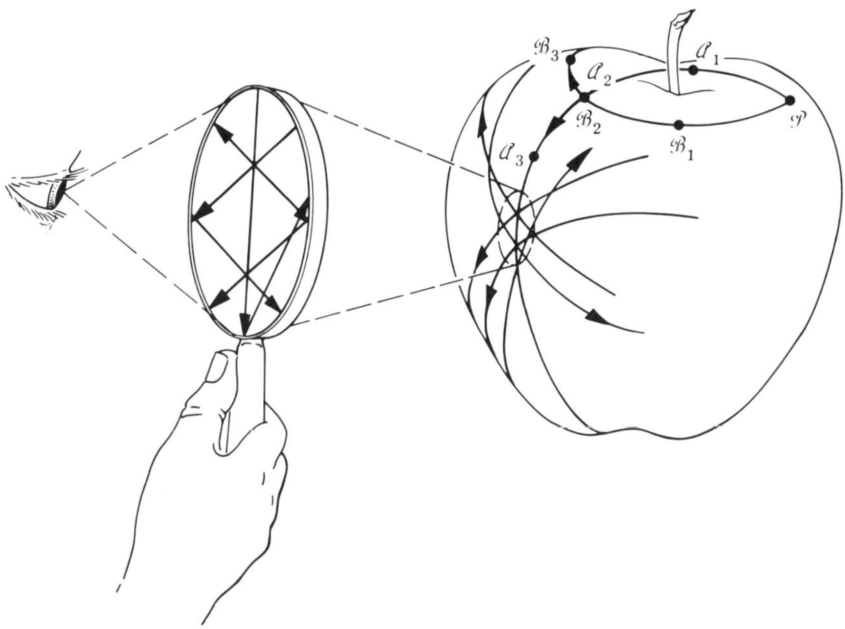
$\mathcal{B}_3$
$\mathcal{A}_1$
$\mathcal{A}_2$
$\mathcal{B}_2$
$\mathcal{P}$
$\mathcal{A}_3$
$\mathcal{B}_1$

*Plate 41*

# Descriptions of Plates

PLATE 1. Existence viewed as a self-synthesizing system. The universe starts small (thin U at upper right), grows (loop of U), and in time gives rise (upper left) to observer-participancy—which in turn determines all we have the right to say about even the earliest days of the universe. (Figure and description adapted from C. M. Patton and J.A.W., "Is Physics Legislated by Cosmogony?" in C. J. Isham, R. Penrose, and D.W. Sciama, *Quantum Gravity: An Oxford Symposium*, Clarendon, Oxford, 1975, Fig. 9.4, p. 565.)

PLATE 2. Critical deformations leading to nuclear fission according to the liquid-drop model. A small distortion of the spherical form (1) leads to stable oscillations about the globular form, unless the droplet is so highly charged electrically (charge $= [10 \times$ surface tension $\times$ volume$]^{\frac{1}{2}}$) that even the slightest deformation makes it possible for the electrostatic repulsions to rupture the droplet. If the charge has less than its limiting value (as corresponds to the case of all actual nuclei) a finite deformation will be required to lead to a critical configuration of unstable equilibrium. This deformation is small (2) and the energy which a bombarding neutron must have to bring it about is not great for the most highly charged nuclei, but as the charge is diminished the deformation necessary to lead to rupture is greater (3) and fission is no longer possible for the bombarding energies obtainable. (From N. Bohr and J.A.W., "Mechanism of Nuclear Fission," *Phys. Rev.* 56 [1939]: 426.)

PLATE 3. Schematic representation of construction potential, $V(x, y, z; \alpha)$. The parameter $\alpha$ may describe a change of shape of the potential well, such as spheroidal deformation; or a change in orientation, as in the case of a collective rotation; or a change in extension, as in a dilatational vibration; or there may be several such parameters to describe simultaneously several types of collective motion. (From J. J. Griffin and J.A.W., "Collective Motions in Nuclei by the Method of Generator Coordinates," *Phys. Rev.* 108 [1957], Fig. 1, p. 315.)

PLATE 4 from J.A.W., "On the Mathematical Description of Light Nuclei by the Method of Resonating Group Structure," *Phys. Rev.* 52 (1937): 1118.

PLATE 5. (*a*) Bipyramid alpha-particle model for the nucleus $Ne^{20}$. (*b,c*) A distortion separating alpha-particles 2,4 and bringing 5,1 into contact. The result is an exchange plus rotation through 90 degrees. (From E. Teller and J.A.W., "On the Rotation of the Atomic Nucleus," *Phys. Rev.* 53 [1938], Fig. 2, p. 782.)

PLATE 6 from J.A.W., "The Lesson of the Black Hole," *Proc. American Phil. Soc.* **125** (1981), Fig. 7, p. 35. Symbolic representation of the foam-like structure expected for space at the Planck scale of distances.

PLATE 7. Symbolic representation of the unquantized charge of classical theory. For ease of visualization the number of space dimensions is reduced from three to two. However, the two-dimensional curved and multiply connected space is imbedded in a three-dimensional Euclidean space. The third dimension, measured off the surface, has no physical meaning. Of course the topology and geometry of the 2-space receives its best mathematical formulation in intrinsic terms, without its imbedding of the manifold in a space of higher dimensionality. The 2-space is multiply connected, but free of all singularities. An imaginary ant crawling over the surface and entering the tunnel or handle or "wormhole" finds there the same two-dimensional space he experienced everywhere else. Electric lines of force that converge on the right-hand mouth of the tunnel continue to obey at each point Maxwell's equation, $\text{div}\mathbf{E} = 0$. The field is everywhere free of singularity. The lines of force have no escape but to continue through the tunnel. They emerge from the left-hand mouth. Outside the tunnel mouths the pattern of lines of force is identical with that due to equal positive and negative charges. An observer endowed with poor vision sees evidence for two point charges. He may even construct a boundary around the right-hand charge, determine the flux through this boundary, incorrectly apply the theorem of Gauss, and "prove" that there is charge inside the boundary. He does not recognize that he has been making tacit and unjustified assumptions about the topology of space. He is not aware that his "boundary" does not bound any region interior to it. He assumes, either that Maxwell's equations fail in the vicinity of the charge, or that there exists some magic substance at which lines of force end and to which he gives the name "electricity." But a closer inspection discloses that the lines of force do not end. Neither is there any violation of Maxwell's equations for charge-free space. Nowhere can one place one's finger and say, "Here there is some charge." Such is the purely topological picture of unquantized electric charge which is adopted in this paper. This classical charge has no *direct* relation whatsoever to quantized electric charge. At this classical level there is a freedom of choice in the strength of the charge, and an individuality about the connection between one charge and another, that must be entirely changed in any proper *quantum* theory of electricity.—The distance along the wormhole from one mouth to the other need have no correspondence whatever with the distance in the open space between the same two mouths. The connection can be as short as the radius of the wormhole itself, for example, even when the openings are very far apart in the upper space, as one sees by bending the upper space to bring the backs of the two holes into coincidence. (From C. W. Misner and J.A.W., "Classical Physics as Geometry. Gravitation, Electromagnetism, Unquantized Charge, and Mass Properties of Curved Empty Space," *Ann. Phys.* 2 [1957]: 525–603.)

PLATE 8. With sufficient increase of the neutrino pressure, all the electrons of all the atoms are converted to mu-mesons. The new all-mu-meson atoms are two hundred times smaller than the original atoms. (From J.A.W., "Nanosecond Matter" in Hans Mark and Lowell Wood, eds., *Energy in Physics, War and Peace*, World Scientific, Singapore, 1987, Fig. 2, p. 109.)

PLATE 9. The laws of physics are not chiseled on a slab of granite that stands from everlasting to everlasting. Those regularities had to come into being at the big bang and be obliterated in the gravitational collapse that takes place at the center of a black hole. (From J.A.W., *Physics and Austerity: Law without Law* [in Chinese], Anhui Science and Technology Publications, Anhui, Hefei, 1982, Fig. 11.)

PLATE 10. Idealized picture of a black hole. Any detail of the infalling matter is washed out: we end up with a configuration uniquely determined by mass, charge, and angular momentum. (From R. Ruffini and J.A.W., "Relativistic Cosmology and Space Platforms," in *Proceedings of an ESRO Colloquium held at Interlaken, Switzerland, 4 September 1969*, Book SP-52, European Space Research Organization, Neuilly-sur-Seine, France, 1971, Fig. 14, p. 86. Logo at lower left identifies place of origin as High Island.)

PLATE 11. Time rules physics. Its highest platform of authority is general relativity. Three cracks suggest that the foundations are falling to pieces and will require reassembly by a new plan: (1) Time ends in big bang and gravitational collapse. (2) "Past" and "future" are interlocked in a way contrary to the causal ordering presupposed by "time." (3) Quantum theory denies all meaning to the concepts of "before" and "after" in the world of the very small. Appreciation is expressed to Ros Lin Chin for making this drawing. (From J.A.W., *Physics and Austerity: Law without Law* [in Chinese], Anhui Science and Technology Publications, Anhui, Hefei, 1982, Fig. 16, pp. 41–42.)

PLATE 12. Harrison-Wakano-Wheeler (HWW) catalogue of all equilibrium configurations of cold, catalyzed matter. HW equation of state. Integrations to central density of $10^{18} g/cm^3$, M.W.; integrations to $10^{22} g/cm^3$, B. K. H. Masses (*after* assembly) in units of the mass of the sun, $M_\odot = 1.987 \times 10^{33} g$ ($M_\odot^* = 1.474 \times 10^5 cm$). *Black dots*: points of change of stability. *L* stands for Landau, who predicted the existence of the two most important of these critical points. The HWW curve agrees closely in the range $10^{14}$ to $10^{17} g/cm^3$ with the 1939 curve of Oppenheimer and Volkoff, who were the first to determine the position of the LOV critical point (model of the ideal neutron gas). From low densities up to $\sim 10^{17} g/cm^3$ the HWW curve for $M(\rho_0)$ agrees in its general run with the curve calculated by Chandrasekhar in 1935 for a substance deriving its pressure from an ideal electron gas, and with an effective mass per electron $\mu_{eff}$ equal to $\frac{1}{26}$ of the mass of an $Fe^{56}$ atom. However, the Chandrasekhar curve shows no maximum; it approaches an asymptotic $M_\infty/M_\odot = 1.26$ as the central density goes to infinity. When one allows for the crunching of electrons into combination with the protons (HW equation of state) one arrives (1958) at the LHWW critical point. MZ, critical points discovered by Misner and Zapolsky (1964); H, by Harrison (1964). The musical scales show which among the acoustical normal modes of oscillation (purely radial; spherical symmetry) are unstable (*black notes*) between one critical point and the next. (From B. K. Harrison, K. S. Thorne, M. Wakano, and J.A.W., *Gravitation Theory and Gravitational Collapse*, University of Chicago Press, 1965, p. 46, Fig. 5.)

PLATE 13 from J.A.W., "On the Nature of Quantum Geometrodynamics," *Ann. of Phys.* 2 (1957): 606.

PLATE 14. Measuring the strength of a gravitational wave. Time of round-trip travel between two geodesics responds to oscillations in the curvature of spacetime (diagram is schematic only; symbolic of laser pulse sent from Earth to a corner reflector on the Moon and back at a time when a very powerful, long-wavelength gravitational wave passes by; the wave would have to be powerful because a direct measure of distance to better than 10 cm is difficult, and such precision produces a much less sensitive indicator of waves than the vibrations in length [$10^{-14}$ cm or less] of a Weber bar). The geodesics are curved toward each other in regions where the relevant component of the Riemann curvature tensor, call it $R_{\hat{x}\hat{0}\hat{x}\hat{0}}$, has one sign, and curved away from each other in regions where it has the opposite sign. The diagram allows one to see at a glance the answer to an often expressed puzzlement: Is not any change in round-trip travel time mere trumpery flummery? The metric perturbation, $\delta h_{\mu\nu}$, of the wave changes the scale of the distances slightly but also correspondingly changes the scale of time. Therefore does not every possibility of any really meaningful and measurable effect cancel out? Answer: (1) The widened separation between the geodesics is not a local effect but a cumulative one. It does not arise from the local value of $\delta h_{\mu\nu}$ directly or even from the local value of the curvature. It arises from an accumulation of the bending process over an entire half-period of the gravitational wave. (2) The change in separation of the geodesics is a true change in proper distance, and shows up as a true change in proper time (see "ticks" on the world line of one of the particles). Note: When one investigates the separation between the geodesics, not over a single period, as here, but over a large number of periods, one finds a cumulative, systematic, net slow bending of the rapidly wiggling geodesics toward each other. This small attractive acceleration is evidence in gravitation physics of the effective mass-energy carried by the gravitational waves. (From C. W. Misner, K. S. Thorne, and J.A.W., *Gravitation*, Freeman, San Francisco, 1973, p. 1014, Fig. 37.3.)

PLATE 15. Above: qualitative diagram of a gravitational wave which comes in from great distances, reaches a stage of maximum contraction at a moment of time symmetry, and re-expands outwards. Middle: qualitative diagram of how metric varies with time in the course of its implosion and explosion, showing Schwarzschild type of asymptotic approach to flatness at great distances. The mass associated with this $1/r$ approach to flatness remains unchanged throughout the entire evolution of the gravitational wave. Below: more detail about such a wave at the moment of time symmetry. A slice through 4-space at this moment has a characteristic three-dimensional geometry of its own. A typical component of the metric associated with this 3-geometry departs from unity the more, the greater is the strength of the gravitational wave at this moment (different problems, not the same problem at different moments). For a certain critical value of the strength of the gravitational wave the space is curved up into a closed universe. For any greater value of the strength the metric is singular and inadmissible. For the critical value of the strength, the 3-geometry is everywhere perfectly regular, and it remains so for a finite proper time. However, it is conjectured here that as this *closed-space* metric evolves with time, however it is originally chosen, it will always develop a singularity in its intrinsic geometry. (From J.A.W., *Geometrodynamics*, Academic, New York, 1962, p. 56, Fig. 13.)

PLATE 16. A 4-cube of dimensions $\Delta y$, $\Delta z$, $\Delta t$ and its eight 3-cube faces depicted as "exploded off" of it. The content of stress-energy tensor in the 3-cube at the upper right (shaded only at its "top," and with diagonal stripes) measures the amount of momentum-energy that passes out of the 4-cube across a surface of area $\Delta x \Delta y$, located at $z + \frac{\Delta z}{2}$, during the time $\Delta t$. To demand conservation of momentum and energy in all eight cubes, taken with due regard to sign, shall add to zero. This demand is fulfilled by representing the amount of momentum-energy in the given 3-cube (a vector or, in its dual representation, a 3-index tensor) as a sum of the "moments of rotation" associated with all six 2-dimensional faces of that 3-cube. When we add up these contributions for all eight 3-cubes we have $6 \times 8 = 48$ 2-dimensional faces to deal with, but their contributions all cancel out in pairs, as required. Thus the contribution of the striped face just considered is canceled by the contribution of the striped face of the "top" or $\Delta x \Delta y \Delta z$ 3-cube. This is the payoff of the principle that the "2-dimensional boundary of a 3-dimensional boundary of a 4-cube" is zero. This is how spacetime grips mass so as to ensure and enforce the law of conservation of momentum and energy. (Figure from C. W. Misner, K. S. Thorne, and J.A.W., *Gravitation*, Freeman, San Francisco, 1973, p. 366, Box 15.1; description from J.A.W., *Physics and Austerity: Law without Law* [in Chinese], Anhui Science and Technology Publications, Anhui, Hefei, 1982, Fig. 31, p. 77.)

PLATE 17. Space (upper left), spacetime (upper right), and superspace (below). The *leaf of history* that curves through superspace includes all the configurations ($A$, $B$, $B'$, . . .) achieved by space in its classical dynamical evolutions in time; that is, all spacelike slices through the given spacetime. A different spacetime (not shown), that is, a classical history of space when the dynamics of space is started off with different initial conditions, corresponds to a different leaf of history (also not shown) cutting through superspace. (Figure adapted, and description taken, from J.A.W., "Frontiers of Time," in G. Toraldo di Francia, ed., *Problems in the Foundations of Physics: Proceedings of the International School of Physics "Enrico Fermi." Course 72*, North Holland, Amsterdam, 1979, p. 438, Fig. 6.)

PLATE 18. Black hole and big crunch seen as part and parcel of the final singularity. The closed model universe is uniquely foliated by a sequence of spacelike hypersurfaces distinguished one from another by the value—constant over any one hypersurface—of the mean extrinsic curvature; that is, the trace of the tensor of extrinsic curvature or the fractional rate of decrease of volume per second. (From J.A.W., "Beyond the Black Hole," in H. Woolf, ed., *Some Strangeness in the Proportion: A Centennial Symposium to Celebrate the Achievements of Albert Einstein*, Addison-Wesley, Reading, Massachusetts, 1980, Fig. 22.3, p. 346.)

PLATE 19. Space grips mass, keeps it going straight when free, and by its power enforces conservation of energy and momentum in a smash. The coupling of mass and geometry, far from being the weakest force in nature, is the strongest. (From J.A.W., *Physics and Austerity: Law without Law* [in Chinese], Anhui Science and Technology Publications, Anhui, Hefei, 1982, Fig. 28, p. 72.)

PLATE 20. Rotations and the 1-boundary of the 2-boundary of a 3-cube. Left: the rotation $A$ experienced by a sample vector on being transported parallel to itself around the indicated circuit. This rotation measures one component of the spacetime curvature or gravitational tide-producing acceleration. Right: the rotations associated with all six faces together add up to zero; the diagram closes. It closes because each edge of the cube is traversed twice, and in opposite directions, in the circumnavigation of the two abutting faces of the cube: $\partial \circ \partial = 0$. (From J.A.W., *Physics and Austerity: Law without Law* [in Chinese], Anhui Science and Technology Publications, Anhui, Hefei, 1982.)

PLATE 21. "Lighting profile": Intensity of lighting up of dust and other scattering substances by the supernova. The point where the intensity of lighting is half maximum moves out with exactly the speed of light. Units of time [$t$] and distance are taken to be the same but otherwise may be chosen at will (cm, km, lyr., kpc, etc.). (From W. Miller and J.A.W., "4-Geodesy," *Il Nuovo Cimento* 8 [1985], Fig. 3, p. 430.)

PLATE 22. Formation and fragmentation of a "pancake" of nuclear matter (schematic; adapted from Ruffini and J.A.W., 1971). A: rotating white dwarf core of a star that has become unstable and starts to collapse. B: resulting "pancake" of nuclear matter, turning much more rapidly. B′: same, seen from observer looking "down" along the axis of rotation. C: fragmentation of this pancake into neutron stars. Eventually these stars lose enough angular momentum by gravitational radiation that they can coalesce stepwise into a black hole. (From J.A.W., "Mechanisms for Jets," *Pontificiae Academiae Scientarum Scripta Varia* 35 [1971], Fig. 8, p. 553.)

PLATE 23. Results obtained for selected real energy values by integrating numerically the exact Dirac wave equation for a spin 1/2 electron in curved space; region of wave running down into the black hole; complex energy value (perpetual leakage and exponential decay). From it follows the listed values of energy levels (center of resonance) and level width (full width at half maximum). (Appreciation is expressed to Jacob Bekenstein for construction of this graph.) (From J.A.W., "Transcending the Law of Conservation of Leptons," in *Atti del* [Cortona 10–12 Giugno 1970] *Convegno Internazionale sul Tema: The Astrophysical Aspects of the Weak Interactions*, Accademia Nazionale dei Lincei, Roma, Quaderno No. 157, 1971, Fig. 5, pp. 158–159.)

PLATE 24. *Mechanism of radiation*. J. J. Thomson's way to understand why the strength of an electromagnetic wave falls only as the inverse first power of distance $r$ and why the amplitude of the wave varies (for low velocities) as $\sin\theta$ (maximum in the plane perpendicular to the line of acceleration). The charge was moving to the left at uniform velocity. Far away from it, the lines of force continue to move as if this uniform velocity were going to continue forever (Coulomb field of point-charge in slow motion). However, closer up the field is that of a point-charge moving to the right with uniform velocity ($1/r^2$ dependence of strength upon distance). The change from the one field pattern to the other is confined to a shell of thickness $\Delta\tau$ located at a distance $r$ from the point of acceleration (amplification of field by "stretching factor" $r\sin\theta\Delta\beta/\Delta\tau$). We thank C. Teitelboim for the construction of this diagram. (From C. W. Misner, K. S. Thorne, and J.A.W., *Gravitation*, Freeman, San Francisco, 1973, p. 111, Fig. 4.6.)

PLATE 25. In geometrodynamics the structure of superspace is to be considered as defined entirely internally; that is to say, by the very form of the "Einstein-Schrödinger equation" itself. We write this equation symbolically as shown. (From J.A.W., "Superspace and the Nature of Quantum Geometrodynamics," in C. M. DeWitt and J.A.W., eds., *Battelle Rencontres: 1967 Lectures in Mathematics and Physics*, Benjamin, New York, 1968, p. 278.)

PLATE 26. Building two 3-geometries into a thin sandwich 4-geometry, by interposing perpendicular connectors between the two, with preassigned lengths and shifts. What would otherwise be flexible thereupon becomes rigid. The flagged point illustrates the following equation:

$$ds^2 = g_{ij}(dx^i + N^i dt)(dx^j + N^j dt) - (Ndt)^2$$

(From C. W. Misner, K. S. Thorne, and J.A.W., *Gravitation*, Freeman, San Francisco, 1973, p. 506, Fig. 21.2.)

PLATE 27. Smoky dragon as symbol of Bohr's elementary quantum phenomenon. This drawing, motivated by the delayed-choice experiment, was made by Field Gilbert of Austin, 1983. (From W. A. Miller and J.A.W., "Delayed-Choice Experiments and Bohr's Elementary Quantum Phenomenon,'' in S. Kamefuchi et al., eds., *Proceedings of International Symposium on Foundations of Quantum Mechanics*, Physical Society of Japan, Tokyo, 1984, Fig. 3, p. 142.)

PLATE 28. Proposed delayed-choice experiment extending over a cosmological reach of space and time. Left, quasar *Q* recorded at receptor as two quasars by reason of the gravitational lense action of the intervening galaxy *G*-1. Middle, schematic design of receptor for delayed-choice experiment: (a) filter to pass only wavelengths in a narrow interval, corresponding to a long wave train, suitable for interference experiments; (b) lense to focus the two apparent sources into the acceptor faces of two optic fibers; (c) delay loop in one of these fibers of such length, and of such rate of change of length with time, as to bring together the waves traveling the two very different routes with the same, or close to the same, phase. Right, the choice. Upper diagram, nothing is interposed in the path of the two waves at the crossing of the two optic fibers. Wave 4a goes into counter I, and 4b into counter II. Whichever of these photo-detectors goes off, that —in a bad way of speaking— signals ''by *which* route, *a* or *b*, the photon in question traveled from the quasar to the receptor.'' Lower diagram, a half-silvered mirror, $\frac{1}{2}S$, is interposed as indicated at the crossing of the two fibers. Let the delay loop be so adjusted that the two arriving waves have the same phase. Then there is never a count in I. All photons are recorded in II. This result, again in a misleading phraseology, says that ''the photons in question come by *both* routes.'' However, at the time the choice was made whether to put in $\frac{1}{2}S$ or leave it out, the photon in question had *already* been on its way for billions of years. It is not right to attribute to it a route. No elementary phenomenon is a phenomenon until it is a registered phenomenon. (From J.A.W., *Physics and Austerity: Law without Law* [in Chinese], Anhui Science and Technology Publications, Anhui, Hefei, 1982.)

PLATE 29. Quantum mechanics promotes the mere ''observer of reality'' to ''participator in the defining of reality.'' It demolishes the view that the universe sits ''out there.'' (From J.A.W., ''Genesis and Observership,'' in R. E. Butts and K. J. Hintikka, eds., *Foundational Problems in the Special Sciences*, Reidel, Dordrecht, 1977.)

PLATE 30. We note that the *three coupling constants determined quite independently agree with one another within the limits of error of experiment and theory*. (From J. Tiomno and J.A.W., ''Charge-Exchange Reaction of the Mu-Meson with the Nucleus,'' *Rev. Mod. Phys.* 21 [1949], unnumbered figure from p. 157, text from p. 156.)

PLATE 31. An object is connected to its surroundings by elastic threads (eight are shown here; any number could be used). Rotating the object through 720° and then following the procedure outlined (Edward McDonald) in frames 2–8 (with the object remaining fixed), one finds that the connecting threads are left disentangled, as in frame 9 (lower right). (From C. W. Misner, K. S. Thorne, and J.A.W., *Gravitation*, Freeman, San Francisco, 1973, p. 1149, Fig. 41.6.)

PLATE 32. Pictorial description of classical scattering of alpha particles by nuclei. (a) Assumed form of real part of potential. (b) Classical deflection function versus angular momentum, *l*, ignoring absorption. Depicted are the characteristic features of rainbow angle and spirial scattering. If the actual nuclear absorption is large, the attenuation should fall from large to small values within the interval $l_1 \leq l \leq l_r$. (c) Geometry of classical orbits. In particular, three different orbits, *a*, *b*, and *c*, with the same scattering angle are shown.

Quantum mechanically these must interfere. However, $c$ should be strongly damped by absorption, while the larger area associated with $a$ will make its contribution to the scattering amplitude at this angle much greater than the contribution from $b$. (From K. W. Ford and J.A.W., "Application of Semiclassical Scattering Analysis," *Ann. of Phys.* 7 [1959], Fig. 5, p. 299.)

PLATE 33. A squeezed state is obtained by first 'squeezing' the ground state in one (here $x$-) variable, at the expense of the other ($p$-) variable and then displacing it by $\sqrt{2}\alpha$. For strong squeezing the contour lines in phase space (lines where $P_{sq}$ is constant), have the shape of a long, thin ellipse ("Gaussian cigar"). Therefore, a band defined by the radii $r_m$ and $r_{m+1}$ with $m$ sufficiently larger than $\alpha^2$ "intersects" this cigar twice in the two symmetrically located dark areas. In one zone the oscillator is moving to the right, in the other, to the left. The total probability amplitude, $\sqrt{W_m}$ is the sum of contributions $\sqrt{A_m}$ exp $(\pm i\phi_m)$ from the dark areas. The phase $\phi_m$ is fixed by the shaded area. No simpler illustration is given for "interference in phase space." (From W. Schleich and J.A.W., "Oscillations in Photon Distribution of Squeezed States and Interference in Phase Space," *Nature* 326 [1987]: 574–577.)

PLATE 34. The construction of rope that falls apart upon being pulled symbolizes the elaborate theorem that, conformant to the inexorable demand of mathematics, dissolves into identity on being analyzed [1987 addition; analog for the structure of existence]. The author thanks William Wootters for engineering this photograph. (From J.A.W., "Beyond the Black Hole," in Harry Woolf, ed., *Some Strangeness in the Proportion: A Centennial Symposium to Celebrate the Achievements of Albert Einstein*, Addison-Wesley, Reading, Mass., 1980, Fig. 22.11, p. 360.)

PLATE 35. The paradox of advance effects. Does the pellet strike $X$ at 6 P.M.? If so, the advanced field from $A$ sets $B$ in motion at 1 P.M., and $B$ moves $A$ at 8 A.M. Thereby the shutter $TS$ is set in motion and the path of the pellet is blocked, so it cannot strike $X$ at 6 P.M. If it does not strike $X$ at 6 P.M., then its path is not blocked at 5:59 P.M. via this chain of actions, and therefore the pellet ought to strike $X$. (From J.A.W. and R. P. Feynman, "Classical Electrodynamics in Terms of Direct Interparticle Interaction," *Rev. Mod. Phys.* 21 [1949], Fig. 1, p. 426.)

PLATE 36. *Commutator without commutator.* The fabric of history is unrolled differently at I and II. Histories themselves therefore have different topologies in the two cases, as witness two different histories that lead from the same initial configuration to the same final configuration. Let $I_H$ denote the classical action for the history II of a harmonic oscillator. Let $S_H$ denote the operation of summing over all histories, with equal weight for each, and with such a normalization that the sum $\langle x''t''|x't'\rangle = S_H \exp(iI_H/\hbar)$ is unitary. Then this expression is the propagator. It has the same value whether the sum goes over the histories of the type I or histories of the type II. Not the same, however, are the values of the sum

$$\langle x''t''|x(t_b)x(t_a)|x''t''\rangle = S_H x(t_b)x(t_a) \exp(iI_H/\hbar),$$

for the two classes of histories. The difference is equal to the commutator of the operators $x(t_b)$ and $x(t_a)$ of the operator formalism, despite the fact that $x(t_b)$ and $x(t_a)$ are ordinary c-numbers, not operators. (From J.A.W., *Geometrodynamics*, Academic, New York, 1962, p. 84, Fig. 18.)

PLATE 37. Meaning circuit as a model of existence; a loop tied into closure by observer-participancy. (From J.A.W., "How Come the Quantum," in D. M. Greenberger, ed., *New Techniques and Ideas in Quantum Measurement Theory, Ann. New York Acad. Sci.* 480 [1987]: 121–141—as augmented from Fig. 9 of J.A.W., "Bits, Quanta, Meaning," in A. Giovannini, F. Mancini, and M. Marinaro, *Problems in Theoretical Physics*, University of Salerno Press, 1984, pp. 121–141, by the following description.) *The Meaning Model.* Physics gives light and sound and pressure, tools of communication. It gives biology and communicators. Communication between communicators gives meaning. Meaning calls for the asking of questions, but the asking of one question stands in a complementary relation to the asking of another. The reception of an answer demands distinguishability. Mathematical analysis of distinguishability demands probability amplitudes. Complementary demands that these probability amplitudes be complex. A complex probability amplitude has a phase. The change of phase around a closed loop can be regarded as the definition and measure and even the sole form of existence of the "flux of a field" through that loop. Fields—electrodynamic, geometrodynamic, and chromodynamic—give rise to particles and physics.

PLATE 38. "Mutability." Each tread of the staircase of physics registers a law (for example, the fixity of nuclear charge number and mass number). Each riser marks the transcendence of that law (nuclear transmutation in this example), the imposition of conditions so extreme that that law no longer applies. The condition most extreme of all is gravitational collapse. There is no law of physics that has not required space and time for its statement. With the complete collapse of space, the framework falls down for everything one ever called a law. (Illustration: J.A.W., "From Relativity to Mutability," in J. Mehra, ed., *The Physicists' Conception of Nature*, Reidel, Dordrecht, 1973, Fig. 6, p. 217; the description comes from J.A.W., "Conference Summary: More Results Than Ever in Gravitation Physics and Relativity," in G. Shaviv and J. Rosen, eds., *General Relativity and Gravitation: Proceedings of the Seventh International Conference (GR7). Tel-Aviv University, June 23–28, 1974*, Wiley, New York, 1975, Fig. 5, p. 334.)

PLATE 39. The multisheeted character of superspace. Spin 1/2, if it occurs naturally in the context of quantum geometrodynamics, can hardly occur in any other sense than that in which Pauli spoke of spin from the very beginning, as a "nonclassical two-valuedness." Moreover, a nonclassical two-valuedness is already inescapable in the formalism. There are separate probability amplitudes for a 3-geometry with descriptor $w_k = +1$ and for an otherwise identical 3-geometry with descriptor $w_k = -1$. (From J.A.W., "Superspace and the Nature of Quantum Geometrodynamics," in C. M. DeWitt and J.A.W., eds., *Battelle Rencontres: 1967 Lectures in Mathematics and Physics*, Benjamin, New York, 1968, Fig. 8, p. 289, original one-sentence caption supplemented by three sentences of original text.)

PLATE 40. Reversible *versus* irreversible transformation of a black hole. Starting with an extreme Kerr black hole in this example ($Q = 0$; $J = M^2 = 2M_{ir}^2$), one can in principle reduce its mass from $M - 1.414M_{ir}$ to $M - M_{ir}$ by a sequence of reversible transformations (Christodoulou, 1970) of the Penrose type, with the energy thereby set free being taken away by the outgoing particles. If one proceeds less efficiently, via irreversible transitions, one ends up with a dead black hole with $M - M_{ir}$ now greater ($1.2M_{ir}$ in the example) than the original irreducible mass (diagram adapted from Christodoulou). (From J.A.W., "The Black Hole," in R. Debever, ed., *Astrophysics and Gravitation: Proceedings of the Sixteenth Solvay Conference on Physics*, Editions de l'Université de Bruxelles, Brussels, 1974, Fig. 9, p. 300.)

PLATE 41. The Riemannian geometry of spacetime of general relativity is here symbolized by the two-dimensional geometry of the surface of an apple. The geodesic tracks followed by the ants on the apple's surface symbolize the world line followed through spacetime by a free particle. In any sufficiently localized region of spacetime, the geometry can be idealized as flat, as symbolized on the apple's two-dimensional surface by the straight-line course of the tracks viewed in the magnifying glass ("local Lorentz character" of geometry of spacetime). In a region of greater extension, the curvature of the manifold (four-dimensional spacetime in the case of the real physical world; curved two-dimensional geometry in the case of the apple) makes itself felt. Two tracks A and B, originally diverging from a common point P, later approach, cross, and go off in very different directions. In Newtonian theory this effect is ascribed to gravitation acting at a distance from a center of attraction, symbolized here by the stem of the apple. According to Einstein a particle gets its moving orders locally, from the geometry of spacetime right where it is. Its instructions are simple; to follow the straightest possible track (geodesic). Physics is as simple as it could be locally. Only because spacetime is curved in the large do the tracks cross. Geometrodynamics, in brief, is a double story of the effect of geometry on matter (causing originally divergent geodesics to cross) and the effect of matter on geometry (bending of spacetime initiated by concentration of mass, symbolized by effect of stem on nearby surface of apple). (From C. W. Misner, K. S. Thorne, and J.A.W., *Gravitation*, Freeman, San Francisco, 1973, p. 4, Fig. 1.1.)